SOLAR VARIABILITY AND CLIMATE

Cover illustration: composite of the solar Blue Continuum image recorded by the Precision Solar Photometric Telescope on 14 February 1999 at Mauna Loa Solar Observatory (upper half, from White *et al.*, pp. 75–82) with an image of the Earth taken by the crew of Apollo-17 (lower half, NASA).

Space Sciences Series of ISSI

Volume 11

The International Space Science Institute is organized as a foundation under Swiss law. It is funded through recurrent contributions from the European Space Agency, the Swiss Confederation, the Swiss National Science Foundation, and the University of Bern. For more information, see the homepage at http://www.issi.unibe.ch/.

The titles in this series are listed at the end of this volume.

SOLAR VARIABILITY AND CLIMATE

Proceedings of an ISSI Workshop,
28 June–2 July 1999, Bern, Switzerland

Edited by

E. FRIIS-CHRISTENSEN
Danish Space Research Institute,
Dk-2100 Copenhagen, Denmark

C. FRÖHLICH
Physikalisch-Meteorologische Observatorium Davos,
World Radiation Center,
CH-7260 Davos, Switzerland

J. D. HAIGH
Blackett Laboratory,
Imperial College of Science
Technology and Medicine
London SW7 2BZ, U.K.

M. SCHÜSSLER
Max-Planck-Institut für Aeronomie,
31791 Katlenburg-Lindau, Germany

R. VON STEIGER
International Space Science Institute,
CH-3012 Bern, Switzerland

Reprinted from *Space Science Reviews*, Volume 94, Nos. 1–2, 2000

Springer Science+Business Media, B.V.

INTERNATIONAL SPACE SCIENCE INSTITUTE
Space Sciences Series of ISSI

A C.I.P. Catalogue record for this book is available from the Library of Congress.

ISBN 978-94-010-3809-6 ISBN 978-94-010-0888-4 (eBook)
DOI 10.1007/978-94-010-0888-4

Printed on acid-free paper

TABLE OF CONTENTS

INFLUENCES

CLIMATE

Solar Variability and Climate
ISSI Workshop, 28 June – 2 July 1999, Bern, Switzerland

Group Photograph

1. Nigel Weiss
2. George Reid
3. Jürg Beer
4. Giles Harrison
5. Mrs. Reid
6. Don Farley
7. Reindert Haarsma
8. Mrs. Farley
9. Linton Floyd
10. Oskar Steiner
11. Jeffrey Kuhn
12. Werner Mende
13. Gary Rottman
14. Judit Pap
15. Paul Evenson
16. Karin Labitzke
17. Joel Norris
18. Ulrich Cubasch
19. Galina Bazilevskaya
20. Ms. Tinsley
21. Steve Tobias
22. Claus Fröhlich
23. Brian Tinsley
24. Ari Laaksonen
25. Alfred Bürgi
26. David Parker
27. Keith Shine
28. Neil Arnold
29. Joanna Haigh
30. Mrs. Evenson
31. Henrik Svensmark
32. Jasper Kirkby
33. Oran White

Not in this picture: Raymond Bradley, Robert Cess, Len Fisk, Marcel Fligge, Eigil Friis-Christensen, Johannes Geiss, James Hansen, Erkki Kyrölä, Judith Lean, Werner Schmutz, Henk Spruit, Manfred Schüssler, Sami Solanki, Bernhard Stauffer, Thomas Stocker, Peter Stott, Yvonne Unruh, Heinz Wanner

Picture taken at Spiez Castle by Rudolf von Steiger

Foreword

The variability of the Sun is well established, as well as that of the Earth's climate. To what extent the two are connected, in the sense that solar variability drives climate, is the subject of considerable research and, in some cases, controversy.

After an earlier workshop at the International Space Science Institute (ISSI) on *Solar Composition and its Evolution*, two of the participants came up with the idea to initiate a similar project on the topic of *Solar Variability and Climate,* a workshop aimed at obtaining an overview of the current knowledge of the variability of the Sun and of the Earth's Climate, and of their possible connections. A further, equally important objective was the strengthening of the interaction between the two, often diverse communities of solar physicists and climatologists.

ISSI took up this idea and invited six convenors, E. Friis-Christensen, C. Fröhlich, J. Haigh, J. Hansen, M. Schüssler, and S. Solanki, who subsequently formulated the aims and goals of the workshop, nominated a list of invitees, drafted a programme of introductory talks, and structured the workshop into three sections. For each section there was a concluding discussion session moderated by two co-chairs. Moreover, there was a number of contributed poster papers for which there were two viewing sessions. The main intent of this format was to leave ample time for open, informal discussions, which is one of the principal aims of ISSI.

The workshop was held at ISSI on 28 June – 2 July 1999, and the present volume, number 11 in the *Space Sciences Series of ISSI* is the collection of all introductory papers, most contributed papers (some of which were integrated with an introductory paper), a summary of the discussion sessions, and concluded by a workshop summary. Each of the papers was reviewed by an external referee and by one of the editors. This has lead to a coherent volume giving the intended overview while still conveying the lively spirit that prevailed at the workshop.

As the topic of this workshop is a politically sensitive one there was some desire to address the general public with some of the issues discussed at the workshop. A "public forum" was organised jointly with ProClim–, the Forum of Climate and Global Change of the Swiss Academy of Sciences, where some of the workshop participants summarised the main points of the deliberations during the workshop. The event was attended by some 70 participant and concluded by a lively discussion. A press release was also issued (reproduced at the end of this volume) that summarised the main result with a single sentence: "The Sun's influence is an important factor in global climate change; however, the rapid warming during the last 20 years is dominated by other factors".

Finally it is a pleasure to thank all those who have contributed to this volume and to the workshop in general. We thank the authors for writing up their contributions and the reviewers for their critical reports, which have contributed significantly to the quality of the volume. We also thank the directorate and staff of ISSI for selecting this topic for a workshop and for their support in making it happen, in particular J. Geiss, G. Indermühle, V. Manno, U. Pfander, X. Schneider, D. Taylor, and S. Wenger. The production of the camera-ready papers for this volume was greatly facilitated by U. Pfander and X. Schneider, for which they deserve special thanks.

August 2000
E. Friis-Christensen, C. Fröhlich, J. Haigh, M. Schüssler, R. von Steiger

SOLAR VARIABILITY AND THE EARTH'S CLIMATE: INTRODUCTION AND OVERVIEW

GEORGE C. REID
CIRES, University of Colorado and Aeronomy Laboratory, NOAA, Boulder, Colorado, U.S.A.

Received: 22 October 1999; Accepted in final form: 3 February 2000

Abstract. Numerous attempts have been made over the years to link various aspects of solar variability to changes in the Earth's climate. There has been growing interest in this possible connection in recent years, spurred largely by the need to understand the natural causes of climate change, against which the expected global warming due to man's activities will have to be detected. The time scale of concern here is that of decades to centuries, and excludes the longer millennial scale in which orbital variations play a dominant role. The field has long been plagued by the lack of an acceptable physical mechanism by which solar variability can affect climate, but the discovery of variability in the Sun's total irradiance (the solar "constant" of meteorology) by spacecraft instruments has pointed to a direct mechanism. Other less direct mechanisms that have been suggested involve variations in the Sun's ultraviolet flux and in the plasma outflow of the solar wind. The purpose of this paper is to summarize the current state of the field, emphasizing the proposed mechanisms as an introduction to the more detailed papers that follow. The particular case of sea-surface temperature data will be used as an illustration.

Keywords: solar variability, Earth's climate

1. Introduction

The Earth's climate has undergone dramatic variations during the planet's history, and the cause of these variations has long provided a fertile field for speculation. The Sun's output of electromagnetic radiation and energetic particles also varies, and since the Sun is the ultimate driver of the climate system, it seems natural to link the two together and to look for the source of climate variability in the Sun itself. Many tantalizing correlations between variations in climate and in some aspect of solar variability have been discovered, but the correlations have usually had difficulty in meeting the standards of statistical validity, or have had the awkward property of disappearing as time goes by. As a result, the possible connection between solar variability and global climate change has long been the most controversial area of solar-terrestrial relations.

This situation has changed markedly in recent years for several reasons. First, the work of Eddy (1976) pointed out the rough coincidence between the coldest period of the so-called Little Ice Age in the late 17th and early 18th centuries and the Maunder Minimum of sunspot activity, when no sunspots were detected for a period of about 45 years (Maunder, 1894). Eddy went on to use the radiocarbon record to show that there was also a rough coincidence between earlier periods of warmth and cold in the northern hemisphere and periods of unusually high and low solar activity respectively, and to suggest that the cause might lie in a variation in the Sun's total radiative output (the solar "constant") that was in phase with the long-term average envelope of the 11-year sunspot cycle. While such variations in the Sun's total irradiance had been suggested from time to time by ground-based measurements of the solar constant, contamination by the atmosphere could not be ruled out, and high-precision measurements had to await the development of spacecraft radiometers. These instruments clearly showed that the Sun's total irradiance does indeed vary on a wide range of time scales, from the lifetime of individual sunspots and faculae to that of the 11-year activity cycle (see, e.g., Willson and Hudson, 1991).

Further impetus was given to Sun-climate studies by the growing concern in recent years with possible anthropogenic forcing of climate change through the increasing atmospheric content of greenhouse gases, especially the carbon dioxide generated by the burning of fossil fuel. The need to estimate the magnitude of this effect led to an increased focus on the natural causes of climate change, against whose background any anthropogenic contribution would have to be detected. The availability of reliable climate data for more than a century from many parts of the world has made it possible to search for a solar signal with more hope of success than ever before, and as a result solar variability is now becoming accepted as a serious potential, if not dominant, contributor to global-scale climate variability during the pre-industrial period, and as possibly a significant contributor even today.

By its nature, the study of Sun-climate relationships is highly interdisciplinary, and practitioners of the two principal disciplines involved, climatology and solar physics, rarely communicate with each other on the same plane and in mutually understandable language. The succeeding papers in this volume represent an attempt to remedy that situation. The present paper is intended to give a fairly superficial introduction and overview of the current state of the field to help set the stage for the more detailed presentations that follow.

2. Evidence from Sea-Surface Temperature Variations

Figure 1a shows an example of the kind of correlative evidence that has often been used to support claims of a relationship between solar activity and global climate (see, e.g., Reid, 1987). In this case, the solar parameter is the 11-year running mean of the annual sunspot number (a measure of the envelope of the 11-year solar cycle), and the departure of the globally averaged sea-surface temperature from the long-term average. Sea-surface temperatures (SSTs) have been measured on a routine basis by ocean-going ships since the middle of the 19th century. The data have now been compiled and published in atlas and digital form by Bottomley

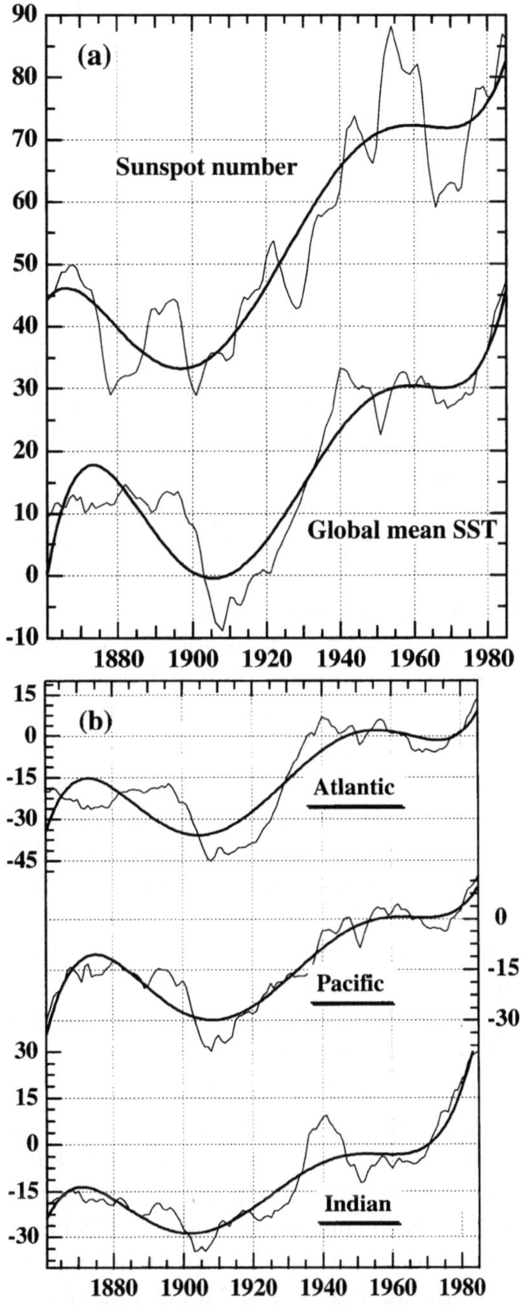

Figure 1. (a) Eleven-year running mean of the annual sunspot numbers (upper light curve) and of the annual mean global sea-surface temperature shown as departures from the 1951–80 average in units of 0.01K (lower light curve). Heavy curves are least-squares 7th degree polynomial fits to the data. (b) Same as (a) for the three major ocean basins.

et al. (1990), and the reader is referred to that publication and others (see, e.g., Folland et al., 1984; Parker et al., 1994) for details of the extensive and complex procedures used to take account of changes in measurement techniques and global coverage over the period of a century and a half. While most attempts to correlate solar activity with some climate parameter have used land surface temperatures, or a combination of land and sea-surface temperatures, the sea-surface temperature itself is in many ways a better measure of global climate, since it represents over 70 percent of the planet's surface area, and is much more spatially and temporally homogeneous than the land surface, as well as being free from such potential problems as urban warming.

The heavy curves in Fig. 1a are least-squares polynomial fits to the data, and emphasize the similarity of the long-term variation in both time series. This similarity alone, however, can only be taken as fairly weak circumstantial evidence for a physical connection. SSTs are known to be influenced by many factors other than direct insolation, such as surface wind patterns, El Niño events, which redistribute warm waters across the tropical Pacific Ocean, and upwelling of cold deep water associated with the global-scale thermohaline circulation (see, e.g., Broecker, 1997). While a global average would be expected to reduce these effects, there remains a suspicion that the global average might not represent a truly global signal, but be strongly influenced by what is going on in some localized region.

Figure 1b shows the 11-year running mean SST data averaged over each of the three main ocean basins, and the overall similarity of the least-squares polynomials is a strong indication that the temperature variation is truly global in nature. The signal is weaker in the Indian Ocean than in the other two, possibly because of its relatively sparse coverage by ship tracks. The cooling around the turn of the century appears everywhere, and is followed by a warming from 1920 through about 1950, and a levelling off until the start of a further warming beginning in the 1970s. The short-lived warming in the 1940s, particularly pronounced in the Indian Ocean record, is probably an artifact related to changes in operating conditions during World War II. No attempt has been made to extend the data beyond 1990, the end of the data set published by Bottomley et al. (1990), and the 11-year averaging involves the loss of 5 years at each end of the time series.

A world-wide variation in SST of this kind puts fairly severe constraints on a possible forcing mechanism. Changes in the greenhouse-gas concentration in the atmosphere would cause a global variation in SST, but the variation in the early 20th century occurred before anthropogenic increases were significant, and no natural increases have been suggested. Tropical volcanic eruptions are known to inject enough new aerosol into the stratosphere to produce significant global cooling, as was seen after the eruption of Mount Pinatubo in 1991 (Dutton and Christy, 1992), but the effects of a single volcano are over within a year or two. A prolonged period of intense volcanism would be needed to affect SSTs, given the large thermal inertia of the ocean, and no such event occurred in the early 20th century. Changes in the thermohaline circulation of the oceans would also produce global-scale temperature changes, but such changes are unlikely to be of the same sign and similar magnitude in all of the world oceans. Furthermore, climate models have frequently shown random variations in temperature that have been interpreted as stochastic fluctuations caused by the highly nonlinear and chaotic nature of the climate system itself, and not as artifacts of the models. Again it would be difficult to imagine a stochastic fluctuation that would cause all three oceans to show the apparently synchronous behavior seen in Fig. 1b.

These arguments compel us to consider a forcing mechanism external to the Earth, i.e., in the Sun itself or in the interplanetary medium. Passage of the solar system through an interstellar dust cloud could be one such mechanism, but there is no astronomical evidence to suggest that this has occurred. Solar variability of some kind appears to be the most likely explanation of at least the early part of the record seen in Fig. 1b, with the anthropogenic greenhouse effect possibly becoming an important contributor in the later part. We shall return to this SST evidence later, after a discussion of the various mechanisms of solar variability that have been suggested as possibly being the source of terrestrial climate change.

3. Mechanisms of Sun/Climate Interaction

The major stumbling block in the way of acceptance of solar variability as a cause of climate change has long been the lack of a generally acceptable mechanism for the interaction. The possible mechanisms that have been suggested fall roughly into three categories: variations in the Sun's total irradiance (the solar "constant"), variations in the spectral irradiance in the ultraviolet, and variations in the solar wind and the flux of energetic particles. Each of these has attractions, but each also has substantial drawbacks. Here we shall simply outline the underlying basis for the mechanisms and their difficulties. Many of the details will be described in later papers in this volume.

3.1. Variations in Total Irradiance

The possibility that the Sun's effective luminosity might vary with time has a long history, and formed the rationale for the long series of ground-based measurements of the solar constant by C.G. Abbott and his collaborators (Hoyt, 1979, and references therein). Substantial variability was seen in these measurements, but was not accepted as being solar in origin, since changing atmospheric transmissivity could not be excluded as a cause.

As mentioned above, the first real breakthrough in solar total irradiance measurements came with the advent of radiometers on spacecraft, notably the Earth Radiation Budget instrument on Nimbus-7, launched in late 1978 (Hickey et al., 1980), and the Active Cavity Radiometer for Irradiance Monitoring (ACRIM) instrument on the Solar Maximum Mission, launched in 1980 (Willson, 1981). The short-term variations seen by these instruments and their successors (see, e.g., Pap and Fröhlich, 1999) are related to solar oscillations and to the passage of individual sunspots, sunspot groups, and faculae across the solar disk. Their short duration coupled with their relatively small magnitude (a few tenths of a percent of the total irradiance) precludes any significant effect on climate, but the variation with the 11-year cycle of order 0.1 percent (see, e.g., Willson and Hudson, 1991) is marginally significant, especially if it is amplified during the larger-amplitude solar cycles, e.g., cycle 19 with its maximum in 1957–58. The effect of this 11-year variation in total irradiance has in fact been measured in sea-surface temperatures (White et al., 1997), thus making the first direct measurement of a climatic effect of solar variability. We shall return to the implications of this measurement later. A strong association between the 11-year cycle and tropospheric pressures and temperatures, primarily over the oceans, has been described by Labitzke and van Loon (1989), but the quantitative relationship between these observations and the measured changes in SST remains uncertain.

To place the 11-year variation in perspective, the Sun's total irradiance is about 1370 Wm^{-2} in space. Since the intercepted radiation is distributed over the surface of the Earth, the average solar radiation at the top of the atmosphere is one-quarter of this, or about 340 Wm^{-2}, and a variation of 0.1% corresponds to 0.34 Wm^{-2}. Planetary albedo scattering reduces this further to about 0.24 Wm^{-2}. This is small, but not totally insignificant, compared with the current estimated radiative forcing by anthropogenic greenhouse gases of 2 to 2.5 Wm^{-2} (Schimel et al., 1996), and comparable with the estimated forcing by such other anthropogenic sources as sulfate aerosols or tropospheric ozone. The cyclical nature of the solar forcing, however, substantially reduces its climatic impact, since the thermal inertia of the ocean is large enough to dampen an 11-year cycle considerably.

The spacecraft measurements have not yet covered a sufficient period of time to detect variations with much longer periods than the 11-year cycle, but those variations, if they exist, will be the most important ones from a climatic point of view. In particular, the variation with the envelope of the solar cycle suggested by Eddy (1976) and Reid (1987, 1997) has not been seen, perhaps at least partly because solar cycles 21 and 22, during which the spacecraft measurements have been made, were of closely similar magnitude, with very little change in the envelope. Despite this lack of direct observation, several attempts have been made to estimate the change in the Sun's radiative output since the time of the Maunder Minimum and the impact of the change on global climate (see, e.g., Lean et al., 1992, 1995; Nesme-Ribes et al., 1993; Hoyt and Schatten, 1993; Reid, 1997; Lean and Rind, 1998, 1999; Cubasch and Voss, 2000). The general conclusion of these estimates is that the Sun's total irradiance has probably varied by somewhere between 0.2 and 0.6% over the 300-year period, and that this would have been sufficient to account for the cooling of the Little Ice Age, which has been estimated to have been about 0.5 to 1C on a global basis.

3.2. Variations in UV Spectral Irradiance

The solar-cycle variation in the ultraviolet wavelengths responsible for the production and loss of ozone in the stratosphere are now known to be considerably larger than the variations in total irradiance, but it is not immediately obvious how the resultant changes in stratospheric chemistry and dynamics can affect surface climate. The important wavelengths (roughly 200–300 nm) are almost entirely absorbed in the stratosphere itself, and their 11-year solar cycle variation is now known to be in the range of a few percent at the shorter wavelengths to a few tenths of a percent at the longer end (see, e.g., Rottman, 1999), which is enough to cause a significant change in stratospheric conditions. The question of stratospheric influences on the troposphere and hence on surface climate, however, remains debatable and somewhat controversial. Model results from both GCMs and simpler models do show changes in tropospheric winds and storm tracks resulting from realistic variations in stratospheric ozone and solar ultraviolet fluxes (Kodera et al., 1990; Haigh, 1996, 1999; Shindell et al., 1999). These results will be described in detail in succeeding papers in this volume, and will not be discussed further here. It is worth mentioning here, however, that the question of spectral irradiance variations in the visible part of the solar spectrum also needs to be considered in the context of climate, since the depth profile of ocean heating varies markedly across the spectrum. Radiation at the blue end of the spectrum penetrates to depths of 100 meters or more, while radiation in the near infrared is absorbed in the outermost skin of the sea surface (Apel, 1987). The planetary albedo is also a function of wavelength, so that differences in spectral irradiance could have a significant impact on climate, particularly at the longer time scales.

3.3. Variations in the Solar Wind and Energetic Particles

The most dramatic manifestations of solar variability occur in the extreme ultraviolet (EUV) spectral range, in the outflow of hot plasma known as the solar wind, and in the energetic particles resulting from solar flares and coronal mass ejections. The EUV radiation does not penetrate significantly below the thermosphere, and while variations in its intensity have a major impact on the ionosphere, there is no obvious way in which they can affect the surface climate. Similarly, solar energetic particles are known to deplete ozone and to cause other chemical changes in the upper stratosphere and mesosphere (see, e.g., Crutzen et al., 1975; Reid et al., 1991), but the changes are rather small and are not likely to influence climate significantly.

The solar wind, however, does have the potential for bringing about climate change through a rather indirect mechanism, first suggested by Ney (1959). When solar activity is low, the solar wind flow is relatively smooth, and when activity is high waves and turbulence in the wind scatter incoming galactic cosmic-ray particles back out of the heliosphere, thus reducing their flux at the Earth. The net result is an 11-year cycle in cosmic-ray flux that amounts to as much as 20–30% at high geomagnetic latitudes, decreasing to near zero at the magnetic equator, which is accessible only to particles with very high energies that are not significantly affected by the interplanetary magnetic fields. Cosmic-ray fluxes are thus anticorrelated with the sunspot cycle, with maximum flux at sunspot minimum and vice versa. Galactic cosmic-ray particles are generally considerably more energetic than those of solar origin, and substantial fluxes can penetrate to the upper troposphere. Ney (1959) suggested that the resultant ionization could affect lightning in the troposphere, and give rise to an 11-year cycle in thunderstorm activity.

A broader potential impact of cosmic-ray ionization can come about through the formation of ice condensation nuclei, which could be dependent on the local concentration of ions in the upper troposphere or lower stratosphere, leading to a relationship between solar activity and cloudiness, which would be most pronounced at high latitudes where the solar-cycle modulation of cosmic rays is most intense. This basic idea was discussed and extended by Dickinson (1975), who suggested that the nucleation of sulfuric acid aerosols in the lower stratosphere and upper troposphere could be enhanced by the presence of ions, and that they could in turn act as cloud condensation nuclei. This possibility was criticized by Mohnen (1990) on the grounds that the effect of ionization would be negligibly small. Recently, however, Svensmark and Friis-Christensen (1997) claimed to have discovered a significant relationship between cloud cover on a global scale and the cosmic-ray flux over the solar cycle. This claim, however, has been criticized in turn by Kernthaler et al. (1999), based on a careful analysis of the occurrence of different types of clouds in the ISCCP data base. The net result is that the entire issue of a cosmic-ray influence on cloud radiative forcing, while intuitively attractive, remains in the area of speculation.

A different influence of solar-wind variation on climate has been proposed by Tinsley and his colleagues (see, e.g., Tinsley and Deen, 1991; Tinsley and Heelis, 1993). Their mechanism invokes an assumed effect of ionization or electric fields associated with solar activity on the freezing of supercooled water droplets in high clouds. Supercooled water can exist in the free atmosphere to temperatures as low as –40°C before spontaneously freezing and releasing the latent heat of liquefaction. If this process were to be made to occur at higher temperatures as a result of changing electrical parameters, an effect of solar activity on weather and climate could be a possibility.

The whole area of solar-wind and cosmic-ray effects on cloud formation is probably the most controversial aspect of Sun-climate relationships, and will be discussed at length in some of the accompanying papers. The outline given here is simply intended as an introduction to the current background state of the field, hopefully assisting the reader to place the papers that follow in their proper perspective.

4. Discussion

The three broad types of suggested Sun-climate mechanism have been discussed above in the order of their current state of development. Variations in solar total irradiance certainly affect climate, since the Sun's total irradiance is ultimately responsible for driving the climate system, while the effect of the 11-year cycle in total irradiance seems to have been directly observed in global sea-surface temperatures (White et al.,1997). In the case of spectral irradiance variations in the ultraviolet, general circulation models using observed 11-year variations in stratospheric ozone show that tropospheric circulation may well be affected, thereby influencing surface climate, but the linkage will probably be difficult to verify observationally. The suggested cosmic-ray and solar-wind effects remain controversial, and are also likely to be difficult to verify.

When thinking about climate change on the time scale of decades to centuries, it is important to recognize the dominant effect of the oceans, which cover over 70% of the planet's surface, and are largely responsible for the transport of heat from low to middle and high latitudes, particularly during the winter season. For a mechanism to be viable as a Sun-climate link, it should be capable of explaining the SST variations of the kind illustrated in Fig. 1b, occurring with the same magnitude and direction in all of the Earth's ocean basins. Only in the case of the total irradiance variations is the physical link clear, and only in this case is it possible to estimate the quantitative effect. White et al. (1997) used a 40-year series of bathythermograph records of the temperature profile of the upper ocean to deduce the temperature variations on decadal and interdecadal time scales. In each case they found a signal that decreased in amplitude with increasing depth, but did not penetrate through the main pycnocline, indicating that earlier model calculations (see, e.g., Reid, 1991) probably overestimated the loss of heat to the underlying deep ocean. They estimated that the sensitivity on an interdecadal time scale of the globally averaged SST was about 0.07 K for a change in solar total irradiance at the top of the atmosphere of 0.5 Wm^{-2}. Corresponding values for the decadal time scale were about 0.05 K for 0.5 Wm^{-2}. Applying the decadal sensitivity to the data in Fig. 1a implies that the cooling of about 0.2 K between 1895 and 1910 could be caused by a change in solar total irradiance at the top of the atmosphere of about 2 Wm^{-2}, or about 0.15%, while applying the interdecadal sensitivity value to the warming of about 0.4 K between 1910 and 1940 leads to a corresponding total irradiance increase of about 3 Wm^{-2}, or about 0.2%.

It is hardly necessary to say that these numbers are highly speculative. They are comparable with the total irradiance variation associated with the 11-year activity cycle, implying that there are long-term variations in total irradiance that are comparable with the variation seen between high solar activity and the very quiet solar activity of sunspot minimum. How such a variation could arise is a matter for solar physics to explain, but meanwhile continued measurements of the Sun's total irradiance from space are critically important if we are ever to have direct evidence that variations of this magnitude do in fact occur. If they do, we will have taken a major step toward explaining past changes in global climate, and also in being able to estimate the natural background of climate change against which anthropogenic change will have to be detected.

References

Apel, J.R.: 1987, Principles of Ocean Physics, *International Geophysics Series* **38**, Academic Press, London.

Bottomley, M., Folland, C.K., Hsiung, J., Newell, R.E. and Parker, D.E.: 1990, Global Ocean Surface Temperature Atlas, U.K. Meteorological Office and Massachusetts Institute of Technology.

Broecker, W.S.: 1997, Thermohaline Circulation, the Achilles Heel of our Climate System: Will Man-Made CO_2 Upset the Current Balance?', *Science* **278**, 1582–1588.

Crutzen, P.J., Isaksen, I.S.A., and Reid, G.C.: 1975, Solar Proton Events: Stratospheric Sources of Nitric Oxide', *Science* **189**, 457–458.

Cubasch, U. and Voss, R.: 2000, 'The Influence of Total Solar Irradiance on Climate', *Space Science Reviews*, this volume.

Dickinson, R.E.: 1975, Solar Variability and the Lower Atmosphere', *Bull. Amer. Met. Soc.* **56**, 1240–1248.

Dutton, E.G. and Christy, J.R.: 1992, Solar Radiative Forcing at Selected Locations and Evidence for Global Lower Tropospheric Cooling Following the Eruptions of El Chichon and Pinatubo', *Geophys. Res. Lett.* **19**, 2313–2316.

Eddy, J.A.: 1976, The Maunder Minimum', *Science* **192**, 1189–1202.

Folland, C.K., Parker, D.E. and Kates, F.E.: 1984, Worldwide Marine Temperature Fluctuations 1856–1981', *Nature* **310**, 670–673.

Haigh, J.D.: 1996, The Impact of Solar Variability on Climate', *Science* **272**, 981–984.

Haigh, J.D.: 1999, Modelling the Impact of Solar Variability on Climate', *J. Atmos. Sol. Terrest. Phys.* **61**, 63–72.

Hickey, J.R., Stowe, L.L., Jacobowitz, H., Pellegrino, P., Maschhoff, R.H., House, F. and Vonder-Haar, T.H.: 1980, Initial Solar Irradiance Determinations form Nimbus 7 Cavity Radiometer Measurements', *Science* **208**, 281–283.

Hoyt, D.V.: 1979, The Smithsonian Astrophysical Observatory Solar Constant Program', *Revs. Geophys. Space Phys.* **17**, 427–458.

Hoyt, D.V. and Schatten, K.H.: 1993, A Discussion of Plausible Solar Irradiance Variations, 1700–1992', *J. Geophys. Res.* **98**, 18,895–18,906.

Kernthaler, S.C., Toumi, R. and Haigh, J.D.: 1999, Some Doubts Concerning a Link Between Cosmic-Ray Fluxes and Global Cloudiness', *Geophys. Res. Lett.* **26**, 863–866.

Kodera, K., Yamazaki, K., Chiba, M. and Shibata, K.: 1990, Downward Propagation of Upper Stratospheric Mean Zonal Wind Perturbation to the Troposphere', *Geophys. Res. Lett.* **17**, 1263–1266.

Labitzke, K. and van Loon, H.: 1989, Association Between the 11-Year Solar Cycle, the QBO and the Atmosphere. Part I: The Troposphere and Stratosphere in the Northern Hemisphere in Winter', *J. Climate* **2**, 554–565.

Lean, J. and Rind, D.: 1998, Climate Forcing by Changing Solar Radiation', *J. Climate* **11**, 3069–3094.

Lean, J. and Rind, D.: 1999, Evaluating Sun-Climate Relationships Since the Little Ice Age', *J. Atmos. Sol. Terrest. Phys.* **61**, 25–36.

Lean, J., Skumanich, A. and White, O.: 1992, Estimating the Sun's Radiative Output During the Maunder Minimum', *Geophys. Res. Lett.* **19**, 1591–1594.

Lean, J., Beer, J. and Bradley, R.: 1995, Reconstruction of Solar Irradiance Since 1610: Implications for Climate Change', *Geophys. Res. Lett.* **22**, 3195–3198.

Maunder, E.W.: 1894, A Prolonged Sunspot Minimum', *Knowledge* **17**, 173.

Mohnen, V.A.: 1990, Stratospheric Ion and Aerosol Chemistry and Possible Links with Cirrus Cloud Microphysics - a Critical Assessment', *J. Atmos. Sci.* **47**, 1933–1948.

Nesme-Ribes, E., Ferreira, E.N., Sadourny, R., Le Treut, H. and Li, Z.X.: 1993, Solar Dynamics and its Impact on Solar Irradiance and the Terrestrial Climate', *J. Geophys. Res.* **98**, 18,923–18,935.

Ney, E.P.: 1959, Cosmic Radiation and the Weather', *Nature* **183**, 451–452.

Pap, J.M. and Fröhlich, C.: 1999, Total Solar Irradiance Variations', *J. Atmos. Sol. Terrest. Phys.* **61**, 15–24.

Parker, D.E., Jones, P.D., Folland, C.K. and Bevan, A : 1994, Interdecadal Changes of Surface Temperature Since the Late Nineteenth Century', *J. Geophys. Res.* **99**, 14,373–14,399.

Reid, G.C.: 1987, Influence of Solar Variability on Global Sea Surface Temperatures', *Nature* **329**, 142–143.

Reid, G.C.: 1991, Solar Total Irradiance Variations and the Global Sea Surface Temperature Record', *J. Geophys. Res.* **96**, 2835–2844.

Reid, G.C.: 1997, Solar Forcing of Global Climate Change Since the Mid-17th Century', *Clim. Change* **37**, 391–405.

Reid, G.C., Solomon, S. and Garcia, R.R.: 1991, Response of the Middle Atmosphere to the Solar Proton Events of August-December 1989', *Geophys. Res. Lett.* **18**, 1019–1022.

Rottman, G.: 1999 , Solar Ultraviolet Irradiance and its Temporal Variation', *J. Atmos. Sol. Terrest. Phys.* **61**, 37–44.

Schimel, D., et al.: 1996, Climate Change, 1995: The Science of Climate Change, Working Group 1 of the Intergovernmental Panel on Climate Change, Cambridge University Press, Cambridge, p. 65.

Shindell, D., Rind, D., Balachandran, N., Lean, J. and Lonergan, P.: 1999, Solar Cycle Variability, Ozone, and Climate', *Science* **284**, 305–308.

Svensmark, H. and Friis-Christensen, E.: 1997, Variation of Cosmic Ray Flux and Global Cloud Coverage - A Missing Link in Solar-Climate Relationships', *J. Atmos. Terrest. Phys.* **59**, 1225–1232.

White, W.R., Lean, J., Cayan, D.R. and Dettinger, M.D.: 1997, Response of Global Upper Ocean Temperature to Changing Solar Irradiance', *J. Geophys. Res.* **102**, 3255–3266.

Willson, R.C.: 1981, Solar Total Irradiance Observations by Active Cavity Radiometers', *Sol. Phys.* **74**, 217–229.

Willson, R.C. and Hudson, H.S.: 1991, The Sun's Luminosity Over a Complete Solar Cycle', *Nature* **351**, 42–44.

Address for Offprints: Dr. George C. Reid, CIRES, University of Colorado and Aeronomy Laboratory NOAA, Boulder, 80303 Colorado, U.S.A; email: reid@al.noaa.gov; Fax: +1 303 497 5373

I: SOLAR VARIABILITY...

OBSERVATIONS OF IRRADIANCE VARIATIONS

CLAUS FRÖHLICH
Physikalisch-Meteorologisches Observatorium Davos, World Radiation Center
CH–7260 Davos Dorf, Switzerland

Received: 1 April 2000; Accepted in final form: 11 May 2000

Abstract. Since November 1978 a complete set of total solar irradiance (*TSI*) measurements from space is available, yielding a time series of 21 years. From measurements made by different space radiometers (HF on NIMBUS-7, ACRIM I on SMM, ACRIM II on UARS and VIRGO on SOHO) a composite record of *TSI* can be compiled. The corrections which are needed for each radiometer, mainly to compensate for degradation, but also for operational influences are described. ACRIM I (1980–1989) and ACRIM II (1992–1995) are taken as reference because their degradation was monitored by inflight comparisons with spare radiometers; since 1996 VIRGO is used as reference. The scale of ACRIM II is adjusted to ACRIM I by comparison with NIMBUS-7 and ERBE. The reliability of this adjustment depends strongly on the stability of the HF and ERBE radiometer during the period of the gap between the ACRIMs which is discussed in detail together with an assessment of the overall precision of the composite.

Keywords: solar variability, total solar irradiance, solar activity

1. Introduction

Total solar irradiance (*TSI*) monitoring from space with electrically calibrated radiometers started with the launch of NIMBUS-7 in November 1978. In early 1980 the solar Maximum Mission Satellite (SMM) was launched and for the last 20 years at least two missions with solar monitors were operating simultaneously (Fig. 1). These more or less continuous measurements are supplemented by point measurements from balloons and rockets, which were very important in the early years when the decrease of *TSI* during the descending cycle 21 had still to be proven to be of solar origin (Willson et al., 1986). The radiometric accuracy of irradiance measurements made by individual instruments is of the order of 0.2%, and is insufficient to determine long term changes of only about 0.1% that are observed during the 11-year solar cycle modulation. While the instrument repeatability is adequate to monitor short term changes, the long term behavior can only be retrieved by careful tracing of one experiment database to the other, incorporating good knowledge of the degradation of radiometers operating in space. Fortunately several time series exist from different platforms made by different radiometers. This allows the construction of a composite time series having improved long term precision, thus yielding an unbiased estimate of *TSI* variability during the last two solar cycles (Fröhlich and Lean, 1998b; Fröhlich and Lean, 1998a). In the

Space Science Reviews **94:** 15–24, 2000.

following a review of *TSI* measurements is presented and the corrections explained which were applied to homogenize the record with special emphasis on the corrections for the HF radiometer during the gap between ACRIM I and ACRIM II data. Moreover, the VIRGO data have been extended over the SoHO 'vacation' period in summer 1998 with the necessary corrections (Fröhlich and Anklin, 2000) and are now updated until November 1999.

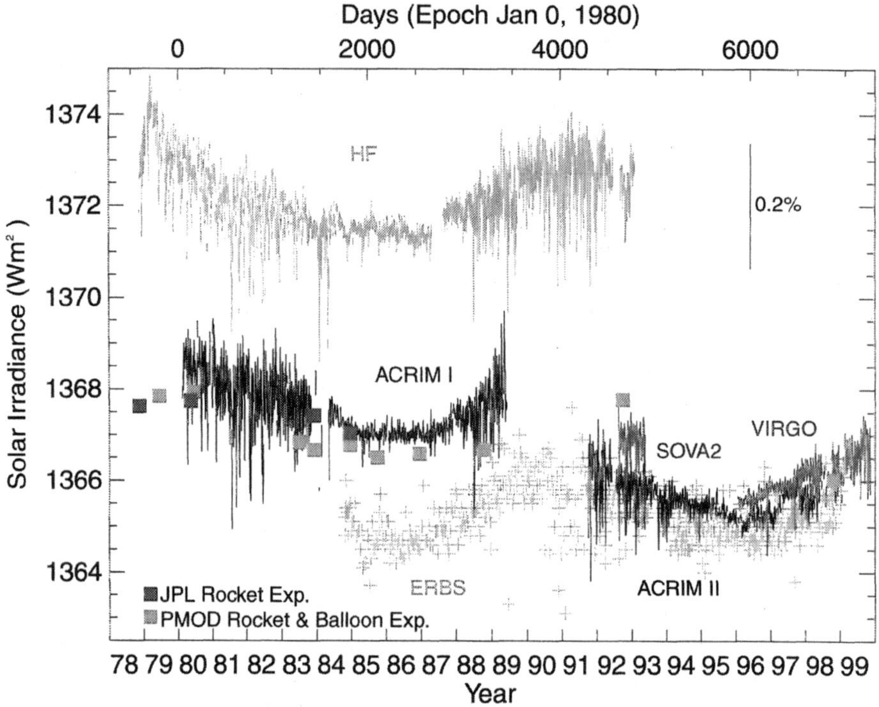

Figure 1. Compared are daily averaged values of the Sun's total irradiance *TSI* from radiometers on different space platforms since November 1978: HF on Nimbus7, ACRIM I on SMM, ERBE on ERBS, ACRIM II on UARS, SOVA2 on EURECA and VIRGO on SoHO. Also plotted are the results from rocket and balloon experiments. The data are plotted as published, that is besides the results from the two ACRIM, the SOVA2 and the VIRGO radiometers the data are not corrected for degradation.

2. Review of Radiometric Corrections

A detailed description of the procedures to construct the composite can be found in Fröhlich and Lean (1998a; 1998b). Besides a summary of the corrections we add new results of a detailed analysis of the glitches of HF during autumn 1989 and spring 1990 and discuss the most recent corrections to the VIRGO data from Fröhlich and Anklin (2000). The radiometer HF on NIMBUS-7 has no back-up

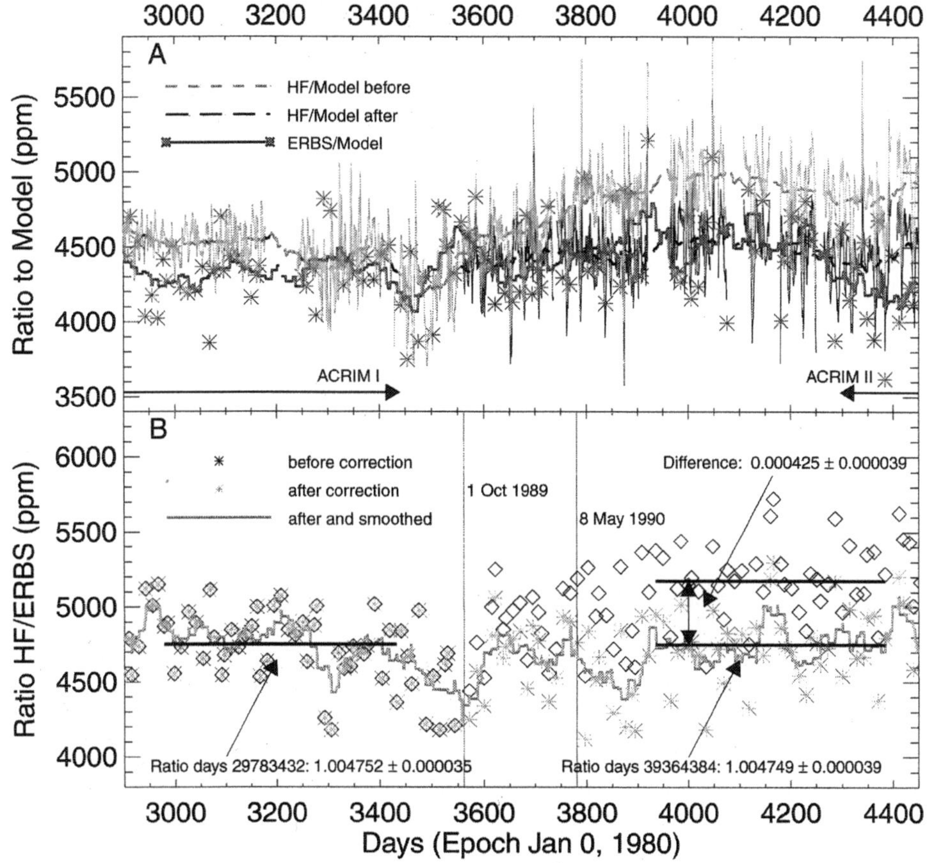

Figure 2. Compared are daily averaged values of the Sun's total irradiance from HF and ERBS with the model from Fröhlich and Lean (1998a)(A) and with each another (B). Also indicated is the range of the ACRIM I and ACRIM II measurements and the dates when the changes in the HF data have occured.

instrument and thus no direct means are available to determine the irradiation-dependent degradation (Hoyt et al., 1992). The cavity of HF has the same characteristics as PMO6-type radiometers, which means it has the same geometry and the same paint is used in the cavity. As seen in Fig. 1 the early data are rather high with a steep increase at the very beginning which immediately suggests that the behaviour could be similar to the one observed by PMO6V on VIRGO/SoHO (Anklin et al., 1998). Thus, we apply corrections for the HF radiometer degradation prior to 1982 by considering the behaviour of PMO6V and by utilizing data from the ACRIM I to determine in the tail an exponential function describing the degradation (although we know from VIRGO that an exponential function is only an approximation). Moreover, we adjust HF data prior to the end of 1980 downwards, corresponding to a slip in the NIMBUS-7 orientation relative to the sun as described in detail in Fröhlich and Lean (1998b). Near 1 October, 1989 and

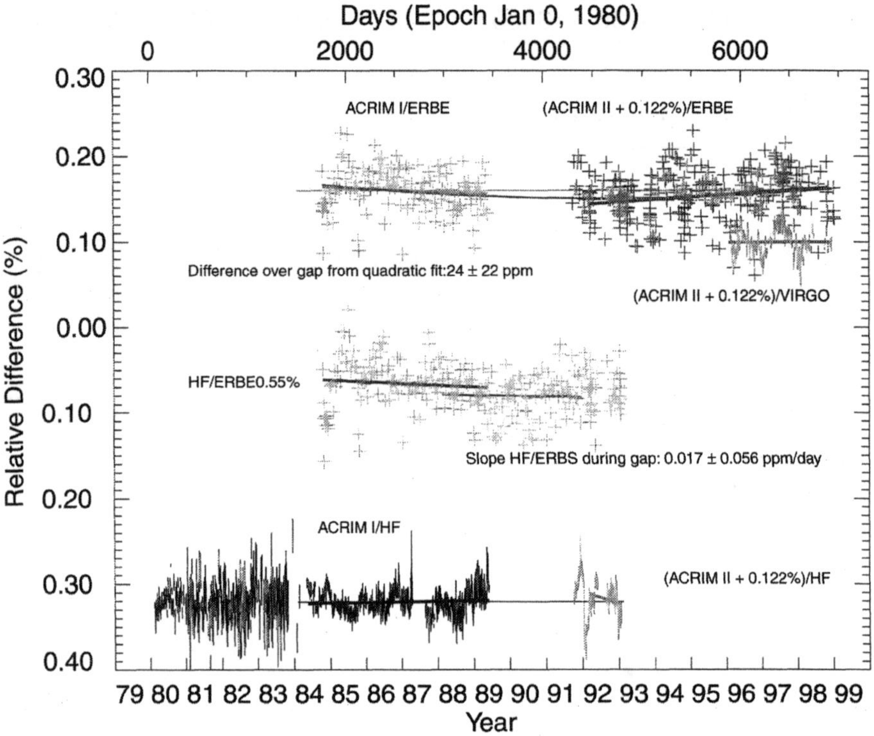

Figure 3. Comparison of different space experiment with ACRIM I and ACRIM II. The adjustment of ACRIM II by 0.122 % and the correction of the time series of HF after the glitches by −0.054 % yield a consistent picture of behaviour of HF and ERBE over the gap between ACRIM I and II.

8 May, 1990 adjustments of −0.31 and −0.37 Wm^{-2} were proposed to correct for glitches detected by comparison with ERBE data (Lee III et al., 1995) and also with models based on ground based observations by Chapman et al. (1996). The total amount of these two changes is crucial for the determination of the long-term changes of *TSI* because they happened just during the gap between the two ACRIM experiments and thus determine the adjustment of ACRIM II to the scale of ACRIM I via comparison with HF and ERBE. Thus these corrections have been revised in detail and the results are presented in Fig. 2. Both comparison show that there was a change in the NIMBUS7 data, although it is not so clear that it happened in two steps. In order to have an independent judgment of the amount of change the total difference is determined by comparing the ratio HF/ERBE before and after the glitches happened. The periods used for the comparison cover roughly 15 months before and after and overlap fully with ACRIM I and partly with ACRIM II, which started only 500 days after the second glitch. The two glitches need to be reduced by 69 ppm each to equalize the ratio before and after. This adjustment turns out to change the data on average (over the second 15 months) by 425 ppm which is slightly smaller than the imposed correction of 542 ppm; this rather important

difference reflects the noisiness of the HF/ERBE comparison in general and may be an indication of the uncertainty of these comparisons. The presently determined correction is about 40 ppm smaller than what was used in earlier evaluations and 138 ppm smaller than what was originally proposed by Lee III et al. (1995) and Chapman et al. (1996).

Before relating ACRIM II to the scale of ACRIM I some adjustments for degradation during the first year of operation and for a glitch in October 1995 have to be made as described in Fröhlich and Lean (1998a). The adjustment of the level of ACRIM II to the one of ACRIM I is performed by comparison to HF and ERBE as shown in Fig. 3. The correction to ACRIM II turns out to be 1.00122 ± 0.00013. As this value depends crucially on the amount of the HF correction the uncertainty of which is not well known we need some further consistency checks: the comparison of HF/ERBE (in the middle of Fig. 3) shows that the adjustment of HF is quite accurate as the slope over the gap is -0.017 ± 0.056 ppm/day, which is not significantly different from zero. The downward trend of HF/ERBE of -0.054 ± 0.051 ppm/day is similar to the one observed by ACRIM I with -0.057 ± 0.044 ppm/day which demonstrates that the corrections for the degradation of HF were properly taken into account. Later during ACRIM II an upward trend of 0.081 ± 0.023 ppm/day is observed. Thus, it seems that ERBE has an irradiation-independent rate of change (with measurements once in 2 weeks, the accumulated dose is essentially zero, even after several years of operation) which corresponds first during ACRIM I to an increase and then during ACRIM II to a decrease. During the critical period of the gap between ACRIM I and II it was turning around and thus was (fortunately) more or less constant; this is confirmed by the quadratic fit shown in Fig. 3 which yields a difference over the gap of -24 ± 22 ppm. All this indicates that (i) the glitches detected in the NIMBUS7 data are real, although less pronounced than originally published, (ii) the adjustment of ACRIM II is known with an uncertainty of about ± 50 ppm, and (iii) during the first several years ERBE increased its sensitivity which turned into a decrease after about 8 years in space.

VIRGO has two types of radiometers, PMO6V and DIARAD mainly to enable a consistent determination of possible changes (e.g. degradation) during the mission. This was always stressed in the objectives of that investigation (Fröhlich et al., 1995), and indeed their different characteristics are reflected in a quite different behaviour as shown in Fig. 4 with level-1 data of the operational and back-up DIARAD-L and -R and PMO6V-A and -B. In the following we give a summary of the procedures used to determine the long-term changes of the four radiometers described in Fröhlich and Anklin (2000). The PMO6V show a much larger degradation than DIARAD, which is astonishing as during the SOVA/EURECA mission they showed a very similar behaviour. Moreover, the PMO6V radiometers increase their sensitivity at the very beginning of the exposure to radiation for a few days before they start to decrease with ≈ 3 ppm per exposure day. A behaviour which is also observed in HF and possibly ERBE. By comparison of PMO6V-B with DIARAD-R a trend of the former of -0.08 ppm/day is found which is interpreted

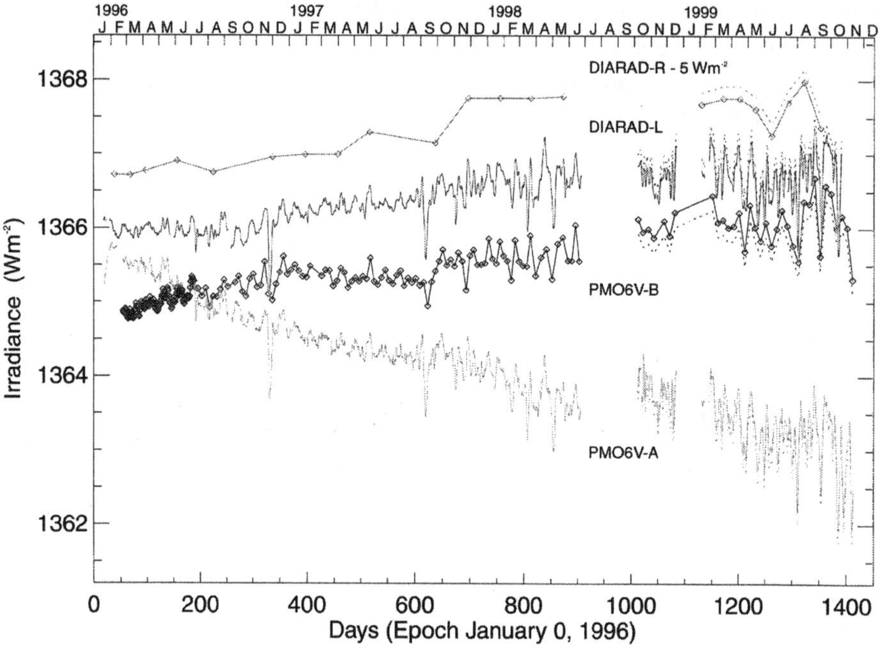

Figure 4. Level-1 radiometer data of DIARAD-R, DIARAD-L, PMO6V-B and PMO6V-A (from top to bottom). The dotted lines after the SoHO interruption (June to September 1998) show the shifts needed to maintain internal consistency over the gap.

as an irradiation-independent change due to exposure to the space environment and for the time being it is assumed to persist through end of 1999. An interesting effect is observed after an accidental switch-off of the VIRGO experiment for a couple of days in September 1996 (Fig. 4): DIARAD-L showed a change after switch-on which recovered in only about 6-8 weeks to the level of before. Moreover, the ratio to DIARAD-R does not show any significant change during that recovery period, indicating that both instruments react in a very similar way to exposure-independent effects, as e.g. a switch off. This observation has been used to argue for a similar change of the DIARADs during the interruption of the SoHO mission, often called SoHO 'vacations', from 28 June until the end of September 1998. The corrections evaluated are shown in Fig. 4 as dotted lines. Comparison with ACRIM II during the 2.5 years before the SoHO gap show a slope of the ratio ACRIM II to VIRGO of -0.09 ± 0.02 ppm/day which may be interpreted as an irradiation-independent (internally undetectable) change of DIARAD-R which was been assumed to be constant up to now. Interestingly enough almost the same trend was allotted to PMO6V-B as a result of the comparison with DIARAD-R. With the corrected DIARAD-L and PMO6V values the VIRGO irradiance is calculated according to the prescriptions given in (Fröhlich et al., 1997). These data are shown in Figure 5 and are available as daily and hourly values from `http://www.pmodwrc.ch/virgo/virgo.html` (version 2.51). An important point

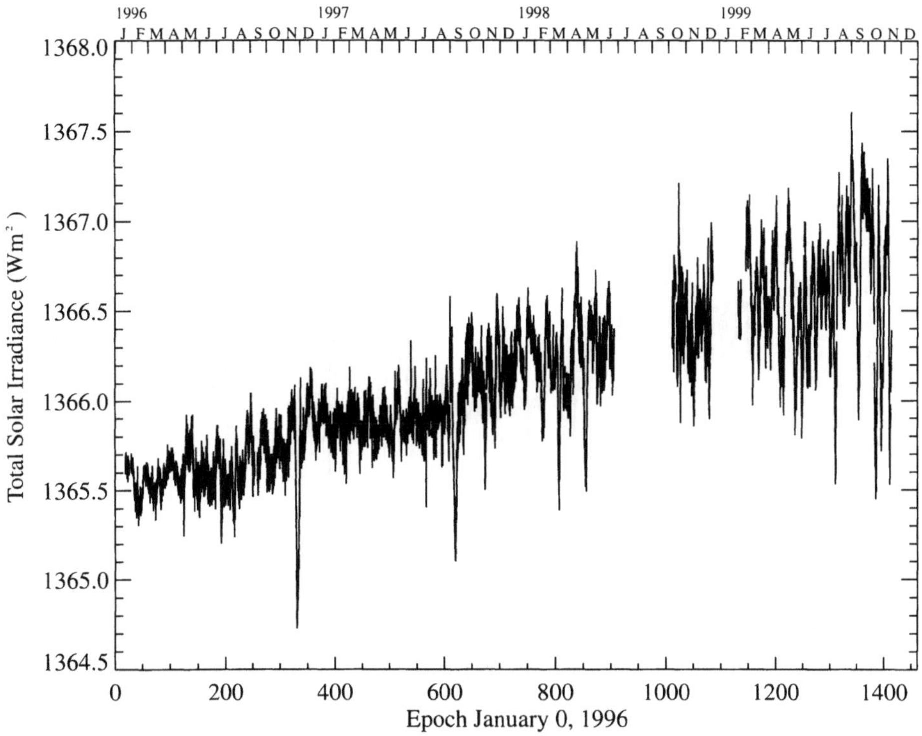

Figure 5. VIRGO *TSI* record (hourly values) from February 1996 until November 1999. For a discussion of the reliability of this record see text.

in the context of this paper is the question how reliable is the early increase observed by VIRGO (during the second half of 1996) for which the model (Fröhlich and Lean, 1998a) would suggest no trend at all. The determined degradation during 1996 is about 120 ± 40 ppm, compared to the increase of about 300 ppm shown in Fig. 5. Even if we assume no correction a substantial increase remains; thus, we believe that the increase is real.

3. Construction of the Composite *TSI*

From the original data of Fig.1 and with all the above-described corrections applied, we construct the composite *TSI* using ACRIM I and ACRIM II as reference in similar way as we did in Fröhlich and Lean (1998a). The result, shown in Figure 6, is then adjusted to Space Absolute Radiometer Reference (SARR; Crommelynck et al., 1995); this does not lower the absolute uncertainty, but is convenient for comparison with other data. In Fig. 7 the composite is compared with the original data for Fig. 1. The data of the composite *TSI* are available from `http://www.pmodwrc.ch/solar_const/solar_const.html`.

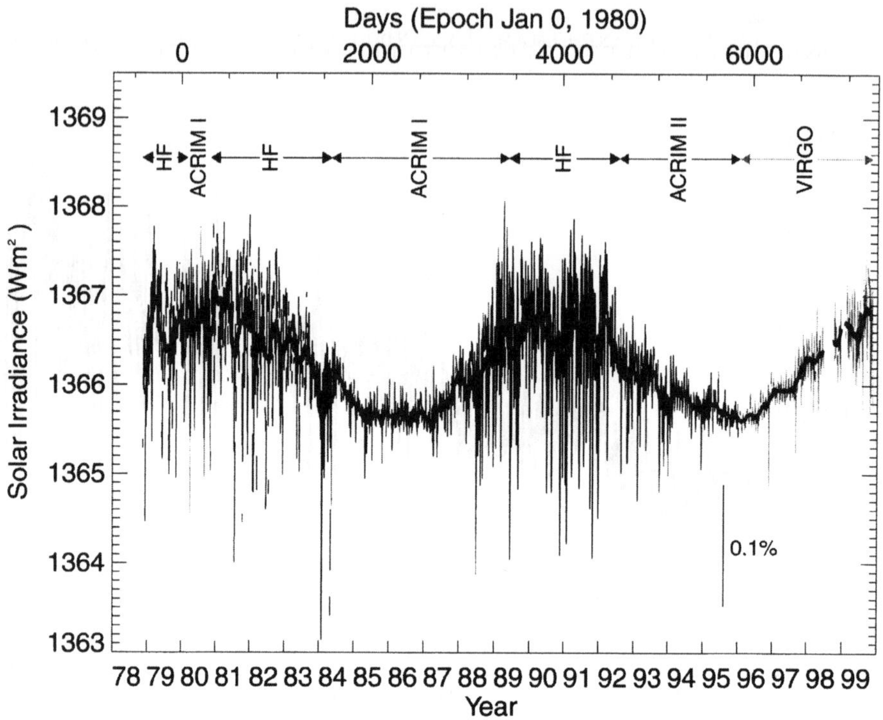

Figure 6. Composite *TSI* with indication of which time series are used at different times (corrected from the data shown in Fig. 1 as described in the text).

4. Conclusions

The analysis of the space radiometry in general and more specifically the VIRGO experiment demonstrates that it is not sufficient to have a less exposed back-up radiometer for the determination of its long-term changes, but at least two types of different construction. As a result of the VIRGO analysis a long-term uncertainty of about 10 ppm/a can be achieved. The difference of *TSI* measured at the two minima (30 day average of the irradiance in April/May 1985 and Jan/Feb 1996) amounts to 60 ppm with the 1996 minimum being higher. The question is how significant is this result from the radiometric point of view and neglecting the fact that the definition of which value corresponds to the solar minimum may be vague and small sunspots can still diminish the irradiance. The above stated radiometric uncertainty may be conservative because the improvement of the long-term reliability from having different experiments simultaneously is obviously very difficult to estimate – from the analysis of the influence of the glitches one might estimate something like 50 ppm over 10 years. More important seems, however, the influence of the changes due to interruptions of the radiometer's operation – especially when no other experiment is in space – which can be of the order of

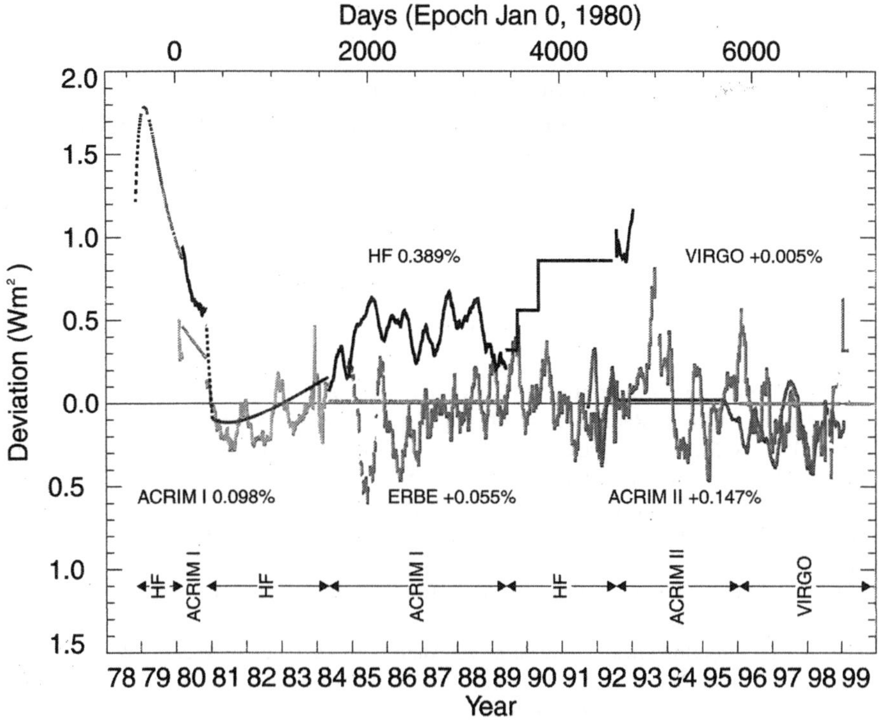

Figure 7. Comparison of the composite *TSI* with the original data of Fig. 1.

more than 100 ppm, as the incident of the SoHO 'vacations' has demonstrated. If we still believe an uncertainty of 10 ppm/a the observed change of 60 ppm over 11 years is not significantly different from zero.

The discussion of uncertainties shows that while the composite appears to give a reliable time series which can be used by solar physicists as well as by climatologists there remains the problem of absolute accuracy and determination of the long-term stability as emphasized in Quinn and Fröhlich (1999). Nevertheless, the reason that we could produce such a reliable time series of *TSI* is that we had always at least two independent experiments simultaneously in space. Although the plans of space agencies for the next 5 years show that we still may have more than one experiment at a time, the plans for the distant future, however, are vague and there may even be a certain reluctance by the agencies to continue such programs. Thus, we need measurements with an accuracy of space radiometry at least a factor of 10 better. Radiometers with such accuracy are available, but have not yet proven to maintain such a level when transferred to space. We need them soon as a benchmark for future measurements which may not be as continuous as needed for accurate traceability.

Acknowledgements

Part of this work is supported by the Swiss National Science Foundation which is gratefully acknowledged. Thanks are extended to the VIRGO/SoHO team for its effort to make this experiment running so well; SoHO is a cooperative ESA/NASA mission launched in December 1995.

References

Anklin, M., C. Fröhlich, W. Finsterle, D. A. Crommelynck, and S. Dewitte: 1998, 'Assessment of Degradation of VIRGO Radiometers Onboard SOHO'. *Metrologia* **35**, 686–688.

Chapman, G. A., A. M. Cookson, and J. J. Dobias: 1996, 'Variations in Total Solar Irradiance During Solar Cycle 22'. *J. Geophys. Res.* **101**, 13541–13548.

Crommelynck, D., A. Fichot, R. B. Lee III, and J. Romero: 1995, 'First Realisation of the Space Absolute Radiometric Reference (SARR) During the ATLAS 2 Flight Period'. *Adv. Space Res.* **16**, (8)17–(8)23.

Fröhlich, C. and M. Anklin: 2000, 'Uncertainty of Total Solar Irradiance: An Assessment of the Last 20 Years of Space Radiometry'. *Metrologia*, in press.

Fröhlich, C., D. Crommelynck, C. Wehrli, M. Anklin, S. Dewitte, A. Fichot, W. Finsterle, A. Jiménez, A. Chevalier, and H. J. Roth: 1997, 'In-Flight Performances of VIRGO Solar Irradiance Instruments on SOHO'. *Sol. Phys.* **175**, 267–286.

Fröhlich, C. and J. Lean: 1998a, 'The Sun's Total Irradiance: Cycles and Trends in the Past Two Decades and Associated Climate Change Uncertainties'. *Geophys. Res. Lett.* **25**, 4377–4380.

Fröhlich, C. and J. Lean: 1998b, 'Total Solar Irradiance Variations: The Construction of a Composite and its Comparison with Models'. In: F. L. Deubner, J. Christensen-Dalsgaard, and D. Kurtz (eds.): *IAU Symposium 185: New Eyes to See Inside the Sun and Stars*. pp. 89–102.

Fröhlich, C., J. Romero, H. Roth, C. Wehrli, B. N. Andersen, T. Appourchaux, V. Domingo, U. Telljohann, G. Berthomieu, P. Delache, J. Provost, T. Toutain, D. Crommelynck, A. Chevalier, A. Fichot, W. Däppen, D. O. Gough, T. Hoeksema, A. Jiménez, M. Gómez, J. Herreros, T. Roca Cortés, A. R. Jones, and J. Pap: 1995, 'VIRGO: Experiment for Helioseismology and Solar Irradiance Monitoring'. *Sol. Phys.* **162**, 101–128.

Hoyt, D. V., H. L. Kyle, J. R. Hickey, and R. H. Maschhoff: 1992, 'The NIMBUS-7 Solar Total Irradiance: A New Algorithm for its Derivation'. *J. Geoph. Res.* **97**, 51–63.

Lee III, R. B., M. A. Gibson, R. S. Wilson, and S. Thomas: 1995, 'Long-term Total Solar Irradiance Variability During Sunspot Cycle 22'. *J. Geoph. Res.* **100**, 1667–1675.

Quinn, T. and C. Fröhlich: 1999, 'Accurate Radiometers Should Measure the Output of the Sun'. *Nature* **401**, 841.

Willson, R. C., H. S. Hudson, C. Fröhlich, and R. W. Brusa: 1986, 'Long-Term Downward Trend in Total Solar Irradiance'. *Science* **234**, 1114–1116.

Address for Offprints: Dr. Claus Fröhlich, PMOD/WRC, Dorfstrasse 33, CH-7260 Davos Dorf, Switzerland; email: cfrohlich@pmodwrc.ch, Fax: +41 81 417 5100

OBSERVATIONS OF VARIABILITY IN COSMIC RAYS

G.A. BAZILEVSKAYA (bazilevs@sci.lebedev.ru)
Lebedev Physical Institute, Russian Academy of Sciences, Leninsky Prospect, 53, 117924, Moscow, Russia

Received: 15 October 1999; Accepted in final form: 28 January 2000

Abstract. Cosmic rays are the main source of ionization in the atmosphere at altitudes below 55–60 km. This circumstance, together with the fact that cosmic ray flux modulation closely mirrors the solar activity time history, makes cosmic rays a good candidate as a possible mediator in the solar variability – climate relationship. The observed cosmic ray flux variations are described with the aim of emphasizing the features which may be useful in the search of correlation between cosmic rays and atmospheric phenomena.

1. Introduction

The Earth is permanently bombarded by energetic charged particles coming from space. These are cosmic rays (CRs). Most of them are generated in interstellar space, but a certain amount is accelerated at the Sun and in the heliosphere. The power brought by CRs to Earth ($\sim 10^{-9}$ W/cm^2) is negligible in comparison with the power involved in the atmospheric processes related to weather. However, CRs are the main source of ionization at atmospheric altitudes below about 55–60 km except the near-ground level below 3–4 km. In addition, CR flux modulation is closely associated with the solar activity dynamics. This provides reason to consider CRs as one of possible mediators in the solar variability – climate relationship. To investigate the role of CRs in the Sun–Earth relationship, it is necessary to outline the CR flux variations to aid the search for physical links between solar activity and the processes in the atmosphere.

The paper aims at a phenomenological description of CR flux variations with emphasis on the relation between apparent CR flux changes and solar activity manifestations. Cosmic ray variations are described as observed by the ground-based and balloon-borne apparatus, i. e. at Earth and near Earth. Cosmic ray modulation is also imprinted as cosmogenic nuclei abundance in different natural archives such as tree rings, polar ice cores, stalagmites and others. This opens additional possibilities for exploring the solar-terrestrial relationship in the past. However this method of observing CR variability is not examined here as it is highlighted in other contributions (e. g., Beer, 2000).

Although the time history of sunspot numbers is commonly accepted as an indicator of solar activity dynamics, it should be noted that sunspots are not at

Space Science Reviews **94:** 25–38, 2000.

all a direct reason for CR flux modulation. The latter is a complex phenomenon of particle interactions with the magnetic fields of the solar wind in the heliosphere (e. g., Fisk and Wenzel et al., 1998) and is beyond the scope of this paper.

2. Cosmic Rays in Interplanetary Space and near Earth

The dominant component of CRs is protons, about 10% comprise He nuclei, and other nuclei make less than 1%. Electrons are a minor constituent of CRs (about 1% or lower). Energy spectra of various components are practically similar when intensity is expressed as a function of kinetic energy for protons and electrons and of kinetic energy per nucleon for nuclei.

Figure 1 presents the integral energy spectra for CR main constituents (GOST, 1986). Only the low-energy part of spectra is shown. Actually, the spectra cover the energy from 1 MeV (1.6×10^{-6} erg) to the enormous value of 3×10^{14} MeV (48 J). The particle intensity at these ultra-high energies is very low, ≤ 1 particle per 10 km^2 and year. Solar activity affects the CR fluxes in the low-energy part of the spectrum up to about 10^5 MeV. It is seen in Fig. 1 that CR fluxes are lower during periods of high solar activity. However the solar CRs, which are mainly produced in high solar activity periods, sporadically give an additional particle flux plotted with the dashed lines. Figure 1 shows CR energy spectra as they are observed in interplanetary space.

Near the Earth, the geomagnetic field reduces the intensity of charged particles and operates as a separator on the arriving cosmic particles according to their rigidity. The magnetic rigidity is defined as $R = cP/Ze$ (here c is the speed of light, P is the particle momentum, Z is the charge number, and e is the unit charge) and it is a key parameter for the particle motion in the magnetic field. The particles with equal rigidities move in a similar way in a given magnetic field. Very roughly speaking, each geomagnetic latitude may be characterized by a cut-off rigidity, R_c, such that particles with less rigidity cannot arrive at this latitude. Most protected are equatorial latitudes; only particles with $R > 15$ GV (kinetic energy of protons $> 1.4 \times 10^4$ MeV) are allowed to reach equatorial regions.

The Earth's atmosphere also acts as a particle energy analyzer because of CR particle interactions with the air. Primary CR particles with energies below several hundreds of MeV/nucleon are just absorbed in the atmosphere due to ionization, whereas the higher energetic primaries produce nuclear-electromagnetic cascades involving electrons, gamma-rays, X-rays, mesons, and nucleons. The total incident flux of secondary CRs first increases from the top of the atmosphere to altitudes of ~17–27 km (depending on R_c and solar activity level) due to dominating cascade generation and then decreases due to absorption. The lower the altitude is the higher the effective energy of primary CRs contributing to the flux of secondaries. Therefore CR intensity near the Earth and at Earth depends not only on the solar activity level but also on the geomagnetic latitude and the altitude in the atmosphere.

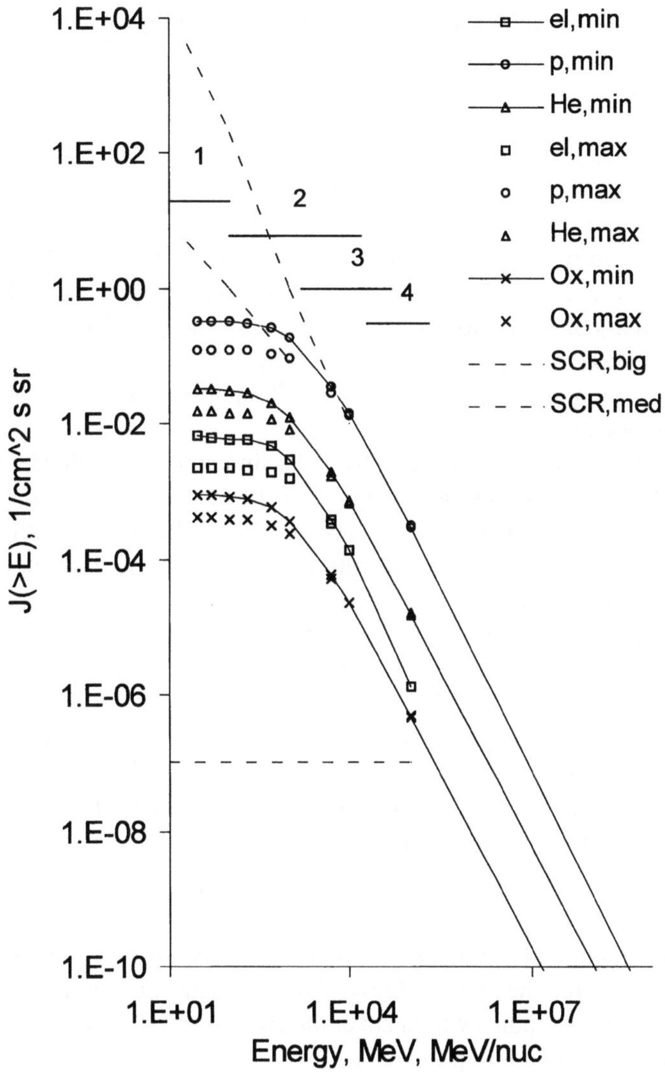

Figure 1. The integral energy spectrum of comic rays at energies $\leq 10^9$ MeV for protons and electrons and $\leq 10^9$ MeV/nucleon for nuclei. The horizontal dashed line indicates the region where cosmic ray intensity is modulated by solar activity. All the components are presented as observed in the minimum and maximum phases of solar activity. The two upper dashed lines relate to solar proton events of different amplitudes. The solid horizontal bars cover the energy regions of the following observations: 1– spacecraft, 2– balloon, 3– neutron monitor, 4– muon telescope.

The low-energy fraction of CRs which affects the upper atmosphere can only be recorded at high latitudes and altitudes. This is observed onboard spacecraft, rockets (< 100 MeV) and balloons ($10^2 - 1.5 \times 10^4$ MeV). On the other hand, at energies above 10^3 MeV CR modulation is well observed at the Earth's surface with neutron monitors and muon telescopes.

3. Cosmic Ray Monitoring with Neutron Monitors and Balloon-borne Detectors

The essential features required of any installation observing CR flux variations are the statistical accuracy of the measurements and the stability of the efficiency during many years of observations. It is necessary to measure CR flux with standard installations so that variations observed at various sites on Earth are comparable. The longest homogeneous time series of CR intensity are those obtained with neutron monitors and balloons.

The neutron monitor was invented by J. A. Simpson, who initiated a worldwide network of standard installations which have been in operation since the 1950s (Simpson, 1948; Simpson et al., 1953; Hatton, 1971). In the 1960s the network of neutron monitors comprised more than 60 stations. Today there are more than 20. A neutron monitor detects neutrons (nuclear component of CRs) that are progenies of the particles incident at the atmosphere with energies higher than $\sim$1000 MeV. The effective energy of CRs for a given neutron monitor depends on the geomagnetic latitude (cut-off rigidity R_c) and the altitude of the station.

There is a direct influence of the atmospheric condition on the neutron monitor count rate. The most significant is the pressure effect, the dependence of the count rate on the air mass above the detector. The uncorrected neutron monitor data, N_{obs}, are processed according to the expression

$$N_{\mathrm{cor}} = N_{\mathrm{obs}} / \exp[\beta(P - P_0)],$$

where β is negative and about -1%/mm Hg, P is the current pressure, and P_0 is the mean pressure for the station. Only corrected neutron monitor data, N_{cor}, are used for CR variation studies which cover the characteristic time-scale from several minutes (and even seconds) up to years.

The monitoring of CR fluxes in the atmosphere with detectors carried by balloons was started by S.N. Vernov and A.N. Charakhchyan in 1957, during the International Geophysical Year (Charakhchyan, 1964; Charakhchyan et al., 1976; Bazilevskaya et al., 1991; Bazilevskaya and Svirzhevskaya, 1998) and has been continued up to now by the Lebedev Physical Institute (LPI, Russia). Standard sounding balloons with Geiger counters are sent up practically every day at several geographic sites of Russia including polar latitudes in the Arctic and Antarctic. The detectors are carefully calibrated to maintain the homogeneity of the time series. A device returns the flux values for protons with energies above several MeV and

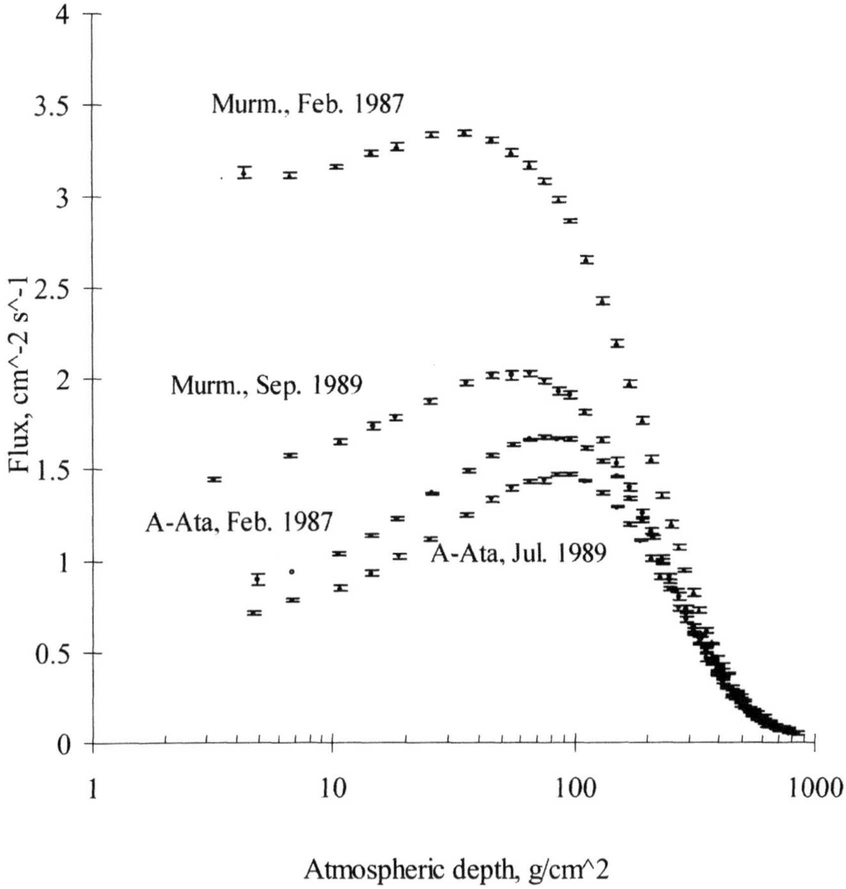

Figure 2. Secondary cosmic ray fluxes in the atmosphere (transition curves) at Murmansk (polar station, cutoff rigidity $R_c = 0.6\,\text{GV}$) and Alma-Ata (mid-latitude station, $R_c = 6.7\,\text{GV}$). Monthly averages are given for the periods of minimum solar activity (1987) and maximum solar activity (1989). Data of the balloon flights during solar energetic particle events are excluded.

electrons above hundreds of keV (global flux) at atmospheric depths from ground level up to ~5 g/cm^2 (altitude ~35 km). These particles are progenies of primaries that are allowed by the geomagnetic field to reach the top of the atmosphere at the site of observation. While ascending in the atmosphere a sounding balloon records the secondary CR flux versus atmospheric depth, which is a so-called transition curve. Figure 2 demonstrates the dependence of transition curves on geomagnetic latitude and solar activity. A significant difference in fluxes at polar and mid-latitudes is seen, as well as a decrease in fluxes during periods of maximum solar activity.

Observations of ion production rate in the atmosphere were performed by H.V. Neher (1967, 1971) up to the end of the 1960s. The relationship between secondary CR fluxes, J, and ionization rate, q, in the atmosphere is shown in Fig. 3. The

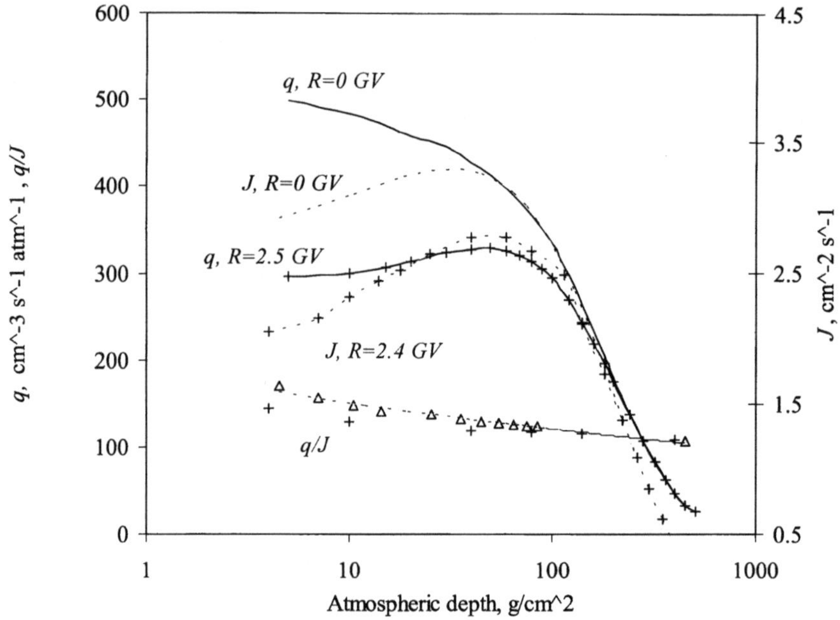

Figure 3. Comparison of results of CR balloon observations in the minimum of solar activity: (1) q - ion production rate normalized to one standard atmosphere (Neher, 1967, 1971) (left axis) at polar ($R_c = 0\,\mathrm{GV}$) and mid-latitudes ($R_c = 2.5\,\mathrm{GV}$), and (2) J - omnidirectional flux of charged particles (ionizing component, right axis) measured by Lebedev Physical Institute group practically at the same latitudes. Below is the ratio q/J (left axis): triangles and dashed line for $R_c = 0\,\mathrm{GV}$, plus signs for $R_c = 2.4 - 2.5\,\mathrm{GV}$.

results of observations at polar and mid-latitudes in the minimum of solar activity and the ratio q/J versus atmospheric depth are presented. The ratio is only weakly dependent on latitude and increases slightly with altitude due to the presence of low-energy, high-ionizing particles.

The geomagnetic field gives rise to a latitude effect in CR fluxes that is controlled by cut-off rigidity R_c and depends also on the altitude because of the effective energy of primary CRs related to each altitude. The response of various instruments (such as a neutron monitor measuring the nucleon component of CRs and LPI balloon-borne Geiger counter mainly recording electrons and muons) to the energy of primary CRs can differ. That leads to different latitude effects for different instruments. Figure 4 shows the results of latitudinal surveys performed with neutron monitors (Stoker, 1994) and with balloon-borne Geiger counters (Golenkov et al., 1990). Both sets of data were taken during periods of minimum solar activity. It is seen that the R_c dependence of neutron monitor data at sea level is equal to that taken with a Geiger counter at the altitude of ~9 km (300 g/cm^2). For comparison, the results of neutron monitor observations at the altitude of 9 km and those of balloon detectors at ~32 km (8 g/cm^2) are also plotted.

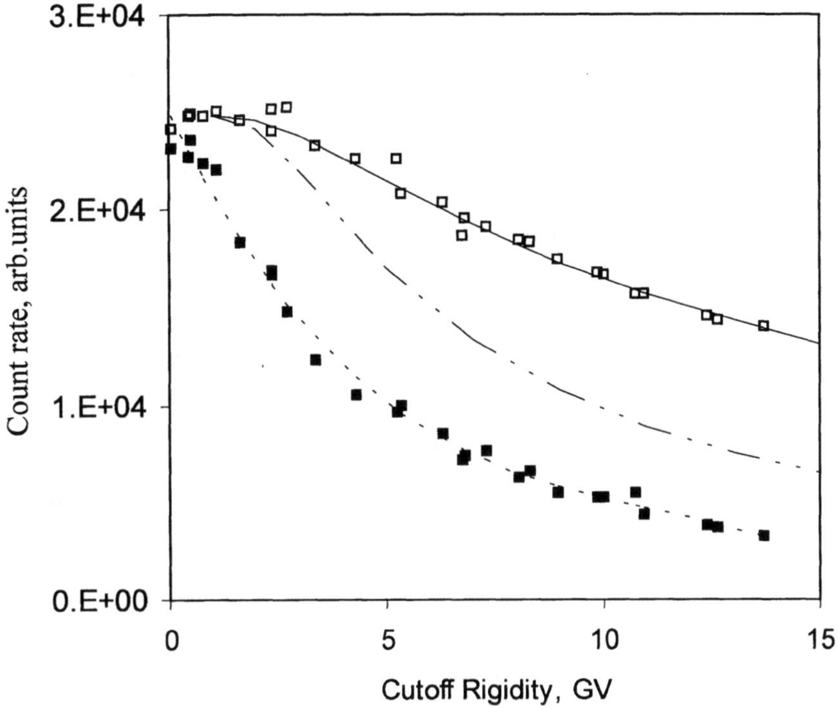

Figure 4. Latitudinal effect of the ionizing and nucleon components of cosmic rays as observed by a neutron monitor at sea-level (solid line) and at an altitude of 9 km (dashed-dotted line); and as observed by a Geiger counter at 9 km (light squares) and at 32 km (dark squares and dashed line). All data are normalized at polar latitudes ($R_c = 0$ GV).

4. Long-term CR Flux Modulation

Cosmic ray flux modulation caused by varying solar activity manifests itself as particle intensity variations on a wide range of time scales. The most impressive modulation is the 11-year variation that is in opposite phase with the sunspot number changes. The time history during the last four solar cycles of CR flux changes relative to the average flux in 1965 is presented in the lower panel of Fig. 5 for the two neutron monitor stations Climax, $R_c = 3$ GV, and Huancayo/Haleakala, $R_c = 13$ GV (Chicago University Web site), as well as for the two CR balloon stations Murmansk, $R_c = 0.6$ GV, and Moscow, $R_c = 2.4$ GV. The modulation strongly depends on R_c and kind of observation. It is ~50% in the transition maximum (at the altitude of ~20–25 km) in the polar stratosphere and around 10% at the equatorial neutron monitor Huancayo/Haleakala. The deeper in the atmosphere the smaller is the amplitude of the 11-year cycle in CRs. At the altitude of 12 km the modulation of the global flux at polar latitudes is about two times smaller than at 25 km. The sunspot number is given for comparison in the upper panel of Fig. 5. A long-term trend can be noticed in the CR data during the last 3 cycles.

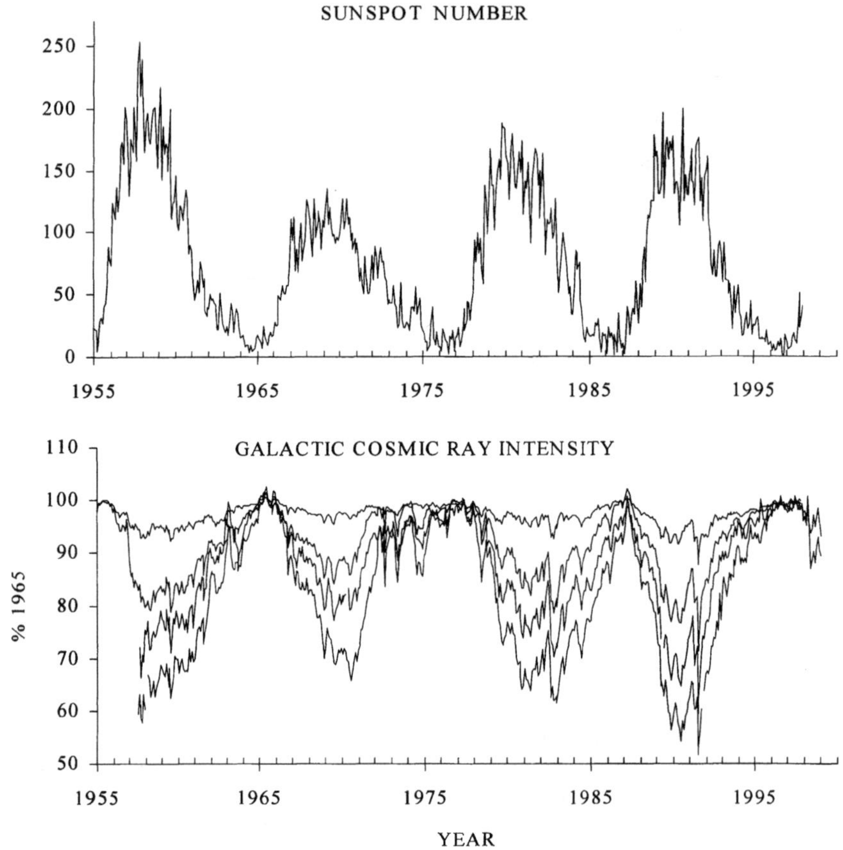

Figure 5. Long-term changes in solar activity (upper panel) and galactic cosmic ray intensity (lower panel) as measured by balloon detectors (BD) in the maximum of the transition curve (at the altitude ~20–25 km) and by neutron monitors (NM). In the lower panel monthly data in % of the yearly average in 1965 are given (from top to bottom) for: NM Huancayo/Haleakala ($R_c = 13\,\text{GV}$), NM Climax ($R_c = 3\,\text{GV}$), BD Moscow ($R_c = 2.4\,\text{GV}$), and BD Murmansk ($R_c = 0.6\,\text{GV}$). The marks on the X-axis relate to the start of a year.

The modulation in solar activity maximum becomes stronger even if the sunspot number maximum in the 1990s is not higher than that in the 1980s. This is an indication of growing solar magnetic activity that governs the conditions in the heliosphere. Some signatures of the 22-year CR modulation can also be seen in Fig. 5. The clearest one is an alternate shape of the 11-year cycle maxima in CR intensity, being peak- or plateau-like depending on the polarity of the large-scale solar magnetic field. In the epochs of the positive magnetic field in the northern hemisphere of the Sun (the 1970s and 1990s) the 11-year cycle in CR intensity had a flat top, whereas during the epochs with the opposite polarity, the 11-year cycle had a sharp top (in 1965 and 1987).

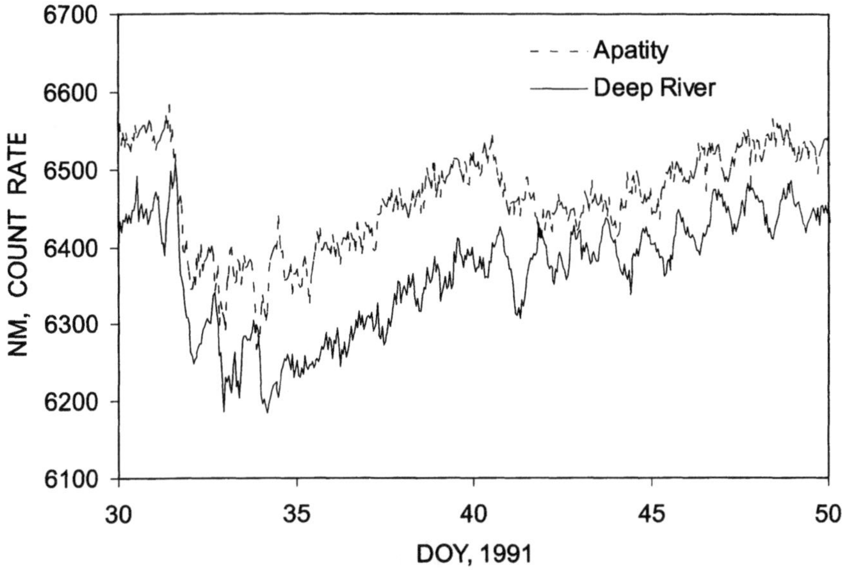

Figure 6. Pressure corrected hourly count rates of the two neutron monitors Apatity ($R_C = 0.6$ GV, latitude 67.33 N, longitude 33.2) and Deep River ($R_C = 1$ GV, latitude 46.1 N, longitude 282.5). For details see text.

5. Forbush Decreases and 27-day Recurrent Variations

Fast releases of solar energy that manifest themselves as flares and coronal mass ejections (CMEs) are usually accompanied by strong magnetic disturbances in the solar wind that modulate CR intensity in space. The CR response that appears as a relatively sharp decrease of CR intensity followed by a gradual recovery is called a Forbush effect. The encounter with an interplanetary disturbance affects the Earth's magnetosphere; therefore, Forbush decreases on the Earth are highly correlated with geomagnetic disturbances. Examples of strong Forbush decreases are shown in Fig. 6, where the data of the Apatity and Deep River neutron monitors are given. The stations are not very different in R_c (0.6 GV and 1 GV, correspondingly) but are widely separated in longitude. It is seen that a Forbush effect started simultaneously at the two stations on 31 January, 1991 (DOY 31), and another one, not as strong, was observed on 9 February, 1991 (DOY 40). A distinct diurnal variation is also seen in Fig. 6 that is dependent on longitude.

Figure 7 demonstrates an example of the 27-day variation in CR intensity as observed by the Apatity neutron monitor and by the balloon-borne Geiger counter at the altitude of ~25 km. It is known that this variation is due to long-lived activity regions on the Sun and the solar rotation. Interplanetary magnetic field (IMF) strength and solar wind velocity are given in the lower part of Fig. 7. Cosmic ray 27-day variation started with the Forbush effect on 26 March, 1976 (DOY 86), and lasted during almost 8 solar rotations. There is an increase in IMF strength B

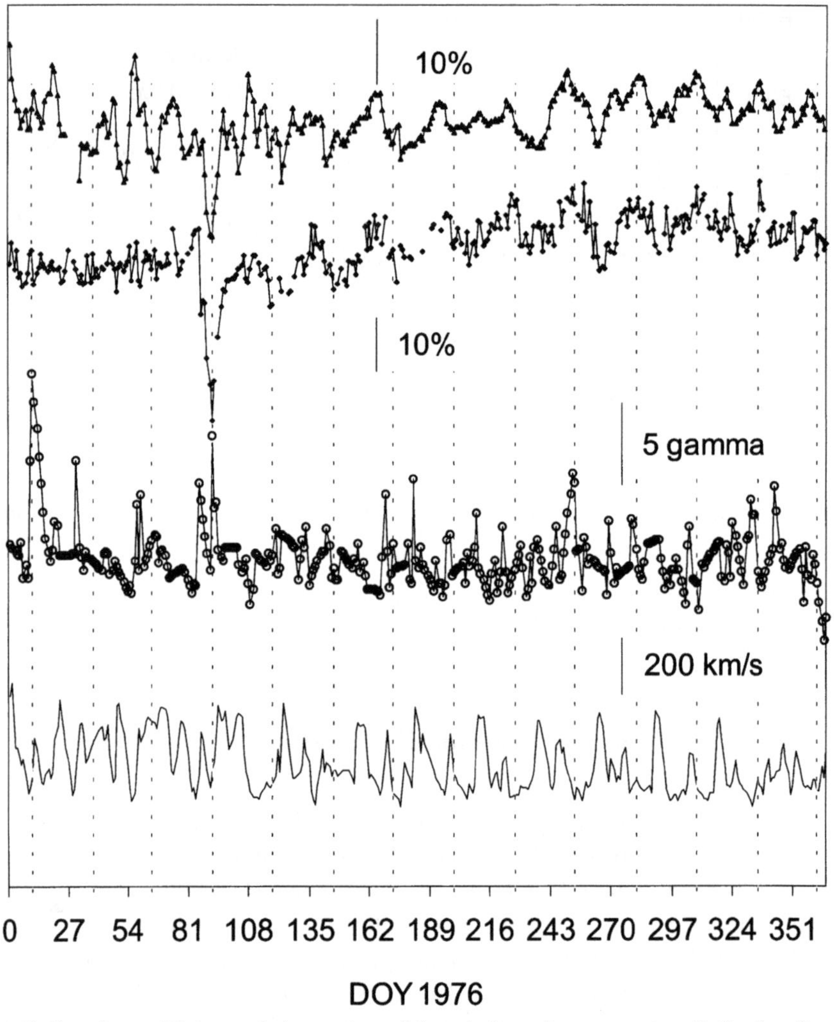

Figure 7. Cosmic ray 27-day variation and conditions in interplanetary space. Daily data from top to bottom: count rate of the neutron monitor Apatity (triangles), count rate of a Geiger counter at the altitude of ~25 km (rhombs), interplanetary magnetic field strength (circles), solar wind velocity (solid line).

that correlates with the Forbush decrease, but the following recurrent CR variation correlates with the solar wind velocity rather than with B.

Forbush decreases are more often observed during periods of high solar activity, whereas 27-day CR variations are usually more distinctive and longer-lasting during the descending or minimum phases of the solar activity cycle. The amplitudes of Forbush effects, 27-day variations, and other CR variations in the range of timescales between ~10 days and ~3 months vary, in general, in phase with the solar activity cycle, i. e. they reach the maximum values during solar activity maximum phase (in the opposite way than the CR intensity itself). However, periods of ~1 year duration usually occur around the maximum of solar activity when CR variability is depressed and no strong Forbush effects are detected at all (Storini, 1997; Bazilevskaya et al., 1998).

6. Solar Cosmic Rays and Electron Precipitation

As seen in Fig. 1 the energy of solar CRs is mainly lower than several GeV/nucleon. Abrupt increases of solar energetic particle (SEP) flux occur after fast and powerful energy releases on the Sun, most often after flares. Therefore, the rate of such events is higher during periods of high solar activity (contrary to galactic CR intensity). Although the elemental composition changes from one SEP event to another, a great majority (> 90%) of solar CRs are protons. The 100 MeV protons cannot penetrate into the atmosphere deeper than ~32 km (8.6 g/cm^2 of atmospheric depth). During SEP events the form of transition curves of CR intensity in the atmosphere changes in such a way that CR intensity increases with altitude sometimes up to 2 orders of magnitude in comparison with quiet time conditions. Neutron monitors are sensitive only to SEP events with relativistic protons (so-called ground-level enhancements, GLEs). The majority of SEP events are not observed in the atmosphere and near Earth because their size distribution decreases with increasing SEP energy. Figure 8 presents the rate of SEP observations by balloon-borne detectors. Each SEP event is plotted as a vertical bar with a height proportional to the maximum intensity of > 100 MeV solar protons. For the period 1957-1995 ninety-five events were observed in the stratosphere and forty-eight GLEs occurred. Both magnitude and duration of SEP events depend on particle energy as illustrated by Fig. 9 (Bazilevskaya and Svirzhevskaya, 1998). It is seen that the time-scale of a GLE is several hours, whereas an event in the stratosphere usually lasts from 1 to 5 days. The relative increase in neutron monitor count rate may be as large as hundreds of percent (more than 4500% during the extraordinary SEP event of 23 February, 1956). Powerful SEP events often happen in series as is seen in Fig. 9. Thus the CR intensity in the stratosphere can be enhanced during a week or longer.

The solar wind disturbance that causes Forbush decreases often originates in the same active region where flares generating SEPs occur. Therefore, GLEs may be observed during Forsbush decreases, or they can be followed by a Forbush decrease. Due to the SEP energy limitation, GLEs are usually detected by neutron monitors situated at rather high latitudes. The energy dependence of Forbush decreases, on the contrary, is rather weak. Therefore, high latitude NMs may detect a GLE, whereas NM at lower latitudes observe a Forbush decrease. Thus, the observed CR intensity may be enhanced at polar stations and depressed at equatorial locations. Such a situation took place in October, 1989. A GLE was observed in polar regions and at mid-latitudes on 19–20 October, followed by a CR intensity

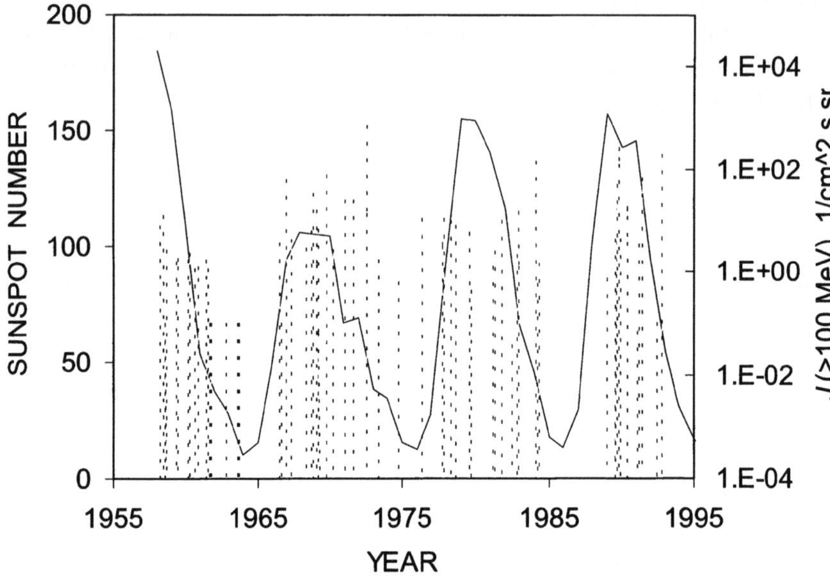

Figure 8. Solar proton events as recorded in the balloon measurements of the Lebedev Physical Institute (Russia). The length of the vertical bars (dashed line, right axis) corresponds to solar proton intensity $J(> 100\,\text{MeV})$ at the altitude of ~30 km. Data before 1967 are lower limits of $J(> 100\,\text{MeV})$. The solid line presents the yearly means of sunspot number (left axis).

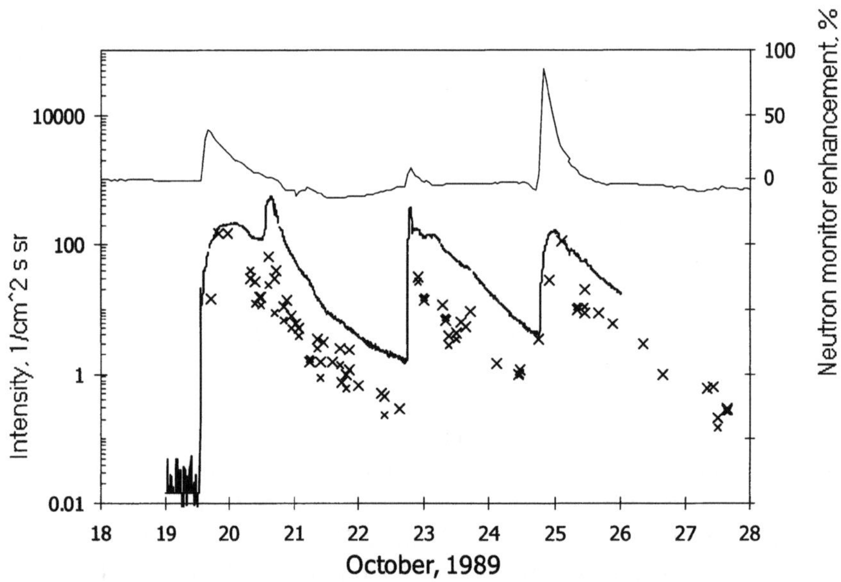

Figure 9. Solar energetic particle events in October, 1989. The upper line gives the Apatity neutron monitor measurements; the lower line gives the > 100 MeV proton intensity as measured by GOES satellite; crosses give > 150 MeV proton intensity observed in the stratosphere. The galactic cosmic ray background is removed (Bazilevskaya and Svirzhevskaya, 1998).

decrease on 21 October. At equatorial stations the CR intensity started decreasing already on 20 October.

Another radiation intensity enhancement, which is occasionally observed at polar stations in the atmosphere, is due to electron precipitation (the electrons are mostly of magnetospheric origin). Electrons are absorbed in $\sim$1 g/cm^2 of atmospheric depth, but they generate X-ray emission that penetrates the atmosphere down to $\sim$20 km or even deeper (Makhmutov et al., 1995; Bazilevskaya and Svirzhevskaya, 1998). Electron precipitation tends to occur in the auroral zone in the form of narrow spikes, about 15 – 100 km wide. Its duration, depending on the event, varies from a few seconds to a few tens of hours. The flux of X-rays at the altitude of $\sim$30 km can be well above 100 cm^{-2}s^{-1}, which is several times more than the X-ray background during quiet times. It can be occasionally observed by LPI balloon-borne detectors. However, daytime may not be favorable for its detection. According to LPI observations, the rate of electron precipitation also demonstrates an 11-year cycle. Its maximum is on the descending branch of the solar activity cycle, rather close to solar minimum.

7. Conclusion

Cosmic ray intensity is strongly modulated by solar activity dynamics. This is successfully used to reconstruct solar-terrestrial links of the past by analyzing cosmogenic isotope abundances. However, that does not actually mean that CRs influence the global Earth's atmospheric processes. The physical relation between CRs and global Earth atmospheric processes should be studied using peculiarities of CR flux modulation such as the latitude effect, the opposite phase of the CR 11-year cycle and the solar activity 11-year cycle, the alternate shape of the CR 11-year cycle, the cases of enhancements of CR flux at the polar latitudes with simultaneous Forbush decreases at the low latitudes, etc. Hopefully, CRs will find their place in the solar variability – climate relationship.

Acknowledgements

I would like to express my gratitude to the Organizing Committee of the ISSI workshop on Solar Variability and Climate for their invitation. Thanks are due to my colleagues from Lebedev Physical Institute with whom the data on the cosmic ray fluxes in the atmosphere were obtained during the many-years' joint work, to the Chicago University group and WDC-C2 for the neutron monitor data available on the Web site and on CD. In addition, I thank H. Debrunner for helpful comments and Louise Wilson for correcting the English manuscript. This work was partly supported by the Russian Foundation for Basic Research, grants No. 98-02-16420, 99-02-18222, and 99-02-31015.

References

Bazilevskaya, G. A., Krainev, M. B., Stozhkov, Yu. I., Svirzhevskaya, A. K., and Svirzhevsky, N. S.: 1991, 'Long-Term Soviet Program for the Measurement of Ionizing Radiation in the Atmosphere', *J. Geomag. Geoelectr. Suppl.* **43**, 893–900.

Bazilevskaya, G. A. and Svirzhevskaya, A. K.: 1998, 'On the Stratospheric Measurements of Cosmic Rays', *Space. Sci. Rev*. **85**, 431–521.

Bazilevskaya, G. A., Krainev, M. B., Makhmutov, V. S., Sladkova, A. I., Storini, M., and Flueckiger, E.: 1998, 'The Gnevyshev Gap in Cosmic Ray Physics', in: J. Medina (ed): *rayos cosmicos 98, Proc. 16th European Cosmic Ray Symposium*, Departamento de Fisica, Universidad de Alcala, Spain, 83–86.

Beer, J.: 2000, 'Long Term, Indirect Indices of Solar Variability'. *Space Sci. Rev.*, this volume.

Charakhchyan, A. N.: 1964, 'Investigation of Stratosphere Cosmic Ray Intensity Fluctuations Induced by Processes on the Sun', *Soviet Physics Uspekhi* **7**(3), 358–374.

Charakhchyan, A. N., Bazilevskaya, G. A., Stozhkov, Yu. I., and Charakhchyan, T. N.: 1976, 'Cosmic Rays in the Stratosphere and in the Near-Earth Space During 19th and 20th Cycles of Solar Activity', *Trudy FIAN*, Nauka, Moscow, **88**, 3–50, (in Russian).

Fisk, L. A. and Wenzel, K. -P., et al.:1998, 'Global Processes that determine Cosmic Ray Modulation', *Space Sci. Rev.* **83**, 179–214.

Golenkov, A. E., Svirzhevskaya, A. K., Svirzhevsky, N. S., and Stozhkov, Yu. I.: 1990, 'Cosmic Ray Latitude Survey in the Stratosphere During the 1987 Solar Minimum', *Proc. 21st Intern. Cosmic Ray Conf.* Adelaide, **7**, 14–17.

GOST, 1986, *Galactic Cosmic Rays. Energy Spectra.* Gosudarstvenny Standard 25645.122-85 - 25645.125-85, Moscow, 1986 (in Russian).

Hatton, C. J.: 1971, 'The Neutron Monitor'. In: J. G. Wilson and S. A. Wouthuysen, (eds.):*Processes in Elementary Particle and Cosmic Ray Physics*, North-Holland Publ. Comp., Amsterdam-London, 3–100.

Makhmutov, V. S., Bazilevskaya, G. A., Podgorny, A. I., Stozhkov, Yu.I., Svirzhevsky, N. S.: 1995, 'The precipitation of electrons into the Earth's atmosphere during 1994', *Proc. 24th Intern. Cosmic Ray Conf.* Rome, **4**, 1114–1117.

Neher, H. V.: 1967, 'Cosmic Ray Particles That Changed from 1954 to 1958 to 1965', *J. Geophys. Res.* **72**(5), 1527–1539.

Neher, H. V.: 1971, 'Cosmic Rays at High Latitudes and Altitudes Covering Four Solar Maxima', *J. Geophys. Res.* **76**(7), 1637–1651.

Simpson, J. A.: 1948, 'The Latitude Dependence of Neutron Densities in the Atmosphere as a Function of Altitude', *Phys. Rev.* **73**, 1389–1391.

Simpson, J. A., Fonger, W. H., and Tremian, S. B.: 1953, 'Cosmic Radiation Intensity-Time Variations and Their Origin: I. Neutron Intensity Variation Method and Meteorological Factors', *Phys. Rev.* **90**, 934–950.

Stoker, P. H.: 1994, 'Relativistic Solar Proton Events', *Space Sci. Rev.* **73**, 327–385.

Storini, M.: 1997, 'Cosmic rays for solar-terrestrial physics', *Nuovo Cim.* **20**(6), 871–880.

Address for Offprints: G.A. Bazilevskaya, Lebedev Physical Institute, Russian Academy of Sciences, Leninsky Prospect, 53, 117924, Moscow, Russia, bazilevs@sci.lebedev.ru

SHORT TERM, DIRECT INDICES OF SOLAR VARIABILITY

J. L. LEAN
E. O. Hulburt Center for Space Research
Naval Research Laboratory, Washington, DC 20375, U.S.A.

Received: 20 October 1999; Accepted in final form: 25 February 2000

Abstract. Indices of solar activity relevant for understanding and modelling solar irradiance variability are identified, and their temporal characteristics compared. Reproducing observed solar irradiance variability requires a minimum of two different types of indices — an index for irradiance depletion by sunspots and an index for global irradiance enhancement by faculae and network. When combined with appropriate wavelength-dependent parameterizations of sunspot and facular contrasts and center-to-limb functions, these indices permit the construction of empirical models of daily, monthly and annual solar total and spectral irradiances. The models are compared with observations at selected wavelengths and for the total irradiance. While the models replicate much of the rotational and 11-year cycle variance in contemporary irradiance databases, differences exist because of either the presence of variability mechanisms additional to solar magnetism, or of unresolved instrumental effects in the databases. The reconstruction of solar irradiance in the past requires speculation about the extent of intercycle fluctuations in the global facular index, or in other, as yet unspecified, variability mechanisms.

1. Introduction

A balance between incoming shortwave solar radiation (which peaks in the visible spectrum) and outgoing longwave terrestrial radiation (which peaks in the vicinity of 10 μm) establishes the equilibrium surface temperature of the Earth. The Sun's irradiance at the top of the Earth's atmosphere, shown in Figure 1, is therefore a critical determinant of Earth's climate. Both the solar spectral irradiance variability, also shown in Figure 1, and the processes which facilitate climate response to solar radiative forcing, are strongly wavelength dependent. Comparison in Figure 1 of the spectrum of radiation incident at the top of the Earth's atmosphere with that incident at the surface (0 km altitude) indicates considerable atmospheric absorption in the ultraviolet (UV) and near infrared (IR) spectral regions. Climate model simulations of Earth's surface temperature response to solar variability in past centuries require knowledge of historical irradiance. Irradiance variability models based on solar activity indices estimate these inputs. Past climate change simulations have used reconstructions of total solar irradiance, alone; future global change studies are planned using reconstructions of the spectral irradiance changes that comprise total irradiance variability, at wavelengths from the UV to the far infrared IR.

Space-based observations of total solar irradiance exist for only about twenty years. A composite record compiled recently from multiple, independent measure-

Space Science Reviews **94:** 39–51, 2000.

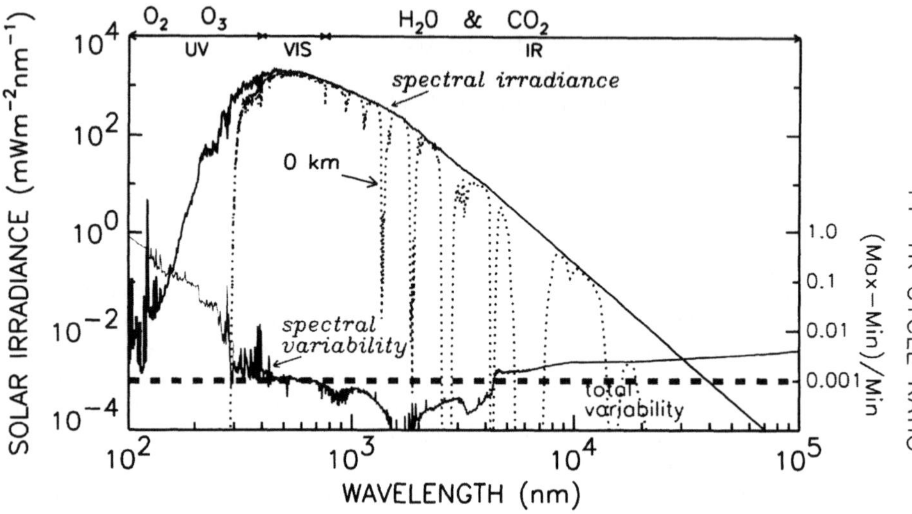

Figure 1. Shown are the Sun's spectral irradiance at the top of the Earth's atmosphere and its estimated variability during the 11-year solar activity cycle. Also shown is the irradiance spectrum at the Earth's surface (0 km). Gases in the Earth's atmosphere (O_2, O_3, H_2O, CO_2) absorb solar radiation, especially in the UV and IR spectrum.

ments (Fröhlich and Lean, 1998; Fröhlich, 2000) exhibits a prominent 11-year cycle of amplitude 0.085% (difference between September 1986 and November 1989 monthly means). Larger fluctuations occur regularly during the Sun's 27-day rotation on its axis. During epochs of high solar activity these shorter term fluctuations correspond to decreases in irradiance of a few tenths percent whereas near solar minimum the decreases are much smaller. Levels of the composite irradiance record are approximately the same during the two minima of 1986 and 1996. But the database is too short, by far, for the reliable detection of long-term irradiance trends such as Willson (1997) has speculated.

Space-based observations of solar spectral irradiance exist primarily at UV wavelengths and are of limited duration compared with those of total solar irradiance. The longest and most reliable data, from the Upper Atmosphere Research Satellite (UARS), have been made only since October 1991. The UARS data suggest solar cycle irradiance changes of 20%, 8% and 3% near 140, 200 and 250 nm, respectively (Lean *et al.*, 1997; Rottman, 2000), as Figure 1 indicates. Observations of solar cycle spectral irradiance variability at visible and infrared wavelengths are lacking because space-based radiometers have had neither the sensitivity nor long-term stability necessary for their detection. For this reason, the solar spectrum changes shown in Figure 1 at wavelengths longward of 400 nm are theoretical estimates (Solanki and Unruh, 1998; Unruh *et al.*, 2000).

Indices of solar activity, such as those compared in Figure 2, permit the development of models of solar irradiance variability that extend beyond the observational database, itself too short for most climate change applications. Prior to the devel-

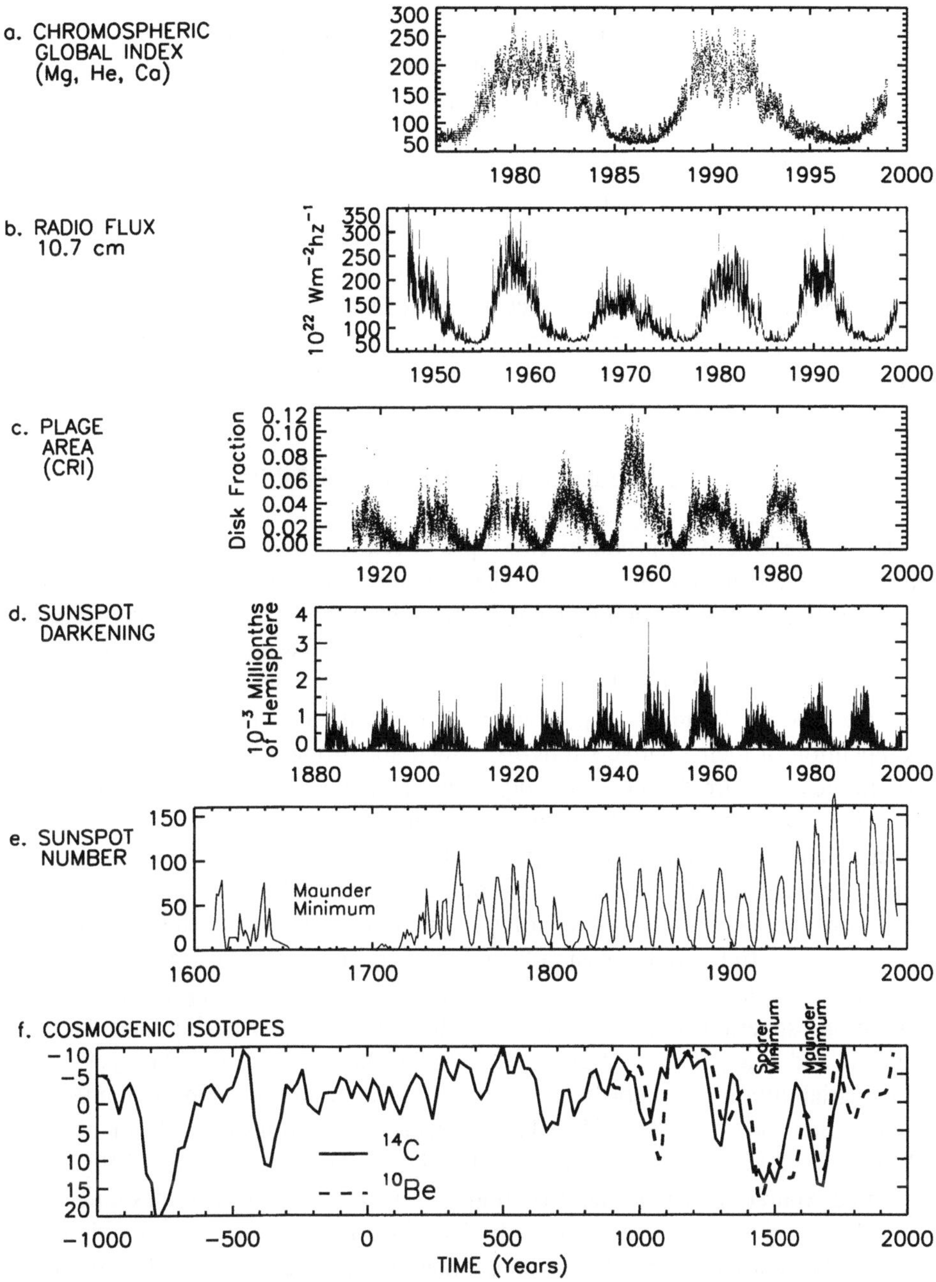

Figure 2. Solar activity produces a range of phenomena from which assorted indices are derived. The indices in panels a)-e) are based on direct solar observations whereas the indirect indices in f) are obtained from terrestrial archives.

opment of such models, the lack of knowledge of long-term irradiance variability constrained many Sun-climate investigations to empirical studies that simply correlated indices of solar activity directly with assorted climate parameters (Hoyt and Schatten (1997), and Lean and Rind (1998), provide overviews). Irradiance reconstructions permit more physical approaches, including general circulation model investigations of the actual processes by which climate responds to forcing by changing solar radiation. The irradiance variability models parameterize selected indices of solar activity that reflect the primary sources of irradiance variability but that exist for much longer than the space-based observations. Of the indices in Figure 2, those in panels a) to e) are obtained from various types of solar observations and are therefore direct indices of solar variability. For comparison, panel f) of Figure 2 shows the cosmogenic isotopes that are indirect indices of solar variability, obtained from terrestrial rather than solar archives (Beer, 2000).

2. Sources of Irradiance Variability

Solar magnetic activity produces two types of features that cause the primary variations observed in solar irradiance. These features are sunspots, which are cooler and darker than the surrounding solar atmosphere, and faculae, which are hotter and brighter. Sunspots and faculae respectively deplete and enhance local solar radiation, thereby altering the net radiation projected in the direction of the Earth, i.e., the irradiance. Since sunspots and faculae usually occur on the Sun together, a competition of their relative strengths, which are strongly wavelength-dependent, determines solar irradiance variability. For example, sunspots have minimal effect on UV radiation but a significant role at visible wavelengths.

Irradiance modulation by sunspots and the largest, contiguous facular regions (those associated with chromospheric plages) accounts for most of the irradiance variations that occur during solar rotation, but only a part of the change during the 11-year cycle. Enhanced emission from bright magnetic features that have smaller scales and are more dispersed on the solar disk (typically tracing the boarders of the chromospheric network) contribute additional variability during the solar cycle (Foukal and Lean, 1988; Fontenla *et al.*, 1999). Other sources of solar cycle irradiance variability are also postulated, such as a change in photospheric temperature (Kuhn and Libbrecht, 1991). Whether 'nonfacular' variations result from, or are independent of, solar magnetic features (thermal shadows, active network) is not yet clarified. Sources and amplitudes of irradiance variations on time scales longer than the 11-year cycle are speculative, only. Evidence from Sun-like stars and cosmogenic solar indices (Beer, 2000) suggests that the Sun is capable of a broader range of activity than witnessed during the past two solar cycles, thereby implying the existence of a long-term irradiance variability component. Changes in the global emission from the "quiet" sun, by a reduction in either faculae brightness or area in the network borders or in the basal emission from the centers of the network cells,

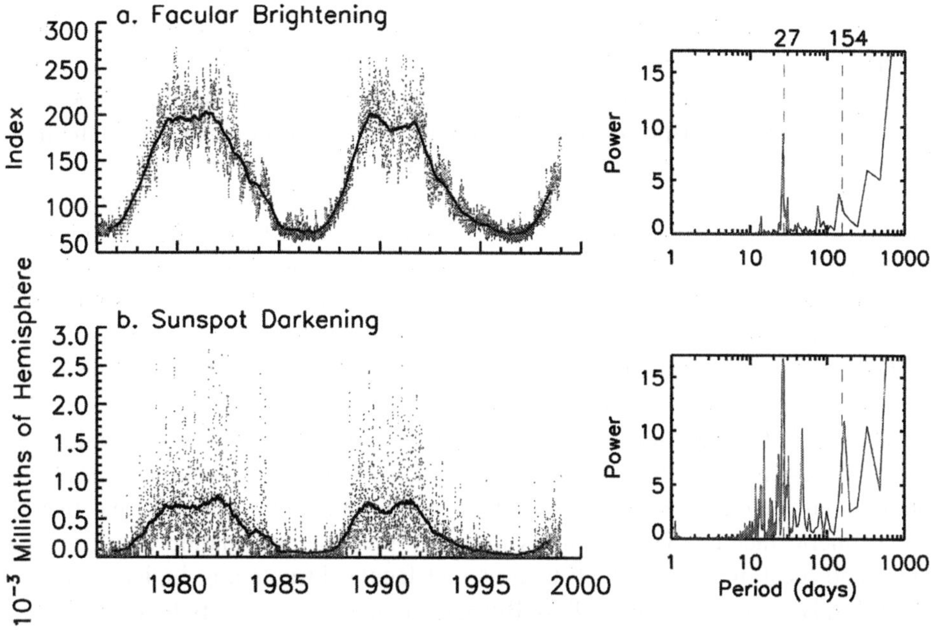

Figure 3. Indices of facular brightening (a) and sunspot darkening (b) are compared during two solar cycles. The periodograms on the right compare the variance in the daily time series.

is one possible long-term irradiance variability mechanism (White *et al.*, 1992). Long-term changes in solar diameter (Sofia and Fox, 1994) and convective strength (Hoyt and Schatten, 1993) are also proposed.

3. Indices of Irradiance Variability

A variety of solar observations provide information about sunspots and faculae (associated with both plages and active network) independently of direct irradiance measurements. The daily facular brightening, P_F, and sunspot darkening, P_S, time series shown in Figure 3 since 1976 are two such indices that are used widely for irradiance variability modelling. The facular brightening index is a composite of directly measured flux ratios of emission from the center of Fraunhofer features (CaII H & K, MgII h & k) relative to emission in the line wings (de Toma *et al.*, 1997; White *et al.*, 1998) whereas white light solar images that record the locations and areas of sunspots provide the primary inputs for the sunspot darkening index.

Sunspot darkening in Figure 3 is calculated explicitly as $P_S = \sum \Delta F_S / F_Q = \sum 5\mu A_{\mathrm{WDC}}(C_S - 1)R(\mu)/2$ where ΔF_S is the reduction in radiation relative to the quiet Sun, F_Q, for a sunspot of area A_{WDC} (in fractions of the solar hemisphere, as reported by the World Data Center) at location μ in heliocentric coordinates. C_S is the sunspot's contrast (ratio of spot emission to the background photosphere) and $R(\mu)$ is the center-to-limb function. The summation is over all spots on the solar disk at a specific time. Ground-based white light images made from 1882 to 1976 by the Greenwich Observatory, and most recently by the U.S. Air Force operational SOON sites (provided to the NOAA NGDC), supply the basic time-dependent information about sunspot areas and locations. A correction is applied to the time series in Figure 3 that reduces the Greenwich areas by 20% to account for differences in sunspot areas between the Greenwich and SOON datasets (Fligge and Solanki, 1997). The sunspot darkening in Figure 3 utilizes bolometric contrast and center-to-limb functions (Lean *et al.*, 1998) and is therefore directly appropriate for modelling total solar irradiance variations. Sunspots actually comprise a very dark central umbra and a less dark surrounding penumbra but present versions of P_S do not include this distinction because the SOON data lack sufficient information about these specific sunspot features.

The explicit calculation of a facular index analogous to P_S is also possible. In solar images made in the core of Ca and other Fraunhofer lines the emission is preferentially enhanced in chromospheric plage and bright network that overlie photospheric faculae. These images therefore provide information about facular areas and locations from which to construct facular brightening indices (Lean *et al.*, 1998; Harvey and White, 1998). But such indices are generally less satisfactory for irradiance variability modelling than is the global flux index in Figure 3. It is difficult to quantify the areas and contrasts of small-scale faculae whose spatial scales and modest brightnesses are comparable to that of the network that composes the quiet Sun. Also, physical characteristics of faculae and plages such as their center-to-limb variations and contrasts are less well known than for sunspots. In comparison, chromospheric flux ratios offer a number of advantages: the technique of ratioing the core and wing fluxes minimizes instrumental drifts, the line cores originate in the chromosphere where bright features have greater contrast than their photospheric counterparts (factor of 2 versus a few percent for an active region), and the flux measurements automatically integrate emission over the full disk. One disadvantage is that the chromospheric and photospheric radiations can have somewhat different center-to-limb dependences of emission and contrast.

Knowledge of long-term sunspot darkening exists since the 1880s, as Figure 2 shows, but there are no flux measurements of the Sun in the Fraunhofer line cores prior to 1975. An index of global chromospheric variability can be constructed since about 1950 by combining flux measurements of the sun in the radio spectrum (the 10.7 cm flux, shown in Figure 2) with the McMath plage index to account for solar cycle and solar rotation modulation, respectively. Digitized scalings of Mount Wilson Ca K spectroheliograms enable the construction of an index of plage emission since 1905, also shown in Figure 2 (Foukal, 1998). The correlation of the two nominally equivalent Ca K plage indices (from McMath and Mount Wilson images) from 1945 to 1985 is only 0.83, implying either real or instrumental differences. Future analyses of the Mount Wilson spectroheliograms may yield improved facular indices for this century.

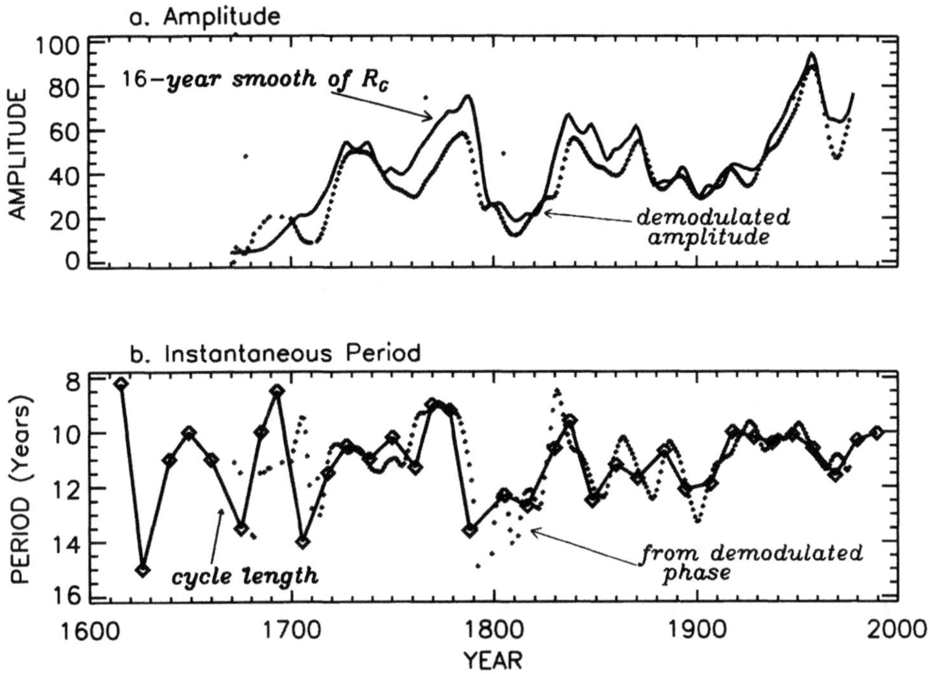

Figure 4. Shown in the upper panel are the demodulated amplitude of the group sunspot number compared with 16-year smoothed values. In the lower panel the instantaneous period of the 11-year cycle and the cycle length are compared. Both cycle amplitude and length have been used as indices for long-term solar variability.

The only direct solar index prior to 1882 is the sunspot number, recently revised (Hoyt *et al.*, 1994) to the sunspot group number, R_G, shown in Figure 2. Indices have been developed from R_G to simulate postulated sources of long-term irradiance variability on multi-decadal to centennial time scales. Such indices derive from either the smoothed amplitude of the 11-year R_G cycle or its instantaneous period (i.e., cycle length). As Figure 4 shows, the temporal structure of these two quantities are rather different from each other, and their correlation is poor (−0.34). Which of the two time series in Figure 4 better represents long-term irradiance variability, if it exists, is unknown.

4. Empirical Irradiance Variability Models

Empirical models of solar irradiance variability over multiple time scales utilize as inputs three specific solar indices, namely sunspot darkening, P_S, facular brightening, P_F, and an index of postulated long-term variability, I_L. The models are constructed in different ways, depending on whether or not suitable irradiance observations are available at the wavelengths of interest, over the time scales of interest. Models of total and UV spectral irradiance variability, for which adequate

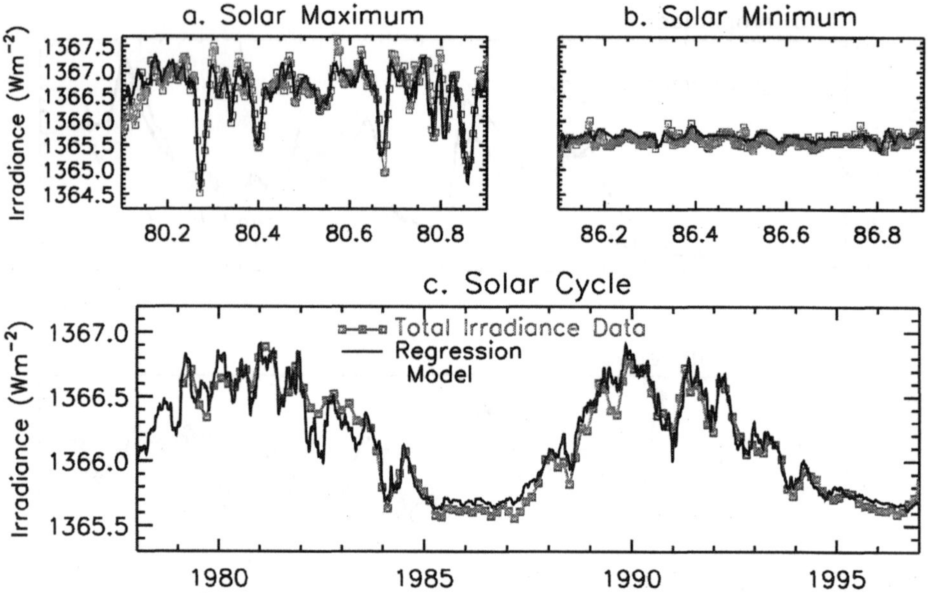

Figure 5. An empirical model of total solar irradiance variability based on two indices — sunspot darkening and facular brightening from Figure 3 — is compared with the composite observational irradiance record during solar rotation at (a) high solar activity and (b) lower solar activity. Compared in (c) are the smoothed model and measurements during two solar cycles.

observational databases exist, are developed in one of two ways. Either the P_F and P_S time series are regressed together against the observed irradiance, S, or the observations are first corrected for sunspot influences then the residual time series, $S - P_S$, linearly regressed against P_F (Lean *et al.*, 1997; Fröhlich and Lean, 1998). Both approaches yield an expression of the form $S(t) \propto a_\lambda P_S(t) + b_\lambda P_F(t)$ for solar irradiance at time t, where the indices contain time dependent information about sources of the irradiance variability and the coefficients a_λ and b_λ adjust the relative contributions of these sources for radiation at different wavelengths, or for total irradiance. In practice, the observations may or may not be detrended prior to the regression, depending on the confidence (or lack of) in their long-term stability.

Empirical models typically account for more than 80% of observed total irradiance variability. Figure 5 compares the composite total solar irradiance record with an empirical variability model developed from multiple regression of P_S and P_F against the daily data since 1978. Some improvement in the model's reproduction of short-term total irradiance variations is achieved from multiple regression of the observations against three indices, namely P_S, P_F and $\langle P_F \rangle$ where $\langle P_F \rangle$ is smoothed over about 3 months. This approach better accommodates differences in center-to-limb functions of the visible spectrum compared with that of the chromospheric emissions that comprise the P_F index (Fröhlich and Lean, 1998), and increases the correlation of the model and observations from 0.91 to 0.92.

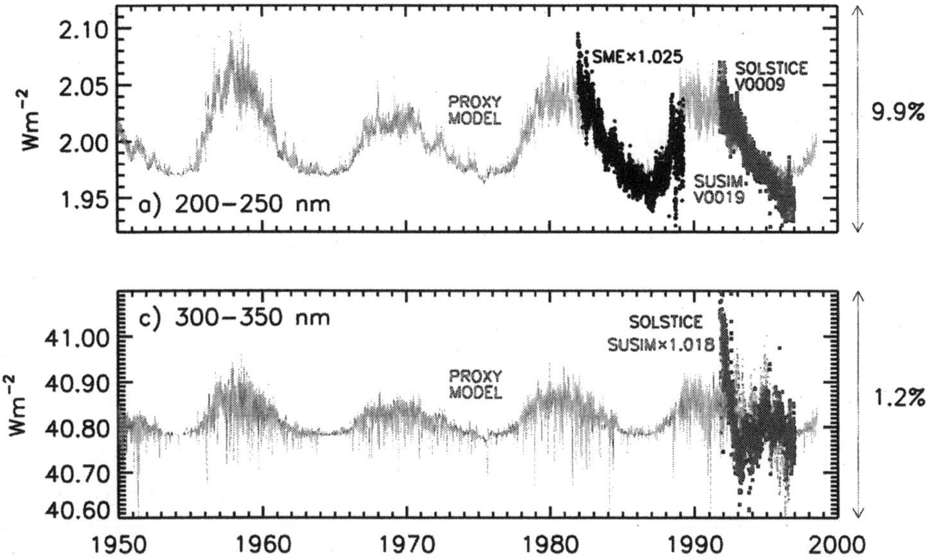

Figure 6. Empirical models constructed from combinations of the facular brightening and sunspot darkening indices in Figure 3 are compared with observations in two UV bands.

At UV wavelengths, empirical models have been developed by regression of the sunspot and facular indices against observations (Lean *et al.*, 1997) and from facular indices determined directly from Ca K images (Lean *et al.*, 1998; Worden *et al.*, 1998). Figure 6 compares models and observations in two UV bands. The models generally reproduce observed variability better at shorter wavelengths where the changes are larger but instrument long-term stability comparable. For example, a model developed from regression of the indices in Figure 3 with detrended SOLSTICE data correlates better with the data at 200 nm (0.88) than at 250 nm (0.7). At some wavelengths, differences between the two UARS datasets can be as large as differences between the data and the model. One cause of discrepancies may be wavelength-dependent instrumental effects remaining in the observations, whose long-term stabilities are still being revised. If the UV irradiance trends present in the observations are assumed to be real (rather than instrumental), an additional component of long term variability is needed. However, if the larger-than-predicted solar cycle changes evident in the observations at some wavelengths are indeed real — such as continued irradiance declines even while solar activity increases after solar minimum — then the empirical models will require additional components unrelated to solar magnetism. What these might be is unclear.

Models of solar irradiance variability at wavelengths longer than 400 nm, for which adequate observational databases are lacking, also utilize the sunspot darkening and facular brightening indices in Figure 3 but rely on theoretical solar models (Unruh *et al.*, 2000) to provide information about the coefficients a_λ and b_λ. The wavelength dependencies of the contrasts of faculae and sunspots, shown

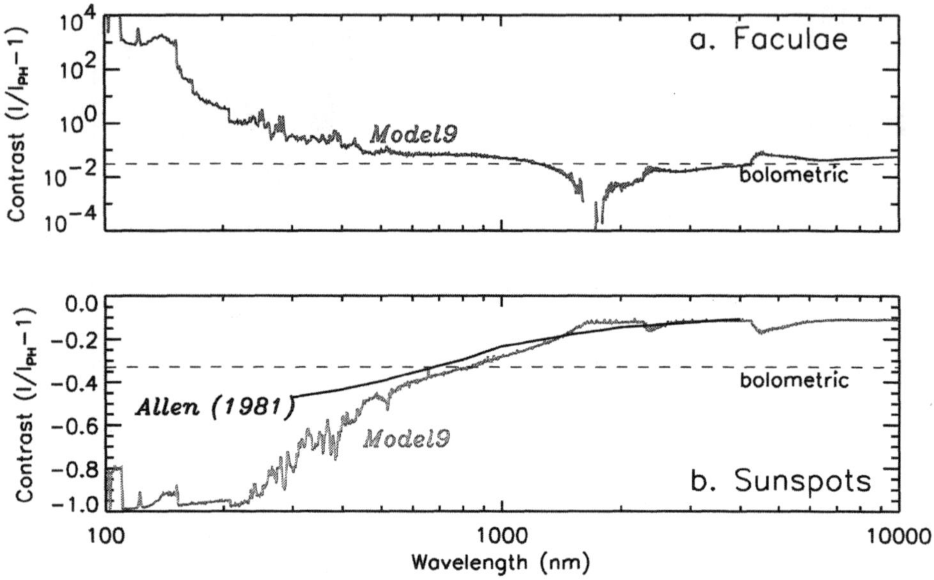

Figure 7. The contrasts of bright faculae and dark sunspots are shown as functions of wavelength from the UV to the IR spectral regions. The contrasts are ratios of fluxes to the background photosphere, determined from theoretical models of the solar atmosphere (Unruh *et al.*, 2000) and limited observations (Allen, 1981).

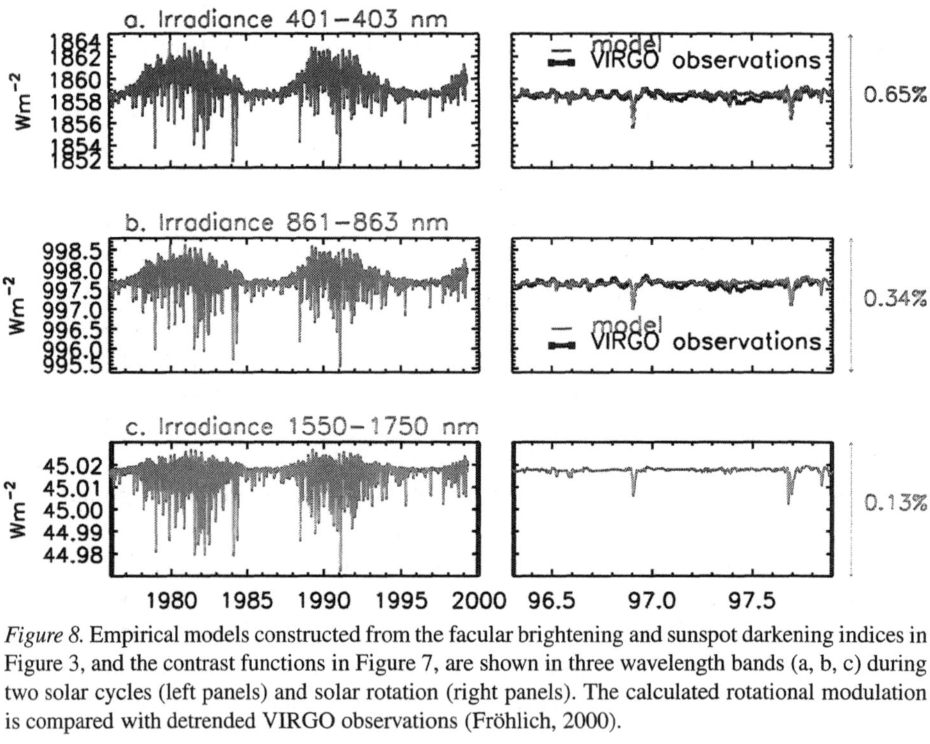

Figure 8. Empirical models constructed from the facular brightening and sunspot darkening indices in Figure 3, and the contrast functions in Figure 7, are shown in three wavelength bands (a, b, c) during two solar cycles (left panels) and solar rotation (right panels). The calculated rotational modulation is compared with detrended VIRGO observations (Fröhlich, 2000).

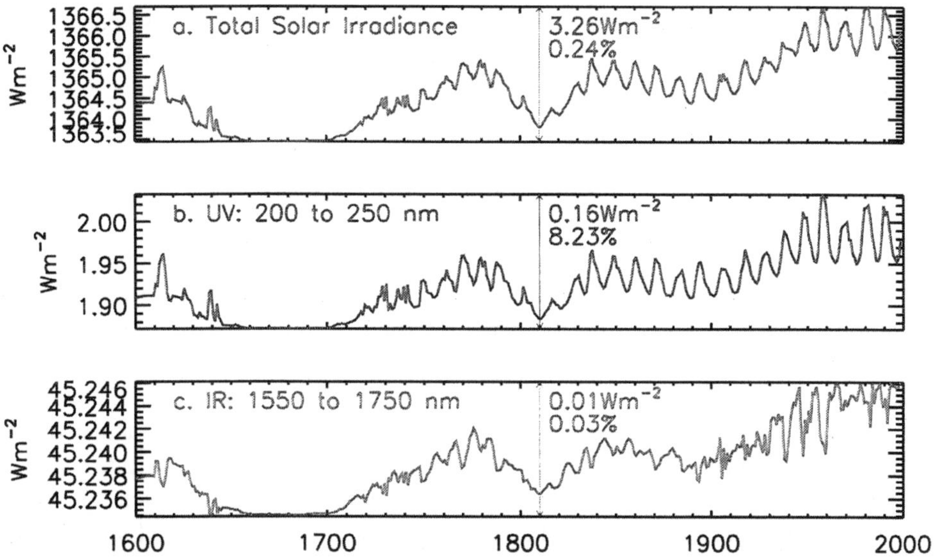

Figure 9. Model estimates of long-term variability in the total solar irradiance and in two spectral bands in the UV and IR. In these calculations, the adopted amplitude of the long-term facular component is approximately equal to its contemporary solar cycle amplitude.

in Figure 7, determine appropriate combinations of P_S and P_F for reproducing irradiance variabilities at different wavelengths. As the model time series in Figure 8 show, the short-term irradiance dips caused by sunspots are more pronounced in the 1550 – 1750 nm band than at shorter wavelengths because the facular influence is less at IR than at UV wavelengths. Comparisons of model calculations with observations at visible and infrared wavelengths are only possible in three narrow wavelength bands, and only during solar rotation. The comparisons in Figure 8 with rotational modulation of the SOHO/VIRGO spectral bands (Fröhlich, 2000) suggest that the infrared region is modelled better than the UV region. However, data covering only a few sunspot events are inadequate for detailed statistical evaluation of the model's capabilities or shortcomings. Note that since the VIRGO spectral channels lack adequate long-term stability for solar cycle comparisons, detrended data are shown in Figure 8.

Simulation of long-term irradiance variability consistent with trends inferred from Sun-like stars and cosmogenic isotopes requires the adoption of an index of long-term solar variability, I_L, in addition to the sunspot darkening and facular brightening indices in Figure 3. Figure 9 presents a scenario in which solar irradiance is determined as $S(t) \propto a_\lambda P_S(t) + b_\lambda P_F(t) + c_\lambda I_L(t)$ where I_L is the smoothed sunspot group number from Figure 4, and c_λ accounts for the wavelength-dependent amplitude of the long-term component. The long-term component is scaled such that the irradiance change from 1650 to 1986 is approximately equal to the solar cycle variation in the facular component of the variability.

Estimating the magnitude of long-term solar irradiance variability is speculative, at best, since it is presently not known whether the long-term irradiance variations actually do occur.

5. Summary and Future Research

Indices derived from a variety of solar activity phenomena are essential components of empirical models of solar irradiance variability that climate models use to simulate terrestrial response to solar forcing. The indices permit the construction of models that aid in interpreting contemporary irradiance data and facilitate estimates of past variability. Improvements are possible in the indices themselves and in the models developed from them. Knowledge of sunspot and facular properties (areas, contrasts, center-to-limb functions) is presently rather rudimentary, and limits model reproduction of day-to-day, monthly and solar cycle changes. Improved sunspot darkening and facular brightening indices obtained with inputs of higher quality and greater photometric precision will help to better specify the magnetic components of solar variability, and aid in the detection of nonmagnetic variability components. Future programs of ground-based observations such as SOLIS, ISOON and RISE and at BBSO are planning to generate data needed for improved indices. However, a critical component of the models, the long-term variability, will likely remain speculative and highly uncertain until improved understanding and longer time series of observations are achieved. Future space-based monitoring programs, including SORCE, PICARD and the SOLSPEC and SOVIM instruments on the ISS, will monitor total and spectral irradiance with improved precision, long-term stability and wavelength coverage during the early part of the next decade. Solar irradiance monitoring must be continued indefinitely to extend the historical record to sufficient duration for the detection of long-term irradiance variability trends.

Acknowledgements

The models and indices described in this paper have been developed over the past decade in collaboration with Peter Foukal, Claus Fröhlich, Dick White and Bill Livingston, among others. Gary Rottman and Dick Willson supplied UARS observations and Yvonne Unruh provided the contrast models in Figure 7. The VIRGO team produced the observational time series in Figure 8. NASA and ONR supported this work.

References

Allen, C. W.: 1981, *Astrophysical Quantities*, 3rd Edition, Athlone, London.

Beer, J.: 2000, 'Long-Term Indirect Indices of Solar Variability', *Space Sci. Rev.*, this volume.

de Toma, G., White, O. R., Knapp, B. G., Rottman, G. J., and Woods, T. N.: 1997, 'Mg II core-to-wing index: Comparison of SBUV 2 and SOLSTICE time series', *J. Geophys. Res.* **102**, 2597.

Fligge, M., and Solanki, S. K.: 1997, 'Inter-cycle variations of solar irradiance: Sunspot areas as a pointer', *Solar Phys.* **173**, 427.

Fontenla, J., White, O. R., Fox, P. A., Avrett, E. H., and Kurucz, R. L.: 1999, 'Calculation of Solar Irradiances. I. Synthesis of the Solar Spectrum', *Astrophys. J.* **518**, 480.

Foukal, P.: 1998, 'Extension of the F10.7 Index to 1905 using Mt. Wilson Ca K Spectroheliograms', *Geophys. Res. Lett.* **25**, 2902.

Foukal, P. and Lean, J.: 1988, 'Magnetic Modulation of Solar Luminosity by Photospheric Activity', *Astrophys. J.* **328**, 347.

Fröhlich, C.: 2000, *Space Sci. Rev.*, this volume.

Fröhlich, C. and Lean, J.: 1998, 'The Sun's Total Irradiance: Cycles, Trends and Climate Change Uncertainties since 1976', *Geophys. Res. Lett.* **25**, 4377.

Harvey, K. L., and White, O. R.: 1998, 'Spectral Irradiances and Magnetic Structures', *Astronomical Society of the Pacific Conference Series* **140**, 247.

Hoyt, D. V., and Schatten, K. H.: 1993, 'A Discussion of Plausible solar irradiance variations', *J. Geophys. Res.* **98**, 18,895.

Hoyt, D. V., and Schatten, K. H.: 1997, 'The Role of the Sun in Climate Change', Oxford University Press, Oxford.

Hoyt, D. V., Schatten, K. H., and Nesme-Ribes, E.: 1994, 'The One Hundredth Year of Rudolf Wolf's Death: Do we have the Correct Reconstruction of Solar Activity?', *Geophys. Res. Lett.* **21**, 2067.

Kuhn, J. R., and Libbrecht, K.: 1991, 'Nonfacular Solar Luminosity Variations', *Astrophys. J.* **381**, L35.

Lean, J., and Rind, D.: 1998, 'Climate forcing by changing solar radiation', *J. Climate* **11**, 3069.

Lean, J. L., Rottman, G. J., Kyle, H. L., Woods, T. N., Hickey, J. R., and Puga, L. C.: 1997, 'Detection and parameterization of variations in solar mid and near ultraviolet radiation (200 to 400 nm)', *J. Geophys. Res.* **102**, 29939.

Lean, J. L., Cook, J., Marquette, W., and Johannesson, A.: 1998, 'Magnetic modulation of the solar irradiance cycle', *Astrophys. J.* **492**, 390.

Rottman, G.: 2000, *Space Sci. Rev.*, 1991-1999', this volume.

Sofia, S., and Fox, P.: 1994, 'Solar Variability and Climate', *Climate Change* **30**, 1.

Solanki, S. K., and Unruh, Y. C.: 1998, 'A model of the wavelength dependence of solar irradiance variations', *Astron. Astrophys.* **329**, 747.

Unruh, Y. C., Solanki, S. K., and Fligge, M.: 2000, 'Modelling Solar Spectral Irradiance Variations', *Space Sci. Rev.*, this volume.

White, O. R., Skumanich, A., Lean, J., Livingston, W. C., and Keil, S. L.: 1992, 'The Sun in a Non-cycling State', *Publications of the Astronomical Society of the Pacific* **104**, 1139.

White, O. R., Livingston, W. C., and Keil, S. L.: 1998, 'Variability of the Solar CaII K Line over the 22 Year Hale cycle', *Astronomical Society of the Pacific Conference Series* **140**, 293.

Willson, R. C.: 1997, 'Total Solar Irradiance Trend During Solar Cycles 21 and 22', *Science* **277**, 1963.

Worden, J. R., White, O. R., and Woods, T. N.: 1998, 'Evolution of chromospheric structures derived from CaII K spectroheliograms: implications for solar ultraviolet irradiance variability', *Astrophys. J.* **496**, 998.

LONG-TERM INDIRECT INDICES OF SOLAR VARIABILITY

JÜRG BEER (beer@eawag.ch)
Environmental Physics, EAWAG, CH-8600 Dübendorf, Switzerland

Received: 4 February 2000; Accepted in final form: 27 April 2000

Abstract. Continuous direct records of solar variability are limited to the telescopic era covering approximately the past four centuries. For longer records one has to rely on indirect indices such as cosmogenic radionuclides. Their production rate is modulated by magnetic properties of the solar wind. Using a parameterisation of the solar activity and a Monte Carlo simulation model describing the interaction of the cosmic rays with the atmosphere, the production rate for each cosmogenic nuclide of interest can be calculated as a function of solar activity. Analysis of appropriate well-dated natural archives such as ice cores or tree rings offers the possibility to reconstruct the solar activity over many millennia. However, the interpretation of the cosmogenic nuclide records from these archives is difficult. The measured concentrations contain not only information on solar activity but also on changes in the geomagnetic field intensity and the transport from the atmosphere into the archive where, under ideal conditions, no further processes take place. Comparison of different nuclides (e.g. ^{10}Be and ^{14}C) that are produced in a very similar way but exhibit a completely different geochemical behaviour, allows us to separate production effects from system effects.

The presently available data show cyclic variability ranging from 11-year to millennial time scale periodicities with changing amplitudes, as well as irregularly distributed intervals of very low solar activity (so called minima, e.g. Maunder minimum) lasting typically 100 years.

1. Introduction

Solar variability is an important issue from the solar physics as well as from the terrestrial climate point of view. Solar variability is commonly related to magnetohydrodynamic processes within the convective zone. Most phenomena observed at the solar surface such as sunspots, magnetic network, coronal mass ejections, flares and fluctuations in the amount and the spectral distribution of the emitted radiation are connected in various ways to these processes. At present the theoretical understanding is not yet good enough to model the relationship and the dynamics of these solar variability indices based on physical processes.

The importance of solar variability for the Earth is based on the fact that the Sun is by far the most important source of energy. Geothermal and tidal energy are negligible compared to solar radiation. Therefore, any change in solar radiation or its spectral composition is potentially important for the terrestrial climate (Beer et al., 2000; Mende and Stellmacher, 2000). Terrestrial phenomena related to solar variability are aurorae, the magnetic aa-index and the cosmic ray induced neutron flux in the atmosphere (Beer et al., 1996). All solar indices mentioned so far are based on observations or measurements. Their historical records are therefore lim-

Space Science Reviews **94:** 53–66, 2000.

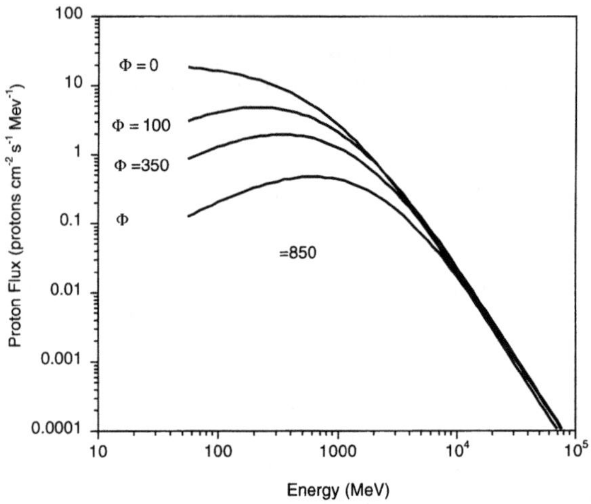

Figure 1. Energy spectrum of the primary cosmic ray proton flux at the top of the atmosphere as a function of the solar modulation parameter Φ. $\Phi = 0$ MeV corresponds to a completely inactive Sun, $\Phi = 850$ MeV to a rather active one.

ited to the period during which observations have been carried out. The longest continuous record is the sunspot record which goes back to the beginning of the 17th century, the time when the telescope was invented. Cosmogenic radionuclides on the other hand offer the unique opportunity to reconstruct solar variability over much longer time scales.

2. Solar Modulation of Cosmogenic Nuclei Production

Cosmogenic nuclei are produced continuously by the interaction of cosmic ray particles with the atmosphere. The primary galactic cosmic rays consist of protons (87%), helium nuclei (12%), and heavier elements (1%). In Fig. 1 the effect of solar modulation on the energy spectrum of primary protons is shown. Modulation of the differential energy spectrum of galactic cosmic rays by solar activity takes place within a radius of about 100 AU around the Sun. During active periods, more coronal mass ejections cause large disturbances in the interplanetary magnetic field, which in turn result in enhanced scatter of galactic cosmic rays away from the inner heliosphere. As a consequence, the net flux reaching the top of the atmosphere is reduced especially at the low energy side of the spectrum. This process has been parameterised using the solar activity parameter Φ. A value of $\Phi = 0$ MeV corresponds to no solar modulation (quiet Sun) and reflects the estimated interstellar spectrum, while $\Phi = 850$ MeV corresponds to a rather active Sun (see Fig. 1).

Using these primary cosmic ray spectra the nuclear interactions with N, O and Ar in the atmosphere can be simulated by Monte Carlo calculations (Masarik and

TABLE I

Some characteristics of three long-lived abundant cosmogenic radionuclides.

Nuclide	Target	Half-life [y]	Production Rate [atoms cm^{-2}s^{-1}]	Inventory [tons]
^{14}C	N	5730	2	62
^{10}Be	N, O	1.51×10^6	0.018	105
^{36}Cl	Ar	3.08×10^5	0.0019	8

Beer, 1999). Depending on the original energy, a cascade of secondary particles develops. Mainly the neutrons produced in such cascades generate – by spallation and other nuclear reactions – a variety of new nuclei most of which are stable or very short-lived. Assuming that the interstellar galactic cosmic ray flux is constant in time, the instantaneous production rate of these nuclides at a certain point in the atmosphere depends only on the solar activity and the geomagnetic field intensity. Averaging over the whole atmosphere provides the mean global production rate (Tab. I). Based on those numbers the steady state global amount of each radionuclide is calculated. The production rate of ^{10}Be is significantly lower than that of ^{14}C because it is only produced by high-energy spallation reactions. In the case of ^{36}Cl, the production rate is even lower because it can only be produced from ^{40}Ar which is quite rare (1%) in the atmosphere. We do not make use of ^{36}Cl in this paper. However, since it is sometimes used in literature instead of ^{10}Be we include it in Tab. I.

Two indices that are especially interesting from the climate point of view are the solar diameter (Gilliland, 1981) and the solar irradiance (Lean, 1997; Pap and Fröhlich, 1997). While the first is very difficult to determine with the necessary precision and therefore does not provide conclusive results so far, the latter shows changes over an 11-year Schwabe cycle in the order of 0.1%. To monitor such small changes over decades, extremely stable radiometers are required and careful corrections for degradation processes must be applied. As a consequence, there is no general agreement with regard to the question if there is a trend during the past 2 cycles (Willson, 1997; Beer et al., 2000) or not (Fröhlich and Lean, 1997; Fröhlich, 2000).

The dependence of the production rate of a particular radionuclide on the geomagnetic field intensity and the solar modulation parameter Φ can be calculated using the proton and neutron fluxes obtained from Monte Carlo calculations (Masarik and Beer, 1999). As an example, the results of such calculations are shown in Fig. 2 for ^{10}Be. The geomagnetic field intensity is given in units relative to its present value. The solar modulation parameter Φ varies in the range from 0 MeV (quiet

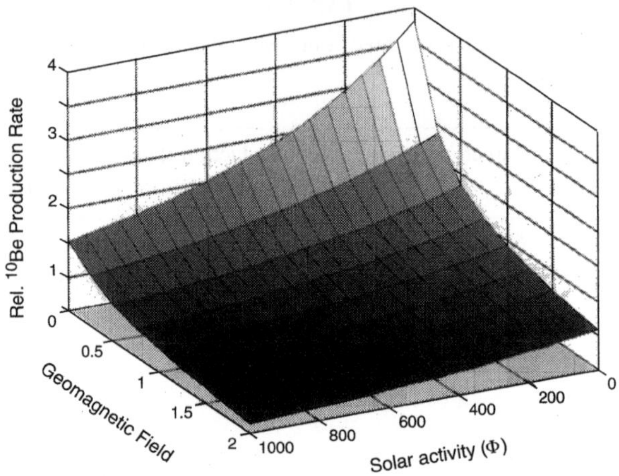

Figure 2. Relative global mean production rate of ^{10}Be as a function of solar activity Φ and geomagnetic field intensity M. M is normalised to the present value. The production rate 1 corresponding to $M = 1$ and $\Phi = 550$ MeV accounts to 0.081 atoms $cm^{-2}s^{-1}$.

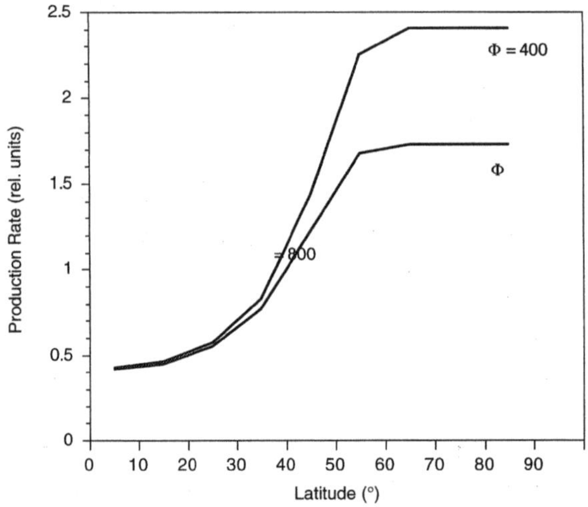

Figure 3. Relative production rate of ^{10}Be as a function of geomagnetic latitude for the present geomagnetic field intensity ($M = 1$) and two different solar activity levels. Note that the solar modulation is effective mainly at high latitudes.

Sun) to 1000 MeV (very active Sun). The relative units of the ^{10}Be production rate are atoms $cm^{-2}s^{-1}$. For an average $\Phi = 550$ MeV and $M = 1$ the mean global ^{10}Be production rate is 0.018 atoms $cm^{-2}s^{-1}$ (Tab. I).

As a consequence of the geomagnetic dipole field which cuts off low energy cosmic ray particles mainly at low latitudes, the solar modulation effect is strongest in polar regions. Figure 3 shows the effect of the latitude on the amplitude of the ^{10}Be signal assuming that Φ fluctuates between 400 and 800 MeV. For a quantita-

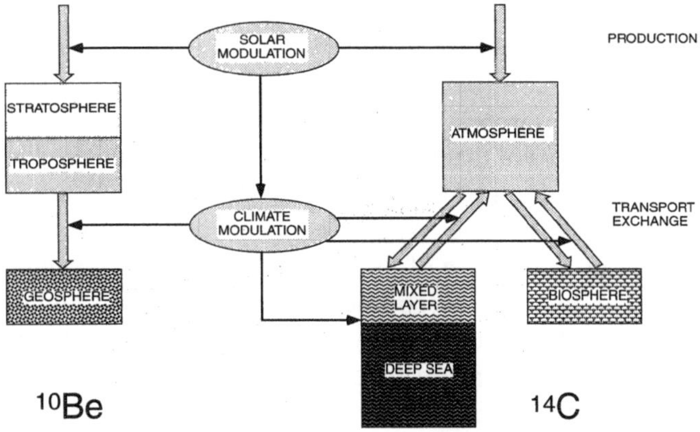

Figure 4. Description of the ^{10}Be and ^{14}C systems by simple box models. Solar modulation affects the production rate of both nuclides in a very similar way. Climate modulation (transport, atmospheric mixing, and deposition), however, leads to completely different effects in the two systems. After its removal from the atmosphere ^{10}Be becomes stored in the geosphere. In contrast ^{14}C continuously exchanges between the different reservoirs resulting in a much more complex behaviour.

tive interpretation of e. g. ^{10}Be in rain or snow it is crucial not only to understand the spatial and temporal variations of the production rate but also the subsequent atmospheric mixing and transport processes. Typical residence times are in the order of years for the stratosphere and weeks for the troposphere. Since about half of the production takes place in the stratosphere, ^{10}Be measurements with a time resolution of less than 1 year mainly reflect atmospheric transport processes. In contrast to the production rate variations that can be modelled quite well, no models based on physico-chemical processes are available today to describe the atmospheric pathways of cosmogenic nuclides from their initial position where they are produced to their final destination where they are deposited. Beside some promising attempts using general circulation models (Brost et al., 1991), most authors rely on rather simple box models which, fortunately, appear to be adequate in many cases when considering cosmogenic nuclide variations on time scales longer than one year (Scheffel et al., 1999).

Figure 4 shows examples of box models generally used to describe the behaviour of ^{10}Be and ^{14}C. In the case of ^{10}Be, the atmosphere is divided into the stratosphere and the troposphere. It is assumed that all the ^{10}Be removed from the troposphere stays in the geosphere and that no recycling takes place. In the case of ^{14}C, $^{14}CO_2$ is formed which exchanges between the four main reservoirs: the atmosphere, the biosphere, the mixed layer, and the deep sea. The shorter the periodicity the more the amplitudes of production variations are attenuated due to the large sizes of these reservoirs. For instance, in the case of the 11-year Schwabe cycle the amplitude of the production rate is damped in the atmospheric ^{14}C concentration by about two orders of magnitude. At the same time, the carbon system causes a periodicity dependent phase lag between production and atmospheric concentration signal (Siegenthaler et al., 1980). Therefore, a quantitative comparison between ^{14}C and other cosmogenic radionuclides such as ^{10}Be and ^{36}Cl is only feasible if based on an appropriate carbon cycle model.

To reconstruct changes of the production rate in the past we need archives where the produced radionuclides are stored in stratigraphically undisturbed layers. Such archives are e. g. ice sheets, glaciers, and sediments in the case of ^{10}Be, tree rings, sediments, and corals in the case of ^{14}C. These archives are assumed to be closed systems with neither internal transport nor any other processes taking place. These assumptions are not always completely fulfilled. ^{10}Be seems to migrate and to become attached to dust grains on time scales of 10^5 years (Baumgartner et al., 1997). Chlorine can become expelled from ice in the form of HCl or from the sea as methyl chloride (Wagnon et al., 1999). In spite of all these disturbing processes the effects on time scales of millennia are rather modest with the possible exception of chlorine.

3. Results

In the following, we will discuss some existing ^{14}C and ^{10}Be data regarding the long-term solar variability.

3.1. ^{14}C

The potential of ^{14}C as a dating tool for organic material was recognised in the early forties, initiating an extensive study of its geochemical cycle and its production history. It was soon obvious that the atmospheric $^{14}C/^{12}C$ ratio exhibits temporal changes of up to about 10% over the last 10'000 years, but that its spatial changes are small in the order of a few permil (tenths of a percent) (McCormac et al., 1998).

This indicates that the atmosphere is globally well mixed regarding ^{14}C. Figure 5 shows the present $\Delta^{14}C$ record (Stuiver et al., 1998) (the relative deviation in permil of the atmospheric $^{14}C/^{12}C$ ratio from a standard value). The $\Delta^{14}C$ curve is characterised by a long-term trend of about 100‰ superimposed to which there are short-term fluctuations of up to 20‰. The long-term trend is mainly caused by the long-term changes in the production rate, the properties of the carbon cycle and the half-life. It is therefore important to keep in mind that the $\Delta^{14}C$ value at any time is not only determined by the actual production rate but also by the production history during the preceding ca. 2 half-lives and by carbon system effects. The short-term $\Delta^{14}C$ fluctuations are characterised by peaks with a typical duration of about 100 years, the so-called Suess-wiggles.

The period 1510–1950 is shown in detail in Fig. 5 (Stuiver and Braziunas, 1993). Sunspot and aurora data strongly suggest that the wiggles can be attributed to periods of low solar activity. The most pronounced of these periods is the Maunder Minimum from 1645–1715 AD. Similarly, less pronounced periods of low

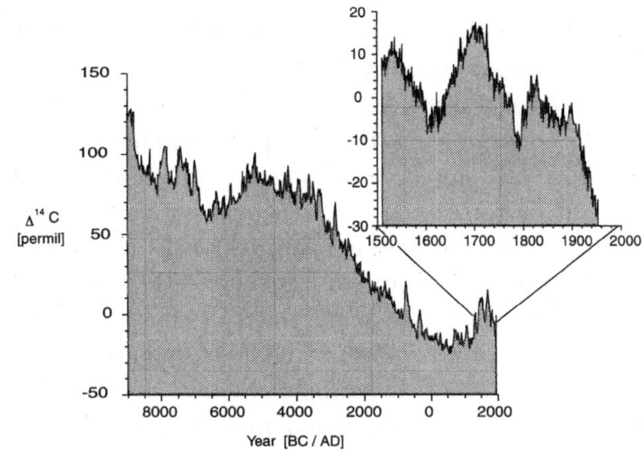

Figure 5. $\Delta^{14}C$ measured in tree rings for the past 11 ky. The record is characterised by short-term variations superimposed to a long-term trend. The short-term variations can be attributed mainly to solar modulation whereas the long-term trend is mainly the result of variations of the geomagnetic field intensity during the past 20 ky and possibly some changes within the carbon system.
The period from 1510 AD to 1950 AD is shown in detail. The peak at 1700 AD corresponds to the maximum of the Maunder minimum, a period famous for its almost complete lack of sunspots. The steep decrease during the 20th century is due to the Suess effect (dilution of the atmospheric ^{14}C content by the growing combustion of fossil fuel) and an increase in solar activity.

solar activity occurred around 1800 (Dalton Minimum) and 1900. The decreasing trend in $\Delta^{14}C$ as from about 1900, however, is to a large extent of anthropogenic origin. The increasing rate of fossil fuel consumption led to an increasing amount of ^{14}C-free CO_2 reducing the $^{14}C/^{12}C$ ratio considerably.

The $\Delta^{14}C$ record covering the past 11500 years (Fig. 5) is based on tree rings. Comparison of the ring width between different trees enabled to match the individual records to a continuous chronology. Before 11500 BP, during the deglaciation, the vegetation was rather different and the number of trees much sparser. Therefore, no continuous record is available yet although several floating chronologies are under construction. Instead of tree rings, other material such as varved sediments (Hajdas, 1993; Hughen et al., 1998; Kitagawa and Van der Plicht, 1998), corals (Bard et al., 1990), or stalagmites have been used. The absolute age can be determined by counting the annual sediment layers or using the U/Th dating technique. However, all these techniques lead to somewhat larger uncertainties in $\Delta^{14}C$ due to dating difficulties or the fact that they do not record directly the atmospheric $\Delta^{14}C$ (corals, stalagmites).

Since solar variability is at least partly quasi-periodic it is interesting to search for periodicities. The most famous periodicity of solar variability is the 11-year Schwabe cycle. Although this periodicity varies considerably between 7 and 18 years since 1750 it seems quite stable and close to 11 years if averaged over longer time periods. Due to the dampening effect of the carbon cycle system, the Schwabe cycle is attenuated strongly and difficult to detect in the $\Delta^{14}C$ (Fig. 5) (Siegenthaler et al., 1980). In addition, only a small part of the total $\Delta^{14}C$ record has been measured with a time resolution of better than 10 years. A longer periodicity, the existence of which has also been detected in the aurorae records, is the 88-year Gleissberg cycle (Gleissberg, 1965). It is also present in the $\Delta^{14}C$ record, however, its amplitude changes with time (Stuiver and Braziunas, 1993). The most prominent periodicity in the $\Delta^{14}C$ record is the 205-year De Vries cycle. Unfortunately, the observational records are too short to unambiguously attribute this cycle to solar variability. However, the fact that solar minima periods such as Maunder, Spoerer and Wolf occur roughly every 200 years clearly points to a solar origin of this cycle. Its amplitude is also time dependent and its frequency seems stable averaged over long periods of time. Another periodicity often attributed to the Sun is the 2100/2300-year Hallstatt cycle (Damon and Jirikowic, 1992). Although still well in the range of potential solar variability cycles there are no strong arguments yet in favour of a solar origin. Other periodicities found in the power spectrum could be of solar or geomagnetic origin, or related to processes within the carbon cycle system.

3.2. ^{10}Be

^{10}Be can be found in different archives such as ice cores, sediments, and loess (Beer and Sturm, 1995). For studies of solar variability, ice cores are the best-suited archives because they sample the atmospheric ^{10}Be content in the most direct way with a time resolution of one year. One year corresponds to the mean atmospheric residence time and therefore represents the upper limit. The disadvantage of the relatively short atmospheric residence time is, however, that the atmosphere is not necessarily well mixed regarding ^{10}Be and that changes in the removal process from the atmosphere may induce a local component in the ^{10}Be signal measured in an ice core. The fact that ^{36}Cl and ^{10}Be at high latitudes reflect nicely the solar and the geomagnetic modulation of the production rate and the good agreement between these two nuclides with ^{14}C point to a relatively good mixing within the hemispheres. Another difficulty is dating ice cores, especially at greater depth where the annual layer thickness decreases by the thinning effect. On the other hand, ^{10}Be has several advantages compared to ^{14}C. Its production signal is practically not attenuated and is therefore well suited to detect the 11-year Schwabe cycle (Beer et al., 1990). Due to its longer half-life (1.51×10^6 years compared to 5730 years), the time range to investigate is considerably longer (at least 10^5 years). Finally, the ^{10}Be system has no memory effect and therefore stores the actual production rate. In Fig. 6 several ^{10}Be records from Greenland are shown covering different periods with different time resolution. Panel a shows the ^{10}Be concentration from Dye 3 (South Greenland) after applying a low pass filter to remove periodicities shorter than 6 years. The ^{10}Be fluxes in panel b and c have been calculated from the measured ^{10}Be concentration according to the formula

$$F = \rho a c$$

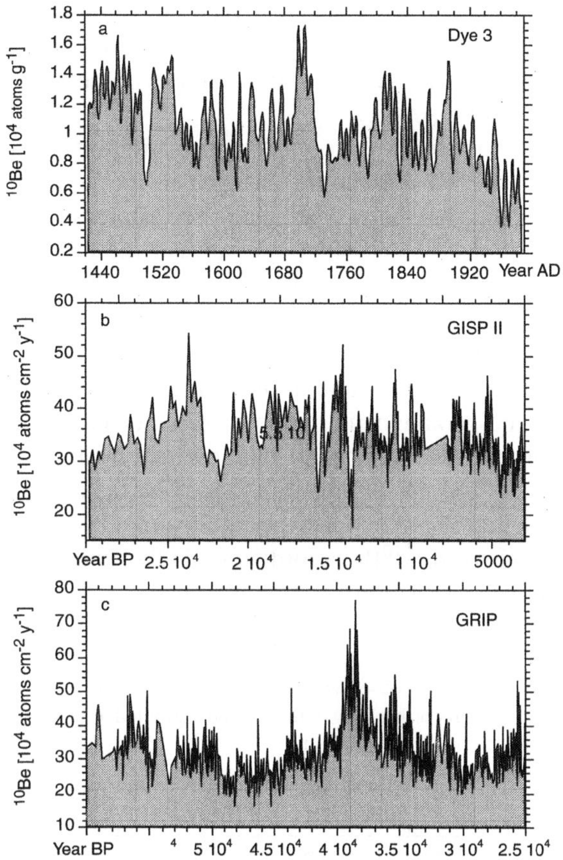

Figure 6. ^{10}Be records with different time resolutions for different time intervals.
a) Annual ^{10}Be concentrations from Dye 3 after removing the short-term fluctuations. b) ^{10}Be flux covering the transition from the last Glaciation into the Holocene . The sharp peaks correspond probably to quiet Sun periods (e.g. Maunder Minimum).
c) ^{10}Be flux during Marine Isotope Stage 3 mainly. The peak close to 40 ky BP and the long-term changes are consistent with changes of the global mean production rate induced by the geomagnetic field intensity.

with the density ρ (g cm^{-3}), the accumulation rate a (cm y^{-1}), and the concentration c (10^4 atoms/g). Since ρ of ice is constant, the flux F is mainly determined by the product of concentration and accumulation rate. During the Holocene the accumulation rate is relatively constant leading to a flux F that looks much the same as the concentration c. During glacial times, however, the accumulation rate changes by up to a factor of 2. As has been shown in earlier studies the ^{10}Be flux can be considered as proportional to the mean global production rate (Wagner et al., 2000a).

As can be seen in Fig. 6 the ^{10}Be records vary on very different time scales. The short-term variations of the ^{10}Be concentration in Fig. 6a reflect the 11-year

Schwabe cycle. Periods of high ^{10}Be concentration correspond to periods of low solar activity such as the Maunder minimum (1645–1715) and partly the Spoerer minimum (before 1540). The general ^{10}Be decrease in the 20th century is mainly due to an increase in solar activity.

Fig. 6b covers the transition from the last glaciation into the Holocene (Finkel and Nishiizumi, 1997). To correct for changes in the accumulation rate at the transition, the ^{10}Be flux has been calculated. The decreasing data density is the consequence of the constant sampling interval and the thinning effect. The narrow peaks most likely correspond to solar minima comparable to the Maunder minimum in panel a.

Fig. 6c shows the ^{10}Be flux of the GRIP ice core between 25 and 65 ky BP (Marine Isotope Stage 3). The main feature is the peak at 38 ky BP that can be attributed to an excursion of the geomagnetic dipole field (so called Laschamp event). The long-term trend can also be explained to a large extent by geomagnetic modulation (Baumgartner et al., 1998). Due to the thinning effect the time resolution given by the sampling interval of 55 cm varies between 30 and 60 years.

A spectral analysis of the ^{10}Be records reveals the following periodicities that can be attributed to solar variability:

TABLE II

Some observed periodicities in the ^{10}Be records of Fig. 6 that can be attributed to solar variability.

Time Interval	Periodicities [y]
1440–1980 AD	10.8, 88
4000–30'000 BP	205
25'000–50'000	205

In the case of ^{10}Be, the most prominent periodicity is the 11-year Schwabe cycle. Its presence has been confirmed for the full length of the Dye 3 annual record back to 1423 AD (Beer et al., 1994). Even during the Maunder minimum when hardly any sunspots could be observed the cycle is still clearly visible in the ^{10}Be data indicating that the solar dynamo was operational during this time (Beer et al., 1998). The Gleissberg cycle was already discovered in the sunspot record (Gleissberg, 1965). Its amplitude changes with time. The 205-year De Vries cycle can also be found with a varying amplitude similar to the case of Δ^{14}C. Recently, it has was found for the first time during the glacial period between 25000 and 50000 BP (Wagner et al., 2000b). This nicely shows the potential of cosmogenic radionuclides to extend our indirect indices of solar variability over much longer time scales. Other periodicities found in these records cannot yet be associated unambiguously with solar variability.

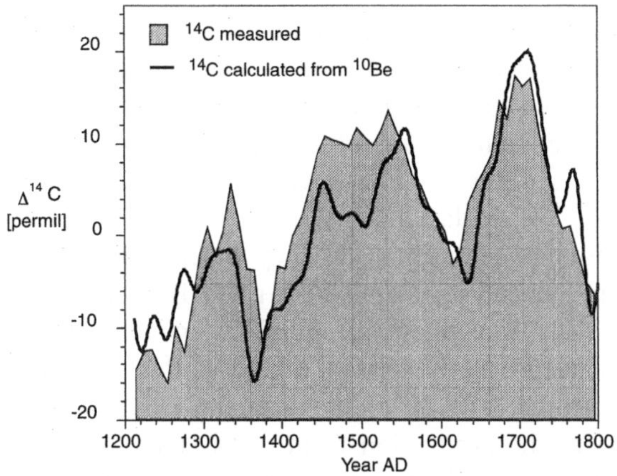

Figure 7. Comparison of the $\Delta^{14}C$ record measured on tree rings with the $\Delta^{14}C$ record calculated based on a combined ^{10}Be record of two ice cores from Milcent (Greenland) and the South Pole. The calculation was performed using the carbon cycle model of Fig. 4 assuming that the combined ^{10}Be record reflects directly the atmospheric production rate. The three periods of low solar activity (Maunder, Spoerer and Wolf) are clearly visible.

4. Comparison ^{14}C-^{10}Be

As discussed earlier, the interpretation of cosmogenic radionuclide records is not straightforward because the signals are composed of a production and a system component (Fig. 4). Since we are interested here in solar variability, we need to separate the production component containing the solar signal from the system component. One solution to this problem is to compare ^{14}C and ^{10}Be. Both nuclides are produced by cosmic ray induced interactions. However, their geochemical behaviour is completely different (see Fig. 4). Therefore the common signal of $\Delta^{14}C$ and ^{10}Be must be attributed to production, the rest to system effects. The comparison can be made in two ways:

1. From the measured atmospheric ^{14}C concentration the production rate is calculated using a carbon cycle model. This calculated production rate is then compared with ^{10}Be data.
2. The ^{10}Be signal is considered as the true production signal and the corresponding atmospheric ^{14}C concentration is calculated based on a carbon cycle model. This calculated data is then compared with the measured ^{14}C record.

We prefer the second approach which has been successfully used already earlier (Beer et al., 1984; Bard et al., 1997). As an example of such a comparison Fig. 7 shows the period 1200 to 1800 AD. The combined ^{10}Be concentration data from Milcent (Greenland) and South pole have been used to calculate the expected $\Delta^{14}C$ record (Beer et al., 1991).

The good agreement with the measured tree ring $\Delta^{14}C$ record clearly indicates that the dominant source of variability in the Holocene is the production rate, most likely caused by solar modulation. A similar good agreement is obtained for the Holocene periods for which detailed ^{10}Be data from Greenland and Antarctica are available. From this we conclude that the atmosphere seems to be relatively well mixed for ^{10}Be as well and that therefore the local component of the ^{10}Be signal is quite small. Larger differences between measured and calculated Δ^{14}C are to be expected for the last deglaciation when the climate was very unstable. Δ^{14}C is especially sensitive to changes in the deep-water formation. If the global deep-water formation is reduced as postulated e.g. by Broecker et al. (1999) less ^{14}C enters the oceans. This leads to a build-up of ^{14}C and a higher concentration in the atmosphere. Since deep-water formation does not affect the ^{10}Be concentration the calculated and measured Δ^{14}C would be different.

The separation of the production from the system signal still leaves an open problem. Changes in the production rate can be caused either by solar or by geomagnetic modulation. In principle, these two effects could be separated by comparing two records, one from the equator (reflecting magnetic effects) and the other from a polar region (reflecting solar modulation). However, due to the observed good atmospheric mixing such a separation is not feasible. As a general rule, we can assume that short periodicities (< 1 ky) are most likely of solar origin whereas longer periodicities (> 3 ky) are probably caused by magnetic modulation. For a more reliable separation independent information is required, e. g. in the form of paleomagnetism derived from remanence measurements on sediments (Wagner et al., 2000a).

5. Conclusions

A better knowledge of the long-term solar variability is crucial not only for solar physics, but also for environmental physics and climatology. The observational records of direct indices of solar variability such as sunspots are limited to the telescopic era and cover roughly 400 years. Indirect indices such as cosmogenic nuclides have the potential to extend these records significantly to 100'000 years using accurately dated ice cores. The main difficulty in applying cosmogenic nuclides is that they provide a complex signal containing not only information on the production rate, but also on geochemical system effects. By combining different nuclides (e. g. ^{10}Be and ^{14}C), the separation of these effects is possible. Several cycles related to solar variability can be found in the records available today (11-year Schwabe, 88-year Gleissberg, 205-year De Vries). A fundamental issue to address in the future is establishing a relationship between solar variability as derived from cosmogenic nuclides and solar irradiance. This relationship is crucial to distinguish between natural and anthropogenic climate changes. It requires the identification of the physical processes on the Sun that are responsible for long-term changes of solar irradiance. In particular, a basic question is whether solar irradiance changes are only the result of photospheric processes (sunspot blocking, magnetic network) or whether there are also changes involved in the energy transport through the convective zone.

Acknowledgements

I am indebted to W. Mende, G. Wagner, and R. von Steiger for valuable comments and suggestions. The manuscript was further improved by the help of R. Muscheler, R. Stellmacher and C. Wedema. This work was supported by the Swiss National Science Foundation.

References

Bard, E., Hamelin, B., Fairbanks, R. G., and Zindler, A.: 1990, 'Calibration of the ^{14}C time scale over the past 30,000 years using mass spectrometric U-Th ages from Barbados corals', *Nature* **345**, 405–410.

Bard, E., Raisbeck, G. M., Yiou, F., and Jouzel, J.: 1997, 'Solar modulation of cosmogenic nuclide production over the last millenium: comparison between ^{14}C and ^{10}Be records', *Earth. Planet. Sci. Let.* **150**, 453–462.

Baumgartner, S., Beer, J., Masarik, J., Wagner, G., Meynadier, L., and Synal, H.-A.: 1998, 'Geomagnetic modulation of the ^{36}Cl flux in the GRIP ice core, Greenland', *Science* **279**, 1330–1332.

Baumgartner, S., Beer, J., Wagner, G., Kubik, P. W., Suter, M., Raisbeck, G. M., and Yiou, F.: 1997, '^{10}Be and dust', *Nucl. Instr. Meth.* **B123**, 296–301.

Beer, J., AndrÈe, M., Oeschger, H., Siegenthaler, U., Bonani, G., Hofmann, H., Morenzoni, E., Nessi, M., Suter, M., Wölfli, W., Finkel, R., and Langway Jr., C.: 1984, 'The Camp Century ^{10}Be Record: Implications for Long-Term Variations of the Geomagnetic Dipole Moment', *Nucl. Instrum. Meth.* **B5**, 380–384.

Beer, J., Baumgartner, S., Hannen-Dittrich, B., Hauenstein, J., Kubik, P., Lukasczyk, C., Mende, W., Stellmacher, R., and Suter, M.: 1994, 'Solar Variability Traced by Cosmogenic Isotopes', in *The Sun as a Variable Star: Solar and Stellar Irradiance Variations*, J. M. Pap, C. Fröhlich, H. S. Hudson and S. K. Solanki (eds.), pp. 291-300, Cambridge University Press.

Beer, J., Blinov, A., Bonani, G., Finkel, R. C., Hofmann, H. J., Lehmann, B., Oeschger, H., Sigg, A., Schwander, J., Staffelbach, T., Stauffer, B., Suter, M., and Wölfli, W.: 1990, 'Use of ^{10}Be in polar ice to trace the 11-year cycle of solar activity', *Nature* **347**, 164–166.

Beer, J., Mende, W., R., S., and White, O. R.: 1996, 'Intercomparisons of proxies for past solar variability', in *Climate variations and forcing mechanisms of the last 2000 years*, P. D. Jones, R. S. Bradley and J. Jouzel (eds.), Vol. I41, Springer-Verlag, Heidelberg.

Beer, J., Mende, W., and Stellmacher, R.: 2000, 'The role of Sun in climate forcing', *Quat. Sci. Rev.* **19**, 403–415.

Beer, J., Raisbeck, G. M., and Yiou, F.: 1991, 'Time Variations of ^{10}Be and Solar Activity', in *The Sun in Time*, C. P. Sonett, M. S. Giampapa and M. S. Matthews (eds.), pp. 343-359. Univ. of Arizona press, Tucson.

Beer, J., and Sturm, M.: 1995, 'Dating of lake and loess sediments','*Radiocarbon* **37**, 81–86.

Beer, J., Tobias, S. M., and Weiss, N. O.: 1998, 'An active Sun throughout the Maunder minimum', *Solar Physics* **181**, 237–249.

Broecker, W. S., Sutherland, S., and Peng, T.-H.: 1999, 'A possible 20th-century slowdown of southern ocean deep water formation', *Science* **286**, 1132–1135.

Brost, R. A., Feichter, J., and Heimann, M.: 1991, 'Three-dimensional simulation of ^{7}Be in a global climate model', *J. Geophys. Res.* **96**, 22,423–22,445.

Damon, P. E., and Jirikowic, J. L.: 1992, 'The Sun as a low-frequency harmonic oscillator', *Radiocarbon* **34**, 199–205.

Finkel, R. C. and Nishiizumi, K.: 1997, 'Beryllium-10 concentrations in the Greenland ice sheet project 2 ice core from 3–40 ka', *J. Geophys. Res.* **102**, 26,699–26,706.

Fröhlich, C., and Lean, J.: 1997, 'Total Solar Irradiance Variations', in *IAU Symposium 185*, F. L. Deubner (ed.), Kluwer Academic Publ., Dordrecht.

Fröhlich, C.: 2000, *Space Sci. Rev.*, this volume.

Gilliland, R. L.: 1981, 'Solar radius variations over the past 265 years', *Astrophys. J.* **248**, 1144–1155.

Gleissberg, W.: 1965, 'The eighty-year solar cycle in auroral frequency numbers', *J. Br. Astron. Assoc.* **75**, 227.

Hajdas, I.: 1993, 'Extension of the Radiocarbon Calibration Curve by AMS Dating of Laminated Sediments of Lake Soppensee and Lake Holzmaar', PhD Thesis, ETH Zürich.

Hughen, K., Overpeck, J. T., Lehmann, S., Kashgarian, M., Southon, J., Peterson, L. C., Alley, R., and Sigman, D. M.: 1998, 'Deglacial changes in ocean circulation from an extended radiocarbon calibration', *Nature* **391**, 65–68.

Kitagawa, H., and Van der Plicht, J.: 1998, 'A 40,000-year varve chronology from lake Suigetsu, Japan: Extension of the ^{14}C calibration curve', *Radiocarbon* **40**, 505–515.

Lean, J.: 1997, 'The Sun's variable radiation and its relevance for Earth', *Ann. Rev. Astron. Astrophys.* **35**, 33–67.

Masarik, J., and Beer, J.: 1999, 'Simulation of particle fluxes and cosmogenic nuclide production in the Earth's atmosphere', *J. Geophys. Res.* **104**, 12,099–13,012.

McCormac, F. G., Hogg, A. G., Higham, T. F. G., Lynch-Stieglitz, J., Broecker, W. S., Baillie, M. L. L., Xiong, L., Pilcher, J. R., Brown, D., and Hoper, S. T.: 1998, 'Temporal variation in the interhemispheric ^{14}C offset', *Geophys. Res. Lett.* **25**, 1321–1324.

Mende, W. and Stellmacher, R.: 2000, *Space Sci. Rev.*, this volume.

Pap, J. M., and Fröhlich, C: 1997, 'Total irradiance variations', *J. Atmos. Solar-Terr. Phys.*, 1–13.

Scheffel, C., Blinov, A., Massonet, S., Sachsenhauser, H., Stan-Sion, C., Beer, J., Synal, H. A., Kubik, P. W., Kaba, M., and Nolte, E.: 1999, '^{36}Cl in modern atmospheric precipitation', *J. Geophys. Res.* **26**, 1401–1404.

Siegenthaler, U., Heimann, M., and Oeschger, H.: 1980, '^{14}C Variations Caused by Changes in the Global Carbon Cycle', *Radiocarbon* **22**, 177–191.

Stuiver, M., and Braziunas, T. F.: 1993, 'Sun, Ocean, Climate and Atmospheric $^{14}CO_2$, an evaluation of causal and spectral relationships', *The Holocene* **3**, 289–305.

Stuiver, M., Reimer, P. J., Bard, E., Beck, J. W., Burr, G. S., Hughen, K. A., Kromer, B., McCormac, G., Van der Plicht, J., and Spurk, M.: 1998, 'INTCAL98 Radiocarbon age calibration, 24,000-0 cal BP', *Radiocarbon* **40**, 1041–1083.

Wagner, G., Masarik, J., Beer, J., Baumgartner, S., Imboden, D., Kubik, P. W., Synal, H.-A. and Suter, M.: 2000a, *Nucl. Instr. Meth.*, in press.

Wagner, G., Beer, J., Masarik, J., Muscheler, R., Kubik, P.W., Laj, C., Mende, W., Raisbeck, G. M., and Yiou, F: 2000b, 'Presence of the solar de Vries cycle (205 years) during the last ice age', *Geophys. Res. Lett.*, submitted.

Wagnon, P., Delmas, R. J. and Legrand, M.: 1999, 'Loss of volatile acid species from upper firn layers at Vostok, Antarctica', *J. Geophys. Res.* **104**, 3423–3432.

Willson, R. C.: 1997, 'Total solar irradiance trend during solar cycles 21 and 22', *Science* **277**, 1963–1965.

EXTREME SOLAR CYCLE VARIABILITY IN STRONG LINES BETWEEN 200 AND 400 NM

ORAN R. WHITE (orw@ucar.edu)
High Altitude Observatory, National Center for Atmospheric Research, Box 3000 Boulder, Colorado 80307 USA

JOHN FONTENLA (jfontenla@mho.net)
1874 MacCullen Drive, Erie, Colorado 80516 USA

PETER A. FOX (pfox@ucar.edu)
High Altitude Observatory, National Center for Atmospheric Research, Box 3000 Boulder, Colorado 80307 USA

Received: 23 September 1999; Accepted in final form: 28 April 2000

Abstract. Our study of solar cycle irradiance variability in the UV between 200 and 400 nm requires a detailed knowledge of the composition of the solar spectrum in this wavelength range. We compute the synthetic spectrum from 250 to 300 nm and compare it with ATLAS3 and SOLSTICE observations. Synthetic solar spectra for solar minimum and maximum conditions show large variations in broad, strong UV lines. Strong lines of FeI between 260 nm and 264 nm show increases between 0.4× and 3× in their max/min ratio. Our "broad lines" database shows 167 lines with similar properties between 200 nm and 400 nm. Our results raise issues of the importance of such large variability in narrow bands and the difficulty of detection in measurements with spectral resolutions of 1 nm.

1. Calculation of the Solar Spectrum

In our study of the composition of the solar UV spectrum, we compute synthetic solar spectra with all lines fully resolved in order to determine the degree of variability in the UV between 200 and 400 nm. The upper panel of Figure 1 shows our first computation of the solar spectrum between 250 and 260 nm with the observed spectra from ATLAS3 (Van Hoosier, 1996) and SOLSTICE (Rottman and Woods, 1994). See Fontenla et al. (1999) for details of the computations.

Our computed spectrum agrees well with observations in the cores of strong lines, particularly the MgI and MgII lines at 280 nm and 286 nm. However, our spectra show too much radiation between 260 to 278 nm and 290 to 300 nm. This excess radiation is due to missing opacities in our first calculations. With too small an opacity, we "see" the higher temperatures deeper in our model atmospheres; hence, the emergent intensity is too large. We improved the match to observations by including free-bound continua of Fe, Ca, K, and Na in the calculation. The bottom panel of Figure 1 shows new results for the 260–270 nm band together with the two observed spectra and our earlier computation. The agreement with observation

Space Science Reviews **94:** 67–74, 2000.

is excellent below 263.3 nm. Above that wavelength, the computed spectrum is closer to the observed, but not in good enough agreement with observations. We still have missing opacity sources such as the broad line at 295 nm, completely missing in the synthetic spectrum. This result emphasizes the need for complete energy level information and accurate ionization rates.

In strong lines, computed irradiances agree well with observation because the opacity is determined by a single bound-bound transition. Therefore, here we concentrate on variability of the so-called "winged lines" described by Moore et al. (1982).

2. Calculation of Spectra for Feature in Active and Quiet Regions

We compute the solar spectra for different solar features and combine them to approximate the solar cycle variation. Fontenla et al. (1999) give the necessary model atmospheres and a scheme for combining the resultant spectra.

We use model atmospheres for seven solar structure types in Table I:

TABLE I
Model Atmospheres

Model	Description
A	faint supergranule cell interior
C	average supergranule cell interior
E	average chromospheric network
F	bright network/faint plage
H	average plage
P	bright plage
S	sunspot umbra

Using the latest Avrett models and Kurucz's tabulation of solar lines, we compute the emergent intensity at a particular wavelength by evaluating the total opacity and source function as a function of depth for each particular structure. Line source functions are estimated using detailed computations of non-LTE populations for hydrogen and an approximate form for all elements from helium to zinc in the periodic table.

We compute the angular dependence of the radiation emerging from each structure using a plane parallel atmosphere. Using this center-to-limb variation, we derive irradiance estimates from the average disk intensity using the solid angle of the Sun subtended at the Earth. Thus, we compute the spectral irradiance from seven hypothetical 'Suns' completely covered by one of the seven structures, i.e., plage, network, cell. Given these irradiance estimates for each structure, we weight

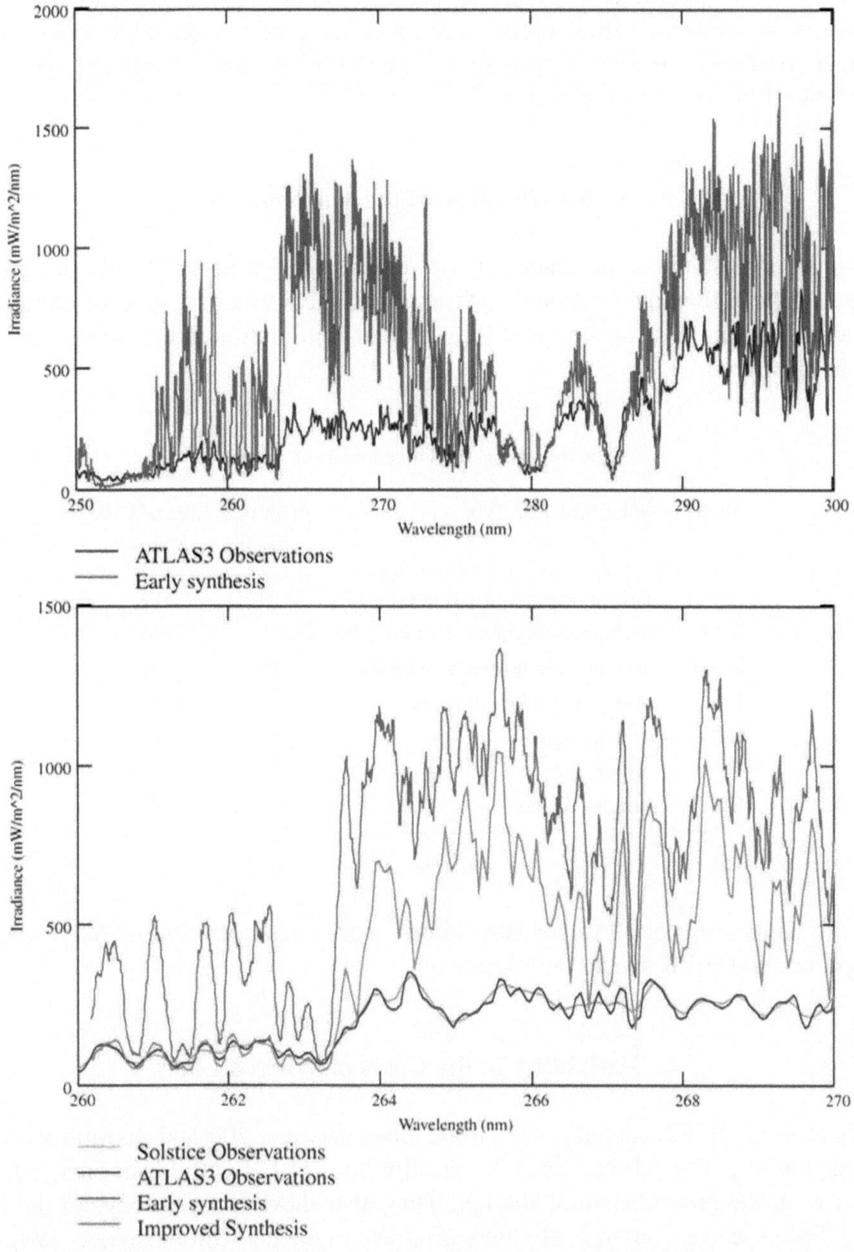

Figure 1. Comparison of Observed and Synthetic Spectra. *Upper panel*: Our first synthesis of the quiet Sun's spectrum between 250 nm and 300 nm at 0.1 nm resolution. *Lower panel:* Our first synthetic spectrum for the 260–270 nm band and the improved spectrum together with observed irradiance spectra from ATLAS3 and SOLSTICE.

them by estimates of the areas for each of the seven structures to estimate the irradiance for a particular level of solar activity. For example, a quiet Sun irradiance comes from using only irradiances for Models A, C, and E in the table above. To estimate irradiances at solar maximum, we weight the model atmospheres according to Table II in the following section.

3. Specification of the Variable Sun

In order to test the solar spectrum for sensitivity to different levels of activity, we use the following fractional disk areas of each structure type in computing irradiance estimates for the quiet Sun at solar minimum and the active Sun near solar maximum.

TABLE II

Structure Areas for Three Levels of Activity

Model	Solar Structure Type	Fractional Area of Disk	
		Quiet Sun	*Active Sun*
A	faint supergranule cell interior	0.10	0.08
C	average supergranule cell interior	0.77	0.67
E	average chromospheric network	0.10	0.10
F	bright network/faint plage	0.03	0.07
H	average plage	0.0	0.05
P	bright plage	0.0	0.026
S	sunspot umbra	0.0	0.004

These areas are typical values determined from decomposition of NSO magnetograms and CaII K spectroheliograms.

4. Variability in the Cores of Winged Lines

Moore et al. (1982) identify 167 strong lines between 200 and 400 nm with extended wings. These broad lines are usually lines of FeI with lower energy levels near or at the ground state of the ion. They also show emission beyond the limb in eclipse spectra; consequently, they originate in the low chromosphere. We computed spectra between 250 nm and 300 nm using our synthesis algorithm with dynamic wavelength spacing, i.e., the point spacing decreases in known lines to resolve the individual line profiles. These very dense spectra are then 'filtered' to give equally spaced irradiance values at resolutions of 0.01 nm and 1 nm for presentation here.

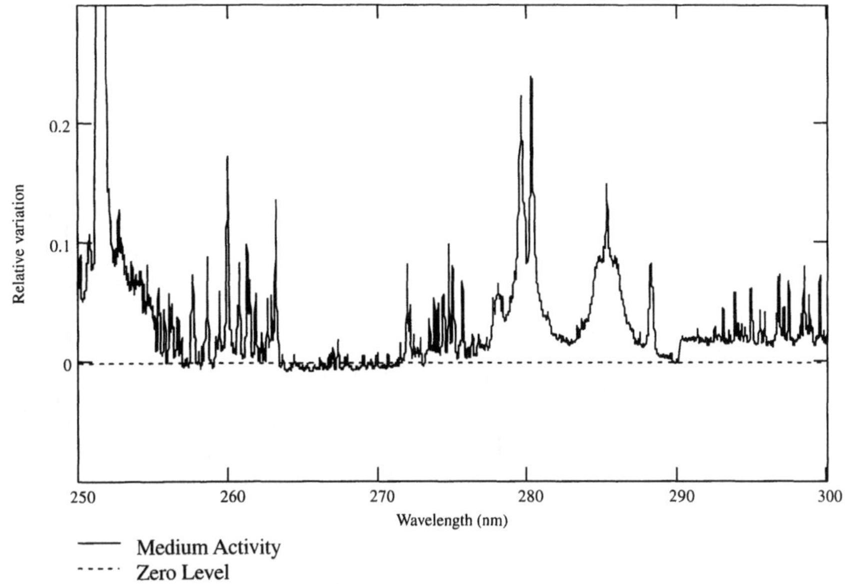

Figure 2. The solar cycle variability spectrum from our first synthesis of the 250–300 nm band. MgI and MgII lines at 280 nm and 285 nm are highly variable because they originate in the chromosphere. Excessive variability at 252 nm is an artifact due to incorrect data in the atomic database. The discontinuity near 290 nm is an artifact of the computation.

Using these spectra corresponding to quiet and active Sun , we compute the fractional irradiance change from solar minimum to near maximum by

$$\text{fractional variability} = V_{\lambda} = [I_{\text{active}} - I_{\text{quiet}}]/I_{\text{quiet}}.$$

Figure 2 shows the radiative variability spectrum for the 250 to 300 nm band at 0.1 nm resolution. The extreme variability at 280 nm and 286 nm comes from the MgI and MgII lines formed in the chromosphere. The high variability near 251 nm is spurious and arises from incorrect data in the atomic database. The discontinuity in the spectrum at 290 nm is an artifact from computing the spectrum in in 10 nm segments.

In Figure 3 we show only the UV band from 260 to 264 nm as an example of the extreme variability we find in the “winged” lines. Here we show the computed irradiance at 0.01 nm resolution in the upper panel and the corresponding variability spectrum, V_{λ}, in the lower panel. At this high resolution we find seven lines with V_{λ} values greater than 0.5. This value of 0.5 corresponds to irradiance increases of 1.5$\times$ from solar minimum to maximum in cores of these lines. This result is typical of the other bands we computed.

Since current observations give variability spectra at 1 nm resolution, we smooth this 260–264 nm band to show the effect of decreased spectral resolution on variability measurements in the UV. At 1 nm resolution, the variability spectrum shown in Figure 4 loses almost all of its spectral structure and gives V_{λ} ~0.02 between

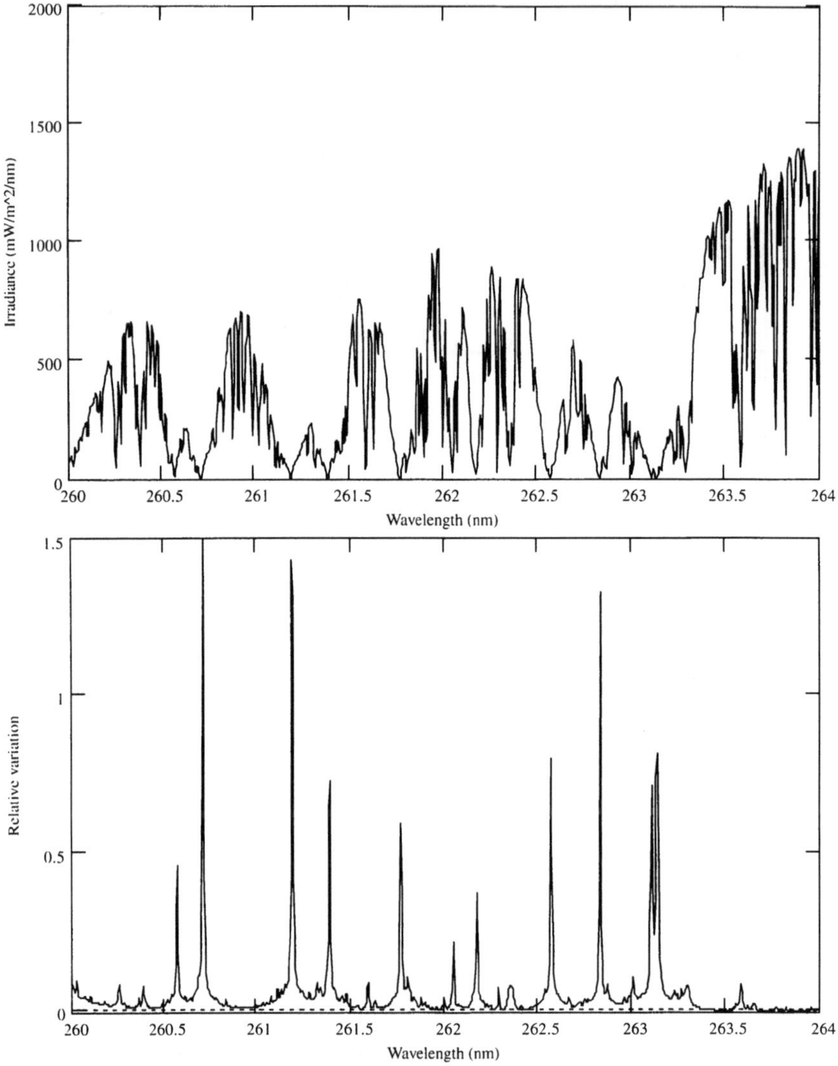

Figure 3. High resolution (0.01 nm) synthetic spectrum between 260 and 264 nm band and its variability from minimum to maximum in the solar cycle. *Upper panel:* The synthetic spectrum. *Lower panel:* Solar cycle variability spectrum. Note extreme variability in the cores of the broad lines in this spectral band.

260 nm and 263 nm. Above 263 nm, we see negative variability! Again, this may be due to missing opacity. The shape of the variability spectrum in this band is very similar to that reported by Rottman (2000) at this meeting. In order to detect solar cycle variability unambiguously in this band, our synthesis shows that long-term precision has to be maintained at better than 1 percent.

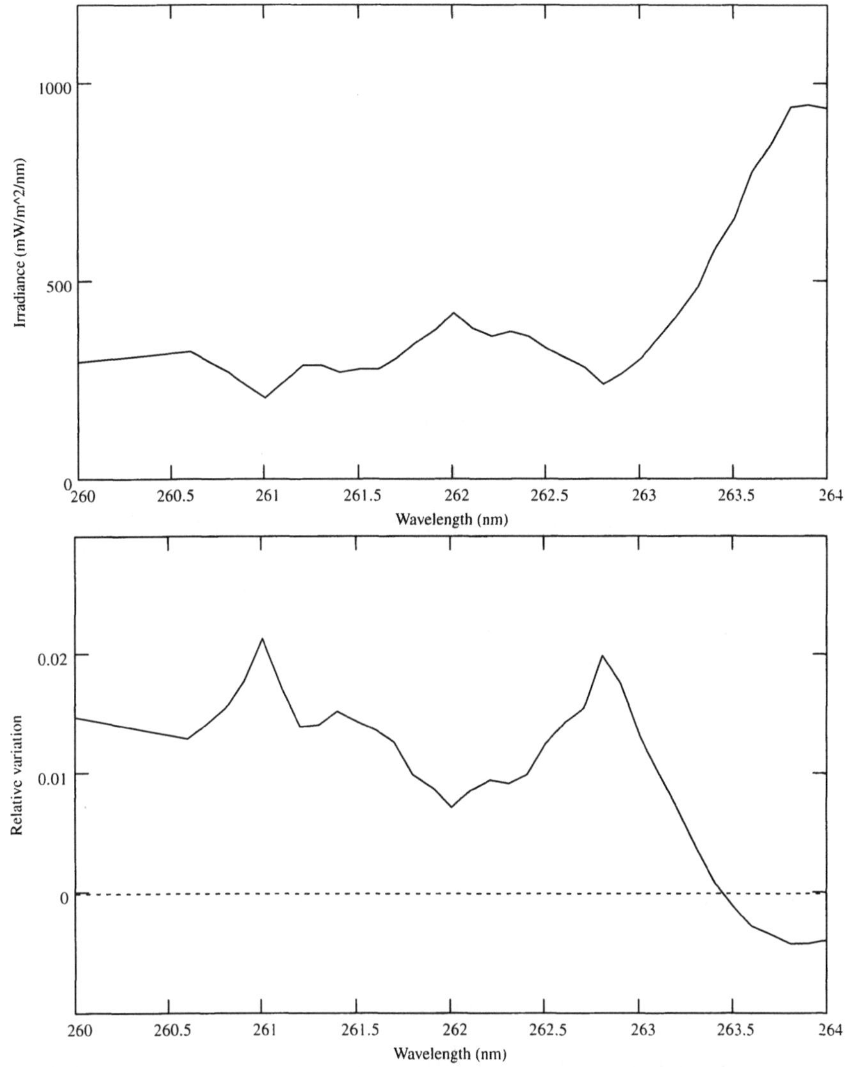

Figure 4. Low resolution (1 nm) synthetic spectrum between 260 and 264 nm and its variability from minimum to maximum in the solar cycle. *Upper panel:* The synthetic spectrum. *Lower panel:* Solar cycle variability spectrum. Note the loss of spectral detail at this resolution. Only the variability due to individual lines near 261 nm and 263 nm is detectable. Negative variability above 263 nm is discussed in the text.

We studied a second band between 306–312 nm (not shown) with the same result. A third band between 259.8 and 260.0 nm (not shown) contains three strong lines of FeI, and V_λ is greater than 0.25 in bands 0.025 nm wide in the line cores. The maximum variability in the line core intensities is $2\times$ to $3\times$.

5. Discussion

Our results raise several questions:

- *How will such strong radiative forcing in narrow bands affect the chemical composition and the heating rate in the upper atmosphere where this radiation is absorbed?*
  Answer: [to be discussed at this workshop]
- *How many strong broad lines showing extreme variability are in the UV between 200 and 400 nm?*
  We find 167 such lines in the Kurucz tabulation, but we already find similar lines in observed spectra not listed in Kurucz tables.
- *What are the implications for UV measurements of the solar cycle variability?*
  If the energy variation in these lines is large enough to affect the physical state of the upper atmosphere, measurement of the solar cycle variation in a small number of well-chosen bands should be undertaken with sufficient precision and resolution to detect extreme radiative variability unambiguously. Our calculations give average values of variability, $\langle V_\lambda \rangle$, of $\sim$0.5% for the 260 nm and 300 nm bands at 1 nm resolution. For higher resolutions of 0.01 nm and 0.1 nm, we expect V_λ values larger than 25% near 260 nm and values larger than 16% near 300 nm.

Acknowledgements

This research is supported by the U.S. National Science Foundation through the RISE program grant, NCAR proposal 1997-014, Application of Spectra Synthesis to the Study of Solar Visible, UV, EUV, and Infrared Variability.

References

Fontenla, J., White, O.R., Fox, P.A., Avrett, E.H., and Kurucz, R.L.: 1999, 'Calculation of Solar Irradiances. I. Synthesis of the Solar Spectrum', *ApJ* **518**, 480–499.

Moore, C.E., Tousey, R., Brown, C.M.: 1982, 'The Solar Spectrum between 3069A and 2095A', *NRL Report* **8653**, Naval Research Laboratory, Washington, D.C., 1–169.

Rottman, G.J.: 2000, 'Variations of the Solar UV Irradiance Observed by the UARS SOLSTICE - 1991 to 1999', *Space Sci. Rev.*, this volume.

Rottman, G.J., and Woods, T.N.: 1994, 'Upper Atmosphere Research Satellite (UARS) Solar Stellar Irradiance Comparison Experiment (SOLSTICE)', *SPIE Proceedings* **2266**, San Diego, 317–327.

VanHoosier, M.E.: 1996, 'Solar Ultraviolet Spectral Irradiance Data with Increased Wavelength and Irradiance Accuracy', *SPIE Proceedings* **2831**, Denver, 57–64.

Address for Offprints: O.R. White, High Altitude Observatory, National Center for Atmospheric Research, 3450 Mitchell Lane, Boulder, Colorado 80301 USA

DATA FROM THE PRECISION SOLAR PHOTOMETRIC TELESCOPE (PSPT) IN HAWAII FROM MARCH 1998 TO MARCH 1999

ORAN R. WHITE (orw@ucar.edu), PETER A. FOX (pfox@ucar.edu), RANDY MEISNER (meisner@ucar.edu), MARK P. RAST (mprast@ucar.edu), ERIC YASUKAWA (yasukawa@ucar.edu), DARRYL KOON (koon@ucar.edu) and CRYSTAL RICE (crice@hao.ucar.edu)
High Altitude Observatory, National Center for Atmospheric Research, Box 3000 Boulder, Colorado 80307 USA

HAOSHENG LIN (lin@nso.noao.edu)
U.S. National Solar Observatory, Sunspot, New Mexico 88349 USA

JEFF KUHN (kuhn@ifa.hawaii.edu) and
ROY COULTER (coulter@ifa.hawaii.edu)
University of Hawaii, 2680 Woodlawn Drive, Honolulu, Hawaii 96822 USA

Received: 13 October 1999; Accepted in final form: 10 March 2000

Abstract. Two Precision Solar Photometric Telescopes (PSPT) designed and built at the U.S. National Solar Observatory (NSO) are in operation in Rome and Hawaii. A third PSPT is now in operation the NSO at Sunspot, NM. The PSPT system records full disk solar images at three wavelengths: K line at 393.3 nm and two continua at 409 nm and 607 nm throughout the observing day. We currently study properties of limb darkening, sunspots, and network in these images with particular emphasis on data taken in July and September 1998. During this period, the number of observations per month was high enough to show directional properties of the radiation field surrounding sunspots. We show examples of our PSPT images and describe our study of bright rings around sunspots.

1. Rationale for Development of a Precision Photometric Solar Telescope

Impetus for development of a solar disk imager for high precision photometry of photospheric structure came from the NSF program to study Radiative Input from the Sun to the Earth (RISE). Under the leadership of Dr. Peter Foukal, the RISE steering committee recommended initiation of this new NSF program for measurement and analysis of solar images to determine basic properties of radiation from the Sun in the direction of the Earth. Program objectives included estimation of irradiance changes from solar images, determination of directional properties of radiation from surface structures, and resolution of the question about changes in the solar luminosity. The nature of energy transport through the convection zone remains a fundamental question that requires precision measurements of radiation from magnetic and non-magnetic regions on the solar disk over a solar cycle.

Space Science Reviews **94:** 75–82, 2000.

Acting on this recommendation, the NSO proposed to build a state-of-the-art telescope/photometry system with a precision of 0.1% per pixel and a spatial resolution of at least 2 arcsec. Dr. Jeff Kuhn led the development with his colleagues Haosheng Lin and Roy Coulter at Sacramento Peak Observatory. In order to fulfil the scientific objectives in the RISE program, the new instrument has to record a daily series of images and make them available immediately for flat field correction and analysis of their information content, i.e., signal/noise ratio, spatial resolution, image artifacts, etc.

The NSO proposal called for two PSPT systems: one to be operated at Sac Peak and the other for a second unknown site. Colleagues at the Rome Observatory studying radiative variability proposed to buy a third instrument. The first operational telescope system was installed in Rome in 1996 and upgraded to the final telescope system as the development proceeded. Ermolli et al. (1998) describe their operation of the first PSPT. The Rome team developed a method for numerically tracing the chromospheric network pattern in their CaII K images. This tool allowed studies of the geometrical properties of the network (Berrilli et al., 1998), its time variation (Berrilli et al., 1997), and its solar cycle variation (Caccin et al., 1998).

HAO proposed to operate the second instrument at Mauna Loa Solar Observatory as part of our daily coronal observing program. Following instructions from Kuhn and his colleagues, the MLSO facility was enlarged to accommodate the second PSPT. Installation of the second PSPT was completed in the fall of 1997, and we began daily operation in March of 1998. We now have over one year of PSPT solar images as well as data on physical conditions at the site plus the operational/engineering record of system performance. The MLSO installation can be seen at `http://rise.hao.ucar.edu/pspt/instrument.html`.
We describe the PSPT system and give an initial result on bright rings around sunspots in the sections below (see also Fox et al., 2000).

2. Description of the Telescope System

The measurement strategy from the Steering Committee called for ‘continuum’ images at wavelengths in the blue and red plus a CaII K image to show locations of concentrated magnetic fields. Folding this requirement into the photometric precision and spatial resolution requirements led to a 15 cm objective of 230 cm focal length imaging on a 2048 × 2048 pixel Thomson CCD detector. The telescope layout can be viewed at `www.noao.edu/PSPT`. Wavelengths and filter band passes are summarized in Table I.

The Blue Continuum wavelength was chosen in a spectral region relatively free of Fraunhofer lines and close to the K line to insure the same focus and solar image size in the two channels. The Red Continuum band shows no strong lines in the filter bandwidth. The filter wheel can accommodate a fourth channel, and we wish

TABLE I
Definition of Wavelength Channels

Channel	Center Wavelength (nm)	Bandpass (nm) FWHM
CaII K	393.3	0.23
Blue Continuum	409.37	0.2
Red Continuum	607.0	0.45

to add a channel as close to 1000 nm as the CCD response allows. The passbands of the blue and red filters isolate the centers of these continua free of strong lines. These choices give continuum images of the distribution of photospheric radiation on the visible solar disk.

The PSPT has the following subsystems running in parallel to the main camera:

TABLE II
Optical Subsystems

Instrument	Function
Servo Guider	X-Y guider for the telescope assembly
Active Fast Mirror	fast mirror to stabilize image on the CCD
Transparency Monitor for each channel	filter photometer to monitor light levels
Seykoura Seeing Monitor	scintillation monitor of the seeing

The Xedar CCD camera is the key element in achieving the 0.1% photometric precision. Electron wells in the Thomson CCD are 50,000 electrons deep and allow accumulation of many photons per read out cycle. The camera reads out every 2 seconds, and the image sequence is accumulated for 10 minutes per channel to dampen the 5 minute oscillation. The seeing monitor runs continuously and can trigger recording of an image sequence.

Correction for fixed patterns in the CCD is a crucial requirement. Since the CCD camera reads out in 1024 $\times$ 1024 quadrants, problems with the flat fielding are usually quite apparent to the critical analyst. We use the Kuhn-Lin technique (Kuhn et al., 1991) with 16 off-axis images recorded at least once per day. The Kuhn-Lin correction algorithm is then used after observing to correct all other images collected during the day.

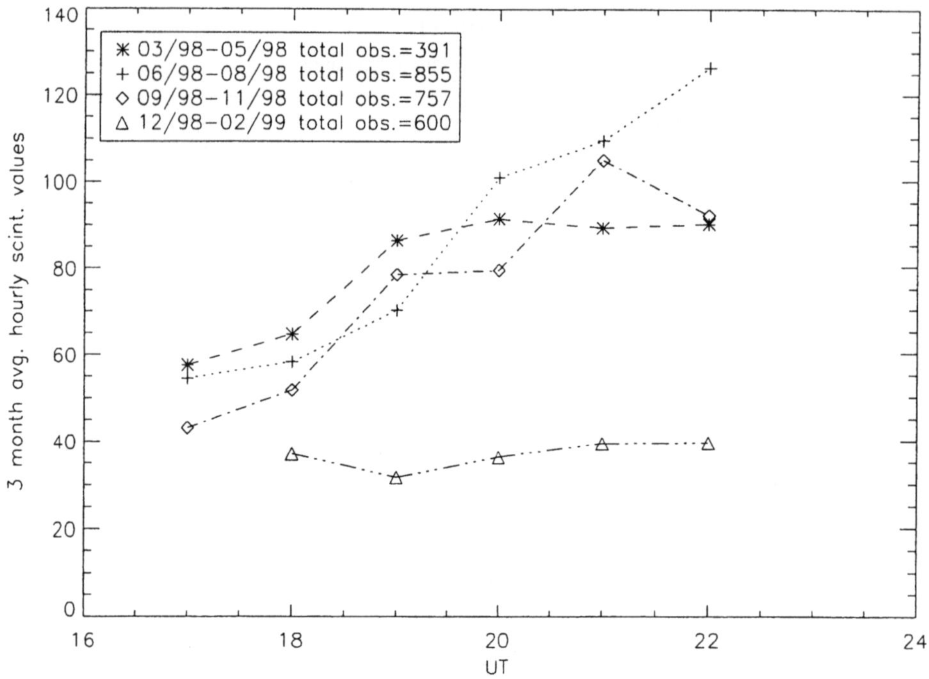

Figure 1. Seasonal averaged seeing monitor output measured every two hours during the MLSO observing day. Lower curve: winter, uppermost curve: summer. This instrument measures image degradation due to atmospheric seeing. Lower values correspond to better image quality. During the winter (lower curve), measurements show good seeing throughout the day.

3. The First Year of Operation at the Mauna Loa Solar Observatory

Regular daily operation of the full PSPT system began in March of 1998 under the supervision of Eric Yasukawa and Darryl Koon. The first year of data are now processed and archived in the NCAR mass storage system. This archive contains two types of images: 1) the basic full disk image corrected for fixed pattern noise; and 2) an image with a smooth limb-darkened photospheric surface subtracted. All images are stored in FITS format, and their headers contain the ancillary information defining the observation: date, time, seeing and transparency output, observers' notes, and processing records. Play back of the full year of data shows disk passages of active regions as they evolve, but one also sees variation in the image quality due to seeing fluctuations.

The scintillation monitor data are separated by season and averaged over each hour in the day to give typical seeing variations during the MLSO observing day. We show this site characteristic in Figure 1.

The lower curve shows that the seeing is best during the winter and lasts all day. In contrast, the three curves for spring, summer, and fall show the steady deterioration of the average image quality over the MLSO observing day. This seeing record from the first year allows definition of an optimum observing scheme during the second year of operation.

Coverage during this first year is not complete because of both instrument problems and weather. Nearly complete sequences containing both active and quiet regions were obtained in July and September of 1998. On 1 July 1998, a modified flat fielding routine was put in the processing stream. By that time we had sufficient experience in both instrument operation and data processing to recognize problems in the data and correct them. The flat field images need to be recorded only once a week unless an instrument failure interrupts the observing program. PSPT image data are available through the HAO web site, `http://rise.hao.ucar.edu/pspt/request_data.html`.

4. Examples of PSPT Images

The two solar images in Figure 2 are examples of our full disk images taken on 14 February 1999 under good seeing conditions. The image sequence is CaII K at the top and Blue Continuum below.

These images are corrected for fixed pattern noise and dark current. The Blue Continuum image has limb darkening removed, but this correction has not been made in the K image.

To appreciate the potential of the PSPT data, in Figure 3 we show subimages in K and the Blue Continuum containing the large sunspot in the southern solar hemisphere in Figure 2.

Again, the image sequence is CaII K (top) and Blue Continuum (bottom). The brighter plage and network in the K image locates regions with concentrations of magnetic field. A faint bright ring can be seen around the smaller spot in the Blue Continuum image in our original data.

We currently study several such bright sunspot rings. One question is whether the ring is of magnetic or convective origin. Rast et al. (1999) argue that the enhanced radiation in the ring is not solely plage but also contains thermal radiation emerging at the periphery of the sunspot penumbra. Since the ring is usually observed around nearly circular spots, we may have a simple flux tube geometry amenable to numerical simulation. Again the fundamental question is energy transport in and near sunspots.

5. Plans for the Second Year

The second year of PSPT operation is important because the Sun will be near the maximum of Cycle 23. We will change the observing program to take better advantage of the seeing profiles shown in Figure 1. Continuous image sequences will be taken during the first two hours of the day when the image quality is the best.

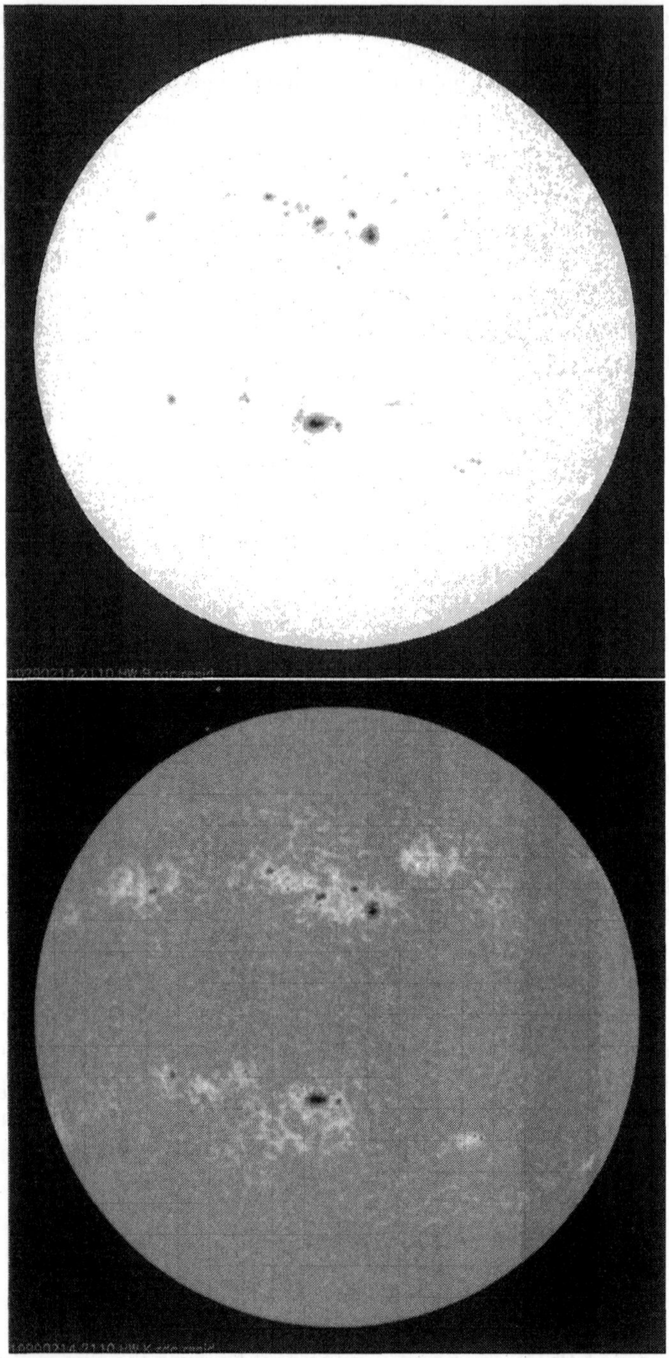

Figure 2. Full Disk CaII K and Blue Continuum images recorded by the PSPT on 14 February 1999 at MLSO. The Blue continuum image has the quiet Sun limb darkening subtracted. The CaII K image is the original observation without any limb darkening removal. The K image shows loci of concentrated magnetic fields in the plage and chromospheric network.

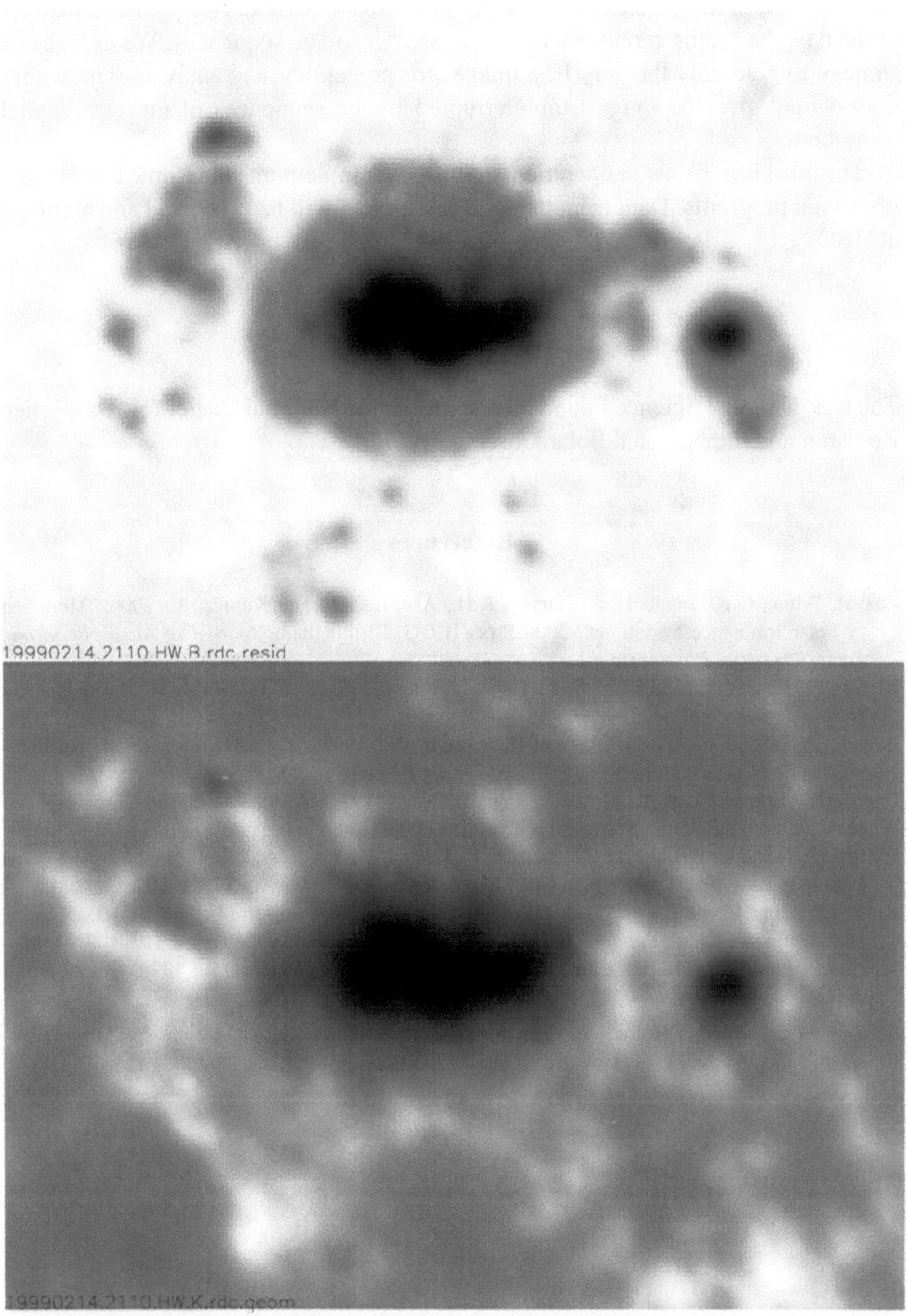

Figure 3. CaII K and Blue Continuum images of the large sunspot in the southern hemisphere extracted from the full disk images in Figure 2. The original images show a faint bright ring at the periphery of the spot penumbra. We frequently see such rings around older, unipolar sunspots.

The daily flat fielding observations will follow this sequence. During the remainder of the day, the seeing monitor will trigger shorter image sequences. We expect this strategy to give only the very best images for processing and analysis. The winter season may give the longest, uninterrupted image sequences of uniform spatial resolution.

The third PSPT now in operation at Sunspot complements the Rome and Hawaii observing programs. Data from the third instrument will be processed and archived at HAO.

Acknowledgements

The U.S. National Science Foundation sponsors the National Center for Atmospheric Research and the National Solar Observatory.

References

Fox, P., White, O.R., Fontenla, J., Harvey, K.H., Avrett, E.H., and Kurucz, R.: 2000, 'Modeling of Solar Irradiance Variations', 2000 Proc. IUGG, Birmingham, *Journal of Atmospheric and Solar-Terrestrial Physics* (in press).

Berrilli, F., Florio, A., and Ermolli, I.: 1998, 'On the Geometrical Properties of Chromospheric Network', *Solar Physics* **180**, 29–45.

Berrilli, F., Ermolli, I., Florio, A., and Pietropaolo, E.: 1997, 'Time Dependence of the Chromospheric Network Properties', *SIF Conferences Proceedings* **58**, 105–120.

Caccin, B., Ermolli, I., Fofi, M., and Sambuco, A.M.: 1998, 'Variations of the Chromospheric Network with the Solar Cycle', *Solar Physics* **177**, 295–303.

Ermolli, I., Fofi, M., Bernacchia, C., Berrilli, F., Caccin, B., Egidi, A., and Florio, A.: 1998, 'The Prototype RISE-PSPT Instrument Operating in Rome', *Solar Physics* **177**, 1–10.

Kuhn, J.R., Lin, H., and Loranz, D.: 1991, 'Gain Calibrating Non-uniform Image-Array Data Using Only the Image Data', *Publications of the Astronomical Society of the Pacific* **103**, 1097–1108.

Rast, M.P., Fox, P.A., Lin, H., Lites, B.W., Meisner, R.W., and White, O.R.: 1999, 'Bright Rings Around Sunspots', *Nature* **401**, 14,678–14,679.

Address for Offprints: O.R. White, High Altitude Observatory, National Center for Atmospheric Research, 3450 Mitchell Lane, Boulder, Colorado 80301 USA

VARIATIONS OF SOLAR ULTRAVIOLET IRRADIANCE OBSERVED BY THE UARS SOLSTICE — 1991 TO 1999

GARY ROTTMAN (gary.rottman@lasp.colorado.edu)
Laboratory for Atmospheric and Space Physics, University of Colorado Boulder, CO 80309 USA

Received: 22 February 2000; Accepted in final form: 7 April 2000

Abstract. The Solar Stellar Irradiance Comparison Experiment (SOLSTICE) is one of ten instruments on the Upper Atmosphere Research Satellite (UARS) — one of two instruments measuring the solar ultraviolet irradiance. The instrument is a three channel spectrometer covering the spectral range 120 to 420 nm with a spectral resolution of approximately 0.2 nm. It has been successfully operating since October 1991, and has now provided more than eight years of data, extending from near the peak of solar cycle 22, through solar minimum and into the new cycle. The data provide time series that display solar variations over time scales from a few days up to the 11-year solar cycle. Quantitative estimates of amplitudes of both rotational modulation and the solar cycle variation in the 1991–1999 epoch are given for the UV spectrum between 119 and 300 nm.

1. Introduction

The Upper Atmosphere Research Satellite (UARS) was launched in September 1991 and includes a payload of ten scientific instruments to study the energy, dynamics and chemistry of the Earth's middle atmosphere (Reber, 1990). Although the spacecraft is three-axes stabilized and Earth oriented for the nadir and limb viewing instruments, it also carries an autonomous, de-coupled platform with three instruments which are pointed at the Sun. One of the instruments, ACRIM, measures the total solar irradiance (TSI) — that is, the total flux density undistinguished in wavelength (Willson, 1994). The other two instruments are spectrometers that measure the solar spectral irradiance in ultraviolet wavelengths between 120 and 420 nm. One of these instruments is the Solar Ultraviolet Spectral Irradiance Monitor, SUSIM, and the other is the Solar Stellar Irradiance Comparison Experiment, SOLSTICE. Both instruments cover basically the same spectral range with similar spectral resolution; however, they employ quite different approaches to determine changes in their sensitivity during the mission. SUSIM uses a system of duplicate and fully redundant optics and detectors with varying duty cycles to observe the Sun, and by comparing these observations from the different "optical channels" it is possible to deduce changes in the instrument response (Brueckner *et al.*, 1993). In contrast, SOLSTICE has the unique capability to observe both the Sun and bright, blue stars using the very same optics and detectors (Rottman and Woods, 1994). The stars then become reference standards that establish long-term changes in the instrument sensitivity. This report will concentrate on SOLSTICE observations and

Space Science Reviews **94:** 83–91, 2000.

results. The agreement between SOLSTICE and SUSIM measurements is quite good in studies where comparisons are available (Woods *et al.*, 1996).

Solar radiation with wavelengths below 300 nm is almost completely absorbed by the Earth's atmosphere and becomes a dominant direct energy input playing a major role in the dynamics, temperature, structure, and chemistry of the mesosphere and stratosphere. This radiation is especially important to the ozone photochemistry and serves both as a major source of ozone through the dissociation of molecular oxygen ($150 < \lambda < 200$ nm) and a major loss of ozone through the direct photodissociation of ozone ($200 < \lambda < 300$). Even small changes in the input solar ultraviolet will have commensurate influence on the atmosphere. Detailed studies of the Earth's upper atmosphere such as ozone variability require precise and accurate measurement of the solar UV flux.

2. SOLSTICE Observations

SOLSTICE is a three channel spectrometer with each channel covering a different, but overlapping portion of the solar spectrum. For historical reasons the channels are designated as "G" for the spectral range 115 to 185 nm, "F" for the range 170 to 320 nm, and "N" for the range 280 to 430 nm. Details of the instrument design and its calibration and characterization can be found in Rottman *et al.* (1993) and Woods *et al.* (1993). Basically the spectrometer has a small entrance slit (10^{-4} cm^2), followed by a plane diffraction grating, a focusing element and finally the exit slit. A pulse-counting photomultiplier detector is positioned behind the exit slit and detects monochromatic radiation with wavelength established from the angular setting and ruling density of the grating, and with a bandpass set by the width of the exit slit. The ruling density of the grating, the coatings on the optics, and the cathode material of the photomultiplier restrict the instrument sensitivity to a limited spectral range — these factors thereby specify the three spectral channels.

On the UARS spacecraft the solar instruments are pointed directly at the Sun, and by rotating the diffraction grating a spectrum is achieved. Figure 1 shows a typical spectrum covering the full spectral range from 119 to 420 nm, and the dashed lines delineate the wavelength boundaries between the three channels.

This spectrum illustrates the general characteristics of the ultraviolet solar spectrum — approximated by a continuum interrupted by absorption lines and free-bound absorption edges. The solar UV spectrum changes from an emission line spectrum below ~200 nm to the Fraunhofer absorption line spectrum above this wavelength. The emission spectrum and its weak background originate in the solar chromosphere, while above ~200 nm the solar radiation comes from the deeper photospheric layers.

The UARS spacecraft has an orbit altitude near 600 km, and orbits the Earth approximately every 96 minutes. The SOLSTICE and the other two solar instruments are pointed at the Sun during roughly half of the sixty minute day portion of

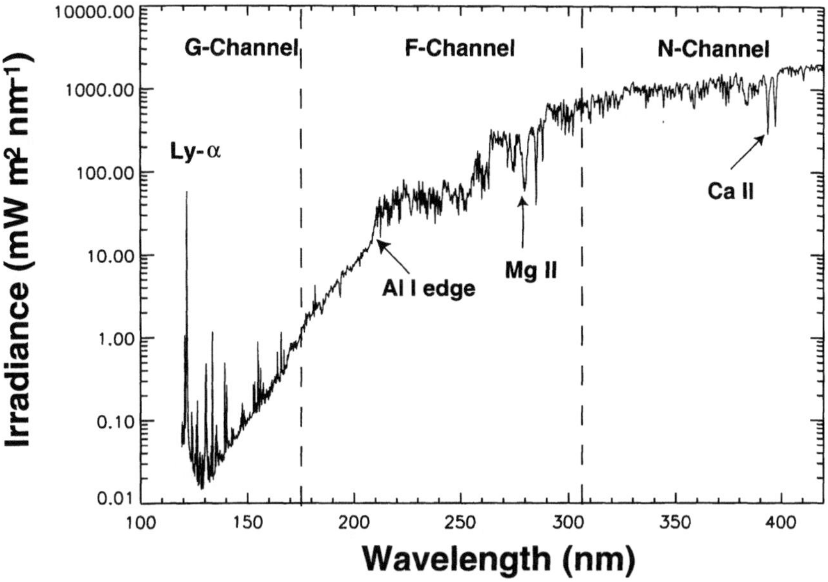

Figure 1. SOLSTICE solar spectrum between 119 nm and 420 nm

each orbit, and the remainder of the day portion is unavailable due to obscurations from various spacecraft structures. SOLSTICE operates autonomously, and during each available Sun view it does a prescribed solar experiment. The primary experiment is a slow spectral scan (2048 grating or wavelength steps each with a one second integration time), but rapid scans (8 times as fast), mini scans, and fixed wavelength experiments are also conducted. The spectrum shown in Figure 1 is the daily average of all of the observations taken during the calendar day, and therefore, combines 10 to 15 independent samples at each wavelength. This high resolution spectrum is then binned to a lower resolution spectrum with one sample every 1 nm (centered on the half mn). This final data product, corrected for all known artifacts and for changes in response of the instrument, is cataloged as the SOLSTICE Level 3BS data product and is available from the EOS Distributive Active Archive Center (DAAC).

As mentioned above, SOLSTICE has the unique capability of observing bright blue stars, and using these as reference standards to first establish and then correct for changes in the instrument sensitivity. During night portions of each orbit the instrument reconfigures itself by changing the entrance slit to an aperture 20,000 times larger (2 cm^2), by increasing the exit slit by a factor of 40 (G channel by 20 and N channel by 60), and by increasing the dwell time of an observation by as much as a factor of 1000. By combining these factors, SOLSTICE is 10^8 to 10^9 times more sensitive for the stellar observations — roughly the order of magnitude difference of the Sun and a bright star in the ultraviolet. Individually these early type stars should be quite stable, varying by a small fraction of a percent over

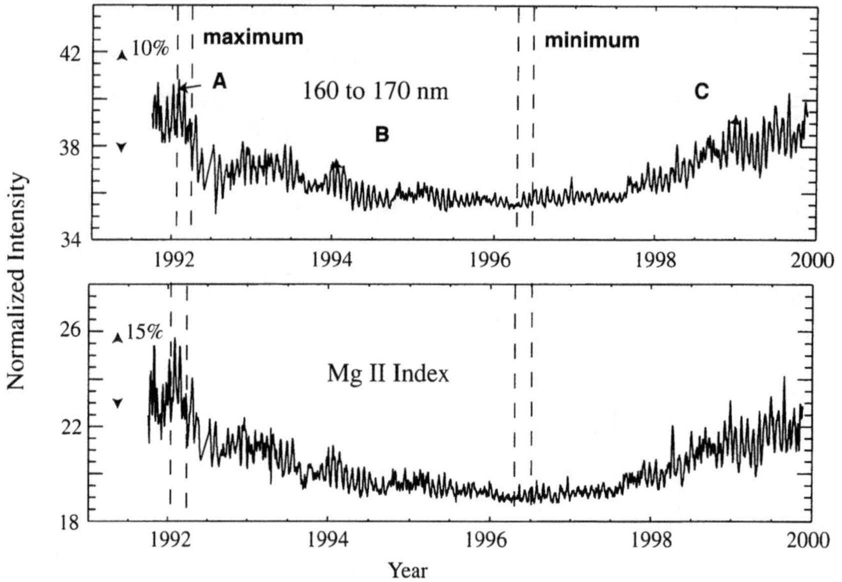

Figure 2. SOLSTICE irradiance (corrected for instrument degradation using the stellar technique) at ~165 nm from the beginning of the UARS mission to October 1999. The bottom panel shows for comparison the SOLSTICE Mg II index. Three periods of clear 27-day rotational modulation (A, B, and C) are identified for later discussion. Pairs of vertical lines mark periods of solar maximum in early 1992 and solar minimum in 1996.

time periods of thousands of years (Mihalas and Binney, 1981). The ensemble average flux from fifteen to twenty stars is an even more stable reference. Eighteen wavelengths (for each of the three channels) have been selected, and repeated measurements of the calibration stars are made whenever possible. Combining all of the stellar data at a given wavelength provides an unambiguous indication of the degradation in the instrument response. Such data also provide a mechanism for identifying stars that may exhibit intrinsic variability, and such stars are removed from the data set. In general, the response of the SOLSTICE degraded at a wavelength dependent rate of 2 to 4% per year. This degradation factor is known with an accumulated uncertainty over the eight year mission of about 1 to 2%. This is only a preliminary estimate, and following SOLSTICE final data processing it may improve to better than 0.5%.

After correcting the daily SOLSTICE spectra for instrument degradation, time series of the irradiance variability can be extracted and evaluated. For example, using the value between 160 nm and 170 nm from each day of the mission, we obtain the time series given in the top panel of Figure 2 . The bottom panel is an index derived from the blended doublet of singly ionized magnesium in the solar spectrum. This is the strong absorption feature identified in Figure 1 as Mg II, where the core of the line is an emission feature arising in the solar chromosphere. Since the core exhibits more variation than the neighboring spectrum, an index derived as the ratio of the core to a reference level from the neighboring photospheric radiation was proposed by Heath and Schlesinger (1986). The ratio is, to first order, free of all instrument degradation. The index is included in Figure 2 to illustrate that the irradiance measurement, 160 nm to 170 nm, follows the expected solar cycle trends.

The time series in Figure 2 show the typical nature of solar variation due to magnetic activity — a clear oscillation with a period near the 27-day rotation period of the Sun and a much slower variation from a high in late 1991, through a minimum in mid 1996, and then regaining strength during the onset of solar cycle 23. The 27-day oscillation is the modulation of solar radiation as active regions containing plage and sunspots cross the visible solar hemisphere. The $\sim$11 year variation follows the level of sunspot activity over the Schwabe cycle.

3. Rotational Modulation of Solar Radiation

As new active regions erupt into the visible solar photosphere, this modulation effect becomes quite clear in the chromospheric emission lines. The eruption and subsequent subsidence of these regions do not occur regularly in time nor in location over the solar disk. Lifetimes of these active centers are typically several months, and the presence of several active regions distributed across the visible solar disk is the norm during periods of high solar activity, just as the absence of activity becomes the norm at later periods. The Sun rotates with a period near 27 days, and this rotation coupled with the non-uniform distribution of activity on the solar disk gives the striking 27-day variation. Variability at 13.5 days may also exist in the time series. For example, two active regions separated in longitude by approximately 180° often gives an oscillation with a quasi 13-day period. Three different epochs are identified as A, B, and C in Figure 2 to illustrate the 27-day modulation of solar irradiance at different phases of the longer term (solar cycle) variation. As a measure of rotational modulation, we take the daily spectrum at the local maximum of the time series and divide by the spectrum of the adjacent minimum. In practice, we average three days near the peak of the time series and three days centered on the adjacent minimum. Since the two time periods are separated by only about one week, the difference of the instrument performance and sensitivity is negligible (or certainly small) — especially when the ratio of two measurements are constructed. Figure 3 gives the amplitude of the 27-day modulation as a function of wavelength in the spectrum for rotation "A" of Figure 2. It shows wavelength-to-wavelength differences in variability associated with many features in the solar spectrum shown in Figure 1. Note for example that each of the strong chromospheric emission lines displays much stronger rotational modulation than does the neighboring continuum. Likewise the aluminum edge (208 nm) shows a strong decrease in variability moving from short wavelength to long. Also notice how the Mg II Fraunhofer line at 280 nm shows much stronger variability

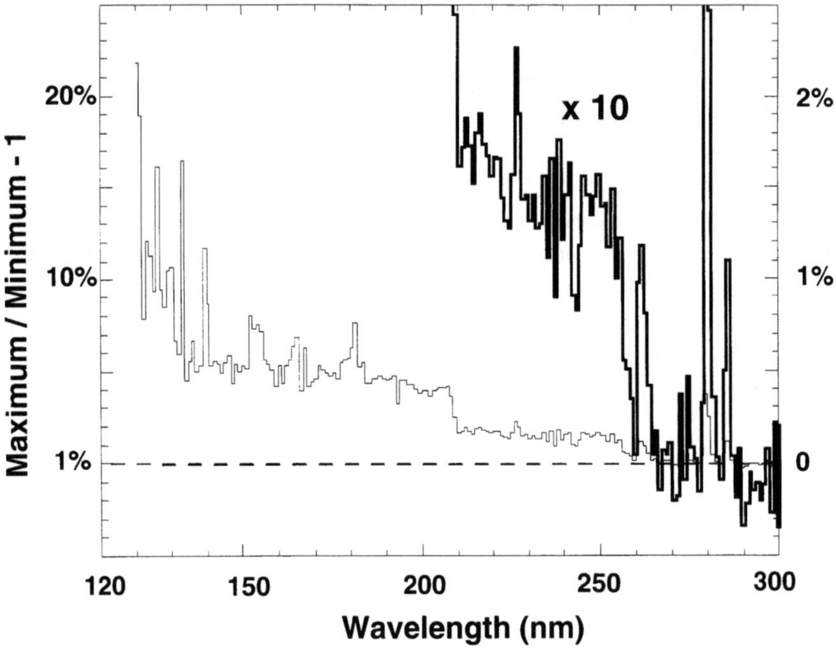

Figure 3. Amplitude of solar rotational modulation as a function of wavelength for the period "A" identified in the time series in Figure 2. The bold curve to the right is an amplification of the curve by a factor of 10 and corresponds to the right ordinate scale.

than the adjacent spectrum — confirming the validity of the Mg II index. (Care must be exercised in comparing Mg II indices from different experiments because its quantitative value depends on the definition used in its construction and on the spectral resolution of the instrument producing the data (DeToma *et al.*, 1997)).

It is beyond the scope of this paper to discuss in detail of the origin of solar rotational modulaton; nevertheless, some general comments are warranted. Although the curve shown in Figure 3 is typical for particular epochs of clear rotational modulation, these ratios are not unique. They do not apply, even with scaling, to other periods of 27-day modulation. For example, the epoch marked "B" in Figure 2 is near solar minimum and shows far lower amplitude of variation. As a general rule a simple scaling of the variation of rotation "A" (Figure 3) will not necessarily provide the variation of rotation "B" or "C". The general spectral shape may be similar, but variance will arise from differences in size and longitudinal distribution of active regions producing variability during the two epochs.

Estimates of solar variability often use information at one particular time period and extrapolate to other epochs. Detailed analyses of the SOLSTICE time series may lead to an improved understanding of the accuracy of such proxy methods for estimating variability in UV irradiances.

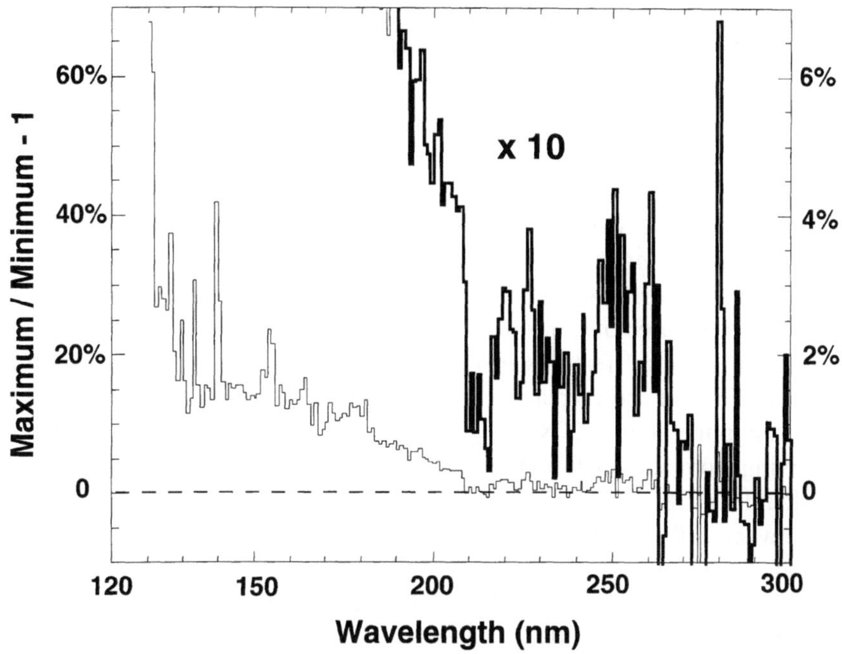

Figure 4. The "solar cycle"variation as a function of wavelength for the periods between "maximum" and "minimum" identified in Figure 2. The bold curve to the right is an amplification of the curve by a factor of 10 and corresponds to the ordinate scale on the right.

4. Solar Cycle Variability from Maximum Levels in 1992 to Minimum in 1996

Most indicators of solar activity show a cyclic variation from maximum to minimum and returning again to maximum in the 11 year Schwabe cycle. Likewise, the ultraviolet irradiance data displayed in Figure 2 show such cyclic behavior between 1991 and 1999. The high activity in late 1991 and early 1992 is near the level of maximum activity occurring perhaps earlier in 1989. The intensity of radiation steadily decreases to the minimum in 1996 (Harvey and White, 1999) then rises toward the maximum of solar cycle 23, expected in mid 2000. The irradiances in Figure 2 continue to rise throughout 1999 as expected.

Although the data of Figure 2 do not represent a single, complete solar cycle, they do cover the descending and ascending phases of two successive cycles. Thus, we estimate the variation in UV irradiance by comparing the two periods of 81 days in Figure 2 (identified by the two pairs of vertical lines) as periods of maximum and minimum activity. We use the same technique employed previously in estimating the amplitude of rotational modulation, except that we use the time series smoothed with a 81 running mean. The amplitude of the solar cycle variation is then the ratio of the 81-day average irradiance for the maximum epoch to that for the minimum epoch for each 1 nm wavelength bin in our SOLSTICE spectrum. Figure 4 shows this maximum-to-minimum variability spectrum. Before considering details of this variability spectrum and interpreting the results, recall from the discussion of the stellar calibration technique in Section 1 that the SOLSTICE analysis presently is limited at 1 to 2% uncertainty for the comparison of any two time periods. This implies that many of the ratios — especially for wavelengths between 210 and 300 nm — are near the uncertainty in the measurements.

Although the SOLSTICE data provided in this report are still preliminary, especially regarding long-term variability shown by our measurements, it is useful to inspect details of Figure 4. It is indeed encouraging that many features of Figure 4 reproduce the features of the rotational modulation in Figure 3, and these in turn correspond to spectral features in Figure 1. The strong chromospheric emission lines below 180 nm show greater variation than the adjacent continuum. The aluminum absorption edge at 208 nm again shows a drop in variation of about a factor of 2 moving from shorter to longer wavelengths, and likewise the variation in the Mg II doublet at 280 nm is strikingly larger than in the neighboring spectrum. The ratios between 210 and 270 nm are in the range of 1 to 3%, values of about the same magnitude as our present uncertainty in these estimates of long-term variability. Nevertheless, narrow bands near 230, 260, and 280 nm show increased variability in both the rotational modulation and cycle maximum-minimum variability data. Of these bands, only the 280 nm band is identified with strong chromospheric radiation, the Mg II doublet, and is expected to show strong variability. The obvious differences in the shapes of the two variability spectra remain to be studied. In our final analysis we expect the SOLSTICE data (stellar calibration data) to provide long-term accuracy of ~0.5%. Such an improvement will significantly refine these estimates of "solar-cycle" radiative variability.

5. Conclusions

We provide preliminary results from the UARS SOLSTICE covering a time period late in 1991 to late 1999, spanning the minimum between solar cycles 22 and 23. Solar variations are seen on all time scales, but are predominated by the 27-day modulation due to rotation of active regions across the visible solar disk, and a much slower solar cycle variation from a peak early in the mission, a minimum in 1996, and a steady increase back to the high values seen early in the mission. This latter long-term variation comes from the 11-year solar cycle variation. Our first UARS observations in late 1991 were at levels of high solar activity common throughout the maximum phase of cycle 22, and the present observations have yet to reach the peak of the new cycle 23. Nevertheless the maximum to minimum variability spectrum from our UARS experiment provides a reasonable preliminary estimate of the variation from solar minimum to maximum.

All aspects of the UARS mission continue to operate in a nominal fashion, and the solar instruments continue producing good data. There is every intent and expectation that the mission may continue for another three to four years — taking the measurement sequence well through the maximum of solar cycle 23. A second generation SOLSTICE instrument is presently being built and will be flown on NASA's Solar Radiation and Climate Experiment, SORCE, to be launched in July of 2002. This second generation SOLSTICE will extend the analysis presented here well through solar cycle 23. Overlap of the UARS and SORCE missions will ensure the reliability of the long-term solar irradiance record across mission boundaries.

Acknowledgements

The UARS SOLSTICE program at the University of Colorado is supported by NASA Contract NAS5-97145 and NASA Grant NAG5-6850.

References

Brueckner, G. E., Edlow, K. L., Floyd, L. E., Lean, J. L., and VanHoosier, M. E.: 1993, 'The Solar Ultraviolet Spectral Irradiance Monitor (SUSIM) Experiment on Board the Upper Atmospheric Research Satellite (UARS)', *J. Geophys. Res.* **98**, 10,695–10,711.

DeToma, G., White, O. R., Knapp, B. G., Rottman, G. J., and Woods, T. N.: 1997, 'Solar Mg II Core-to-Wing Index: Comparison of SBUV2 and SOLSTICE Time Series', *J. Geophys. Res.* **102**, 2597–2610.

Harvey, K. L., and White, O. R.: 1999, 'What is Solar Cycle Minimum?', *J. Geophys. Res.* **104**, 19,759–19,764.

Heath, D. F., and Schlesinger, B. M.: 1986, 'The Mg 280-nm doublet as a monitor of changes in the solar ultraviolet irradiance', *J. Geophys. Res.* **91**, 8672–8682.

Mihalas, D., and Binney, J.: 1981, *Galactic Astronomy Structure and Kinematics*, W. H. Freeman, New York., pp. 135–139.

Reber, C.: 1990, 'The Upper Atmosphere Research Satellite (UARS)', *Geophys. Res. Letters* **20**, 1215–1218.

Rottman, G. J., and Woods, T. N.: 1994, 'UARS SOLSTICE', *SPIE Proceedings* **2266**, 317–327.

Rottman, G. J., Woods, T. N., and Sparn, T. P.: 1993, 'Solar Stellar Irradiance Comparison Experiment 1: 1. instrument design and operation', *J. Geophys. Res.* **98**, 10,667–10,678.

Willson, R. C.: 1994, 'Irradiance observations of SMM, Spacelab 1, UARS, and ATLAS experiments', *in* J. M. Pap, C. Fröhlich, H. S. Hudson, and S. K. Solanki (eds.), *The Sun as a Variable Star*, Cambridge Univ. Press, Cambridge, pp. 54–62.

Woods, T. N., Rottman, G. J., and Ucker, G. J.: 1993, 'Solar Stellar Irradiance Comparison Experiment 1: 2. instrument calibration', *J. Geophys. Res.* **98**, 10,679–10,694.

Woods, T. N, Prinz, D. K., Rottman, G. J., London, J., Crane, P. C., Cebula, R. P., Hilsenrath, E., Brueckner, G. E., Andrews, M. E., White, O. R., VanHoosier, M. E., Flyod, L. E., Herring, L. C., Knapp, B. G. Pankratz C. K., and Reiser, P. A.: 1996, 'Validation of the UARS Solar Ultraviolet Irradiances: Comparison with the ATLAS–1, –2 Measurements', *J. Geophys. Res.* **101**, 9541–9569.

SOLAR VARIABILITY OBSERVATIONS

Discussion Session 1a

ORAN R. WHITE
High Altitude Observatory, National Center for Atmospheric Research, Box 3000, Boulder, Colorado 80307, USA

GARY J. ROTTMAN
Laboratory for Atmospheric and Space Physics, University of Colorado, Boulder, CO 80309, USA

Received: 18 February 2000; Accepted in final form: 13 June 2000

1. Introduction

Variability of the solar output of radiation and particles is currently recorded in experiments that measure the total radiative output, i. e., the Total Solar Irradiance (TSI) and spectral irradiances, and the particulate output of solar wind plasma and energetic particles. Our emphasis is on direct measurement of solar output not on the traditional indicators of solar activity related to global solar variability in its broadest sense. We then distinguish between sunspot numbers, plage indices, magnetic indices, etc., and the direct measurement of the Sun's physical output. Because of the paucity of measurements of the solar output before 1980, climatic significance of the solar output is usually established through these indices of solar activity.

Of these two outputs, the radiation is the most energetic in terms of possible climatic effect. To establish both the total radiative energy input to the terrestrial atmosphere and that absorbed by particular chemical species in the upper atmosphere, we now have measurements of both the Total Solar Irradiance (TSI) and the spectral irradiance in the near UV from 400 nm to 100 nm. The 10.7 cm radio flux measurement is an absolute spectral irradiance of great value as an indicator of radiative variability because the record is so long, i. e., from 14 February 1947 to date. This radio flux record is the basis for estimating radiative output throughout the solar spectrum at times when no absolute measurements are available.

The effects of the Sun's particulate output tend to be short term: minutes to hours following solar flares and coronal mass ejections. Energetically, their effects appear to be small in the troposphere, but we should consider their aggregate effects of recurrent events such as flares and coronal mass ejections. One such effect discussed at this workshop is modulation of the cosmic ray input to the Earth's atmosphere where it affects ionization and dissociation of chemical species. This effect is not a direct effect of the solar output but rather an aggregate effect on

Space Science Reviews **94:** 93–97, 2000.

galactic and solar cosmic rays in the heliosphere. This results in variability of the energetic particle flux at the earth.

2. Measurement of the Total Solar Irradiance

Regular daily measurement of the TSI began with the ERB experiment on Nimbus 7 in 1979 and continues until the present time through a series of experiments, e. g., ACRIMI, ACRIMII, and VIRGO. Fröhlich and Lean (1998) collected the TSI records from these experiment and constructed a useful composite record that shows the basic properties of variability of the Sun's radiative output in the direction of the Earth. First, the TSI output follows the 11 year Schwabe cycle: the TSI increases with solar activity except at time when a large sunspot appears on the solar disk. During the passage of such a spot across the disk, the TSI decreases according to the apparent area of the spot. The radiative variation is then a slow increase and decrease over a solar cycle punctuated by relatively sharp drops lasting approximately 10 days. The amplitude of the solar cycle variation is $\sim$0.2 percent peak to peak while the decreases due to sunspots are typically $\sim$0.1%.

The Fröhlich and Lean (1998) composite begins near the maximum of cycle 21 in 1980, covers all of cycle 22, and shows the onset of cycle 23. In construction of this composite time series, Fröhlich and Lean found it necessary to reconstruct portions of the Nimbus 7 ERB record, particularly in the 1979 to 1981 period. Studies of the TSI record by Lee et al. (1992) using the ERBS time series and by Chapman et al. (1994) using ground-based photometric measurements showed the need for a downward adjustment of the Nimbus 7 record of about 0.6 W/m^2 in the 1989–1990 time period. The Fröhlich-Lean composite contains this correction, and they conclude that the TSI at the time of minimum activity shows no change in the TSI level in the two minima in 1985 and 1996. This implies that the radiative output of the quiet Sun free of activity is constant. In his analysis of ACRMI, ACRIMII, and Nimbus 7 ERB, Willson (1997) finds an increase of $\sim$0.6 W/m^2 in the solar minimum TSI and claims the adjustment used by Fröhlich and Lean (1998) is not justified. The two conflicting claims are important because of the potential to identify a secular change of $\sim$0.5 W/m^2 per decade in the quiet solar atmosphere. Mende (2000) emphasized the accuracy of the ERBS measurement and TSI models from ground-based observations is insufficient for reliable correction of the Nimbus 7 ERB data.

Despite the lack of measurement over another complete solar maximum phase, data available in 1979–1980 and 1990–1991 suggest that average TSI values at solar maximum show no significant systematic increase or decrease.

Our existing TSI records show both differences in absolute calibration as well as slower changes during the life of the instruments in space. For example, ERB and ACRIMI TSI levels differed by 0.12 percent. The differences in absolute level are much smallerr between the DIARAD and PMO-6 radiometers now flying on

SOHO. Consequently, we see improvement in the basic radiometry. One possible source of such differences is accurate measurement of the input aperture of radiometers.

The slower inflight changes may come from degradation in the absorptivity of the radiometer cavities. All TSI experiments have multiple radiometers for inflight detection of instrumental changes. One reference radiometer is kept closed to solar radiation for long periods of time and opened very briefly for observation. All of these problems in the current set of measurements are now addressed in experiments planned for the future. Particular attention is being paid to the metrology of the entrance aperture, radiometer cavity absorptivity, and noise rejection in the measurement electronics. During the twenty year period we see steady improvement in the agreement between measurements by different groups. We expect to see absolute accuracy and precision of 0.01 percent or 100 ppm in future measurements.

3. Measurement of Particulate Output (Input?)

We had no comprehensive discussion of the output of plasma and particles from the Sun. However, Bazilevskaya (2000) reviewed the salient properties of the cosmic ray input to the Earth's atmosphere. In general, the cosmic ray flux at the Earth is a minimum at the maximum of the solar activity cycle. This cosmic ray modulation occurs in the heliosphere where solar wind structure and transients impede galactic cosmic rays. High energy particle(protons)s are generated in solar flares and come directly to the Earth in a matter of minutes to hours. Such proton bursts tend to occur during the descending phase of the solar cycle. Their energy input is small because the bursts are short-lived and infrequent.

Significant effects of cosmic rays in the terrestrial atmosphere occur well above ground level because of the interaction with the geomagnetic field. Solar cosmic rays are important above 15 km while the energy spectrum of the galactic particles ensures that some particles penetrate to lower altitudes. The geomagnetic field deflects the incoming particles in proportion to their magnetic rigidity (momentum per unit charge); therefore, the interaction zone is determined by both the energy spectrum of the particles and the distribution of magnetic field along their paths.

4. Climatic Effects of Solar Variability

Since climate is defined over time scales typically larger than our record of directly measured solar output, solar climate forcing can only be studied through records of solar activity such as the Schwabe sunspot record. Reid reviewed his studies of variability of sea surface temperatures (SST) over the last 150 years. He emphasizes the upward trend in SST that follows the increase in solar activity since 1850.

This increase appears to be the continuation of increasing following the Maunder Minimum that occurred in the 17th century. Similarity in the SST variation in both the northern and southern hemispheres suggests forcing by an external source. The SST record shows a significant decrease since 1990. This also correlates with the 1991 anomaly in the IPCC global temperature record. The 1990 anomalies are associated with an increase in aerosols due to volcanic activity.

Reid also noted the importance of UV variability in understanding variability of stratospheric winds. The planetary wave system also appears to be affected by UV variability. Variability of the solar UV between 310 nm and 150 nm is also important in determining the ozone balance in the upper atmosphere, i. e., ozone concentration decreases with increasing activity. This is due directly to the increase in UV radiation from active regions.

The missing element in understanding these correlations between solar activity and climate is the connection of the solar activity indices to the physical output of the Sun. Baliunas, Reid, and Lean constructed records of TSI output over the last 150 years using contemporary TSI and activity records. Statististical regressions between the contemporary data give linear transformation functions that convert historical data such as the sunspot record to a TSI record extending back in time.

5. Very Long Term Solar Variability

Beer's (2000) discussion of the analysis of cosmogenic isotope records shows a history of coronal activity extending back over 10,000 years. ^{14}C from tree rings and ^{10}Be from ice cores give the most convincing evidence for long term solar variability. The Maunder Minimum is clearly seen in both isotopic records. Other periods of such extended activity are seen throughout these records at intervals of $\sim$250 years. Time resolution in the ^{14}C deposits does not show individual solar cycles, but ^{10}Be clearly shows the 11-year cycle continuing through the Maunder Minimum even though very few sunspots were observed in this 70-year period.

This latter observation illustrates the importance of seeing solar variability and its effects in the terrestrial atmosphere in a global sense. In order to include all the associations between various types of solar activity and climate change, solar and climate variability has to be considered during many successive 11 year solar cycles. We need a more robust and detailed understanding of the effects of the different mechanisms proposed for solar forcing of climate before we can state with certainty the sensitivity of climate to the ensemble of solar outputs. Such a global view must include radiative and particulate outputs as well as solar disturbances in the heliosphere surround the Earth. Nevertheless, the radiative output of the Sun remains the most important driver, energetically. Measurements and study of the Sun's radiative output and its interaction throughout the Earth's atmosphere remains a highest priority scientific objective for understanding the sun-climate relationship.

References

Bazilevskaya, G.: 2000, *Space Sci. Rev.,* this volume.

Beer, J.: 2000, *Space Sci. Rev.,* this volume.

Fröhlich, C. and Lean, J.: 1998, 'The Sun's Total Irradiance: Cycles and Trends in the Past Two Decades and Associated Climate Change Uncertainties', *Geophys. Res. Lett.* **25**, 4377–4380.

Mende, W. and Stellmacher, R.: 2000, *Space Sci. Rev.*, this volume.

Willson, R. C.: 1997, 'Total solar irradiance trend during solar cycles 21 and 22', *Science* **277**, 1963–1965.

PHYSICAL CAUSES OF SOLAR ACTIVITY

N.O. WEISS (now@damtp.cam.ac.uk) and
S.M. TOBIAS (smt11@damtp.cam.ac.uk)
Department of Applied Mathematics and Theoretical Physics,
University of Cambridge, Cambridge CB3 9EW, UK

Received: 18 October 1999; Accepted in final form: 20 January 2000

Abstract. The magnetic fields that dominate the structure of the Sun's atmosphere are controlled by processes in the solar interior, which cannot be directly observed. Magnetic activity is found in all stars with deep convective envelopes: young and rapidly rotating stars are very active but cyclic activity only appears in slow rotators. The Sun's 11-year activity cycle corresponds to a 22-year magnetic cycle, since the sunspot fields (which are antisymmetric about the equator) reverse at each minimum. The record of magnetic activity is aperiodic and is interrupted by episodes of reduced activity, such as the Maunder Minimum in the seventeenth century, when sunspots almost completely disappeared. The proxy record from cosmogenic isotopes shows that similar grand minima recur at intervals of around 200 yr. The Sun's large-scale field is generated by dynamo action rather than by an oscillator. Systematic magnetic cycles are apparently produced by a dynamo located in a region of weak convective overshoot at the base of the convection zone, where there are strong radial gradients in the angular velocity Ω. The crucial parameter (the dynamo number) increases with increasing Ω and kinematic (linear) theory shows that dynamo action can set in at an oscillatory (Hopf) bifurcation that is probably subcritical. Although it has been demonstrated that the whole process works in a self-consistent model, most calculations have relied on mean-field dynamo theory. This approach is physically plausible but can only be justified under conditions that do not apply in the Sun. Still, mean-field dynamos do reproduce the butterfly diagram and other key features of the solar cycle. An alternative approach is to study generic behaviour in low-order models, which exhibit two forms of modulation, associated with symmetry-breaking and with reduced activity. Comparison with observed behaviour suggests that modulation of the solar cycle is indeed chaotic, i.e. deterministically rather than stochastically driven.

1. Introduction

Solar activity is a consequence of magnetic fields that erupt into the atmosphere of the Sun. In this review we first summarize the observed manifestations of activity, both in the Sun and in other similar stars, and then discuss the physical processes that are responsible for generating magnetic fields by dynamo action in their interiors. The solar dynamo is responsible not only for the 11-year Schwabe activity cycle but also for the secular variability with a characteristic timescale of 200 years. Dynamo theory has developed into a rich and complicated subject in its own right but our aim here is to convey only those aspects that are relevant to short and long-term interactions between solar variability and climate.

Space Science Reviews **94:** 99–112, 2000.

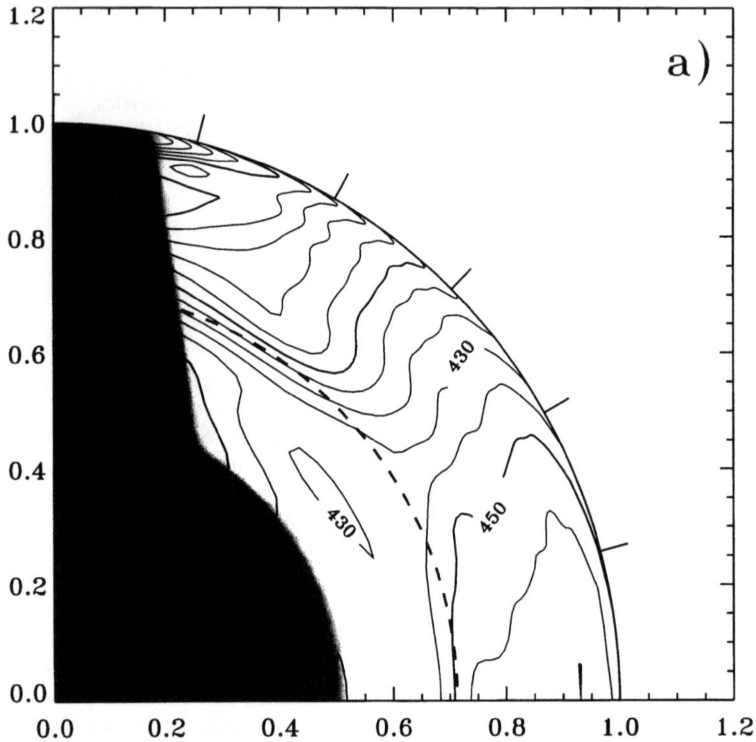

Figure 1. The internal solar rotation rate (from Schou et al. 1998). The frequency in nHz is given by the labels and the shaded area represents regions in which no inference can be made due to insufficient data. The base of the solar convection zone is shown using a dashed line.

The Sun is a typical middle-aged dwarf star on the main sequence (Tayler, 1997), with a radius $R_\odot = 7 \times 10^5$ km, a mass $M_\odot = 2 \times 10^{30}$ kg and a luminosity $L_\odot = 4 \times 10^{26}$ W. Energy is generated by thermonuclear processes (the p-p reaction) in a central core with a radius of $0.2R_\odot$ and is carried outwards by photons through the radiative zone which extends to $0.7R_\odot$. The outer 30% by radius is the convection zone, where energy transport is by turbulent convective motions, which are visible at the solar surface (the photosphere). The central temperature is 1.5×10^7 K, falling to 2×10^6 K at the base of the convection zone and 6×10^3 K at the photosphere. The temperature drops off slightly in the chromosphere but then rises abruptly to values of 1–3 $\times$ 10^6 K in the tenuous corona, which extends outwards for several solar radii and merges into the solar wind. The corona's structure is dominated by magnetic fields, which emerge from the interior of the Sun. They are responsible for coronal heating and thus for driving the solar wind. The great difficulty of solar physics is that, while we have an enormous wealth of data relating to the solar atmosphere, which contains only $10^{-10}M_\odot$, there are very few observations that can tell us about the Sun's interior.

Fortunately, recent developments in helioseismology – the study of global acoustic oscillations (p-modes) – have provided a means of probing the solar interior. From surface measurements it was already known that the equator rotates more rapidly than the poles: the sidereal rotation period at the equator is 25 days (corresponding to the 27 day synodic period that shows up in geomagnetic storms) while that at the poles is about 40 days. Rotational splitting of p-mode frequencies is now yielding information about the internal rotation of the Sun. Figure 1 shows the distribution with radius and latitude of the angular velocity Ω as derived from measurements made in space (Schou *et al.*, 1998). Within the convection zone Ω depends mainly on latitude, rather than on radius, but there is an abrupt transition at, or just below, the base of the convection zone, from differential rotation to almost uniform rotation in the radiative zone. This transition region (the tachocline) plays an essential rôle in the solar dynamo.

In the next section we list the principal features of the solar cycle and relate them to observations of magnetic activity in other stars with deep convective envelopes. Then, in §3, we consider the physical processes that lead to magnetic activity, before going on to review dynamo theory in §4. §5 is devoted to secular modulation of activity cycles, and finally we summarize our conclusions.

2. Observations: Solar and Stellar Magnetic Cycles

Lower main-sequence stars have deep convective envelopes and exhibit magnetic activity but it is only in the Sun that small-scale structures can be resolved. On the other hand, it is only through the solar-stellar connection that the evolution of the Sun's magnetic field can be understood. Stellar magnetic fields can be measured directly and there are active stars with fields of up to 2 kG (0.2 T) over half of their unspotted surfaces; they show photometric variations which are ascribed to starspots that may cover up to 60% of their surfaces, as well as radio emission associated with flaring, and coronal X-ray emission. Chromospheric Ca^+ emission, which is associated with magnetic fields on the Sun, is the most useful indicator of stellar activity. Measurements show that, for stars of given mass, magnetic activity decreases as the rotation period increases. Comparisons between solar-type stars in clusters with different ages show that they spin down (through magnetic braking by stellar winds) as they evolve. Thus both their magnetic fields and their angular velocities decline with age. Moreover, cyclic activity is only found in slow rotators like the Sun (Weiss, 1994; Tayler, 1997; Rosner, 1999).

Many different indicators have been employed as measures of systematic solar activity. Sunspots – which are dark because convective transport is impeded by strong local fields (Thomas and Weiss, 1992; Spruit, 2000) – are the most obvious magnetic features and the sunspot number, though arbitrarily defined, is closely related not only to sunspot areas but also to the 10.7 cm radio flux from the corona, to variations in cosmic ray fluxes at the earth (Beer, 2000) and to the aa-index

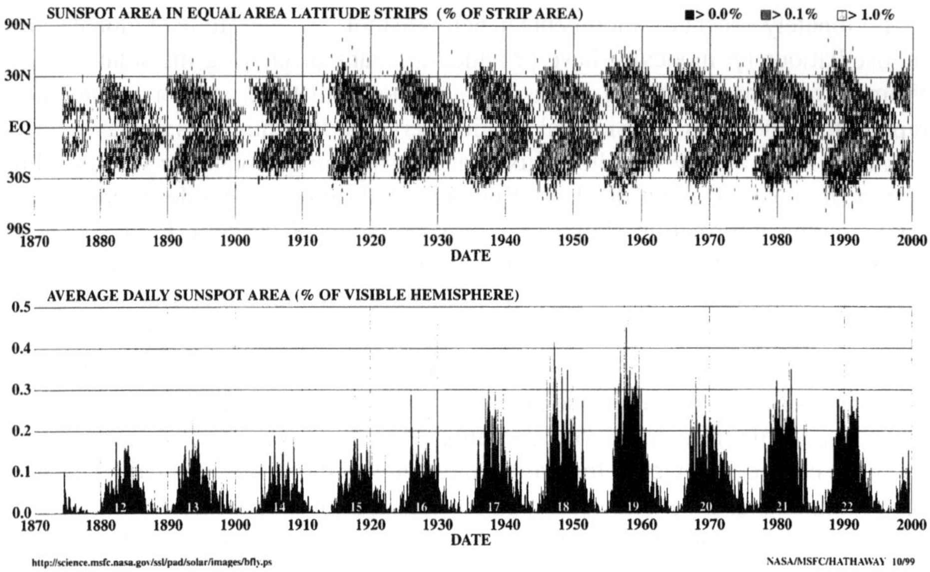

Figure 2. Butterfly diagram showing the positions in latitude of sunspots for each rotation of the sun since May 1874 (upper panel) together with the irregular cyclic variation of the area covered by spots (lower panel). (Courtesy of D.N. Hathaway)

which measures the incidence of geomagnetic storms. Since vigorous cycles are typically shorter, the inverse cycle length has also been used to measure activity. Figure 2 shows the variation of sunspot area with time over the past 120 years. It is apparent that solar activity varies aperiodically but with a well-defined mean period (the Schwabe period) of about 11 years. The butterfly diagram in Fig. 2 displays the location of sunspots in latitude and time: at the beginning of a new cycle spots appear at $\pm 30°$ and the activity zone spreads towards the equator, where it dwindles away as the first spots of the new cycle appear at higher latitudes.

Sunspots appear in active regions, typically as pairs with oppositely directed magnetic fields and oriented roughly parallel to the equator. The polarities of preceding and following spots (referred to the direction of the Sun's rotation) are the same in each hemisphere but opposite on each side of the equator. Moreover, these polarities reverse from one cycle to the next, so the magnetic cycle has a mean period (the Hale period) of 22 years. The magnetic fields in active regions are responsible for flares and for coronal mass ejections. They interact with the solar wind to form the interplanetary magnetic field, which impinges on the earth to cause magnetic storms and also deflects galactic cosmic rays (Bazilevskaya, 2000).

The Sun's magnetic activity is also modulated on longer timescales (Beer, 2000). Sunspots disappeared almost completely during the Maunder Minimum in the seventeenth century (Ribes and Nesme-Ribes, 1993), though the activity cycle persisted at a lower level (Beer, Tobias and Weiss, 1998). The proxy record from cosmogenic isotopes (^{10}Be and ^{14}C) shows that similar episodes of reduced activity (grand minima) have recurred on a characteristic 200-year timescale for the past 10 000 years and there is further evidence of similar behaviour in the interval between 50 000 and 35 000 years BP (Beer, 2000). That this is a normal feature of stellar activity is confirmed by observations of other stars, which suggest that they may be inactive for 30% of the time (Baliunas and Jastrow, 1990).

The magnetic flux that emerges through sunspots corresponds to a systematic azimuthal or toroidal field that is antisymmetric about the equator and reverses after 11 years. The resulting distribution of sunspots, as shown by the butterflies in Fig. 2, is very nearly symmetrical about the equator. This symmetry can, however, be broken at grand minima. For 30 years at the end of the Maunder Minimum, from 1672 to 1712, such activity as did occur was almost entirely confined to the southern hemisphere (Ribes and Nesme-Ribes, 1993).

3. Origin of Magnetic Activity in the Sun

There are two possible mechanisms for reversing the toroidal magnetic field that appears in sunspots. The first relies on a self-excited hydromagnetic dynamo in the interior of the Sun, while the second calls for a hydrodynamically driven oscillator. Dynamo action involves the conversion of kinetic energy into magnetic energy by the inductive effects of fluid motion in an electrically conducting fluid, and the dynamo can explain all the relevant observations: the meridional (or poloidal) component of the field near the poles also reverses, but at sunspot maximum; the basic structure of differential rotation, including the tachocline, does not change appreciably with the solar cycle but such changes as there are (the so-called torsional oscillations) have an 11-year period, as expected if they are driven by the Lorentz force, which is quadratic in the magnetic field **B**; and the whole process is at least qualitatively understood. The oscillator model, on the other hand, requires some unknown mechanism that would reverse the sense of differential rotation every 11 years and generate an alternating toroidal field from a *steady* poloidal field. That is not compatible with the observations.

Although small-scale turbulent convection at the photosphere can act as a dynamo and generate a disordered field (Cattaneo, 1999), the systematic behaviour of the solar cycle requires a large-scale dynamo that relies on rotation to produce an ordered field. The time-scales and spatial dimensions of active regions indicate that they rise from deep in the convection zone, and there is a general consensus that the dynamo is located in a region of weak convective overshoot between the radiative and convective zones, coinciding with the tachocline, where Ω varies rapidly with radius.

In a highly conducting plasma the field lines tend to follow the fluid motion, and there are three different physical processes that operate in this region (Parker, 1979). An initially poloidal field – for example, a dipole field derived from a potential that is symmetric about the equator – will be drawn out by *differential rotation* to produce an antisymmetric toroidal field. As this field grows it may eventually become unstable to modes driven by *magnetic buoyancy*, which can liberate Ω-shaped flux loops that rise up through the convection zone to emerge in active regions. These loops are themselves associated with rising plumes, which expand horizontally owing to the density stratification and therefore rotate through the action of the Coriolis force. These *cyclonic eddies*, with a persistent sense of helicity, create components of the field in meridional planes which can merge through diffusion to generate a reversed poloidal field. Thus the cycle can proceed.

In analysing this process it is important to distinguish the different timescales that are involved. The rotation period is comparable with the convective turnover time τ_c in the deep convection zone (about a month) and both are very much smaller than the resistive decay time $\tau_\eta \approx R_\odot^2/\eta$ for a large scale field, where η is a typical value of the magnetic diffusivity. For the Sun, $\tau_\eta \approx 10^{10}$ yr, which is comparable to its life on the main sequence, and very much greater than the 22-year period of the solar cycle. In the earth's core, on the other hand, $\tau_\eta \approx 10^4$ yr and a dynamo is required to explain the persistence of the geomagnetic field over the earth's history.

4. Dynamo Theory

The dynamo problem raises many subtle issues and is mathematically complicated (Moffatt, 1978; Roberts, 1994; Childress and Gilbert, 1995). In its simplest linear (or kinematic) form, the problem is to find a velocity field **u** such that the induction equation

$$\partial \mathbf{B}/\partial t = \text{curl}\,(\mathbf{u} \times \mathbf{B}) + \eta \nabla^2 \mathbf{B} \tag{1}$$

possesses exponentially growing solutions. There are now many examples of velocities that allow dynamo action. For astrophysical applications we are particularly interested in "fast" dynamos, for which the growth rate becomes independent of the diffusivity η in the limit as the magnetic Reynolds number $R_m = UL/\eta \to \infty$, where L, U are a typical lengthscale and velocity, while the field develops an increasingly fine structure; such dynamos require the fluid motion to be sufficiently complicated (exhibiting Lagrangian chaos).

For a solar dynamo model, the velocity **u** should include the effects of differential rotation and helicity. With R_m sufficiently small any magnetic perturbation will decay but as R_m is increased dynamo action sets in as a magnetic instability, at an oscillatory (or Hopf) bifurcation from the field-free state. Growth of the field is then limited by nonlinear effects. Behaviour in the weakly nonlinear regime is further complicated by magnetic buoyancy, which only becomes effective after the toroidal field exceeds a critical value. Thus we expect to find subcritical behaviour, leading to a bifurcation diagram resembling that shown qualitatively in Fig. 3*a*. Any self-consistent nonlinear model must include the back-reaction of the magnetic field on

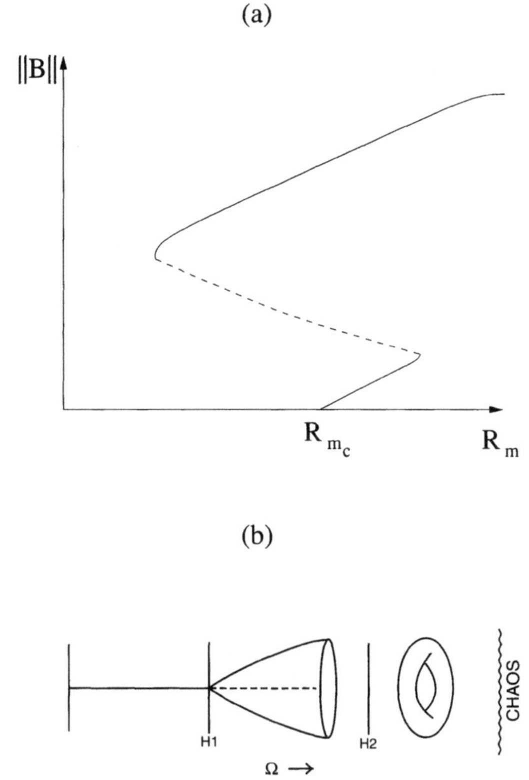

Figure 3. Schematic bifurcation diagrams for dynamo action. In (a) weak dynamo action sets in at $R_m = R_{m_c}$, but magnetic buoyancy can act to produce subcritical behaviour. The bifurcation sequence in (*b*) shows the appearance of periodic cycles, corresponding to a limit cycle, at the first Hopf bifurcation (H1), followed, at the second Hopf bifurcation (H2), by periodically modulated cycles and a two-torus, which then breaks down to yield a chaotically modulated signal. (After Tobias *et al.*, 1995.)

the motion, through the Lorentz force. Thus it is necessary to solve the equation of motion

$$\rho[\partial \mathbf{u}/\partial t + (\mathbf{u} \cdot \nabla)\mathbf{u} + 2\Omega \times \mathbf{u}] = -\nabla p + \mathbf{j} \times \mathbf{B} + \rho\nu\nabla^2\mathbf{u} + \mathbf{F} \quad (2)$$

in addition to (1). Here ρ is the density, p the pressure, $\mathbf{j} = \text{curl}\,\mathbf{B}$ the electric current and ν the viscous diffusivity, while Ω is the angular velocity of the rotating frame. Motion is driven by an external force $\mathbf{F}$ which is either imposed or, in the case of thermal buoyancy, calculated from additional equations. Gilman (1983) solved these equations numerically for a simplified spherical model and demonstrated that the dynamo process does indeed work, though his cycles did not resemble those in Fig. 2. Although computers are much more powerful now it would be premature to attempt a detailed simulation of the solar dynamo until the causes of the observed pattern of differential rotation have been understood.

Instead, most attention has been devoted to mean field dynamo models, which describe the evolution of the large-scale field by averaging over the turbulent motion. This procedure can be justified in certain circumstances but not for stellar convection zones, so it is best regarded as a parametrization that captures the essential physics of the solar dynamo. The azimuthally averaged field is split into a toroidal component $\mathbf{B}_\mathrm{T} = B_\phi \mathbf{e}_\phi$ and a poloidal part $\mathbf{B}_\mathrm{P} = \mathrm{curl}\,(A\mathbf{e}_\phi)$, where $\mathbf{e}_\phi$ is a unit vector in the ϕ-direction, referred to cylindrical polar co-ordinates (s, ϕ, z). Then the effect of the cyclonic eddies in regenerating $\mathbf{B}_\mathrm{P}$ from $\mathbf{B}_\mathrm{T}$ is represented by a parameter α so that

$$\partial A/\partial t = \alpha B_\phi + \eta(\nabla^2 - 1/s^2)A, \tag{3}$$

while differential rotation creates a toroidal field satisfying

$$\partial B_\phi/\partial t = s\mathbf{B}_\mathrm{P} \cdot \nabla\omega + \eta(\nabla^2 - 1/s^2)B_\phi, \tag{4}$$

where ω is the *local* value of the angular velocity and η must now be regarded as a turbulent diffusivity (Weiss, 1997). These linear mean field (or $\alpha\omega$) dynamo equations have been solved in cartesian and spherical geometries and for many assumed spatial distributions of α, ω and η. The behaviour found depends on a dimensionless control parameter, the dynamo number $D = \alpha\omega' L^4/\eta^2 \approx (\Omega\tau_c)^2$. As D is increased, the trivial solution ($\mathbf{B} = 0$) undergoes a Hopf bifurcation giving rise to dynamo waves that propagate towards the equator if ω decreases inwards and α is positive in the northern hemisphere. So the essential features of the butterfly diagram in Fig. 2 are readily reproduced.

In order to limit growth of the field some nonlinear saturation mechanism must be introduced, whether microscopic (through limitation of the α-effect by the small-scale field) or macroscopic (through the effect of the large-scale field on differential rotation). The torsional oscillations that are observed on the Sun suggest that it is appropriate to extend the linear system by adding an equation of the form

$$s\partial\omega/\partial t = (\mu_0\rho)^{-1}\mathrm{curl}\,\mathbf{B}_\mathrm{T} \times \mathbf{B}_\mathrm{P} \cdot \mathbf{e}_\phi + \nu(\nabla^2 - 1/s^2)(s\omega). \tag{5}$$

5. Secular Modulation of Cyclic Activity

Both Ω and the dynamo number D have *de*creased during the Sun's lifetime on the main sequence but in order to explain the origin of grand minima it is convenient to follow solar evolution backwards by studying the effects of *in*creasing Ω and D instead. Then we may expect to find the bifurcation sequence shown schematically in Fig. 3*b*, with successive transitions from a non-magnetic state to periodic cyclic activity (with trajectories attracted to a limit cycle in the phase space of the system) and then to periodically modulated cyclic activity (with a two-torus in phase space), followed by chaotically modulated cycles such as are apparently observed.

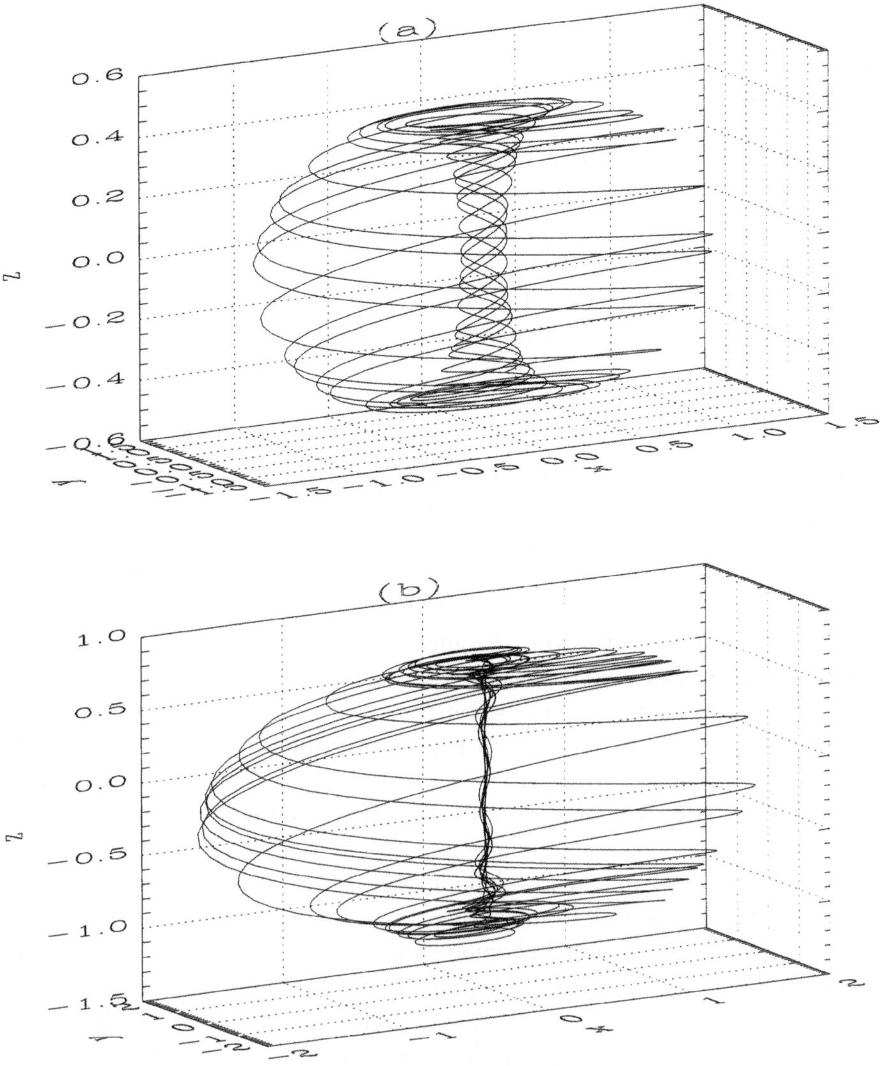

Figure 4. Attractors in the phase space for the system given by equations (6–8)). The trajectory in (*a*), which corresponds to a periodically modulated solution, lies on a two-torus while that in (*b*) is on a chaotic attractor. (After Tobias *et al.*, 1995.)

This bifurcation structure can be demonstrated for a toy model, consisting of three coupled ordinary differential equations (the minimum necessary for chaos to appear). Tobias, Weiss, and Kirk (1995) investigated the system

$$\dot{x} = f\left[\lambda x - \omega y + azx + d\,z(x^2 + y^2)\right], \tag{6}$$

$$\dot{y} = f\left[\lambda y + \omega x + azy\right], \tag{7}$$

$$\dot{z} = f\left[\mu - z^2 - (x^2 + y^2) + cz^3\right]. \tag{8}$$

where the variables $x(t)$ and $y(t)$ represent the toroidal and poloidal fields respectively and all the hydrodynamics is collapsed into the single variable $z(t)$. As the appropriate parameter is varied solutions do indeed follow the expected bifurcation sequence: Figure 4*a* shows a two-torus corresponding to a periodically modulated cycle, while aperiodic modulation yields trajectories that lie on a chaotic attractor like that in Fig. 4*b*. Since these are in fact the normal form equations for a saddle-node/Hopf bifurcation, it follows that their properties are generic and therefore robust. So similar behaviour is to be expected in a wide range of problems. It is this system that has been used to model the solar dynamo as an input to the climate system to demonstrate the power of resonant effects (Tobias and Weiss, 1999, 2000).

To demonstrate this, Tobias (1996) explored behaviour for a two-dimensional mean field dynamo model in slab geometry, with the toroidal field constrained to be antisymmetric about the equator. Once again the same bifurcation sequence appeared. When the symmetry constraint is relaxed there is the possibility of further symmetry-breaking bifurcations (Tobias, 1997). The dynamo equations allow two families of solutions with different symmetries: the first, with toroidal fields that are antisymmetric and poloidal fields with dipole symmetry, corresponds to the current behaviour of the Sun's magnetic field; the second has toroidal fields that are symmetric and poloidal fields with quadrupole symmetry. Linear theory suggests that the dipole solutions become unstable first, but that the quadrupole solutions follow very closely after. In Fig. 5 we compare the mean square fields for the low order model and for the mean field partial differential equations with the variations in ^{10}Be abundance described by Beer (2000)in this volume. All three timeseries show chaotically modulated activity cycles, though the pattern grows richer as one proceeds from the low-order model to the partial differential equations and then to the proxy record. Similar behaviour has since been found for nonlinear mean field dynamos in spherical geometry (Küker, Arlt, and Rüdiger, 1999; Pipin, 1999). Figure 6*a* shows a butterfly diagram with modulated cyclic activity from a nonlinear mean field dynamo model (Beer, Tobias, and Weiss, 1998). When there is vigorous activity, the field has dipole symmetry but that symmetry is broken in grand minima. On emerging from deep minima activity is much stronger in one hemisphere – which is what happened at the end of the Maunder Minimum on the Sun. Furthermore, there is the possibility that symmetry may actually flip in a grand minimum, as illustrated in Fig. 6*b*. These transitions are also supported by low-order model calculations (Knobloch, Tobias, and Weiss, 1998). So there may have been episodes in the past when the Sun's magnetic field had quadrupole, rather than dipole, symmetry.

This account has relied on deterministic behaviour to produce chaotic modulation of activity cycles. It can also be argued that turbulent convection provides stochastic input that affects the solar dynamo and may be responsible for modulation (e.g. Schmitt, Schüssler, and Ferriz Mas, 1996). The vigour of solar activity indicates that the dynamo must be far from the marginal state and so large stochas-

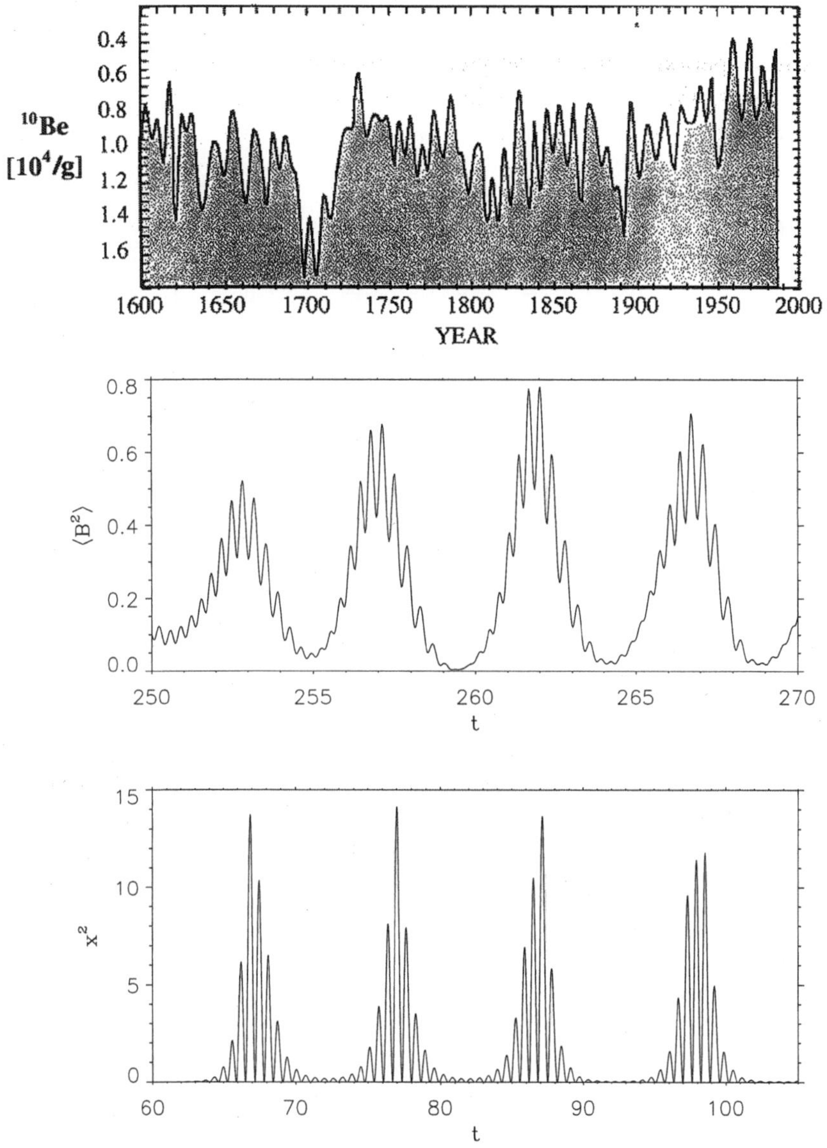

Figure 5. Comparison of proxy and model data. The top panel shows the ^{10}Be data, whilst the lower two panels show the chaotically modulated output from two solar dynamo models. In the central panel the magnetic energy from a two-dimensional mean field model governed by partial differential equations, whilst the lower panel displays the output from the system of ordinary differential equations (6–8).

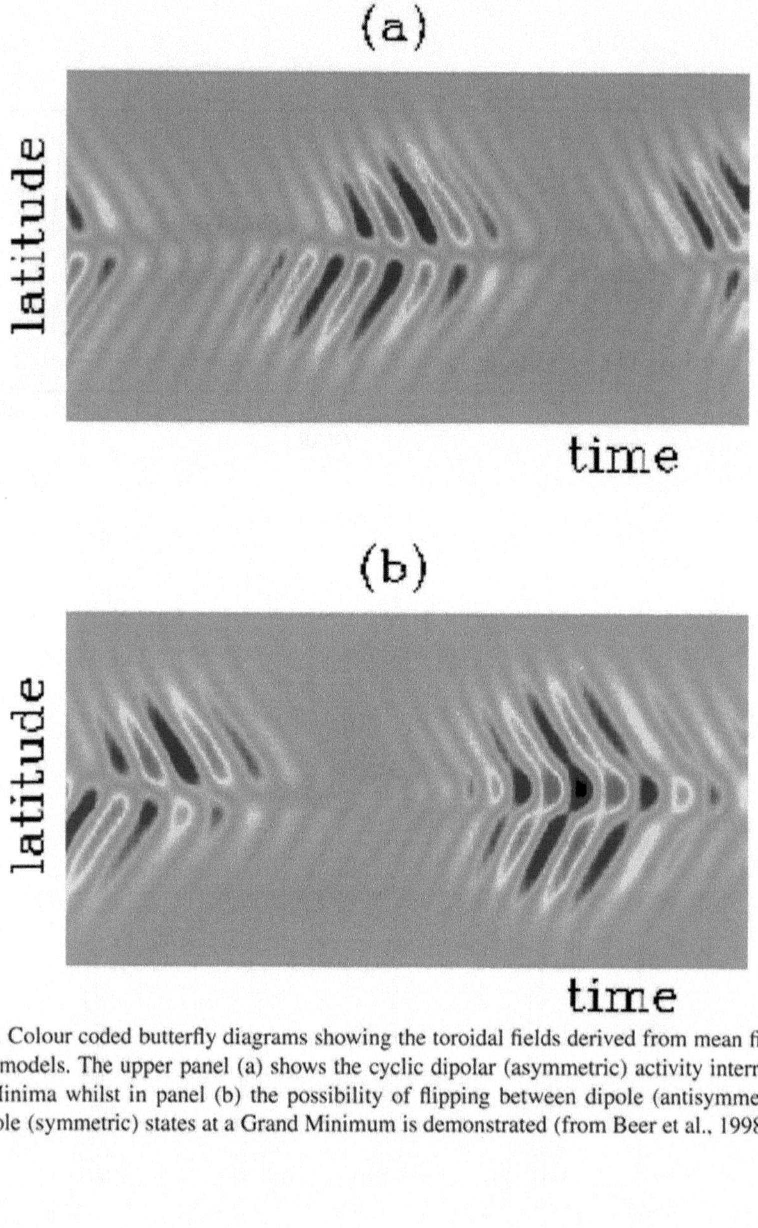

Figure 6. Colour coded butterfly diagrams showing the toroidal fields derived from mean field solar dynamo models. The upper panel (a) shows the cyclic dipolar (asymmetric) activity interrupted by Grand Minima whilst in panel (b) the possibility of flipping between dipole (antisymmetric) and quadrupole (symmetric) states at a Grand Minimum is demonstrated (from Beer et al., 1998)

tic fluctuations would be needed to produce a serious effect. On the other hand, chaotic modulation explains the recurrent features of grand minima, most notably the persistent 200-year timescale in the proxy record. Any weak stochastic fluctuations need not have any qualitative effect (trajectories would 'shadow' motion on the chaotic attractor) except in deep grand minima when they would certainly be important.

6. Summary

This brief account shows that dynamo theory can explain the *qualitative* features of solar activity, though current models lack predictive power. The time is ripe for the development of a more *quantitative* treatment, which must rely on massive computation. The first stage is to establish the relationship between convection and differential rotation, and to clarify the origin of the tachocline; work on these aspects of the problem is already in progress (Miesch, 1999). It will be then be possible to add a magnetic field and to produce a convincing model of the solar dynamo – though even then the effects of small-scale turbulence will have to be appropriately parametrized (Rosner, 1999). Meanwhile, we may conclude that the basic physical processes are understood, and that we must expect to find aperiodically modulated magnetic cycles, with episodes of reduced activity, in stars as active as the Sun. There is also the possibility of symmetry changes and of flipping from dipole to quadrupole symmetry and back in deep grand minima.

In conclusion, we emphasize again that the whole structure of the solar atmosphere is controlled by magnetic fields that are generated by dynamo action in the interior of the Sun. Those fields erupt through the surface to form a suite of magnetic features, ranging from active regions through spots and pores to small magnetic elements. The presence of magnetic activity modifies the total solar irradiance and (to a greater extent) the ultraviolet output; magnetic fields are also responsible for heating the corona (through some combination of tiny flares and wave dissipation) and so for producing X-ray emission, as well as driving the solar wind, which carries the interplanetary magnetic fields that impinge on the Earth.

Acknowledgements

We thank ISSI for inviting us to this stimulating meeting. SMT holds a Research Fellowship from Trinity College, Cambridge.

References

Baliunas, S. and Jastrow, R: 1990, 'Evidence for long-term brightness changes of solar-type stars', *Nature* **348**, 520–523.

Bazilevskaya, G.A., 2000: 'Observations of Variability in Cosmic Rays', *Space Sci. Rev.*, this volume.

Beer, J.: 2000, 'Long Term, Indirect Indices of Solar Variability', *Space Sci. Rev.*, this volume.

Beer, J., Tobias, S.M., and Weiss, N.O.: 1998, 'An active Sun throughout the Maunder Minimum', *Solar Phys.* **181**, 237–249.

Cattaneo, F.: 1999, 'On the Origin of Magnetic Fields in the Quiet Photosphere', *Astrophysical Journal* **515**, L39–L42.

Childress, S. and Gilbert, A.D.: 1995, 'Stretch, Twist, Fold: The Fast Dynamo', Springer, Berlin.

Gilman, P.A.: 1983, 'Dynamically consistent nonlinear dynamos driven by convection in a rotating spherical shell. II - Dynamos with cycles and strong feedbacks', *Astrophysical Journal Supp. Ser.* **53**, 243–268.

Knobloch, E., Tobias, S.M. and Weiss, N.O.: 1997, 'Modulation and symmetry changes in stellar dynamos', *Mon. Not. R. Astron. Soc.* **297**, 1123–1138.

Küker, M., Arlt, R., and Rüdiger, G.: 1999, 'The Maunder minimum as due to magnetic Lambda -quenching', *Astron. Astrophys.* **343**, 977–982.

Miesch, M.: 1999, 'The Coupling of Solar convection and rotation', preprint

Moffatt,H.K., 1978: '*Magnetic field generation in electrically conducting fluids*', Cambridge University Press, Cambridge.

Parker, E.N., 1979: '*Cosmical Magnetic Fields: their origin and their activity*', Oxford, Clarendon Press.

Pipin, V.V., 1999: 'The Gleissberg cycle by a nonlinear alpha Lambda dynamo', *Astron. Astrophys.* **346**, 295–302.

Roberts, P.H., 1994: 'Fundamentals of Dynamo Theory' in *Lectures on Solar and Planetary Dynamos*, M.R.E. Proctor and A.D. Gilbert (eds), Cambridge University Press, pp. 1–58.

Rosner, R. : 1999, 'Magnetic fields of stars', *Proc. Roy. Soc. Lond. A.*, in press.

Schmitt, D., Schüssler, M., and Ferriz-Mas, A.: 1996, 'Intermittent solar activity by an on-off dynamo' *Astron. Astrophys.* **311**, L1–L4.

Schou, J. et al: 1998, 'Helioseismic Studies of Differential Rotation in the Solar Envelope by the Solar Oscillations Investigation Using the Michelson Doppler Imager', *Astrophysical Journal* **505**, 390–417.

Spruit, H.C.: 2000, 'Theory of Solar Irradiance Variations', *Space Sci. Rev.*, this volume.

Tayler, R.J.: 1997, '*The Sun as a star*', Cambridge University Press, Cambridge.

Thomas, J.H. and Weiss, N.O.: 1992, *The theory of sunspots*, in *Sunspots: Theory and observations; Proceedings of the NATO Advanced Research Workshop on the Theory of Sunspots*, J.H. Thomas and N.O. Weiss (eds), Kluwer, Amsterdam, pp. 3–59.

Tobias, S.M.: 1996, 'Grand minima in nonlinear dynamos', *Astron. Astrophys.* **307**, L21–L24.

Tobias, S.M.: 1997, 'The solar cycle: parity interactions and amplitude modulation', *Astron. Astrophys.* **322**, 1007–1017.

Tobias, S.M. and Weiss, N.O.: 1999, 'Resonant Interactions between solar activity and climate', *J. Climate*, submitted.

Tobias, S.M. and Weiss, N.O.: 2000, 'Resonance in a coupled solar-climate model', *Space Sci. Rev.*, this volume.

Tobias, S.M., Weiss, N.O., and Kirk, V.: 1995, 'Chaotically modulated stellar dynamos', *Mon. Not. R. Astron. Soc.* **273**, 1150–1166.

Weiss, N.O.: 1994, 'Solar and stellar dynamos', in *Lectures on Solar and Planetary Dynamos*, M.R.E. Proctor and A.D. Gilbert (eds), Cambridge University Press, pp. 59–95.

Address for Offprints: N.O. Weiss & S.M. Tobias, Department of Applied Mathematics and Theoretical Physics, University of Cambridge, Silver St., Cambridge CB3 9EW, UK. now@damtp.cam.ac.uk smt11@damtp.cam.ac.uk

THEORY OF SOLAR IRRADIANCE VARIATIONS

H. SPRUIT (henk@mpa-garching.mpg.de)
Max-Planck-Institut für Astrophysik, Postfach 1523, 85740 Garching, Germany

Received: 7 September 1999; Accepted in final form: 23 November 1999

Abstract. The theory of stellar structure can be used to identify the most plausible mechanisms for the irradiance variations associated with the solar cycle. Changes in surface emissivity, i.e. the reduced cooling in spots and enhanced emission by small scale magnetic fields, are the most effective mechanisms and account for most of the observed variation. New views of stellar surface convection developed from realistic numerical simulations changes the physical description of thermal perturbations of the solar envelope. Helioseismology will soon be able to test the consequences of changes in surface emissivity, and distinguish them from other scenarios for irradiance variability.

1. Introduction

If a sufficiently complete record of solar irradiance measurements were available, it could be used directly for searching possible correlations with the Earth's climate. In reality, the short length of the record makes this impossible so that proxies like the sunspot record are used in practice. The use of such proxies, however, involves a model for the causes of the irradiance variations, hence requires some form of theory. In addition, the cause of these variations is of course an interesting problem in its own right in solar physics.

Until the advent of helioseismology, our knowledge of the internal structure of the Sun was based on the theory of stellar structure and evolution. Since this theory is reasonably simple, confidence in the models was high even before helioseismological tests. The main uncertainty had to do with the exact values of the opacity of stellar plasma. Increasingly accurate values for this were computed from first principles (atomic physics) by extensive numerical work. The basic correctness of the theory was then confirmed by helioseismology, to such an extent even that one is now confident that the 'solar neutrino problem' has its cause in neutrino physics, not inaccuracies in the solar models.

Though these solar models are spherical and only slowly evolving in time, the theory behind it can also be applied to time dependent and non-spherical disturbances. In this form, the theory makes definite predictions about the thermal behavior of the Sun, that is, the evolution in time and position inside the star of perturbations that might be related to the observed irradiance variations. In the following I present my view how this theory can be used to locate the most plausible source of irradiance variations. After this, I discuss how the new helioseismological

Space Science Reviews **94:** 113–126, 2000.

measurements currently being made can be used to test the predictions of proposed mechanisms of irradiance variation.

Conceptually, one can separate the thermal effects of magnetic fields into three types:

1. 'Sources and sinks'. The generation of a magnetic field involves the conversion of energy of motion into magnetic energy. Since the motions in the solar envelope are thermally driven, this ultimately means conversion of thermal into magnetic energy: building up a magnetic field produces a thermal sink somewhere. The opposite happens when the field decays: magnetic energy is converted into heat. These thermal effects exist only during *changes* in the magnetic energy content of the envelope.

2. 'Shadows', or changes in the heat transport coefficient. Magnetic fields interfere with convection, causing a reduction in the efficiency of heat transport in the envelope. In contrast to (1), these changes last as long as the magnetic field itself is present.

Both these sources and sinks and changes in the transport coefficient cause thermal perturbations, varying with magnetic activity, which propagate through the envelope and cause variations in surface energy flux. Related to the second class of perturbations are:

3. The effects of magnetic fields at the surface of the star. Sunspots, being dark, radiate less than the surrounding photosphere, while the small elements that make up plages and the network have an excess emission. In addition, it is conceivable that the magnetic elements have an indirect effect on surface emissivity by modifying the convective flow in their surroundings slightly (for which there is some observational evidence, see section 3.7).

2. Time Scales

The response of the Sun to thermal perturbations is not governed by a single time scale, but by a wide range of time scales. The longest of these is the thermal time scale of the Sun as a whole, called the Kelvin-Helmholtz time scale. It is about 10^7 years, namely the thermal energy content of the Sun, of the order 10^{41} Joule, divided by its luminosity of $4\,10^{26}$ W. Hence, on time scales of human interest, the Sun has a very large thermal inertia. If the central heat source of the Sun were switched off, the internal structure and the luminosity would start to change on this time scale. It is a global time scale for the Sun as a whole; more generally, we can define the thermal time scale $\tau_{\rm t}$ as a function of depth:

$$\tau_{\rm t}(z) \equiv U(z)/L(z) \approx \frac{1}{L}\int_{R-z}^{R} 4\pi r^2 u \,{\rm d}r, \qquad (1)$$

where L is the luminosity at depth z, U the thermal energy of the envelope down to a depth z, and u the thermal energy (enthalpy) per unit volume, approximately (for an ideal gas of constant γ) given by $u = \gamma P/(\gamma - 1)$. This is the time scale on which the structure of the envelope, and the observed luminosity, would start changing when the heat flux in the star were interrupted, by some magical means, at depth z. Some rough values for this quantity are $\tau_t \sim 10^5$yr at $z = 2 \times 10^5$ km (depth of the convection zone), 10 yr at 20 000 km (the size of a supergranule), 10 hrs at 2000 km (size of a granule). This strong dependence on depth is a consequence of the very strong stratification of the solar envelope (the pressure varies by a factor 10^8 between the surface and the base of the convection zone). As a result, the thermal response of the Sun also depends critically on the location of the disturbance. Near-surface disturbances have much larger effects than ones originating near the base of the convection zone.

A second kind of time scale involved in thermal readjustments is the so-called *diffusive* time scale. This is the time scale on which differences in entropy between different parts of the convection zone are ironed out by the convective flows. This is easily computed in a 'mean field' or diffusion model for convection, in which the local convective heat flux is assumed to be a function of only the local conditions (thermodynamic state of the gas and the temperature gradient). The validity of this model is questionable, since it is now known that stellar envelope convection is in fact extremely non-local, with almost all flows generated by cooling at the stellar surface (e.g. Nordlund, 1986; Stein and Nordlund,1989; for reviews see Nordlund and Stein, 1996; Spruit, 1997). The diffusion model has the advantage, however, that it is simple to apply, and is (unfortunately) still almost universally accepted. For these historical and practical reasons, I still use it for the estimates in this section. Estimates in the context of the new nonlocal view of envelope convection have not yet been developed to the same level of detail. I return to this in section 4, where the examples of heat flux blocking by sunspots and 'thermal shadows' are considered in the nonlocal convection picture. I argue there that the new nonlocal view only strengthens the conclusions based on the diffusion model.

For small perturbations, the heat flux in the local model is proportional to the entropy gradient, and time dependent perturbations are governed by a diffusion equation. The 'turbulent' diffusion coefficient in this equation is $\kappa_t \approx \frac{1}{3} l_c v_c$ where l_c and v_c are the convective length scale and velocity. In mixing length models of the solar convective envelope, this quantity varies only weakly with depth, at a value of the order 10^{13} cm^2/s. In this model thermal inhomogeneities (more precisely: entropy inhomogeneities) of length scale d are smoothed by turbulent diffusion on a time scale

$$\tau_d = d^2/\kappa_t. \tag{2}$$

For $d = 2\,10^5$ km this is about 1 yr, for $d = 2000$ km about 1 hr. Comparing τ_t and τ_d, we see that they are of similar magnitude close to the surface (to be precise: in the surface boundary layer where convection is not efficient enough to keep the stratification close to adiabatic). In deeper layers, the thermal time scale is much *longer* than the diffusive time scale, by a factor of up to 10^5.

The two time scales measure different types of thermal adjustment processes. These same processes appear in the thermal behavior of, say, a chunk of metal, aluminum for example, heated from the inside and suspended in space. The thermal time scale is the time scale on which its temperature adjusts to a change in the heat input, such that the heat radiated from the surface into space balances the heat input again. It is determined by the heat capacity [U in Eq. (1)] and the power level (L). The time scale on which different parts of the chunk equilibrate to the same temperature is governed by a different process, namely the thermal conduction (the equivalent of the turbulent diffusion in the Sun). The diffusion (conduction) time scale is much shorter than the thermal time scale, because of the large heat conductivity of AL. In the Sun, it is the very large turbulent diffusivity in the bulk of the convection zone that causes the very short diffusive time scale compared with the thermal time scale.

How do these different time scales come into play when the convection zone is thermally perturbed by, say, the storage of energy in a growing magnetic field? Such perturbations can be computed in detail, either by numerical methods (Endal et al., 1985; Gilliland, 1988) or more analytically. We can, for example, consider the initial value problem in which a perturbation is allowed to evolve in time by thermal transport in the convection zone. In general this evolution has components on all the time scales of the problem, including the very long thermal time scale. Detailed calculations of this problem have been discussed elsewhere (Spruit, 1982ab, 1991; Arendt, 1992). In the following, the basic conclusions of these analyses are summarized.

3. Expected Level of Luminosity Variations

For quantitative estimates, the strength of the magnetic field and its filling factor in the convection zone have to be specified. Assume that we have a layer of field with strength of the order of 10 000 G (equipartition with the convective flows as estimated by a mixing length model), one pressure scale height deep, near the base of the convection zone (where most of the magnetic flux is probably located).

3.1. Sources-and-Sinks

If the energy needed to build up the solar magnetic field during one half of the solar cycle is taken out of the thermal energy near the base of the convection zone, the calculations show that a surface luminosity variation of only $\delta L/L \sim 10^{-7}$ results. This is due to the very large heat capacity of the lower convection zone. The effect is stronger if the source of the magnetic field is assumed to be closer to the surface, but is still much smaller than the observed irradiance effect. Recent models for the emergence of magnetic flux from the base of the convection zone (D'Silva and Choudhuri, 1993; D'Silva and Howard, 1993; Caligari et al., 1998) indicate that the actual field strength at the base of the convection zone is probably about 10 times higher than the equipartition estimate. The magnetic energy per unit of magnetic flux is then also 10 times higher, but this still does not lead to a significant luminosity effect.

3.2. Shadows

If magnetic fields interfere with convection, a thermal perturbation develops as well. This effect depends crucially on the 'covering factor'. If a reduction of convective efficiency is assumed that uniformly covers a horizontal surface at some depth z, magnetic fields comparable to equipartition with convective flows can have stronger effects than the source-and-sink perturbations (Gilliland, 1988), though a measurable effect is predicted only if the field is located close below the surface. At face value, this would make changes in convective efficiency a reasonable candidate for irradiance variations. In practice, the effect is very strongly reduced, however, if the change in convective efficiency does not extend over an entire horizontal surface. This is because the turbulent heat conductivity in the convection zone is so high that the heat flux is easily 'shunted' past blocking objects below the surface (Spruit, 1977). Compare this with efforts to thermally insulate homes. To be effective, it is not enough that *most* of the building's outer surface is very well insulated. A more modest insulation but without any leaks is more effective. For this reason, thermal shadows of subsurface fields are probably not very important.

3.3. Surface effects

By far the most effective way in which a magnetic field influences the irradiance is by its effect on the *net surface emissivity*. The reduced emission from a sunspot area shows up directly in the irradiance records as a dip tracking the passage of the area across the disk. The only complication is that one might expect, depending on one's prejudices, that part of this reduction could be compensated by a brightening elsewhere, for example in the form of a 'bright ring' surrounding the spot. Evidence of such bright rings is absent for most spots that have been studied for this effect, and where observation of a ring is reported, it compensates for only a small fraction of the spot blocking (e.g. Rast et al., 1999). Real brightening around a spot is hard to measure because it is usually masked facular emission. Such faculae are part of all active regions. Though their excess emission is roughly of the same order as the spot deficit, this is a coincidence that does not hold any more on closer inspection. For example, faculae are equally present in active regions with or without sunspots, and their contribution to the irradiance variation does not correlate very well with spot areas (Foukal and Lean, 1986; Lean et al., 1998).

The absence of bright rings is understood in terms of a turbulent diffusion model for the heat flux in the convection zone (Spruit, 1977, 1982b; Foukal et al.,1983; Chiang et al., 1984). The 'blocked heat flux', for the most part, does not reappear elsewhere on the surface, but stays inside the convection zone, being stored/released on the very long thermal time scale of the convective envelope. This conclusion holds, in the diffusion model, as long as the blocking effect of the spot extends to a depth of at least 1000 km, a mild requirement given that the observed Wilson depression of the umbra of a spot is already of the order of 500 km.

3.4. Spot blocking: stove top analogy

The absence of bright rings in the diffusion model can be understood by a kitchen analogy. Imagine a stove top with an electric heater plate. Assume this plate to be made of a massive block of a metal with a high thermal conductivity, such as aluminum. Such a block, like the solar convection zone, has two time scales, the thermal time scale, given by the heat content divided by the heating rate, and the thermal conduction time scale given by the diffusion time across the block. If the thermal conductivity is large, the diffusion time scale can be much shorter than the thermal time scale. In the solar convection zone this is the same except that the ratio of time scales ($\sim 10^5$) is much more extreme. Suppose the heater plate has been on and has reached a steady temperature, so that heat input balances the conduction/convection losses at the top. Suppose also that the ratio of time scales is sufficiently large. Then put a piece of thermally insulating material on top, covering part of the surface of the plate (as a model for the effect of a spot). Where does the blocked heat flux go? The answer is of course that it does not go anywhere, the blocked heat stays in the block, heating it up slowly, instead of being radiated at some other part of the surface. On account of the large thermal conductivity, the temperature in the block stays uniform, and almost no 'bright ring' appears around the spot. When the insulating spot is removed after a while, the heat flux returns to its old value instantaneously. The temperature has increased only slightly during the blocking phase, and this slight enhancement decays again on a thermal time scale.

3.5. Excess emission by small scale magnetic fields

Apart from the highly visible sunspots, the Sun has a small scale magnetic field consisting of a very large number of tiny magnetic elements. They are called faculae if their density is high, and 'network' at lower density, and collectively contain much more magnetic flux than sunspots. They are visible especially near the solar limb, showing that they radiate more isotropically than the limb-darkened normal solar atmosphere. This excess emission has two components, one originating in somewhat higher atmospheric layers, in the form of line- and UV-continuum emission. This component probably is due to dissipation of mechanical (wave) energy carried by the magnetic field. In addition, there is a component seen in white light, originating in the surface layers of the magnetic structures. This second component is understood theoretically as a radiative transfer effect (Spruit, 1976, 1977): the radiating surface in a magnetic element is a little 'dimple' in the surface, whose side walls radiate an extra flux compared with a flat surface. The large number of small magnetic elements effectively corrugates the surface, and makes it radiate both more effectively and more isotropically, just as a rough surface has a higher emissivity than a polished surface. This effect increases the radiative flux from any part of the surface where small scale magnetic fields are present. It is a net effect (predicted before observation, Spruit, 1977) that is not balanced by a deficit somewhere else on the surface. Since the magnetic elements are so small, they are hardly resolved in the observations. Recent high-resolution observations (Topka et al., 1997), however, verify the picture predicted by flux tube models (Spruit, 1976, figure 11; Steiner et al., 1997) for the appearance of small magnetic elements.

Since the effect of small scale fields is an increased *emissivity*, it increases the *cooling rate* at the surface.

3.6. Surface Effects: Discussion

In the above I have outlined

- How the observed excess emission can be explained in terms of the local effects of small scale magnetic fields.
- That this local excess is not expected to be compensated by deficits elsewhere at the surface: the solar surface has a higher effective emissivity in regions of small scale magnetic fields and a lower one in sunspots.

Together with the observed correlation between the irradiance signal and the magnetic indicators (e.g. Foukal and Lean, 1986; Solanki and Unruh, 1998; Lean et al., 1998), this shows that local surface emissivity changes due to magnetic fields (a reduction in spots, enhancement in small scale fields) account for at least a significant fraction of the observed irradiance variation. It is much harder to prove that there could not be some contribution from other effects as well, for example the photospheric temperature enhancements proposed by Kuhn et al. (1988); Kuhn and Stein (1996). It makes clear, however, that in order to find such other effects one first has to remove the magnetic surface effects. Measurements such as those of Dicke et al. (1987) and Kuhn et al. (1988), in which the contamination by magnetic surface effects is ignored, can not be used to claim the existence of photospheric temperature changes. Whether there is anything left to be explained after the surface effects are taken out is not clear at the moment. Lean et al. (1998) and Solanki and Unruh (1998) argue that no photospheric temperature increase is needed to explain the available irradiance data.

The situation is similar for measurements of (variations of) the solar radius and a possible oblateness of the Sun. At the current sensitivity (a small fraction of an arc second) the high brightness contrast of small scale magnetic fields near the solar limb seriously interferes. In these measurements one typically determines (Kuhn et al., 1998) an average brightness profile across the limb, and variations of the limb position are then measured as offsets (in time or as a function of position around the Sun) of this given profile. Bright areas near the limb cause spurious outward offsets. Since the measured effects are very small, and the brightness of magnetic areas near limb is large (faculae immediately stand out on a solar image) detailed information on any magnetic brightness changes near the limb is necessary to correct for this. As the history of Dicke's oblateness measurements shows, limb positions that ignore the effect of these inhomogeneities are not believable.

A possible way of distinguishing true photospheric temperature changes from small scale surface emissivity effects is through the wavelength dependence of the irradiance signal. The excess emission from magnetic fields is concentrated in the magnetic elements, which have a small surface filling factor. For a given (bolometric) excess, it has to have a higher color temperature than excess emission from something with a large filling factor. This has been tested by Solanki and Unruh (1998, see also Unruh et al., 1999), who conclude that the observed wavelength dependence of the irradiance excess signal favors the small scale magnetic interpretation rather than a diffuse surface temperature increase.

3.7. CHANGES IN GRANULATION IN ACTIVE REGIONS

The changes of surface emissivity by magnetic fields discussed above are due to the different radiative properties of small magnetic elements ('flux tubes'). More indirect effects may play a role as well. The shape of granules appears to be different in magnetic regions (Macris and Roesch, 1983; Muller and Roudier, 1984; Muller, 1986; Title et al., 1992). They are smaller, more irregular, and the measured flow speeds are lower. These effects are present *in addition* to the magnetic elements themselves, which also change the appearance of granulation by filling in the intergranular lanes, especially when seen at low spatial resolution. The changes may be the result of the geometric constraints the magnetic tubes put on the convective flow outside them.

Since the flow appears to be different, it would seem possible that the heat flux it carries is also different. This might contribute to the observed solar cycle variations of irradiance (Muller, 1986; Kuhn and Libbrecht, 1988). Limits on this contribution can be put by observations of the *colors* of the solar cycle variation signal. As discussed above, Solanki and Unruh (1998) find that the wavelength dependence of the irradiance signal argues against a dominant contribution from small changes of surface temperature over a larger areas, but do not quote quantitative limits. In order to contribute in the right sense to the solar cycle variation, the constraints imposed on the flow by the magnetic flux elements would have to lead to an *increase* in the heat flux carried by granulation. Direct (spatially resolved) measurements of a heat flux change in granulation in magnetic regions are probably difficult, since the effect would be small and hard to separate from the enhanced emission from the magnetic elements. Numerical simulations (Nordlund, personal communication) indicate, however, that an effect of the right sign may be present.

Changes in surface heat flux by modified granulation, if they exist, will affect the convective envelope in the same way as the excess emission from the small scale magnetic field. They can also be represented by an effective change of the emissivity of the solar surface. In particular, one does not expect these changes to be 'compensated' by opposite changes elsewhere on the surface (except, as before, on the 10^5yr thermal time scale of the envelope).

4. Improving on the Turbulent Diffusion Picture

In the results quoted a diffusion model for convection was used. It assumes that convection can be modeled by a turbulent viscosity for momentum and a turbulent diffusivity for the transport of heat. While this was a simple and somewhat justifiable model of convection in the absence of detailed knowledge of the convective flows in a stellar envelope, we now know that it does not represent stellar convective envelopes well. Numerical simulations (Nordlund, 1982, 1985ab, 1986, 1990; Nordlund and Dravins, 1990; Nordlund and Stein, 1990, 1991, 1996; Stein and Nordlund, 1989, 1991, 1998; Steffen et al., 1989; Steffen, 1993; Ludwig et al., 1999) show an extremely *nonlocal* picture. The convective flows are driven almost exclusively by cooling at the surface, with narrow fast moving downdrafts between slow almost isentropic upflows (for a discussion see also Spruit, 1997). In the present context, the most important property is that the convective flow at all depths is driven by cooling at the surface rather than by a local overturning process. The material in the cool downdrafts survives to large depths below the surface with little mixing into the upflows.

Given this extremely nonlocal picture, it is appropriate to ask how the thermal behavior of perturbations in the convection zone can be estimated without appealing to a turbulent diffusion model. I do this below for two of the effects discussed, the spot blocking and the 'thermal shadow' problems. In both cases, the conclusion is that thermal perturbations of the normal solar surface are even smaller than in the diffusion picture. This has to with the fact that the flows in the new non-local picture are caused by cooling at the surface, and that much of the changes are one-way only (downward). The diffusion picture on the other hand communicates perturbations equally well in all directions.

4.1. Spot Blocking

Below the spot (modeled as a region of reduced heat transport efficiency extending to some depth below the surface) the upflows have exactly the same temperature as upflows in the unspotted surroundings at the same level, namely that given by the entropy at the base of the upflows. In this sense, there is no 'pile up of heat below the spot'. Because of the reduced heat loss at the surface, however, the downflows below the spot will be less vigorous. The unspotted surface notices nothing of the spot's presence (except for an extremely narrow ring where lateral radiative exchange takes place, and except for the presence of a moat flow, see below). It continues to cool upwellings into downdrafts as before, since the entropy in the upflows has not changed. Thus, we expect again that bright rings will be absent around spots, but the reason is even simpler than in the diffusion model. The spot is a region at the surface where less heat is radiated away, and this is now independent of the depth of the spot below the surface (in contrast to the diffusion model, where the spot has to extend to a minimum depth of 1000 km for the explanation to work).

Still, one may wonder what happens to the amount of heat generated in the solar interior that now fails to be emitted at the surface. This part of the problem is the same as in the diffusion model: the imbalance causes a secular increase of the entropy in the entire convection zone until a new thermal equilibrium is reached. Because of the very long thermal time scale of the convection zone (10^5 yr), the effect is negligible on observable time scales. In a steady state, when the average number of spots does not change, the convection zone does not heat up, because its mean temperature is higher than it would be without spots (Spruit and Weiss, 1986). Episodes of larger than average spot coverage cause heating, those of less than average spot coverage cause cooling on this time scale.

4.2. Thermal Shadows

The thermal shadow expected from a blocking object below the surface (Spruit, 1977; Goode and Kuhn, 1990), which is a small effect already in the turbulent diffusion model, is likely to be even smaller in the new picture of stellar envelope convection. Since all upward flowing gas has nearly the same entropy, a blocking object below the surface has no effect on what happens above it, as long as there is enough room above it for upflows to comfortably reach the solar surface. It may interfere somewhat with downflowing gas, but this has no consequence for the surface temperature above the blocking object, since only a small fraction of the downward moving gas is carried back up to the surface. As before, this is a consequence of the non-local nature of the convective flow.

5. Helioseismological Tests

In the above I have concentrated on surface effects, and argued that temperature enhancements with origins deep inside the convection zone are less likely to contribute. Such deep origins have been advocated by Kuhn et al. (1988), and Kuhn and Stein (1996) who propose that a temperature enhancement with origin at the base of the convection zone causes a temperature enhancement of 0.5K in the active latitudes. Potentially exhausting theoretical debates for or against these models can probably be circumvented since the models make different predictions for the associated perturbations below the surface that can be tested by observation. The excess radiation from small scale magnetic fields comes about through an increased surface emissivity, associated with the increased ‘surface roughness’. This increases the average cooling rate at the surface and creates somewhat stronger and/or cooler downdrafts. It therefore causes slightly *lower* horizontally averaged temperatures below the surface. The two models thus predict opposite thermal effects below the surface.

The prospects for a test of these predictions are good, since helioseismological data are now quite sensitive, with the data obtained by SOHO/MDI and the GONG network (Kosovichev, 1996; Kosovichev and Schou 1997; Duvall et al. ,1998; Duvall and Kosovichev, 1999). Changes in the p-mode frequencies related to the solar cycle were reported by e.g. by Libbrecht and Woodard (1990), Dziembowski et al. (1998). The dependence of the signal on wavenumbers (l, m) shows that most of the effect is due to a change in wave propagation in a layer very close to the solar surface, at the latitudes of magnetic activity. The sense of the effect, an increase of the frequencies, is the *opposite* (Goldreich et al., 1991) of what would be expected if the temperature in these latitudes were higher. Though a higher temperature increases the propagation speed (increasing the frequencies), it also expands the solar envelope at these latitudes, by vertical hydrostatic balance. This lengthens the path traveled and decreases the mode frequencies. The net effect turns out to be dominated by the the second effect, the path length increase, because it is linear in T, while the propagation speed increases only as $T^{1/2}$. In addition to the thermal effect, the presence of the magnetic field has an effect, since it increases the 'stiffness' of the gas (in terms of propagation, the relevant speed is the fast mode speed, which is larger than the sound speed). Since the relative contribution of this effect scales with the ratio of magnetic to gas pressure, it is important only in a thin layer near the surface. The observed changes in mode frequency can thus be due either to a somewhat lower temperature, or to the effect of the magnetic field in the atmosphere. The observation that most of the effect is localized at the surface argues for a major contribution from the magnetic stiffness effect (Goldreich et al., 1991; Woodard et al., 1991). In any case, the observed sign of the effect is incompatible with enhanced temperatures being the main change in envelope structure in the active latitudes.

A second way of measuring variations in propagation conditions below the solar surface is through time-distance seismology (Duvall et al., 1996; Kosovichev, 1996; Braun et al., 1997), in which one does not measure the mode frequencies but their travel time between points on the surface (much like seismology is used to measure irregularities in the Earth's interior). These measurements show shorter travel times in the active regions (Duvall et al., 1998; Duvall and Kosovichev, 1998; Giles et al., 1997; Bogdan et al., 1998; Goode and Dziembowski, 1999), and downflows below the surface. The shorter travel times are compatible with a contribution of the magnetic field to the propagation speed, and with *lower* subsurface temperatures (Brüggen and Spruit, 2000). The results from time-distance seismology have shown some internal inconsistencies, however, and reliable methods for interpretating the data arc apparently still in the development stage.

The lower temperatures expected by enhanced surface cooling in the small scale magnetic field has further observational consequences. In the absence of rotation, the increased negative buoyancy would just cause a downdraft below active regions. Associated with it a horizontal flow converging on the active region is expected. Effects like these are beginning to be explored by time-distance seismology.

5.1. TORSIONAL OSCILLATIONS

The effect of rotation on these flows can probably not be ignored, since the relevant Rossby number (rotation period divided by circulation time scale) can be of order unity. In addition to the circulation, one then expects the Coriolis force to produce a 'thermal wind' (e.g. Pedlosky, 1982) in the active latitudes. A lower average subsurface temperature in these latitudes causes a reduced pressure on equipotential surfaces, which is balanced by horizontal flows in the cyclonic sense. The average effect over longitude would be a contribution to the differential rotation, such that the lower edge of the active latitudes rotate faster, and the upper edge slower than average. The maxima of the disturbance would be at the latitudes where the latitudinal gradient of the magnetic activity has its maxima. Such a pattern of rotation variations associated with the active latitudes is indeed observed in the form of the so-called 'torsional oscillations' (e.g. Ulrich, 1998). A temperature enhancement in the active latitudes such as proposed by Kuhn et al. would cause a thermal wind of the same order of magnitude but of the opposite sign. It must be stressed, however, that other plausible explanations for the oscillation exist, such as the effects of magnetic forces in the deep convection zone (Schüssler, 1981).

References

Arendt, S.: 1992, *Astrophys. J* **389**, 421.

Bogdan, T. J., Braun, D. C., Lites, B. W. and Thomas, J. H.: 1998 *Astrophys. J.* **492**, 379.

Braun, D.C.: 1997, *Astrophys. J.* **487**, 447.

Brüggen, M. and Spruit, H.C., 2000, in prepapration.

Caligari, P., Schuessler, M. and Moreno-Insertis, F.: 1998 *Astrophys. J.* **502**, 481.

Chiang, W. H. and Foukal, P. V.: 1984, *Solar Phys.* **97**, 9.

Dicke, R. H., Kuhn, J. R. and Libbrecht, K. G.: *1987*, *Astrophys. J.* **318**, 451.

D'Silva, S. and Choudhuri, A. R., 1993: *Astron. Astrophys.* **272**, 621.

D'Silva, S. and Howard, R. A.: 1993, *Solar Phys.* **148**, 1.

Duvall, T. L., JR., Kosovichev, A. G., Scherrer, P. H., Bogart, R. S., Bush, R. I., De Forest, C., Hoeksema, J. T., Schou, J., Saba, J. L. R., Tarbell, T. D., Title, A. M., Wolfson, C. J. and Milford, P. N.: 1997, *Solar Phys.* **170**, 63.

Duvall, T., JR.; Kosovichev, A. G. and Scherrer, P. H.: 1998 in *Sounding solar and stellar interiors*, eds. J. Provost, F.-X. Schmider (IAU symposium 181, Nice, France, September 30 - October 3), Dordrecht: Kluwer Academic Publishers.

Duvall, T.L., Jr. and Kosovichev, A.G.: 1999, in SOHO-9 Workshop *Helioseismic Diagnostics of Solar Convection and Activity*, Stanford, California.

Dziembowski, W.A., Goode, P.R., Di Mauro, M.P., Kosovichev, A.G. and Schou, J.: 1998, *Astrophys. J.* **509**, 456.

Endal, A. S., Sofia, S. and Twigg L. W.: 1982, *Astrophys. J.* **290**, 748.
Foukal, P. V., Fowler, P. and Livshits, M.: 1983, *Astrophys. J.* **267**, 863.
Foukal, P. V. and Lean, J.: 1986, *Astrophys. J.* **302**, 826.
Giles, P.M., Duvall, T.L., Jr., Scherrer, P.H. and Bogart, R.S.: 1997, *Nature* **390**, 52–54.
Gilliland, R. L.: 1988, in *Solar radiative output variation*, P. Foukal, ed., Cambridge Research and Instrumentation Inc., 21 Erie st. Cambridge, MA 02139, p. 239.
Goldreich, P., Murray, N., Willette, G., Kumar, P.: 1991, *Astrophys. J.*, **370**, 752.
Goode, P.R. and Kuhn, J.R.: 1990, *Astrophys. J.* **356**, 310.
Goode, P.R. and Dziembowski, W.A.: 1999, in SOHO-9 Workshop "Helioseismic Diagnostics of Solar Convection and Activity", Stanford, California, July 12-15: 1999.
Kosovichev, A.G.: 1996, *Astrophys. J.* **461**, L55.
Kosovichev, A.G. and Schou, J.: 1997, *Astrophys. J.* **482**, L207.
Kuhn, J.R., Libbrecht, K.G. and Dicke, R.H.: 1988, *Science* **242**, 908.
Kuhn, J.R. and Stein, R.F.: 1996, *Astrophys. J.* **463**, L117.
Kuhn, J. R., Bush, R. I., Scherrer, P. and Scheick, X.: 1998, *Nature* **392**, 155.
Lean, J.L., Cook, J., Marquette, W. and Johanneson, A.: 1998, *Astrophys. J.* **492**, 390.
Libbrecht, K. G. and Woodard, M. F.: 1990, *Nature* **345**, 779.
Ludwig, H.-G., Freytag, B. and Steffen, MA.: 1999 *Astron. Astrophys.* **346**, 111.
Macris, C.J. and Roesch, J.: 1983, *Comptes Rendus, ser II*, **296**, 265.
Muller, R.: 1986 *Solar Phys.*, **119**, 229.
Muller, R. and Roudier, T.: 1984, in *The Hydromagnetics of the Sun* ed. T.D. Guyenne, ESA SP 220, p. 51.
Nordlund, Å.: 1982, *Astron. Astrophys.* **107**, 1.
Nordlund, Å.: 1985a, in *Small scale dynamical processes in quiet stellar atmospheres* ed. W. Keil, Sacramento Peak Observatory, Sunspot, NM 88349, USA.
Nordlund, Å.: 1985b, in *Progress in stellar spectral line formation theory*, eds. J.E. Beckman and L. Crivellari (NATO ASI series **152**), Reidel, Dordrecht, p. 215.
Nordlund, Å.: 1986, *Solar Phys.*, **100**, 209.
Nordlund, Å.: 1991, in *Stellar atmopheres: Beyond classical models*, eds. L. Crivellari, I. Hubeny and D.G. Hummer (NATO ASI series **341**), Kluwer, Dordrecht, p. 61.
Nordlund, Å. and Dravins, D.: 1990, *Astron. Astrophys.* **228**, 155.
Nordlund, Å. and Stein, R.F.: 1990, *Comp. Phys. Comm.* **59**, 119.
Nordlund, Å. and Stein, R.F.: 1991, in *Stellar atmopheres: beyond classical models*, eds. L. Crivellari, I. Hubeny and D.G. Hummer (NATO ASI series **341**), Kluwer, Dordrecht, p. 263.
Nordlund, Å. and Stein, R.F.: 1996, in *Proceedings of the 32nd Liège Int. Astrophys. Colloquium 'Stellar Evolution: What should be done'*, eds. A. Noels et al. p. 75.
Pedlosky, J.: 1982, *Geophysical Fluid Dynamics*, New York: Springer.
Rast, M.P., Fox, P.A., Lin, H., Lites, B.W., Meisner, R.W. and White, O.R.: 1999, *Nature* **401**, 678.
Schüssler, M.: 1981, *Astron. Astrophys.* **94**, L17.
Solanki, S.K. and Unruh, Y.C.: 1998, *Astron. Astrophys.* **329**, 747.
Spruit, H.C.: 1976, *Solar Phys.* **50**, 269.
Spruit, H.C.: 1977, *Solar Phys.* **55**, 3.
Spruit, H.C.: 1982a, *Astron. Astrophys.* **108**, 348.
Spruit, H.C.: 1982b, *Astron. Astrophys.* **108**, 356.
Spruit, H.C.: 1991, in *The Sun in Time*, eds. C. Sonett, M. Giampapa and M.S. Matthews, University of Arizona Press, Tucson, p. 118.
Spruit, H.C.: 1992, in *Sunspots: Theory and Observations*, eds. J.H. Thomas and N.O. Weiss, Cambridge: CUP, p. 163.
Spruit, H.C.: 1997, *Mem. Soc. Astron. It.*, **68**, 397.
Spruit, H.C. and Weiss, A.: 1986, *Astron. Astrophys.* **166**, 167.
Steffen, M., Ludwig, H.-G. and Krüss, A.: 1989 *Astron. Astrophys.* **213**, 317.

Steffen, M.: 1993, in *Inside the stars* (IAU Coll 137), eds. W. Weiss and A. Baglin, Astron. Soc. Pac. Conference series **40**, p. 300.
Stein, R.F. and Nordlund, Å.: 1989, *Astrophys. J.* **342**, L95.
Stein, R.F. and Nordlund, Å.: 1991, in *Challenges to Theories of the Structure of Moderate-Mass Stars*, eds. D.O. Gough and J. Toomre, Lecture Notes in Physics **388**, Springer, Berlin, p. 195.
Stein, R.F. and Nordlund, Å.: 1998, *Astrophys. J.* **499**, 914.
Steiner, O., Grossmann-Doerth, U., Knoelker, M. and Schüssler, M.: 1998, *Astrophys. J.* **495**, 468.
Topka, K.P., Tarbell, T.D. and Title, A.M.: 1997, *Astrophys. J.* **484**, 479.
Title, A.M., Topka, K.P., Tarbell, T.D., Schmidt, W., Balke, C. and Scharmer, G.: 1992, *Astrophys. J.* **393**, 782.
Ulrich, R.K.: 1998, in *New eyes to see....*, F. Deubner et al., eds., IAU Symposium **185**, 59.
Unruh, Y.C., Solanki, S.K. and Fligge, M.: 1999, *Astron. Astrophys.* **345**, 635.
Woodard, M. F., Libbrecht, K. G., Kuhn, J. R. and Murray, N.: 1991, *Astrophys. J.* **373**, L81.

Address for Offprints: H.C. Spruit, Max-Planck-Institut für Astrophysik, Postfach 1523, 85740 Garching, Germany, henk@mpa-garching.mpg.de

RECONSTRUCTION OF PAST SOLAR IRRADIANCE

S.K. SOLANKI[1,2] and M. FLIGGE[1]
[1]*Institute of Astronomy, ETH, CH-8092 Zürich, Switzerland*
[2]*Max Planck Institute of Aeronomy, D-37191 Katlenburg-Lindau, Germany*

Received: 25 October 1999; Accepted in final form: 14 January 2000

Abstract. Accurate measurements of solar irradiance started in 1978, but a much longer time series is needed in order to uncover a possible influence on the Earth's climate. In order to reconstruct the irradiance prior to 1978 we require both an understanding of the underlying causes of solar irradiance variability as well as data describing the state of the Sun (in particular its magnetic field) at the relevant epochs.

Evidence is accumulating that on the time-scale of the solar cycle or less, variations in solar irradiance are produced mainly by changes in the amount and distribution of magnetic flux on the solar surface. The main solar features contributing to a darkening of the Sun are sunspots, while active-region faculae and the network lead to a brightening. There is also increasing evidence for secular changes of the solar magnetic field and the associated of solar brightness variability. In part the behavior of sun-like stars is used as a guide of such secular changes.

Under the assumption that solar irradiance variations are due to solar surface magnetism on all relevant time scales it is possible to reconstruct the irradiance with some reliability from today to around 1874, and with lower accuracy back to the Maunder minimum. One major problem is the decreasing amount and accuracy of the relevant data with age. In this review the various reconstructions of past solar irradiance are presented and the assumptions underlying them are scrutinized.

1. Introduction

Changes in the total or UV irradiance of the Sun have been proposed as a possible cause of climate change. Reliable and (almost) continuous measurements of the total solar irradiance started in 1978 (Willson et al., 1981; Willson and Hudson, 1988; Hoyt et al., 1992; Fröhlich and Lean, 1998). Likewise, UV irradiance has been regularly recorded since 1978 (Heath and Schlesinger, 1986; Labs et al., 1987; Rottman, 1988).

The total solar irradiance has now been collected over almost 2 solar cycles (see Fröhlich and Lean, 1998, for a composite of the measurements made by different instruments). These measurements revealed short-term changes of irradiance caused by the disc-passage and evolution of sunspots and faculae. However, the main discovery was the brightening of the Sun by approximately 1.3 W/m^2 at solar activity maximum relative to activity minimum (corresponding to roughly 0.1% of the total solar irradiance).

Space Science Reviews **94:** 127–138, 2000.

In order to detect the influence of solar irradiance variations on climate a longer record is required since the signature of the 11-year cycle has not been reliably detected in tropospheric temperature.

Hence there is a need to reconstruct past solar irradiance over as long a period as possible. Solar irradiance observations provide only limited guidance on the presence or physical causes of changes in the irradiance on a time-scale longer than a solar cycle. Therefore past irradiance can be reconstructed only under certain assumptions regarding the underlying physical mechanisms. These assumptions are outlined in Sect. 2.

Due to our inability to compute the precise solar irradiance from first principles the modeling often restricts itself to finding a simple relationship between some readily available proxy or proxies of the irradiance and the measured value. Such proxy data therefore play a central role in the reconstruction of past solar irradiance and are briefly introduced in Sect. 3. A far more thorough discussion is given by Lean (2000).

In Sect. 4 various reconstructions are reviewed critically. Finally, in Sect. 5 some ideas on how these reconstructions may be improved are given.

2. Basic Assumption

Many processes can all lead to solar irradiance variations. These include changes in the amount and distribution of magnetic flux at the solar surface (Foukal and Lean, 1986, 1988; Lean et al., 1998), changes in convective efficiency and in the internal temperature gradient of the Sun (Kuhn et al., 1988; Kuhn and Libbrecht, 1991), long period r-modes of oscillation (Wolff and Hickey, 1987), release of energy stored in the toroidal magnetic field at the bottom of the solar convection zone and thermal shadowing by this field.

Of all these mechanisms the influence of magnetic fields at the solar surface (through their manifestations, such as sunspots, faculae, network elements) on the solar irradiance is established best. The passage of sunspots across the solar disc (driven by rotation) causes dips in brightness lasting 1–2 weeks (e.g., Willson et al., 1981; Chapman et al., 1994). Similarly, faculae cause a brightening, which however has a more complex temporal profile since they are brightest near the limb (e.g., Fröhlich et al., 1997). Since faculae are more extended, more common and longer lived than sunspots the total, disc-averaged brightening varies more gradually than the darkening (e.g., Chapman, 1987). The disk-integrated contribution of the network is even more stable, since the network is distributed comparatively evenly over the solar surface, so that rotational modulation is small. The magnetic flux in the network changes slowly with time, and is still quite prominent even at activity minimum when few, if any sunspots or faculae are present.

The importance of these surface magnetic features in reproducing observed solar irradiance variations on time-scales of the solar rotation was pointed out by

Willson et al. (1981), Hudson et al. (1982), Chapman et al. (1994), Fligge et al. (2000), and on time-scales up to the solar cycle by Foukal and Lean (1986, 1988), Chapman et al. (1994, 1996), Fligge et al. (1998), Lean et al. (1998). The physical basis for the irradiance variations introduced by surface magnetic features was worked out by Spruit (1982). Due to the high thermal conductivity and the immense heat capacity of the convection zone any perturbation of the heat loss at the surface (due to spots and faculae) influences the complete convection zone (a result of its high conductivity) and only immeasurably affects its properties (caused by its high heat capacity). Hence a local perturbation like a sunspot leads to an instantaneous change in irradiance (Spruit, 2000).

Models based on the assumption that the irradiance variations on solar cycle time-scales are entirely caused by changes in surface magnetism have been remarkably successful in reproducing not only the total irradiance time series (Foukal and Lean, 1990; Chapman et al., 1996; Lean et al., 1995; Fligge et al., 1998), but also a whole row of other observations. These include the change in the spectrum of the irradiance between activity maximum and minimum (Solanki and Unruh, 1998; Fligge et al., 1998; Unruh et al., 1999), the ratio of facular to sunspot areas (Fligge et al., 1998), the variation of the amount of line-blanketing over the solar cycle (Unruh et al., 1999) and the evolution of UV irradiance (Lean et al., 1998). The variations of the ratios between selected spectral lines over the solar cycle (Gray and Livingston, 1997) is at least roughly reproduced (Unruh et al., 2000), although there is still room for improvement.

The main observation which has not as yet been reproduced by current models are the brightness changes measured by limb photometers (Kuhn et al., 1988; Kuhn and Libbrecht, 1991). Efforts should be made to use these additional observations to better constrain the models and possibly the proportion of the irradiance changes produced by surface magnetism.

Kuhn et al. (1988) have argued that their limb photometer restricts surface magnetism, to 50 % of the contribution to the total irradiance. Due to the success of surface-magnetism based models in reproducing a large number and variety of other observations this now appears unlikely on time-scales of the solar cycle or less (although it cannot be completely ruled out, of course). The relative contribution of the various mechanisms to longer-term variations, however, is as yet unknown.

Hence, when reconstructing past solar irradiance the cyclic and (long-term) secular variations are modeled separately. Both partial reconstructions are then combined to give a final reconstruction. An example will be shown in Sect. 4. Here and in the following we count all variations on the time-scale of the solar cycle or less to the cyclic variations.

It should be noted that the reliability of reconstructions of the cyclic component of the variability is much higher than that of the secular trend. Even the existence of the latter is not certain beyond a doubt. Hence, claims by Willson (1997) that the irradiance at two consecutive activity minima is significantly different have subsequently not been confirmed (e.g., Fröhlich and Lean, 1998). The main evidence for such a component is indirect, coming from the reconstructions of the interplanetary magnetic field (Lockwood et al., 1999) and the comparison of the strength of chromospheric network and plage emission from the Sun with Sun-like stars. Plage are the chromospheric counterparts of faculae. It still needs to be established, however, how strongly the interplanetary magnetic field is connected to irradiance variations.

Baliunas and Jastrow (1990) revealed that the flux in the cores of the Ca II H&K lines in a sample of Sun-like stars exhibits a much wider scatter than the Sun does in the course of a typical activity cycle. Significant is the fact that many of these stars show much weaker Ca II emission than the Sun, even at its activity minimum. Moreover, the stars with the weakest such emission are in a non-cyclic state akin to the Maunder Minimum, a period in the 17th century when the Sun exhibited very few sunspots, a sign of low magnetic activity.

Now, Ca II core emission is sensitive to the temperature in the chromospheric layers of the solar, respectively, stellar atmosphere. It is thought that a minimum chromospheric temperature (i.e. Ca II flux) is maintained by the dissipation of acoustic waves (basal flux, Schrijver, 1987; Ulmschneider et al., 1999) while higher Ca II flux values signal the presence of surface magnetic fields. On the Sun Ca II brightness is well correlated with the network and with faculae, i.e. features that are also bright in white light. Since over the last few solar cycles the magnetic network has been present even at sunspot activity minimum the minimum Ca II flux has been distinctly higher than the basal value. Therefore, if the Ca II flux of the Sun had in the past decreased below the current minimum level the Sun would have been less bright at that time.

Using the stellar measurements it is possible to estimate how bright the Sun was during the Maunder Minimum and thus obtain an estimate of how large the longer-term variation of solar brightness is (White et al., 1992; Lean et al., 1992). Note that the longer-term variation obtained in this manner is still due mainly to the magnetic field, since only it significantly affects the strength of the Ca II emission. It is thought that the strength of the magnetic network changes on time-scales longer than the solar cycle. At present we cannot, however, rule out small changes in the convective energy flux, which should leave the Ca flux relatively unaffected. These could in principle produce even larger secular fluctuations in solar irradiance which would, however, have gone undetected by the proxies available to us.

Most reconstructions in recent years have made the assumption that all variations of solar irradiance in the past centuries have been driven by changes in the amount and distribution of magnetic flux on the solar surface. There are, however, exceptions. More details are given in Sect. 4.

3. Historical Records of Proxy Data

It is currently not possible to predict or reconstruct the precise amount or distribution of magnetic flux on the solar surface from first principles. Hence, all reconstructions of solar irradiance are based on measured time-series of the magnetic flux (magnetograms) or other proxies of either the magnetic field or the irradiance.

A variety of records exist which are in some way related to solar magnetism and thus to irradiance. In addition to the direct proxies of solar activity, indirect indicators, caused by the modulation of cosmic-ray flux by solar activity, are stored in terrestrial archives (e.g., ^{10}Be and ^{14}C records). These are discussed in detail elsewhere (Beer, 2000).

These proxy records are not without their problems. For one, fewer records are available at earlier times. Also, they are often of variable quality, with the older data often being less precise and having longer gaps. Efforts to uncover and incorporate further historical observations are ongoing (Foukal, 1993; Hoyt et al., 1994; Hoyt and Schatten, 1998). Often even recently acquired data need to be scrutinized closely, particularly when multiple observatories are involved. For example, when the daily recording of sunspot areas passed from Royal Greenwich Observatory to other stations there was a 20% jump in the average recorded areas, although this was not obviously visible from the time-series (Fligge and Solanki, 1997).

Another problem is how these proxy records relate qualitatively to the brightness fluctuations caused by sunspots, faculae and the network. In particular the network turns out to be elusive and none of the 4 direct proxy records lasting longer than a century is directly sensitive to it. It is therefore not straightforward to constrain the exact behavior of the secular trend exhibited by the irradiance, whose presence has been deduced, e.g., from the behavior of other stars (see Sect. 2).

Various secondary time-series can be constructed from the primary time-series. Examples are the length, the amplitude and the total strength of the solar cycle (the latter defined as the integral of the record over a cycle). The decay time of sunspots is another secondary proxy that has been employed for irradiance modelling (Hoyt and Schatten, 1993). These secondary proxies are often used to describe the slowly varying, secular component of the irradiance.

4. Reconstruction of the Past Total Solar Irradiance

Early reconstructions of solar irradiance only considered cyclic variations. Thus Foukal and Lean (1986, 1988, 1990), Lean and Foukal (1988) took into account sunspots and faculae (the network was counted to the faculae) to reconstruct solar irradiance to as far back as 1874. Sunspot areas, A_s, in combination with other parameters were used to calculate the darkening due to sunspots. Different proxies were used to reconstruct the brightening due to faculae because the best proxies have been available for only a relatively short time.

The construction of relations between the employed proxies and the irradiance form a central part of the model building exercise. Such relations are calibrated using the direct irradiance measurements of the past 20 years. For earlier times it is generally assumed that these calibrations are equally valid (but see Fligge and Solanki, 1998; Solanki and Fligge, 1998). Separate (although not always ideal) proxies of the sunspots and faculae are available since 1874. From this year on the active-region (cyclic) contribution to irradiance change can be reconstructed with a fair measure of reliability. The remaining uncertainty is largely introduced by inaccuracies in the historical proxy observations and the assumption that the conversion relation between the proxy and the irradiance is time independent.

The various reconstructions of the irradiance for this period differ, depending on how carefully the proxy data were chosen, on the level of sophistication of the calibration between proxy and irradiance, etc. For example, Lean et al. (1995) assumed a linear relation between the group sunspot number, R_g, of Hoyt and Schatten (1998) and the irradiance residuals. In contrast, Foukal and Lean (1990) as well as Solanki and Fligge (1998) treated spots and faculae separately. After removing the influence of spots using measurements of their total area, they calibrated the facular proxies against irradiance measurements using linear and quadratic relations, respectively. Further major improvements are expected from the use of images of proxies of facular brightening, e.g. Ca K images, from the time from which they are available.

Prior to 1874 the reliability of the reconstructions decreases. This is mainly due to the lack of independent proxies of sunspot and facular contributions. Recently, Solanki and Fligge (1999) found a way of achieving reasonable, if not perfect, reconstructions in spite of this handicap by making use of the fact that sunspots and faculae evolve on somewhat different time-scales.

A composite of reconstructed cyclic irradiance is plotted in Fig. 1. The quality of the reconstruction improves in steps: It is lowest in the 1700–1750 period, when only yearly values of R_Z and R_g were available and is highest since 1874, when separate proxies were used for spots and faculae.

Averaged over the solar cycle the rise in solar irradiance since 1880 due to the contribution of active regions and the cyclic variations of the network alone is only 0.5 W/m^2, much smaller than the 1.3 W/m^2 that the irradiance changes in the course of a cycle. However, as pointed out in Sect. 2, evidence has emerged from the study of the Earth's atmosphere and of Sun-like stars, that there is probably a secular component in solar variability. The magnitude of this component is assumed to be 2.5 W/m^2 by Lean et al. (1992) and White et al. (1992), 2–6 W/m^2 by Zhang et al. (1994), 3.5 W/m^2 by Hoyt and Schatten (1993) and 2.5 W/m^2 by Solanki and Fligge (1998). This component thus dominates the total irradiance changes.

Not only is there considerable disagreement between different researchers regarding the amplitude of the secular change in irradiance, there is equal disparity

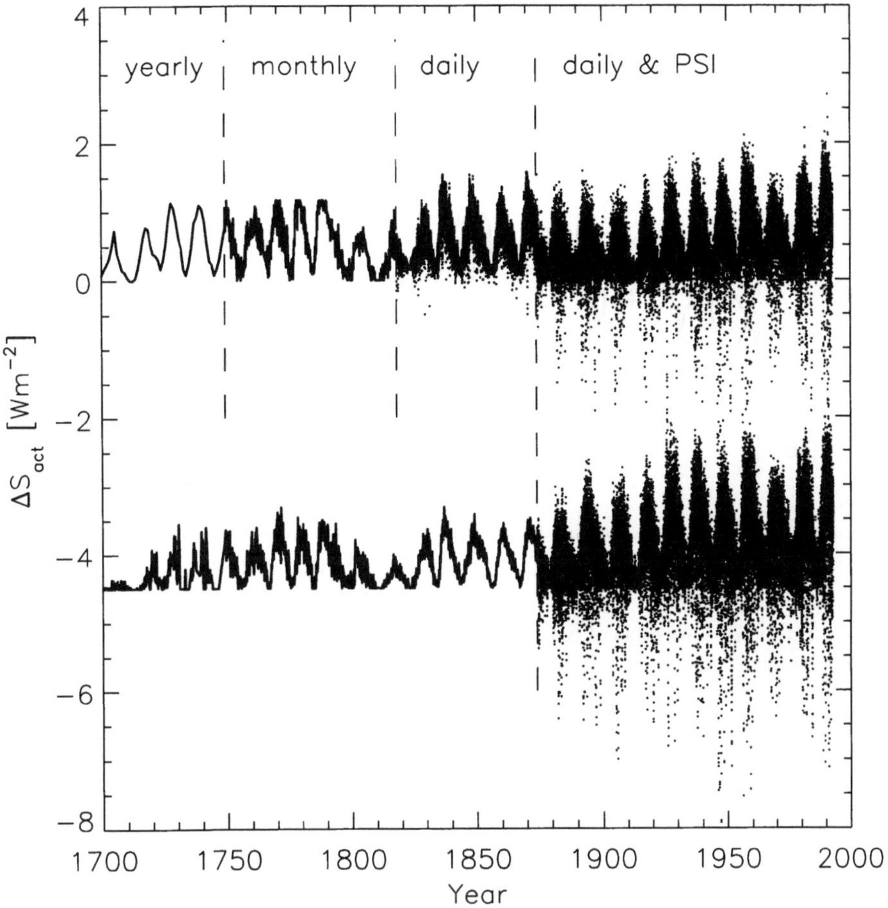

Figure 1. Reconstructed excess irradiance due to active regions between 1700 and 1999. The upper curve corresponds to the composite irradiance record based on Zurich sunspot relative number, R_Z, using different reconstruction methods while the lower curve (shifted by -4.5 W/m^2) is based on the group sunspot number, R_g. The type of data available for the reconstruction is indicated in the figure. PSI denotes 'Photometic Sunspot Index' (From Solanki and Fligge, 1999).

regarding its time variation. The stellar data themselves give conflicting indications. On the one hand they have been interpreted by Lean et al. (1992, 1995) and White et al. (1992) to suggest that the irradiance changes linearly with the strength of the activity indicator, respectively with the cycle amplitude. On the other hand, Baliunas and Soon (1995) present evidence that the irradiance may correlate with the length of the cycle. Which of these indicators is better, or indeed if either of them is adequate, is at present unknown. One reason why it is difficult to distinguish between them is that there is a correlation between the cycle amplitude and length (e.g., Hoyng, 1996).

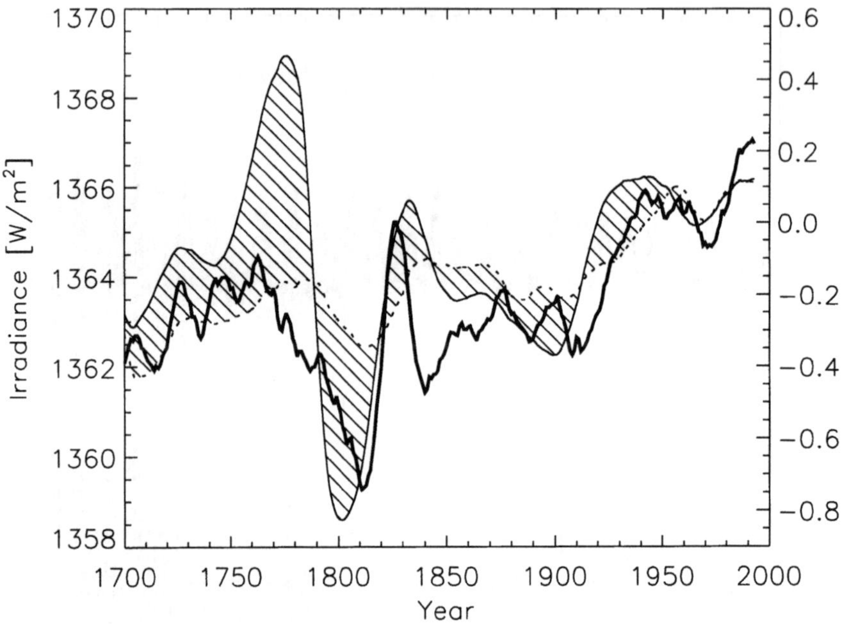

Figure 2. 11-year running mean of two reconstructions of total solar irradiance, including the contribution of both the cyclic and secular components. Thin solid curve: the secular trend is represented by the solar cycle length; thin dashed curve: The amplitude of R_g is used as a proxy of the secular trend. The hatched area between these curves thus gives a rough indication of the uncertainty in the reconstructions. Also plotted is the northern hemisphere land temperature (thick solid curve) according to Groveman and Landsberg (1700 – 1880) and the IPCC (1880 – 1990) (From Solanki and Fligge, 1999).

It is nevertheless important to distinguish between these two possibilities, since the correlation coefficient between cycle length and strength is increased considerably if a phase lag of approximately a solar cycle is introduced in the sense that the cycle length runs ahead of the cycle amplitude by approximately one solar cycle. This timing is crucial when comparing with climate measurements, since the cycle length runs ahead of global climate change (or correlates well with air temperatures above northern hemisphere land-masses) while the cycle amplitude appears to slightly lag behind some climate indicators.

Solanki and Fligge (1998, 1999) have employed both the cycle length and the amplitude, thus effectively creating two reconstructions of solar irradiance that can be directly compared to each other. The largest difference between the two proxies is seen at around 1750 – 1800 (see Fig. 2). The cycle amplitude shows a decreased value around this epoch, while the cycle length exhibits large excursions in both directions. This epoch is interesting because it coincides with an anomaly in the Earth's temperature. One problem with the cycle length as a long-term proxy is that it is ill defined over prolonged periods of low activity such as the Maunder

Minimum. This is particular unfortunate since it is exactly the Maunder Minimum which is usually used to scale the amplitude of the longer-term irradiance changes.

In the only reconstruction that does *not* assume that surface magnetism is responsible for all irradiance variations Hoyt and Schatten (1993) use a host of indicators of solar activity as proxies of long-term irradiance changes. These include the solar equatorial rotation rate, the amplitude, length and decay rate of the solar cycle and the structure (umbral to penumbral area ratio) and decay rate of sunspots. By combining all these proxies in a suitably weighted manner they obtain a secular variation that looks rather similar to the cycle length. Their approach still awaits empirical validation whether most of their proxies are indeed related to irradiance variations.

In summary, the secular variability is obviously the most uncertain (but a very significant) part of the irradiance reconstruction. This is the part which in future deserves the most attention.

5. Outlook

Over the last 15 years our understanding of the underlying causes of irradiance variations as well as the quantitative reconstruction of solar irradiance have improved greatly, although the basic premise, namely that solar surface magnetism is the main agent, has remained unchanged.

Nevertheless, there are considerable further grounds for improvement. These include making use of larger, more sophisticated data sets for the reconstruction of the cyclic component of irradiance variations (such as magnetograms or Ca II spectroheliograms), as well as attempting to close the considerable and serious gaps in our knowledge concerning the secular variation.

The obvious next major step as far as the cyclic variations are concerned is to move from disc-integrated proxies to images of the solar disc, at least for the time that these are available. The case for using such data has been amply demonstrated by Lean et al. (1997) and Fligge et al. (2000), who employed Ca II K images and magnetograms, respectively. These investigations have been restricted to fractions of the solar cycle, however. Extending them to cover a number of solar cycles is a major and time-consuming exercise, but should provide significantly improved estimates of the irradiance variations.

In parallel, the quantitative relation between irradiance, on the one hand, and magnetogram signal, Ca II K brightness (or some other proxy), on the other, needs to be improved. The work of Lean et al. (1997), Unruh et al. (1999) and Fligge et al. (2000) represent steps in this direction. One interesting extension of this work is to incorporate the difference in brightness between magnetic elements in the network and in active regions found by Solanki and Stenflo (1984), Solanki and Brigljević (1992). This difference is probably to a large extent due to the different magnetic filling factors in the two regions - filling factor is the fraction of the solar surface covered by the magnetic field within a given area.

Also, the problems associated with Ca II K brightness (different underlying physical processes from those governing irradiance variations) and magnetograms (decreasing signal toward the limb, since only the line-of-sight component of the vertical magnetic field is measured) need to be dealt with.

We expect that the recent advances in modeling the solar spectral irradiance (Solanki and Unruh, 1998; Fligge et al., 1998; Unruh et al., 1999; Fontenla et al., 1999) and the success in reproducing the temporal evolution of UV irradiance observations (Lean et al., 1997) will in the near future lead to considerable improvements of reconstructions of past UV irradiance.

Particularly worrisome is the fact that most of the evidence for a secular change of the irradiance comes from indirect sources such as stellar observations, activity of the Earth's magnetosphere or cosmogenic isotopes such as ^{10}Be or ^{14}C. It is also troubling that there has been no physical explanation for such a variation. Additional evidence for a secular variation would be very helpful. A more precise estimate of the change in solar irradiance since the Maunder Minimum is particularly needed. Just as important is to uncover the physical causes of secular irradiance variations.

References

Baliunas, S. and R. Jastrow: 1990, 'Evidence for long-term brightness changes of solar-type stars'. *Nature* **348**, 520–523.

Baliunas, S. and W. Soon: 1995, 'Are Variations in the Length of the Activity Cycle Related to Changes in Brightness in Solar-Type Stars?'. *Astrophys. J.* **450**, 896.

Beer, J.: 2000, 'Long Term, Indirect Indices of Solar Variability', *Space Sci. Rev.*, this volume.

Chapman, G. A.: 1987, 'Variations of solar irradiance due to magnetic activity'. *Ann. Rev. Astron. Astrophys.* **25**, 633–667.

Chapman, G. A., A. M. Cookson, and J. J. Dobias: 1994, 'Observations of changes in the bolometric contrast of sunspots'. *Astrophys. J.* **432**, 403–408.

Chapman, G. A., A. M. Cookson, and J. J. Dobias: 1996, 'Variations in total solar irradiance during solar cycle 22'. *J. Geophys. Res.* **101**, 13541–13548.

Fligge, M. and S. K. Solanki: 1997, 'Inter-cycle variations of solar irradiance: sunspot areas as a pointer'. *Sol. Phys.* **173**, 427–439.

Fligge, M. and S. K. Solanki: 1998, 'Long-term behavior of emission from solar faculae: steps towards a robust index'. *Astron. Astrophys.* **332**, 1082–1086.

Fligge, M., S. K. Solanki, and Y. C. Unruh: 2000, 'Modelling irradiance variations from the surface distribution of the solar magnetic field'. *Astron. Astrophys.* **353**, 380–388.

Fligge, M., S. K. Solanki, Y. C. Unruh, C. Fröhlich, and C. Wehrli: 1998, 'A model of solar total and spectral irradiance variations'. *Astron. Astrophys.* **335**, 709–718.

Fontenla, J., O. R. White, P. A. Fox, E. H. Avrett, and R. L. Kurucz: 1999, 'Calculation of Solar Irradiances. I. Synthesis of the Solar Spectrum'. *Astrophys. J.* **518**, 480–499.

Foukal, P.: 1993, 'The curious case of the Greenwich faculae'. *Sol. Phys.* **148**, 219.

Foukal, P. and J. Lean: 1986, 'The influence of faculae on total solar irradiance and luminosity'. *Astrophys. J.* **302**, 826–835.

Foukal, P. and J. Lean: 1988, 'Magnetic modulation of solar luminosity by photospheric activity'. *Astrophys. J.* **328**, 347–357.

Foukal, P. and J. Lean: 1990, 'An empirical model of total solar irradiance variation between 1874 and 1988'. *Science* **247**, 556–558.

Fröhlich, C., B. Andersen, T. Appourchaux, G. Berthomieu, D. A. Crommelynck, V. Domingo, A. Fichot, W. Finsterle, M. F. Gomez, D. Gough, A. Jimenez, T. Leifsen, M. Lombaerts, J. M. Pap, J. Provost, T. R. Cortes, J. Romero, H. Roth, T. Sekii, U. Telljohann, T. Toutain, and C. Wehrli: 1997, 'First results from VIRGO, the experiment for helioseismology and solar irradiance monitoring on SOHO'. *Sol. Phys.* **170**, 1–25.

Fröhlich, C. and J. Lean: 1998, 'The Sun's total irradiance: Cycles, trends and related climate change uncertainties since 1976'. *Geophys. Res. Lett.* **25**, 4377–4380.

Gray, D. F. and W. C. Livingston: 1997, 'Monitoring the solar temperature: Spectroscopic temperature variations of the Sun'. *Astrophys. J.* **474**, 802.

Heath, D. F. and B. M. Schlesinger: 1986, 'The Mg 280-nm doublet as a monitor of changes in solar ultraviolet irradiance'. *J. Geophys. Res.* **91**, 8672–8682.

Hoyng, P.: 1996, 'Is the Solar Cycle Timed by a Clock?'. *Sol. Phys.* **169**, 253–264.

Hoyt, D. V., H. L. Kyle, J. R. Hickey, and R. H. Maschhoff: 1992, 'The NIMBUS-7 Solar Total Irradiance: A New Algorithm for its Derivation'. *J. Geophys. Res.* **97**, 51–63.

Hoyt, D. V. and K. H. Schatten: 1993, 'A discussion of plausible solar irradiance variations, 1700-1992'. *J. Geophys. Res.* **98**, 18895–18906.

Hoyt, D. V. and K. H. Schatten: 1998, 'Group sunspot numbers: A new solar activity reconstruction'. *Sol. Phys.* **179**, 189–219.

Hoyt, D. V., K. H. Schatten, and E. Nesmes-Ribes: 1994, 'The one hundredth year of Rudolf Wolf's death: Do we have the correct reconstruction of solar activity?'. *Geophys. Res. Lett.* **21**, 2067–2070.

Hudson, H. S., S. Silva, M. Woodard, and R. C. Willson: 1982, 'The effects of sunspots on solar irradiance'. *Sol. Phys.* **76**, 211–219.

Kuhn, J. R. and K. G. Libbrecht: 1991, 'Nonfacular solar luminosity variations'. *Astrophys. J., Lett.* **381**, L35–L37.

Kuhn, J. R., K. G. Libbrecht, and R. H. Dicke: 1988, 'The surface temperature of the sun and changes in the solar constant'. *Science* **242**, 908–911.

Labs, D., H. Neckel, P. C. Simon, and G. Thuillier: 1987, 'Ultraviolet solar irradiance measurement from 200 to 358 nm during Spacelab 1 mission'. *Sol. Phys.* **107**, 203–219.

Lean, J.: 2000, 'Short Term, Direct Indices of Solar Variability', *Space Sci. Rev.*, this volume.

Lean, J., J. Beer, and R. Bradley: 1995, 'Reconstruction of solar irradiance since 1610: Implications for climate change'. *Geophys. Res. Lett.* **22**, 3195–3198.

Lean, J. and P. Foukal: 1988, 'A model of solar luminosity modulation by magnetic activity between 1954 and 1984'. *Science* **240**, 906.

Lean, J., G. J. Rottman, H. L. Kyle, T. N. Woods, J. R. Hickey, and L. C. Puga: 1997, 'Detection and parameterization of variations in solar mid- and near-ultraviolet radiation (200–400 nm)', *J. Geophys. Res.* **102**, 29,939–29,956.

Lean, J., A. Skumanich, and O. White: 1992, 'Estimating the sun's radiative output during the Maunder Minimum'. *Geophys. Res. Lett.* **19**, 1595–1598.

Lean, J. L., J. Cook, W. Marquette, and A. Johannesson: 1998, 'Magnetic Sources of the Solar Irradiance Cycle'. *Astrophys. J.* **492**, 390–401.

Lockwood, M., R. Stamper, and M. N. Wild: 1999, 'A doubling of the Sun's coronal magnetic field during the past 100 years'. *Nature* **399**, 437–439.

Rottman, G. J.: 1988, 'Observations of solar UV and EUV variability'. *Advances in Space Research* **8**, 53–66.

Schrijver, C. J.: 1987, 'Heating of stellar chromospheres and coronae: Evidence for non-magnetic basal heating'. In: *Colorado Univ., 5th Cambridge Workshop on Cool Stars, Stellar Systems and the Sun 1 p (SEE N88-13092 04-89).*

Solanki, S. K. and V. Brigljević: 1992, 'Continuum brightness of solar magnetic elements'. *Astron. Astrophys.* **262**, L29–L32.

Solanki, S. K. and M. Fligge: 1998, 'Solar irradiance since 1874 revisited'. *Geophys. Res. Lett.* **25**, 341–344.

Solanki, S. K. and M. Fligge: 1999, 'A reconstruction of total solar irradiance since 1700'. *Geophys. Res. Lett.* **26**, 2465–2468.

Solanki, S. K. and J. O. Stenflo: 1984, 'Properties of solar magnetic fluxtubes as revealed by Fe I lines'. *Astron. Astrophys.* **140**, 185–198.

Solanki, S. K. and Y. C. Unruh: 1998, 'A model of the wavelength dependence of solar irradiance variations'. *Astron. Astrophys.* **329**, 747–753.

Spruit, H. C.: 1982, 'Effect of spots on a star's radius and luminosity'. *Astron. Astrophys.* **108**, 348–355.

Spruit, K.: 2000, 'Physical Causes of Irradiance Variability', *Space Sci. Rev.*, this volume.

Ulmschneider, P., J. Theurer, Z. E. Musielak, and R. Kurucz: 1999, 'Acoustic wave energy fluxes for late-type stars. II. Nonsolar metallicities'. *Astron. Astrophys.* **347**, 243–248.

Unruh, Y. C., S. K. Solanki, and M. Fligge: 1999, 'The spectral dependence of facular contrast and solar irradiance variations'. *Astron. Astrophys.* **345**, 635–642.

Unruh, Y. C., S. K. Solanki, and M. Fligge: 2000, 'Modelling solar spectral irradiance variations'. In: E. Friis-Christensen, C. Fröhlich, J. Haigh, M. Schüssler, and R. von Steiger (eds.): *Solar Variability and Climate*. Dordrecht, p. in press.

White, O. R., A. Skumanich, J. Lean, W. C. Livingston, and S. L. Keil: 1992, 'The sun in a noncycling state'. *Publ. Astron. Soc. Pac.* **104**, 1139–1143.

Willson, R. C.: 1997, 'Total solar irradiance trend during solar cycles 21 and 22'. *Science* **277**, 1963–1965.

Willson, R. C., S. Gulkis, M. Janssen, H. S. Hudson, and G. A. Chapman: 1981, 'Observations of solar irradiance variability'. *Science* **211**, 700–702.

Willson, R. C. and H. S. Hudson: 1988, 'Solar luminosity variations in solar cycle 21'. *Nature* **332**, 810–812.

Wolff, C. L. and J. R. Hickey: 1987, 'Multiperiodic irradiance changes caused by r-modes and g-modes'. *Sol. Phys.* **109**, 1–18.

Zhang, Q., W. H. Soon, S. L. Baliunas, G. W. Lockwood, B. A. Skiff, and R. R. Radick: 1994, 'A method of determining possible brightness variations of the Sun in past centuries from observations of solar-type stars'. *Astrophys. J.* **427**, L111–L114.

MODELLING SHORT-TERM SPECTRAL IRRADIANCE VARIATIONS

M. FLIGGE[1], S.K. SOLANKI[1,2] and Y.C. UNRUH[3]
[1]*Institute of Astronomy, ETH Zentrum, CH-8092 Zurich, Switzerland*
[2]*Max-Planck-Institut für Aeronomie, D-37191 Katlenburg-Lindau, Germany*
[3]*Institut für Astronomie, Universität Wien, Türkenschanzstr. 17, A-1180 Wien, Austria*

Received: 19 October 1999; Accepted in final form: 23 December 1999

Abstract. On time-scales of the solar rotation most of the solar irradiance variations are caused by the changing distribution of solar surface magnetic features. We model these short-term irradiance variations using calculations of sunspot and facular contrasts as a function of wavelength and limb angle on the Sun. The position of active regions on the solar disc is derived from the MDI magnetograms. The reconstructed irradiance variations are compared with total and spectral irradiance measurements obtained by the VIRGO experiment on SOHO.

1. Introduction

Within the flux-tube picture of the magnetic field (Solanki, 1993), the average brightness of solar magnetic features is mainly a function of two parameters: the size of the magnetic feature and its position on the Sun (Knölker and Schüssler, 1988; Solanki, 1993 and references therein). For observations that cannot resolve individual small flux tubes (and the full-disk maps used in this study are of this type), the flux-tube size can be replaced by the magnetic filling factor (Solanki and Brigljević, 1992) which is simply the fraction of the solar surface within a given pixel of the map covered by magnetic flux tubes. Solar irradiance variations are then basically explained by the changing surface area coverage and distribution of magnetic features like sunspots and faculae, at least on the time-scales considered here.

The irradiance reconstruction presented here makes use of sunspot and facular contrasts calculated as a function of wavelength and limb angle on the Sun (Unruh et al., 1999; henceforth called USF99) using Kurucz's ATLAS9 spectral synthesis code (Kurucz, 1992a, 1992b, 1992c). The position and size of magnetic features on the solar disk are extracted from full-disk magnetograms and continuum intensity images obtained by the Michelson Doppler Interferometer (MDI) onboard the Solar and Heliospheric Observatory (SOHO) spacecraft.

In the discussions to follow we first briefly introduce our 3-component model of solar surface magnetism and sketch how we calculate irradiance variations in Sect. 2. We present the calculated total and spectral irradiance variations in Sect. 3 where we also compare them to measurements obtained by the Variability of Irradi-

Space Science Reviews **94:** 139–144, 2000.

ance and Gravity Oscillations (VIRGO) experiment onboard the SOHO spacecraft. Finally, we summarize our results in Sect. 4.

2. Model

The general modeling procedure is as follows: First, magnetic maps are created by decomposing magnetograms into sunspot, facular and quiet Sun regions. The value of the magnetic flux for each pixel is converted into a corresponding filling factor taking the limited spatial resolution of the magnetogram into account. Then, each pixel is replaced by a corresponding intensity value that depends on the filling factor and position on the solar disk. Finally, by summing the intensities over the full solar disk, the value for the irradiance is obtained.

We divide the solar surface into 3-components, namely quiet Sun I^q, sunspots I^s and faculae I^f, whose intensities I depend on the cosine of the limb angle μ and wavelength λ, but do not otherwise change with time t. The 3-components are employed for each solar surface element, i.e. each pixel of a magnetogram. The intensities are calculated for eleven different limb angles from $\mu = 1.0$ (disk center) to the very limb ($\mu = 0.05$) following USF99.

The intensity $I^{\mathrm{tot}}_{i,j}(\lambda; t)$ of a given element (i,j) on the solar disk at time t and wavelength λ is then given by:

$$\begin{aligned} I^{tot}_{i,j}(\lambda; t) = [1 - \alpha^s_{i,j}(\Phi; t) &- \alpha^f_{i,j}(\Phi; t)] \cdot I^q(\mu(i,j), \lambda) \\ &+ \alpha^s_{i,j}(\Phi; t) \cdot I^s(\mu(i,j), \lambda) \\ &+ \alpha^f_{i,j}(\Phi; t) \cdot I^f(\mu(i,j), \lambda), \end{aligned} \tag{1}$$

where $\alpha^s_{i,j}(\Phi; t)$ and $\alpha^f_{i,j}(\Phi; t)$ are the filling factors per surface element for sunspots and faculae, respectively, which are extracted from MDI magnetograms and continuum intensity images. After correcting for fore-shortening (magnetograms) and center-to-limb variations (continuum intensity images), respectively, a simple threshold criterion is used to select facular and sunspot pixels. Then, the magnetic flux density Φ (measured by the magnetogram) of each selected pixel is converted into a corresponding filling factor α.

Sunspots are well resolved by the full-disk MDI magnetograms and since stray light plays a minor role in the seeing-free environment of MDI we set $\alpha^s_{i,j}(\Phi) = 1$ for each pixel (i, j) within a sunspot.

For faculae, the situation is more complex. Faculae originate from more loosely packed magnetic elements (flux-tubes). At least for small facular regions (with low magnetic flux density) the filling factor is smaller than 1 and an appropriate scheme which converts magnetic flux densities into filling factors must be applied. We stress that the meaning of the filling factor used here differs from the usually employed magnetic filling factor. The latter has a considerably smaller value.

The conversion scheme we have employed is illustrated in Fig. 1. We increase the facular filling factor α^f linearly from zero at Φ_{th} to 1.0 at Φ_{sat}. The threshold

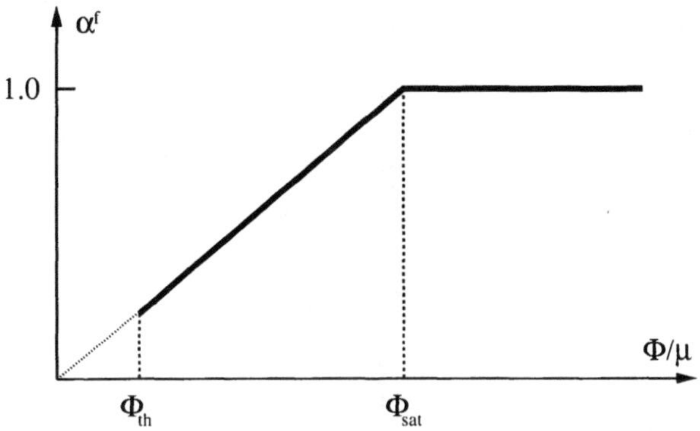

Figure 1. Conversion of magnetic flux Φ measured within a pixel of a magnetogram into a corresponding facular filling factor α^f. The facular filling factor α^f increases linearly from zero at $\Phi_{\rm th}$ to 1.0 at $\Phi_{\rm sat}$.

$\Phi_{\rm th}$ separates the signal from the noise within the magnetogram. For magnetic flux densities larger then $\Phi_{\rm sat}$, however, α^f remains unity. The reason for introducing this saturation lies in the observation that the average temperature enhancement within flux tubes relative to the quiet Sun decreases with increasing filling factor (Hirayama, 1978; Solanki and Stenflo, 1984; Solanki and Brigljević, 1992). Since we describe all faculae by the same thermal structure the simplest, although not particularly accurate way of accounting for the above observations in our model is by introducing a saturation in $\alpha^f(\Phi)$. For the results discussed in the following we employ $\Phi_{\rm sat} = 100$ G which means that already relatively weak faculae originate from densely packed magnetic elements.

As a result we get the filling factor maps $\alpha_s(\Phi; t)$ for sunspots and $\alpha_f(\Phi; t)$ for faculae. The total intensity of each pixel, $I_{i,j}^{tot}(\lambda; t)$, can then be calculated according to Eq. (1). Finally, relative solar irradiance variations are obtained from the relation:

$$\frac{\Delta S_\lambda}{S_\lambda} = \frac{\sum_{i,j} I_{i,j}^{tot}(\lambda) - \sum_{i,j} I^q(\mu(i,j),\lambda)}{\sum_{i,j} I^q(\mu(i,j),\lambda)}. \tag{2}$$

To model VIRGO measurements we constrain the wavelengths to the desired spectral band by multiplying the intensities $I_{i,j}^{\rm tot}(\lambda)$ and $I^q(\mu(i,j),\lambda)$ with the appropriate transmission function for each of the three spectral filters prior to integration over wavelength (Fligge et al., 1998).

3. Results

In the following, we present two reconstructions of solar total and spectral irradiance variations over the periods 15 August 1996–15 September 1996 and 6 November 1996–6 January 1997 and compare them to VIRGO measurements

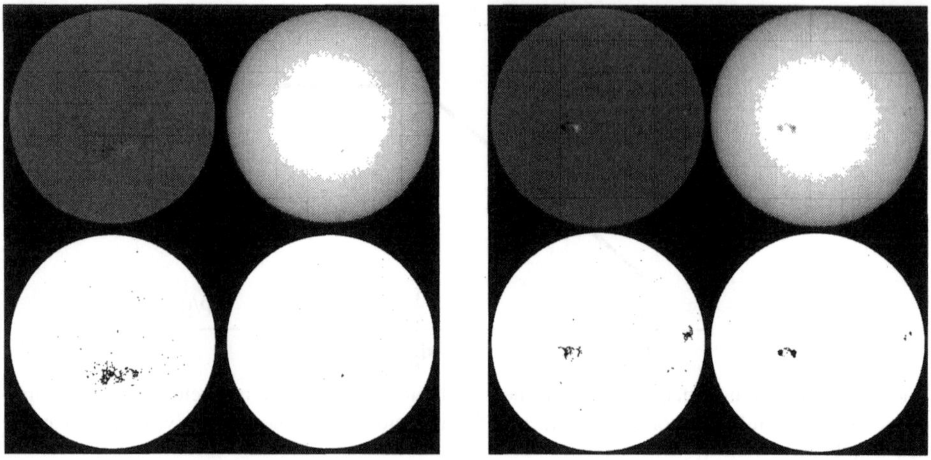

Figure 2. Extraction of sunspot and facular regions from a magnetogram. Left panel: The magnetogram (upper left) and the corresponding continuum intensity image (upper right) were taken on 30 August, 1996 by MDI. Below are the extracted maps for faculae (lower left) and sunspots (lower right). The decaying active region visible on the solar disk is dominated by faculae, although a small spot is also present. Right panel: Same, but for 25 November, 1996.

of total and spectral irradiance at 402 nm, 500 nm and 862 nm, respectively. At this time the Sun was still very close to activity minimum, making it particularly suitable for a detailed study of the influence of single active regions on irradiance variations unhampered by the presence of other active regions.

Fig. 2 shows the extraction of sunspot and facular regions for two days, i.e. 30 August 1996 (left panel) and 25 November 1996 (right panel), respectively. The upper two images of each panel show the magnetogram (left) and continuum intensity image (right) recorded by MDI. Below are the extracted maps for faculae (lower left) and sunspots (lower right), i.e. α_f and α_s. The first period is mainly faculae-dominated while during the second period an active region harboring two large sunspots showed up on the disc.

The reconstructed (solid lines) and measured (dashed lines) irradiance variations are presented in Fig. 3.

The model is able to reconstruct irradiance variations on time-scales of the solar rotation and reproduces the double-peaked structure originating from the center-to-limb variation (CLV) of the facular brightening rather well. The model yields linear correlation coefficients between the observed and modeled time-series of 0.98, 0.92, 0.93 and 0.90 for the total, blue, green and red channels, respectively. For comparison, the dotted curve shows an earlier reconstruction which neglects the CLV of the facular contrast (Fligge et al., 1998). Quite clearly, the new reconstruction, based on magnetograms and the CLV of intensity spectra is far superior.

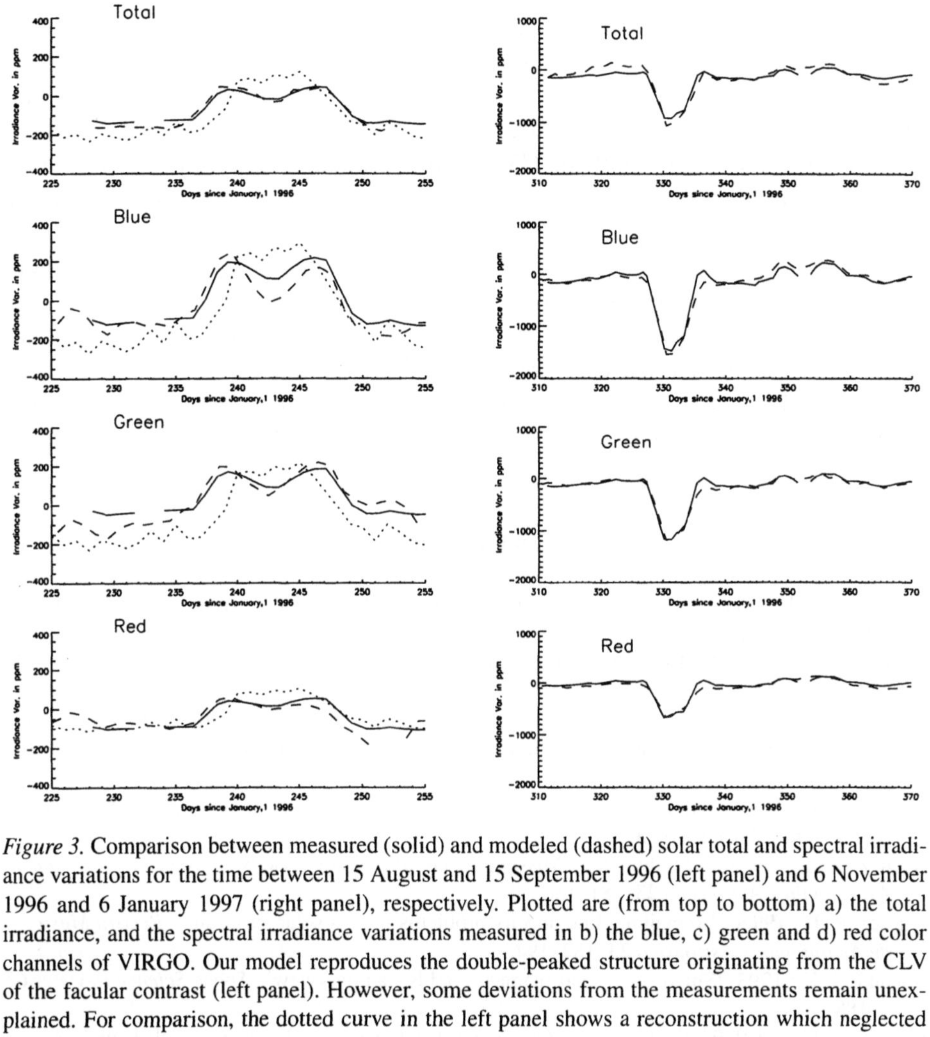

Figure 3. Comparison between measured (solid) and modeled (dashed) solar total and spectral irradiance variations for the time between 15 August and 15 September 1996 (left panel) and 6 November 1996 and 6 January 1997 (right panel), respectively. Plotted are (from top to bottom) a) the total irradiance, and the spectral irradiance variations measured in b) the blue, c) green and d) red color channels of VIRGO. Our model reproduces the double-peaked structure originating from the CLV of the facular contrast (left panel). However, some deviations from the measurements remain unexplained. For comparison, the dotted curve in the left panel shows a reconstruction which neglected the CLV of the facular contrast. The dimming of solar irradiance due to the passage of sunspots (right panel) is also well reproduced.

4. Conclusions

We present a model of solar total and spectral irradiance variations based on a 3-component model which includes the center-to-limb variation of the brightness of magnetic features. It makes direct use of magnetic maps (together with calculated intensity spectra) to reconstruct irradiance rather than proxies, such as Ca II K, Mg II k or He I 10830 Å indices which are formed in chromospheric layers. This requires detailed calculations of intensity spectra for each of the three atmospheric components as a function of wavelength and limb angle described in USF99. This approach is able to significantly improve reconstructions of irradiance variations on time-scales of the solar rotation relative to models based on disk-integrated proxies.

We wish to point out that spectra from the same atmospheric models of quiet Sun, sunspots and faculae as we use, also reproduce a number of other observations, such as the variation of the UV spectral irradiance between solar minimum and maximum, the ratio of facular to spot area, the fraction of irradiance variations due to line blanketing, etc. (see USF and Unruh et al., 2000, for more details).

The success of our model further strengthens the hypothesis that the magnetic field is the dominant driver of solar irradiation variations, at least on the short time scales considered in this paper. In order to further fine-tune the input parameters of our model, increase the statistical significance of the reconstructions and test our basic assumptions, longer time intervals have to be considered which, preferentially, should cover different levels of solar activity.

References

Fligge, M., S. K. Solanki, Y. C. Unruh, C. Fröhlich, and C. Wehrli: 1998, 'A model of solar total and spectral irradiance variations'. *Astron. Astrophys.* **335**, 709–718.

Hirayama, T.: 1978, 'A Model of Solar Faculae and Their Lifetime'. *Publ. Astron. Soc. Jpn.* **30**, 337–352.

Knölker, M. and M. Schüssler: 1988, 'Model calculations of magnetic flux tubes. IV - Convective energy transport and the nature of intermediate size flux concentrations'. *Astron. Astrophys.* **202**, 275–283.

Kurucz, R. L.: 1992a, 'Atomic and molecular data for opacity calculations'. *Revista Mexicana de Astronomia y Astrofisica* **23**, 45–48.

Kurucz, R. L.: 1992b, 'Finding the 'Missing' solar ultraviolet opacity.'. *Revista Mexicana de Astronomia y Astrofisica* **23**, 181–186.

Kurucz, R. L.: 1992c, 'Remaining line opacity problems for the solar spectrum'. *Revista Mexicana de Astronomia y Astrofisica* **23**, 187–194.

Solanki, S. K.: 1993, 'Small scale solar magnetic fields - an overview'. *Space Science Reviews* **63**, 1–188.

Solanki, S. K. and V. Brigljević: 1992, 'Continuum brightness of solar magnetic elements'. *Astron. Astrophys.* **262**, L29–L32.

Solanki, S. K. and J. O. Stenflo: 1984, 'Properties of solar magnetic fluxtubes as revealed by Fe I lines'. *Astron. Astrophys.* **140**, 185–198.

Unruh, Y. C., S. K. Solanki, and M. Fligge: 1999, 'The spectral dependence of facular contrast and solar irradiance variations'. *Astron. Astrophys.* **345**, 635–642.

Unruh, Y. C., S. K. Solanki, and M. Fligge: 2000, 'Modelling solar spectral irradiance variations', *Space Sci. Rev.*, this volume.

Address for Offprints: M. Fligge, Institute of Astronomy, ETH Zentrum, CH-8092 Zurich, Switzerland. fligge@astro.phys.ethz.ch

MODELLING SOLAR IRRADIANCE VARIATIONS: COMPARISON WITH OBSERVATIONS, INCLUDING LINE-RATIO VARIATIONS

Y. C. UNRUH (unruh@astro.univie.ac.at)
Institute of Astronomy, University of Vienna, Türkenschanzstr. 17, 1180 Wien, Austria

S. K. SOLANKI (solanki@linmpi.mpg.de)
Max-Planck-Institut für Aeronomie, D-37191 Katlenburg-Lindau, Germany

M. FLIGGE (fligge@astro.phys.ethz.ch)
Institute of Astronomy, ETH Zentrum, CH-8092 Zürich, Switzerland

Received: 22 October 1999; Accepted in final form: 27 January 2000

Abstract. Solar irradiance variations show a strong temporal and spectral dependence. The progression of the Sun through its activity cycle as well as solar rotation are mirrored in the irradiance variations. The spectral dependence is such that the variations are several magnitudes larger in the EUV than in the visible or infrared.

We present a simple 3-component model that is based on the assumption that changes in the solar flux are exclusively due to changes in spot and facular coverage. We compare our model to observations of the spectral solar irradiance variations. Despite its simplicity, we find that the agreement between our model and the observations is surprisingly good. We also explore the reliability and the limitations of our approach by comparing observations of the solar facular contrast and of the changes in spectral line depths with our calculations.

1. Introduction

Models of total and spectral irradiance variations are important not just for obtaining an understanding of the physical causes of solar irradiance variations, but also for successful reconstructions of such variations for times prior to satellite measurements. We summarize the main results of a simple model, emphasizing the comparison of its predictions with a wide variety of observations. In particular we present the first comparison with the variation of selected line ratios over the solar cycle (Sect. 4.3).

2. The Model

To model the solar irradiance variations, we assume that they are solely due to changes in the spot and facular coverage (Solanki and Unruh, 1998). The facular component here includes any brightenings due to magnetic features, i.e. also the network. The flux of the active Sun is then simply the sum of the three components

Space Science Reviews **94:** 145–152, 2000.

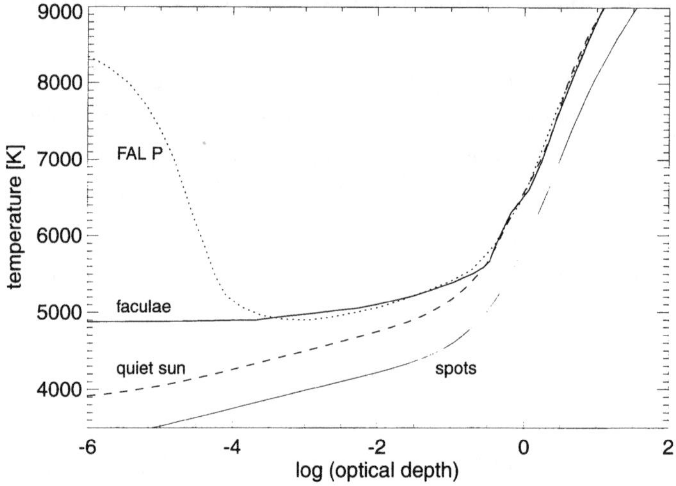

Figure 1. The temperature stratification of the quiet-sun (dashed line), the facular (upper thick solid line) and the spot model (thin solid line) atmospheres as a function of optical depth. The dotted line shows the original facular model by Fontenla et al. (1993).

(quiet-Sun, spots and faculae) weighted according to their filling factors:

$$F_\lambda = (1 - f_F - f_S)F_\lambda^\odot + f_F F_\lambda^F + f_S F_\lambda^S,$$

where f_F and f_S are the facular and the spot filling factors respectively. The filling factors here are defined to be the fraction of the projected solar disk covered by spots or faculae. F_λ, $F_\lambda^\odot$, F_λ^F and F_λ^S are the active-sun, the quiet-sun, the facular and the spot fluxes.

The fluxes are calculated with the ATLAS9 spectral synthesis code written by R. Kurucz that takes into account up to 58 million lines by means of opacity distribution functions. We neglect any non-LTE effects in our calculations. For the quiet-Sun flux we use Kurucz's solar model atmosphere (Kurucz, 1991). This describes a quiet photosphere in radiative equilibrium with an effective temperature of 5777 K. For the sunspot radiative flux we use a model atmosphere with an effective temperature of 5150 K that was interpolated from the Kurucz grid of model atmospheres. This model represents the joint contribution of the umbral and penumbral flux, assuming a ratio of almost 1:3 for the umbral and penumbral areas. We assumed temperatures of approximately 4500 K and 5400 K for sunspot umbrae and penumbrae respectively (see Solanki and Unruh, 1998, and Fligge et al., 1998). The facular model atmosphere was developed from model P of Fontenla et al. (1993) which we used as a starting point. We excluded the temperature inversion in the upper photosphere, thereby creating a "pseudo-LTE" atmosphere. Further changes were introduced to achieve better agreement with the observations. The model atmospheres of the three components are shown in Fig. 1 along with model P of Fontenla et al. (1993). For more details see Unruh et al. (1999).

Note that ATLAS9 also allows us to calculate the intensities (as a function of limb angle) for all three components. We can hence obtain the total flux via a disk integration over all surface pixels (Fligge et al., 2000a, 2000b) and we can also calculate the limb-dependent contrasts of the faculae (see Sect. 4.2).

3. Comparison with Observations: Spectral Irradiance Variations

The total irradiance variation over one solar cycle is on the order of 0.1 % (Willson and Hudson, 1991; Fröhlich and Lean, 1998). The variations are small in the visible and IR, but the solar flux can vary by more than 100 % in the EUV. The spectral dependence of the irradiance variations over a solar cycle has been compiled by Lean (1997) and is plotted in Fig. 2 (dotted line). Note that the data in the UV are measurements, while the behaviour in the visible and IR has largely been estimated.

We obtain the facular and spot filling factors from a fit to the data from Fig. 2 over the wavelength range between 200 and 400 nm. We also require the "total" irradiance variations (integrated from 164 to 100 000 nm) to amount to 0.1 %. Our fit matches the overall behaviour of the data well, although there are differences in detail. We tend to overestimate the variability between about 320 and 380 nm. Part of this is due to the greater sensitivity of the fit to the larger variations between 200 and 260 nm, though we also expect discrepancies at UV wavelengths due to the LTE approach (see Fontenla et al., 1999 and White et al., 1999 for the description of a non-LTE approach). Some of the difference may also arise because the measurements in the range from 300 to 400 nm are associated with relatively large errors.

In order to check whether a facular component with a distinct temperature gradient was really needed to reproduce the data, we also calculated the irradiance variations that one would obtain if the Sun were 1 K hotter at activity maximum compared to activity minimum. These irradiance variations are shown by the dashed lines in Fig. 2. They clearly do not reproduce the spectral dependence of the irradiance and underestimate the variations in the blue part of the spectrum by a large amount while overestimating the variability in the visible and IR. The temporal variation of the total and the spectral variation predicted by our model has been worked out and compared with ACRIM (Willson and Hudson, 1991) and VIRGO (Fröhlich et al., 1997) data by Fligge et al. (1998; 2000a; 2000b). For details we refer the reader to those papers.

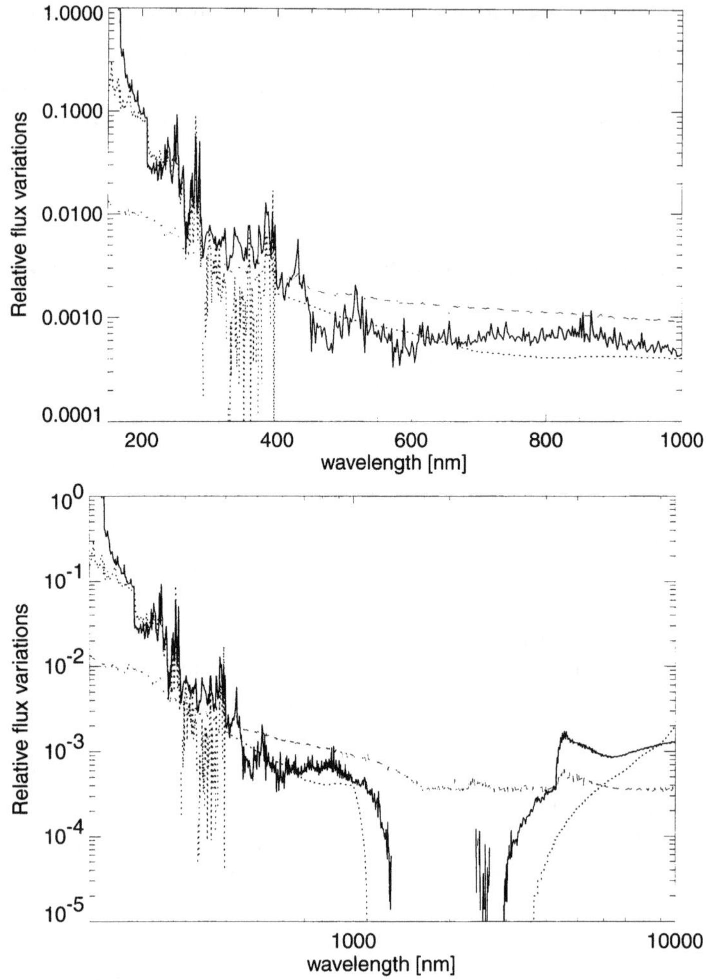

Figure 2. Relative solar irradiance variation in the wavelength ranges between 150 and 1000 nm (top) and between 150 and 10 000 nm (bottom). Plotted is the relative difference between the solar spectrum at activity maximum and minimum. The data compiled by Lean (1997) as well as her estimates are represented by the dotted lines; the solid lines show our model with a facular filling factor of 2.7 % and a spot filling factor of 0.23 %. The predictions of a model whereby the temperature of the Sun at solar maximum has been raised by 1 K relative to activity minimum are indicated by the dashed lines.

4. Comparison with Observations: Further Consistency Checks

4.1. Spot and Facular Coverage

We can check whether the spot and facular filling factors that we obtain are in line with the observed values. Near the maxima of cycles 21 and 22, spots covered about 0.2 % of the solar surface. From observations spanning a 7.5 year-period during cycle 22, Chapman et al. (1997) found the ratio of facular to spot area to be 16.7, although the values varied considerably over the solar cycle.

In order to match the observations plotted in Fig. 2 we need facular filling factors of the order of 2 % to 3 %. The spot filling factor required depends on the facular filling factor that is adopted and ranges from 0.1 % to almost 0.3 %. The corresponding ratios of facular to spot filling factor then range from 20 down to 12 for the lowest and highest filling factors respectively. Our spot and facular filling factors are thus in good agreement with the observations.

4.2. Facular Contrasts

We have calculated the contrasts of the faculae as a function of limb angle and compared them to solar measurements. The contrast is defined as $(I_f - I_q)/I_q$, where I_f and I_q are the intensity of the faculae and the quiet Sun. The spread between different contrast measurements is very large, reflecting the differences in facular structure, wavelength of the observations, magnetic filling factor and size. The scatter in the measurements is evident in Fig. 3, where the linked symbols represent a selection of measurements. The situation is particularly ambiguous close to the solar limb where the measurements by e.g. Libbrecht and Kuhn (1985) indicate a decrease in the contrast beyond about $\mu = 0.2$. Our model is plane-parallel and neglects facular fine structure. We therefore do not expect our calculations to be accurate close to the limb. As the relative effect of the areas close to the limb is small, this should not produce large discrepancies in our irradiance calculations. The three thick lines (solid, dashed and dotted) show our calculations. Given the fact that the observations are inconsistent with each other, our model gives reasonable results (for a more detailed discussion see Unruh et al., 1999).

Our model exhibits excellent agreement with the spectral behaviour of the facular contrast measured by Chapman and McGuire (1977), as shown in Fig. 4. Unruh et al. (1999) also showed that with the same set of free parameters the model gives changes in line blanketing between activity minimum and maximum that are consistent with the findings of Mitchell and Livingston (1991).

4.3. Changes in Line Ratios

As the Sun moves through its cycle, changes in line strengths can be observed in most absorption lines. By considering the ratios between the depths of different lines, Gray and Livingston (1997a) were able to neutralise many of the uncertainties in the measurements. They found the C I (538.03 nm) to Fe I (537.96 nm) ratio ($r_{\rm Fe}$) to vary by about 0.004 and the C I (538.03 nm) to Ti II (538.13 nm) ratio ($r_{\rm Ti}$) to vary by 0.003 between solar activity minimum and maximum. These variations amount to about 1 % of the line ratios. After an empirical temperature calibration of the line ratio changes, based on a comparison with stellar observations (Gray and Livingston, 1997b) they interpreted these variations to be due to a 1.5 K temperature rise of the Sun from activity minimum to maximum. They did

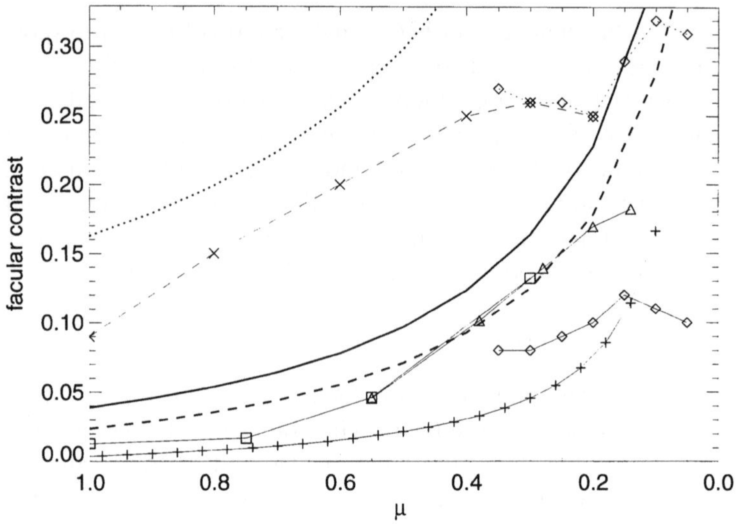

Figure 3. Selected facular contrast measurements, as well as contrasts calculated using our model, plotted vs. limb angle. The squares, crosses, diamonds and triangles indicate data obtained by Frazier (1971), Auffret and Muller (1991) (at 575 nm), Wang and Zirin (1987) (at 386 nm and 525 nm) and Taylor et al. (1998) respectively. The curve linking the plus signs is the parameterisation adopted by Lawrence et al. (1988). Unfortunately the information at what wavelengths the contrasts were measured was not always available. In these cases we assumed that the data were taken around 525 nm. Solid lines link data taken at about 525 nm, dotted lines data below 400 nm and dashed lines data at about 575 nm. The thick curves show our calculations for filters centred at 386 nm (dotted line), 525 nm (solid line) and 575 nm (dashed line). Above about 500 nm the filter width is relatively unimportant. This is not the case in the blue, where e.g. a reduction in the passband from 50 nm to 10 nm predicts an increase in contrast by almost a factor of two.

not consider the influence of changes in the temperature gradient over the cycle, or of the fact that these changes may be concentrated in a small fraction of the solar surface (since they are expected to be associated with the magnetic field).

We have calculated the changes in the line ratios between solar minimum and maximum for our three-component model and for a model where the temperature has been raised by 1.5 K at solar maximum (i.e. similar to Gray and Livingston's model, but based on radiative transfer calculations). Our preliminary findings are summarized in the following table.

change in line ratio	3-component model	ΔT model	Gray and Livingston (data)
Δr_{Fe}	0.0018	0.0008	0.004
Δr_{Ti}	0.0015	0.0006	0.003

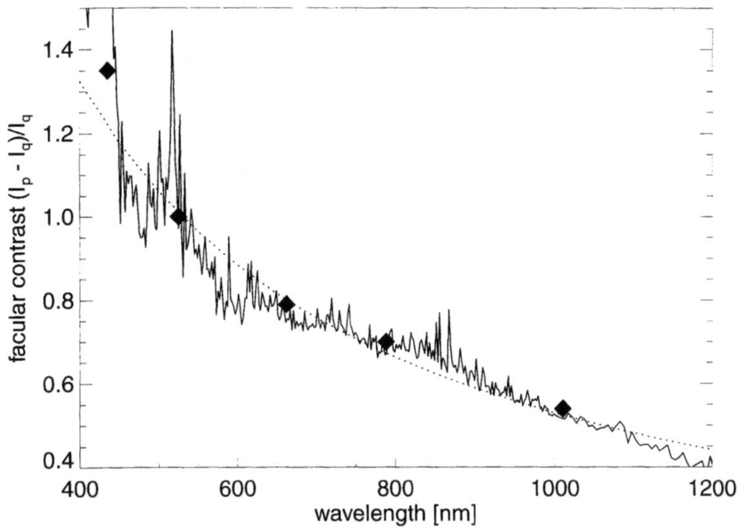

Figure 4. The behaviour of the facular contrast with limb angle as measured by Chapman and McGuire (1977): diamonds. The solid line is the facular contrast obtained with our model. The contrasts have been normalised to be unity at 530 nm. The dotted line shows the inverse wavelength fit suggested by Chapman and McGuire (1977).

Neither of the models is able to satisfactorily reproduce the data. However, our 3-component model yields stronger line-ratio variations than a simple temperature increase, though they are still about a factor of two lower than the measurements. The cause of the discrepancy is still unclear. It may be due to the uncertainties associated with our calculations, not least because of uncertainties in the model atmospheres. For instance, we find that line ratios calculated from the quiet-sun model and from the standard Kurucz model atmospheres for stars between 5500 and 6000 K cannot be reconciled with the observations of solar-type stars in this temperature range by Gray and Livingston (1997b). This poor correspondence may, however, also partly have to do with the simple empirical calibration used by them. For example, we find that the line ratios also depend on gravitational acceleration and not just temperature.

5. Conclusions and Outlook

- The 3-component model matches the measured spectral irradiance variations reasonably well. It can be used to calculate the solar irradiance from solar disk images in a straightforward way. The model also reproduces a number of further observational constraints.
- Some discrepancies remain in the blue, and particularly between 300 and 400 nm. The UV calculations should improve when non-LTE calculations are performed.

- We have presented a preliminary comparison with the cyclic line-ratio variations published by Gray and Livingston (1997a; 1997b). Although our model fares better than a simple increase of the effective temperature, it still does not give a satisfactory fit to the the data.
- Improvements to the model which take into account that quiet-region and active region faculae have distinct temperature stratifications are planned.

Acknowledgements

J. Lean kindly provided us with her compilation of the observational data, which we gratefully acknowledge. YCU would like to acknowledge support from the Austrian Fond zur Förderung der wissenschaftlichen Forschung under grant No. 7302. MF acknowledges support from the Swiss Nationalfonds under grant No. 2000-046894.96.

References

Auffret, H. and Muller, R.: 1991, *A&A* **246**, 264.
Chapman, G. A. and McGuire, T. E.: 1977, *ApJ* **217**, 657.
Chapman, G. A., Cookson, A. M. and Dobias, J. J.: 1997, *ApJ* **482**, 541.
Fligge, M., Solanki, S. K., Unruh, Y. C., Fröhlich, C. and Wehrli, C.: 1998, *A&A* **335**, 709.
Fligge, M., Solanki, S. K. and Unruh, Y. C.: 2000a, *A&A* **353**, 380.
Fligge, M., Solanki, S. K. and Unruh, Y. C.: 2000b, *Space Sci. Rev.*, this volume.
Fontenla, J. M., Avrett, E. H. and Loeser, R.: 1993, *ApJ* **406**, 319.
Fontenla, J. M., White, O.R., Fox, P.A., Avrett, E. H. and Kurucz, R.L.: 1999, *ApJ* **518**, 480.
Frazier, E. N.: 1971, *Solar Phys.* **21**, 42.
Fröhlich, C. and Lean, J.: 1998, *GRL* **25**, 4377.
Fröhlich, C. et al.: 1997, *Solar Phys.* **170**, 1.
Gray, D. F. and Livingston, W. C.: 1997a, *ApJ* **474**, 802.
Gray, D. F. and Livingston, W. C.: 1997b, *ApJ* **474**, 798.
Kurucz, R. L.: 1991, in Stellar Atmospheres: Beyond Classical Models, L. Crivellari, I. Hubeny, D. G. Hummer (Eds), Kluwer, Dordrecht, p. 441.
Lawrence, J. K. and Chapman, G. A.: 1988, *ApJ* **335**, 996.
Lean, J.: 1997, *ARA&A* **35**, 33.
Libbrecht, K. G. and Kuhn, J. R.: 1985, *ApJ* **299**, 1047.
Mitchell, W. E. Jr. and Livingston, W. C.: 1991, *ApJ* **372**, 336.
Solanki, S. K. and Unruh, Y. C.: 1998, *A&A* **329**, 747.
Taylor, S. F., Varsik, J. R., Woodard, M. F. and Libbrecht, K. G.: 1998, *Solar Phys.* **178**, 1.
Unruh, Y. C., Solanki, S. K. and Fligge, M.: 1999, *A&A* **345**, 635.
Wang, H. and Zirin, H.: 1987, *Solar Phys.* **110**, 281.
White, O.R., Fontenla, J. M. and Fox, P.A.: 2000, *Space Sci. Rev.*, this volume.
Willson, R. C. and Hudson, H. S.: 1991, *Nat* **351**, 42.

Address for Offprints: S. K. Solanki, Max-Planck-Institut für Aeronomie, D-37191 Katlenburg-Lindau, Germany

RESONANCE IN A COUPLED SOLAR-CLIMATE MODEL

S.M. TOBIAS (smt11@damtp.cam.ac.uk) and
N.O. WEISS (now@damtp.cam.ac.uk)
Department of Applied Mathematics and Theoretical Physics, University of Cambridge, U.K.

Received: 18 October 1999; Accepted in final form: 20 January 2000

Abstract. The 11–year solar activity cycle is magnetic in origin and is responsible for small changes in solar luminosity and the modulation of the solar wind. The terrestrial climate exhibits much internal variability supporting oscillations with many frequencies. The direct effect of changing solar irradiance in driving climatic change is believed to be small, and amplification mechanisms are needed to enhance the role of solar variability. In this paper we demonstrate that resonance may play a crucial role in the dynamics of the climate system, by using the output from a nonlinear solar dynamo model as a weak input to a simplified climate model. The climate is modelled as oscillating about two fixed points (corresponding to a warm and cold state) with the weak chaotically modulated solar forcing on average pushing the solution towards the warm state. When a typical frequency of the input is similar to that of the chaotic climate system then a dramatic increase in the role of the solar forcing is apparent and complicated intermittent behaviour is observed. The nonlinear effects are subtle however, and forcing that on average pushes the solution towards the warm state may lead to increased intervals of oscillation about *either* state. Owing to the intermittent nature of the timeseries, analysis of the relevant timeseries is shown to be non-trivial.

1. Introduction

The current trend towards global warming appears to be dominated by anthropogenic forcing. In particular the increase in greenhouse gases has led to a sharp rise in the globally averaged temperature over the past 40–50 years. There are, however, many examples of global climate change that must be ascribed to other causes. A growing amount of evidence suggests a link between solar variability and climate change. This evidence is presented in many of the review articles in this volume and so will not be summarised here.

Solar activity which is aperiodic, and has a well-defined mean period of about 11 years, is due to changes in the solar magnetic field. This field reverses after 11 years leading to an overall oscillation with a 22–year period. The basic cycle is modulated on an approximately 200 year timescale. The Maunder minimum of the 17th Century, when sunspots almost totally disappeared, is now established via proxy data as one of many such recurrent events. The solar magnetic field is generated by a self-exciting magnetohydrodynamic dynamo operating at the base of the solar convection zone, and it is possible to construct nonlinear dynamo models that reproduce the modulation of the basic cyclic activity (the secular trend), that is observed on the Sun. More details of the observations of solar magnetic activity

Space Science Reviews **94:** 153–160, 2000.

and of solar dynamo theory can be found in the review by Weiss and Tobias (2000) elsewhere in this volume.

Although recent satellite observations have shown that the total solar irradiance does indeed increase by about 0.15% from sunspot minimum to sunspot maximum, it is not thought that this variation is enough to drive a significant direct response in the climate system, even allowing for systematic increases due to the secular variation in irradiance as magnetic activity increases from Grand Minima to maxima. However there is much evidence supporting the conjecture that solar variability may have a significant influence on climatic change. For this reason amplification mechanisms have been sought and many of them are discussed elsewhere in the volume. They include: modulation of ozone production in the upper stratosphere leading to modification of the Hadley Circulation (Haigh, 2000; Van Loon and Labitzke, 2000, elsewhere in this volume), the effect of cosmic rays, which are modulated by the solar wind, on cloud formation (Marsh and Svensmark, 2000, this volume), and modification of the earth's global electric circuit (Tinsley, 2000, this volume).

We present a complementary process that relies on resonant interactions between solar activity and natural variability of the climate system. The Earth's atmosphere and ocean support oscillations with a wide range of frequencies (Ghil, 1991). Variations in solar activity can have a weak direct effect on climate through some mechanism (e.g. one of those described above) which couples to a natural oscillation of the climatic system, just as palaeoclimate is influenced by the Milankovitch effect. A similar process has been demonstrated by introducing irradiance variations as periodic perturbations in a simplified climatic model (Drijfhout *et al.*, 1999). Here we show that resonance can play an important role in the amplification process *even when the input to the climate is the aperiodically modulated output from a solar dynamo model.* In the next section we present the model equations for both the dynamo and climate systems. We then go on to highlight the important results in section 3, focusing on the role of resonance in modifying the solutions. We conclude by discussing the implications of the results for the impact of solar variability on climatic change.

2. The Dynamo and Climate Models

We represent the chaotically modulated solar cycle by a low-order model of coupled ordinary differential equations (introduced by Tobias, Weiss, and Kirk, 1995). This is the simplest possible model that contains the interaction between magnetic fields and velocity fields and includes nonlinear coupling. More details of this dynamo model (presented below) and its relation to more sophisticated models that include spatial structure are given by Weiss and Tobias (2000) elsewhere in this volume. The equations

$$\dot{x} = f\left[\lambda x - \omega y + azx + d\,z(x^2 + y^2)\right], \quad (1)$$
$$\dot{y} = f\left[\lambda y + \omega x + azy\right], \quad (2)$$
$$\dot{z} = f\left[\mu - z^2 - (x^2 + y^2) + cz^3\right]. \quad (3)$$

govern the evolution of the solar toroidal magnetic field (x), the solar poloidal magnetic field (y) and the hydrodynamic state of the Sun (z). We choose parameter values so that the model reproduces behaviour similar to that of the Sun (Tobias *et al.*, 1995). The basic 11-year magnetic activity cycle is, for these parameter values, modulated on a longer timescale and recurrent grand minima are apparent (see Figure 5 of the review by Weiss and Tobias, 2000, in this volume). The only parameter that is allowed to vary is the frequency ratio f which controls the timescale of magnetic activity relative to that of the climate model described below.

The climate is modelled using the well-known Lorenz equations (Lorenz, 1963). This third-order system was originally derived by truncating the PDEs describing nonlinear Rayleigh-Bénard convection. It is, however, most often used as an example of nonlinear processes that are believed to be important in the climate system. For certain parameter values (see e.g. Sparrow, 1982) the solutions exhibit chaotic oscillations about the non-trivial fixed points, occasionally flipping between the states $T \gtrless 0$. We utilise this model and think of the variable T as representing the global temperature, regarding the non-trivial fixed points T^+ and T^- as warm and cold climatic states. The chaotic Lorenz attractor represents a situation where the temperature naturally oscillates about a fixed point and occasionally flips from one state to the other. We link the solar and climate models by including nonlinear forcing through solar activity in the climate model. Any solar forcing must depend on the level of activity, but not on the sign of the toroidal field, so we add a term proportional to x^2, yielding the system

$$\dot{T} = \sigma\,(U - T), \quad (4)$$
$$\dot{U} = \rho T - U - TV + \delta x^2, \quad (5)$$
$$\dot{V} = -\beta V + TU, \quad (6)$$

where x is given by the output of the solar dynamo model in equation (1). Note that this is a one-way coupling since the Earth's climate does not affect the Sun. It is the behaviour of this coupled system that forms the basis for the rest of this paper.

3. Results

We present the results of a series of integrations of the coupled model given by equations (1–6). We fix the amplitude of the forcing from the solar dynamo model to be very small and investigate the significant effects of allowing the relative

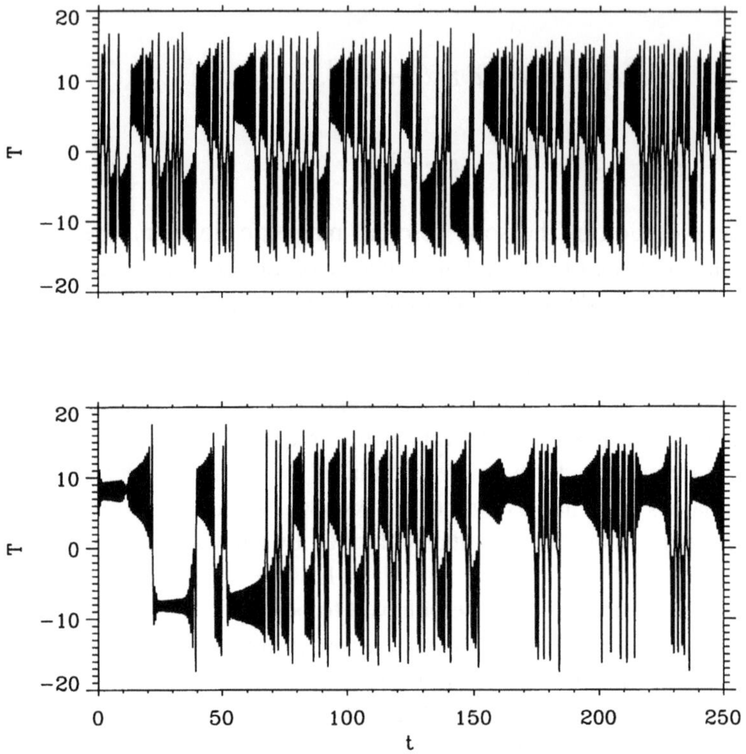

Figure 1. "Non-resonant" and "resonant" behaviour for the weakly forced climate system: timeseries for T for solutions of equations (1)–(3) and (4)–(6) with the parameters chosen so that the input from the chaotic dynamo is very small in both cases (see Tobias and Weiss, 1999, for details). The "relative frequency" is changed from $f = 0.6$ (upper panel) to $f = 0.25$ lower panel. Changing the frequency of the forcing has a dramatic effect on the form of the solutions.

frequency f to vary. In this way we demonstrate that frequency of the forcing plays a crucial role when the aperiodic forcing is very weak. Figure 1 displays the most important result of the paper, by showing a quantitative comparison of the behaviour of the timeseries for T for two different values of f. In Figure 1 the typical timescale for the climate variability (as given by the typical time for the attractor to make one oscillation about a fixed point) is approximately 2–3 times that for the forcing from solar activity. The chaotic behaviour observed, with oscillations about the fixed points and aperiodic flipping, is similar to the unforced system and the role of the forcing is largely undetected. A spectral analysis of the timeseries (Tobias and Weiss, 1999) simply yields a peak at the natural frequency of the climate and no trace of the forcing can be detected.

In contrast, Figure 1(b) shows the timeseries for T when $f = 0.25$ and the timescale for oscillation in the climate system is comparable to that for a typical oscillation of the solar forcing. This figure is significantly different from that in Figure 1(a) and is characterised by two different types of behaviour. For long periods (e.g. between $t \approx 70$ and $t \approx 150$) the climate system exhibits a *normal pattern* – that is it acts in a similar manner to the unforced system. However, there are also long intervals (e.g. between $t \approx 0$ and $t \approx 70$) when the climate system acts quite differently. For this *resonant pattern* the solution typically spends long intervals trapped near the non-trivial fixed points and the time taken to flip is, on average, much longer than in the unforced case. Thus resonance can be seen to be playing a vital role even in the case where the forcing is due to an aperiodic signal. It is interesting also to note that the solution can spend long periods being trapped in *either* the hot or cold state ($T \gtrless 0$) even though the forcing is designed to push the solution towards the hot state ($T > 0$). In Tobias and Weiss (1999) we demonstrate that this resonant behaviour continues to be found for other parameter values that alter the amplitude of the oscillations of either the climate or the forcing *provided the typical frequencies for these oscillations remain comparable*.

We demonstrate that the "resonant behaviour" shown in Figure 1(b) is due to frequency locking by performing a number of calculations at many values of the relative frequency f and comparing the result. As noted above the most striking consequence of the resonance is the intermittent behaviour between intervals of regular flipping and those where the flipping time is greatly increased. A quantitative measure of the role of the relative frequency can therefore be gained by integrating the equations for a long time in order to generate a distribution of flipping times (i.e. the time spent in either $T \gtrless 0$ between flips). The mean and variance of this distribution can then be calculated. A summary of these calculations is shown in Figure 2. The graphs for the mean and variance of the distribution for t_f both show marked peaks at $f \approx 0.25$ where the timescales for the forcing and the climate are similar. For frequencies away from the resonant band, changing the frequency seems to have little effect on the distribution of flipping times. The figure supports our conjecture that the frequency of the chaotic forcing is *crucial* in determining the effect of the forcing on the Lorenz model. It is not possible to predict the consequences of the forcing solely from knowledge of its amplitude.

Frequency analysis is often used as a tool for detecting the presence of a signal of a given period in a complicated timeseries. Indeed power spectra of climate records have often been used to argue for the presence (or absence) of an effect of solar variability on climate. We now therefore discuss how the differences in the nature of the solutions caused by changing the relative frequency of the forcing can be detected in a frequency analysis of the chaotic climate signals. When a spectral analysis of the climate signal shown in Figure 1 is performed, no detectable signal of the aperiodic forcing is visible in the spectra (see Tobias and Weiss, 1999), and the forcing is shown to have a weak effect. Figure 3 shows the result of a similar analysis of timeseries for the resonant case found for the parameters of Figure 1(b). Here three different spectra are computed. Figure 3(a) shows the spectrum for a subset of the full timeseries given by $0 \leq t \leq 75$. For this portion of the timeseries the system exhibits "resonant behaviour". This is reflected in the spectrum which shows two peaks corresponding to the natural frequency of oscillation of the Lorenz

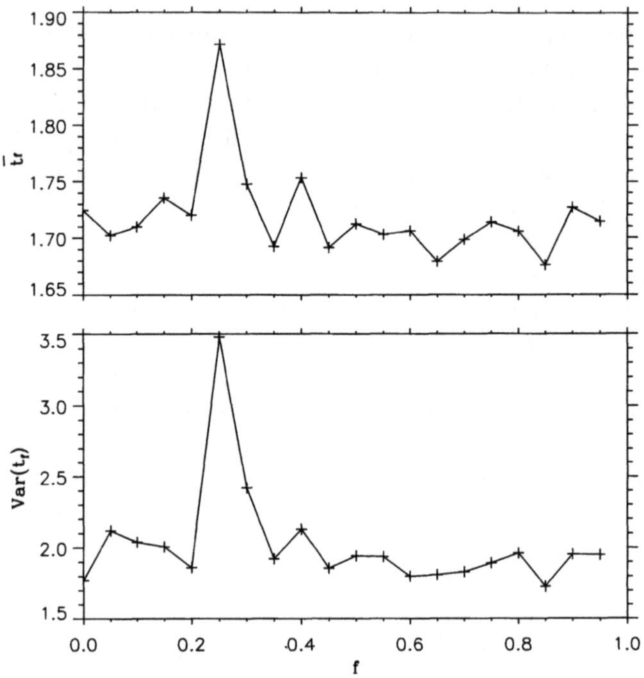

Figure 2. Frequency dependence of flipping time distribution. The upper panel shows the mean of the distribution of flipping times plotted against the frequency ratio, whilst the lower panel shows the variance of the distribution. Both measures show a clear peak (indicating the presence of a significant number of outliers) at a frequency ratio of $f = 0.25$.

system and that of the chaotic input forcing respectively. Figure 3(b) shows the corresponding spectrum for $75 \leq t \leq 150$ – a portion of the timeseries where there is a normal pattern of behaviour. Here only the natural climate frequency can be detected and there is very little evidence of the forcing. The power spectrum for the whole timeseries is shown in Figure 3(c). This timeseries contains intervals both of resonant and of normal behaviour. The primary peak corresponds to the natural climate frequency but the signature of the input is also detectable in the smaller subsidiary peak. The results of a timeseries analysis is therefore dependent on when in the timeseries the spectrum is taken and the length of the timeseries used. This can have extremely important consequences for interpretations of spectra of finite timeseries of climatic data. It is possible that resonant phenomena may be missed if the timeseries utilised is too short and happens not to span a resonant interval. Moreover the resonant signal may not show up clearly if too long a timeseries is analysed and many intervals of resonant and normal behaviours are included. This is due to intermittent switching of the solution between resonant and normal behaviour, which depends on the chaoticity of both the natural climate system and the forcing.

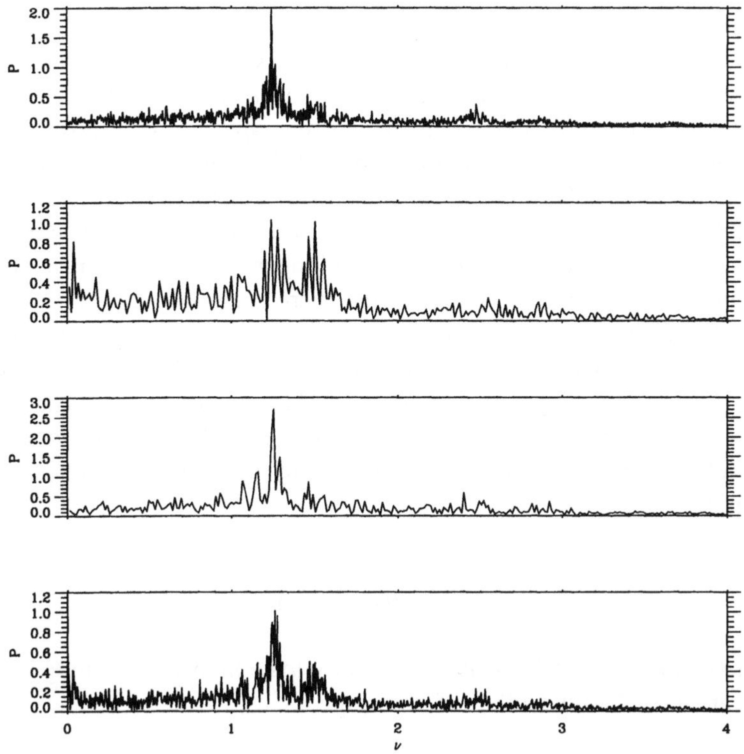

Figure 3. Frequency analysis of the forced Lorenz system with weak "resonant" forcing. (a) Power spectrum of short timeseries ($0 \leq t \leq 75$) containing resonant behaviour. (b) Power spectrum of short timeseries ($75 \leq t \leq 150$) containing normal behaviour. (c) Power spectrum of total timeseries ($0 \leq t \leq 250$). In (a) the forcing frequency can clearly be detected, but it is not visible in the normal timeseries (b) and weaker in the analysis of the long timeseries in (c).

4. Conclusions

In this short paper we have presented the main results that demonstrate that resonant interactions between solar activity and climate can lead to amplification of the solar signal and generate an increased response of the climate to solar forcing. We considered a model consisting of two chaotic weakly-coupled systems to represent solar activity and climate. We utilised a model of solar activity that reproduces not only the basic cyclic activity but also the modulation leading to recurrent grand minima and weakly coupled this to a simple model of climate variability (the Lorenz equations). The amplitude of the coupling is deliberately chosen to be very small compared to the the natural variability of the climate system (the peak value of the input signal from the solar activity is approximately 5% of that of the nonlinear term in the climate model). If the frequency of the forcing is chosen to be similar to that for natural variability of the climate model then the role of the forcing is significantly amplified, and the behaviour of the climate model is very different. The grand minima which are present in the solar activity record also play a significant role. Because the forcing switches off aperiodically, the climate record becomes very intermittent in the resonant case, with intervals where the climate system acts in a resonant manner interspersed with those where it is acting as though unforced. This intermittency of the climate signal means that the results of frequency analysis of the climate signal need to be interpreted with care. Analysis of short timeseries can yield widely differing results depending on the interval that has been chosen, while for much longer timeseries the resonant signal may be diluted. We therefore conclude that there are grave dangers in obtaining power spectra from intermittent timeseries, since the results may be sensitive to the length of the dataset that is available. In summary we have shown that if there exists a coupling (even if it is very weak) between the 11-year activity cycle and a natural periodicity of the climate then resonance provides a mechanism for amplifying the effect of solar forcing regardless of whatever physical process is responsible for the coupling.

Acknowledgements

We wish to thank ISSI for the invitation to participate in this stimulating workshop. SMT is grateful to Trinity College, Cambridge.

References

Drijfhout, S.S., Haarsma, R.J., Opsteegh, J.D., and Selten, F.M.: 1999, 'Solar-induced versus internal variability in a coupled climate model', *Geophys. Res. Lett.* **26**, 205–208.

Ghil, M.: 1991, 'Quaternary glaciations: theory and observations', in *The Sun In Time* ed. Sonett, C.P., Giampapa, M.S., and Matthews, M.S., University of Arizona Press, pp 511–542.

Haigh, J.: 2000, 'The Effect of UV Irradiance Variations on the EarthŠs Atmosphere', *Space Sci. Rev.*, this volume.

Lorenz, E.N.: 1963, 'Deterministic non-periodic flows', *J. Atmos. Sci.* **20**, 131–141.

Marsh, N. and Svensmark, H.: 2000, 'Cosmic Rays, Clouds, and Climate', *Space Sci. Rev.*, this volume.

Sparrow, C.T.: 1982, 'The Lorenz Equations: Bifurcations, Chaos and Strange Attractors', Springer.

Tinsley, B.: 2000, 'Influence of Solar Wind on the Global Electric Circuit, and Inferred Effects on Cloud Microphysics, Temperature, and Dynamics in the Troposphere', *Space Sci. Rev.*, this volume.

Tobias, S.M. and Weiss, N.O.: 1999, 'Resonant Interactions between solar activity and climate', *J. Climate*, submitted.

Tobias, S.M., Weiss, N.O., and Kirk, V.: 1995, 'Chaotically modulated stellar dynamos', *Mon. Not. Roy. Astron. Soc.* **273**, 1150–1166.

Van Loon, H., and Labitzke, K.: 2000, 'The Influence of the 11-Year Solar Cycle on the Stratosphere Below 30 km: a Review', *Space Sci. Rev.*, this volume.

Address for Offprints: S.M. Tobias and N.O. Weiss, Department of Applied Mathematics and Theoretical Physics, University of Cambridge, Silver St., Cambridge, U.K. smt11@damtp.cam.ac.uk now@damtp.cam.ac.uk

A LEAST-SQUARES SOLUTION FOR THE EFFECTIVE CONDUCTIVITY OF THE SOLAR CONVECTION ZONE

J.R. KUHN (kuhn@ifa.hawaii.edu)
Institute for Astronomy, 2680 Woodlawn, Honolulu, HI 96822

D. GEORGOBIANI (dali@pa.msu.edu)
Dept. Physics-Astronomy, Michigan State University, E. Lansing, MI, 48824

Received: 21 February 2000; Accepted in final form: 7 March 2000

Abstract. Here we show how realistic numerical simulations of solar convection can be parameterized with an effective thermal conductivity tensor. We show that this diffusive approximation yields an accurate statistical (in the sense of the χ^2 test) description of the thermal transport properties of a perturbed solar convection zone. This parameterization will allow more accurate large scale solar irradiance and luminosity calculations.

1. Introduction

The problem of heat transport in stellar convective zones has been successfully investigated for several decades using mixing length theory (MLT, cf. Bohm-Vitense, 1958). Although recent theoretical and experimental (cf. Castaing et al., 1989; Siggia, 1994) studies of heat transport in highly turbulent convective fluids imply that the equilibrium heat transport is likely to be a weak power law (e.g. 1/3 exponent) of the temperature gradient, apparently the more complex physical conditions of the Sun's convection zone do not allow this simple behavior. Nonetheless, it is interesting that deviations from the equilibrium convection zone stratification may be simply described. We show here how *perturbations* to the equilibrium convection zone heat transport are described by a diffusive model.

In regions with a high convective efficiency, and where temperature gradients are modest, the heat flux perturbations have been assumed to depend linearly on the temperature gradient: $\vec{q} = -\kappa \cdot \vec{\nabla} T$ (Landau and Lifshitz, 1987). In general, κ is a thermal conductivity tensor, $\kappa = C_p t \tau$ (cf. Kuhn, 1996), where C_p is the specific heat at constant pressure, t is a typical momentum-carrying lifetime of the fluid elements and τ represents the turbulent Reynolds stress tensor, $\tau^{ij} = \overline{\rho v^i v^j}$, with ρ the density and v^i, v^j velocity components in Cartesian directions. The overline indicates time and space (horizontal) averaging. In the mixing length formalism the total stellar flux $F_{\text{total}} = F_{\text{rad}} + F_{\text{conv}}$ is constant (F_{rad} and F_{conv} are radiative and convective fluxes, respectively) and proportional to the superadiabatic temperature gradient, $F_{\text{conv}} = -\kappa \nabla (T - T_{\text{AD}})$. In the convective zone, convective heat transport dominates, and we ignore the radiative flux: $F_{\text{total}} \simeq F_{\text{conv}}$. The convective flux in MLT can be computed from $F_{\text{conv}} = -\alpha\rho C_p \overline{v} H_p \nabla(T - T_{\text{AD}})$ (Cox and Giuli, 1968), where $\overline{v}$ is an average turbulent velocity, and $H_p = P/\rho g$ is the local pressure scale height. Conventional MLT assumes isotropic convection so that κ is the same for transport in vertical and horizontal directions and can be approximated by a depth-dependent, one-dimensional variable (e.g., Spruit, 1977) which we refer to as an effective conductivity.

A realistic physical picture of heat transport, consistent with convection simulations, is needed to understand how the large scale thermal structure seen at the photosphere (Kuhn et al., 1988) comes about. This is particularly troubling since the MLT (cf. Spruit, 1991 and references) predicts a large conductivity throughout the solar convection zone (SCZ) which increases rapidly with depth. In this simple picture the observed thermal variations at the photosphere would be impossible. It is also not clear that the MLT picture is consistent with the observed solar cycle changes in the solar luminosity and flux (Kuhn, 1996).

Another potential problem for the supposed high SCZ thermal conductivity is the tentative observation of bright rings around sunspots. In the basic MLT picture the excess heat blocked by a sunspot (70–75% of the total flux) is redistributed throughout the entire convection zone, to be reradiated at the photosphere over timescales of 10^5 years. Thus, no bright rings are expected. This may be in contrast with recent observations of sunspot bright rings (Rast et al. 1999).

Recent laboratory and numerical experiments (e.g., Kerr, 1996; Stein and Nordlund, 1998) provide new insight on driving mechanisms of convection. In stellar structures, convection is essentially driven by cooling of a thin top boundary layer. Cool, dense threads of fluid descend to the deeper layers, whereas the rest of a fluid slowly flows towards the top boundary replacing the mass in downflows. Descending threads, although turbulent, tend to maintain their identity over many scale heights. Hotter uprising fluid is structureless, isentropic and neutrally buoyant. The convective region is highly anisotropic and horizontal correlation lengths are significantly smaller than vertical ones. Therefore, instead of the popular picture of highly buoyant rising turbulent bubbles, convection consists of strong correlated downflows which penetrate the ambient hotter fluid.

In the light of this new physical picture of turbulent convection, attempts have been made to calculate how entropy inhomogeneities originating in the deep convective zone, affect the outer layers of a star (Kuhn and Stein, 1996; Kuhn, 1996; Georgobiani et al., 1997). On the basis of the Stein and Nordlund numerical code, an experiment has been performed with the incoming entropy flux at the bottom of the simulated volume. The flux was increased by 1% in one half of the lowest layer and decreased by the same amount in the other half. Then the perturbed model was evolved for several turnover times to allow it to relax. From the relative temperature differences between the "hot" and "cold" sides of the simulation volume one can see that the temperature difference peaks in the superadiabatic region, while deeper down, where the turbulent conductivity rapidly increases with depth, there is a decreasing temperature perturbation. Therefore, horizontal mixing is relatively ineffective. From the evolution of the entropy (or temperature) inhomogeneities, it is obvious that vertical heat transfer is significantly more efficient than in horizontal directions.

2. Numerical Model

To obtain a physically realistic picture of the solar convective zone, a three-dimensional hydrodynamic convection simulation has been utilized (cf. Stein and Nordlund, 1998; Stein and Nordlund, 1989; Nordlund and Stein, 1990 and references therein). The numerical code includes LTE radiative transfer. Top and bottom boundaries are transmitting, while horizontal boundaries are periodic. The computational domain represents a region of 6 Mm by 6 Mm horizontally, and its depth ranges from 0.5 Mm above the surface $\tau = 1$ to 2.5 Mm below this surface. While this is a relatively small fraction of the total convection zone depth (about 1%) the simulation volume includes the superadiabatic region and extends into the fully adiabatic convection zone. Three different spatial grid scales, $63 \times 63 \times 63$, $125 \times 125 \times 82$ and $253 \times 253 \times 163$, were used in order to check the influence of resolution on the results of the calculations.

We evolve convective fluid for several turnover times to let it thermally relax. Our calculations are performed for the whole simulation volume. However, the upper part of the volume (approximately 1 Mm) represents the photosphere and the superadiabatic region beneath it. Therefore, the most relevant part of our calculations follow from the lower 2 Mm, where convective energy transport dominates.

3. Calculations

The time dependence of the fluctuating temperature yields a dynamic estimate of the conductivity. We use a heat equation in the form:

$$\frac{\partial}{\partial x_i}\kappa_{ij}\frac{\partial T}{\partial x_j} = \rho C_p \frac{\partial T}{\partial t}, \tag{1}$$

where κ_{ij} are components of a conductivity tensor; ρ, C_p and T are, respectively, the density, specific heat and temperature. Components of κ_{ij} are, in general, coordinate- and time-dependent. Indices i, j run from 1 to 3, with $x_1 = x$, $x_2 = y$, $x_3 = z$ Cartesian coordinates. Summation over repeated indices is implied.

In our least-squares model, we represent the convection zone as a vertically stratified, but horizontally homogeneous medium (κ_{ij} are functions of z only, they do not depend on x and y). Under this assumption, onc can neglect horizontal derivatives of κ. Also, in our calculations $\kappa_{xx} = \kappa_{yy} \equiv \kappa_h$, $\kappa_{xy} = \kappa_{yx} \equiv \kappa_{hh}$, $\kappa_{xz} = \kappa_{yz} = \kappa_{zx} = \kappa_{zy} \equiv \kappa_{hv}$, and $\kappa_{zz} \equiv \kappa_v$.

The presence of significant non-diagonal terms in the conductivity matrix are not a natural feature of a simple convection model, since off-diagonal terms imply that heat transport occurs perpendicular to the temperature gradient. Clearly this can happen naturally over large scales, where off-diagonal terms are caused by rotation (e.g., solar meridional circulation, cf. Kuhn, 1996). Our model does not include large scale mean flows so that *a priori* we expect these terms to be relatively small and of the order of the *physical* uncertainty in the conductivity model. For generality we've retained these terms and shown that they do not change the statistical significance of our model. If they are ignored the remaining vertical and horizontal diagonal terms of the conductivity tensor are sufficient to study the anisotropy of the heat transport in the solar convective zone.

We write the terms of the heat equation, retaining off-diagonal terms for the purposes of comparison. Introducing some new notation for brevity:

$$\kappa_{xx}\frac{\partial^2 T}{\partial x^2} + \kappa_{yy}\frac{\partial^2 T}{\partial y^2} = \kappa_h\left(\frac{\partial^2 T}{\partial x^2} + \frac{\partial^2 T}{\partial y^2}\right) \equiv \kappa_h \nabla_h,$$

$$\kappa_{zz}\frac{\partial^2 T}{\partial z^2} \equiv \kappa_v \nabla_v,$$

$$\kappa_{xy}\frac{\partial^2 T}{\partial x \partial y} + \kappa_{yx}\frac{\partial^2 T}{\partial y \partial x} = \kappa_{hh}\left(\frac{\partial^2 T}{\partial x \partial y} + \frac{\partial^2 T}{\partial y \partial x}\right) \equiv \kappa_{hh} \nabla_{hh},$$

$$\kappa_{xz}\frac{\partial^2 T}{\partial x \partial z} + \kappa_{zx}\frac{\partial^2 T}{\partial z \partial x} + \kappa_{yz}\frac{\partial^2 T}{\partial y \partial z} + \kappa_{zy}\frac{\partial^2 T}{\partial z \partial y} =$$

$$= \kappa_{hv}\left(\frac{\partial^2 T}{\partial x \partial z} + \frac{\partial^2 T}{\partial z \partial x} + \frac{\partial^2 T}{\partial y \partial z} + \frac{\partial^2 T}{\partial z \partial y}\right) \equiv \kappa_{hv} \nabla_{hv},$$

$$\frac{\partial \kappa_{zz}}{\partial z}\frac{\partial T}{\partial z} = \frac{\kappa_{v,\ell+1} - \kappa_{v,\ell-1}}{z_{\ell+1} - z_{\ell-1}}\frac{\partial T}{\partial z} \equiv \left[\kappa_{v,\ell+1} - \kappa_{v,\ell-1}\right]\nabla_z,$$

$$\frac{\partial \kappa_{zx}}{\partial z}\frac{\partial T}{\partial x} + \frac{\partial \kappa_{zy}}{\partial z}\frac{\partial T}{\partial y} = \frac{\kappa_{hv,\ell+1} - \kappa_{hv,\ell-1}}{z_{\ell+1} - z_{\ell-1}}\left(\frac{\partial T}{\partial x} + \frac{\partial T}{\partial y}\right) \equiv \left[\kappa_{hv,\ell+1} - \kappa_{hv,\ell-1}\right]\nabla_x.$$

Here, ℓ represents the depth of the simulated volume and runs from 1 to N_z, where N_z is the number of grid points in the vertical direction. Each of the terms is written for ℓ-th level in depth, unless otherwise specified.

We solve the resulting equation numerically, applying least squares techniques. The numerical model is used to determine the temperature and its time variation which we attempt to describe using the equations above. Our model assumes that the conductivity components depend only on depth. We construct the function χ^2:

$$\chi^2 = \sum_{x,y,t}\left(\sum_{\ell}\left[\kappa_{h,\ell}\nabla_{h,\ell} + \kappa_{v,\ell}\nabla_{v,\ell} + \kappa_{hh,\ell}\nabla_{hh,\ell} + \kappa_{hv,\ell}\nabla_{hv,\ell} + \right.\right.$$

$$\left.\left.\left[\kappa_{v,\ell+1} - \kappa_{v,\ell-1}\right]\nabla_{z,\ell} + \left[\kappa_{hv,\ell+1} - \kappa_{hv,\ell-1}\right]\nabla_{x,\ell} - \nabla_{t,\ell}\right]^2\right). \qquad (2)$$

Here, $\nabla_t = \rho C_p \frac{\partial T}{\partial t}$. The outer summation is performed over the horizontal planes and approximately one hour of solar time. This particular centering requires κ_v and κ_{hv} one zone above the upper boundary as well as one zone beneath the lower boundary. To overcome this at the boundaries we take, $\kappa_{v,2} - \kappa_{v,0} = \kappa_{v,3} - \kappa_{v,1}$, and $\kappa_{v,N_z+1} - \kappa_{v,N_z-1} = \kappa_{v,N_z} - \kappa_{v,N_z-2}$. We use similar arguments for κ_{hv}.

We minimize χ^2 by differentiating it with respect to each of the components of the conductivity tensor, κ_v, κ_{hv}, κ_h and κ_{hh}, and then equating this expansion to zero. As a result, we get a system of linear equations. They establish the connection between the unknown components and some physical characteristics to be extracted from the simulation volume (ρ, C_p and various derivatives of T). Solving the system of linear equations by either constructing a quadratic matrix and inverting it, or constructing and solving a block - pentadiagonal matrix, we obtain values of these conductivity components. We have verified that both solutions are consistent, yielding essentially the same numerical answer. These calculations yield least squares estimates of the conductivity components as functions of density, specific heat at a constant pressure, as well as the temperature derivatives. The simulated convective zone is a self-consistent system, with all depths coupled through the thermal heat transport and the calculations were performed for the whole simulation volume.

One can simplify the problem by reducing the number of unknown conductivity components to κ_v and κ_h - the vertical and horizontal diagonal components of the conductivity tensor. This assumption is simple, yet it captures the important physics and allows us to understand the differences between the vertical and horizontal heat transport in the convective zone. The question arises, what is the statistical significance of this 2-component model? Does it properly represent Eulerian temperature perturbations? Does a model with off-diagonal terms yield a statistically significant improvement in the parametrization of the temperature behavior?

In order to answer these questions, we used a χ^2 test for goodness-of-fit for both 2-component and 4-component models. We construct functions χ_2^2 and χ_4^2 for each level ℓ (these are χ^2 per degree of freedom, cf. Bevington and Robinson, 1992):

$$\chi_i^2 = \frac{1}{N_x}\frac{1}{N_y}\frac{1}{N_t}\sum_{x,y,t}\left(\frac{1}{\sigma^2}[\nabla_t - \kappa\nabla^2 T]^2\right), \tag{3}$$

where i is either 2 or 4 (number of conductivity components in the model), $\sigma = \sqrt{\overline{\nabla_t^2} - \overline{\nabla_t}^2}$, N_x and N_y are total numbers of grid points in x and y directions, N_t is the number of time steps considered, and $\kappa\nabla^2 T$ can be rewritten as $\kappa_h\nabla_h + \kappa_v\nabla_v + [\kappa_{v,\ell+1} - \kappa_{v,\ell-1}]\nabla_z$ for a 2-component model or in the obvious more complex generalization for a 4-component model (cf. Equation (3)). Here, we use our calculated components of the conductivity to check how χ_2^2 and χ_4^2 compare and whether either of them is statistically significant and whether these equations represent the observed temperature fluctuations. The reduced χ^2 should be near unity if this model accurately describes the observed temperature fluctuations and correlations.

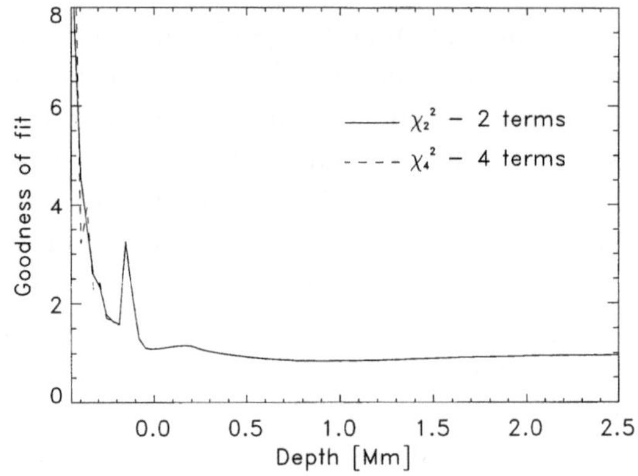

Figure 1. Test for goodness-of-fit: depth dependence of χ_2^2 and χ_4^2.

In fact figure 1 establishes this condition for most of the simulation volume for both the two and four-parameter models. Furthermore, χ^2 is larger than unity only in the uppermost layers, where we expect radiation to be important and our diffusive model to break down.

It is evident that there is no significant difference between χ^2 from the two and four-parameter models, and no empirical justification to include the off-diagonal terms to describe the turbulent conductivity. In the region of pure convection, χ_2^2 and χ_4^2 profiles coincide with each other. Therefore, having more terms in the heat equation does not yield a more accurate physical model of convective transport. Just 2 terms, κ_v and κ_h, are needed to describe heat transport properties of the convective zone. Since χ^2 is close to one in the convective region, the model captures the actual variations in the data, and therefore it is a statistically *complete* model of the correlated solar entropy fluctuations.

4. Results

The four components of the conductivity tensor have been calculated for the lowest resolution runs, $63 \times 63 \times 63$ whereas only two components were computed for the higher resolutions, $125 \times 125 \times 82$ and $253 \times 253 \times 163$. Figure 2 plots the two dominant components, κ_v and κ_h from the lower resolution calculations.

In all three cases the conductivity components are not physically descriptive in the region of pure radiation, as well as in the overshoot region, where the convective energy transport does not play an important role. The magnitudes of κ_v and κ_h vary from $10^{10} - 10^{11}$ at the top of the simulation box to about 10^{15} near the bottom (κ's are expressed in cgs units). Except near the lower boundary, κ_v always dominates over κ_h. The ratio κ_v/κ_h depends on the depth and may be affected by the resolution of the simulation. Since the numerical viscosity of the simulations also depends on the resolution this may not be surprising. For $63 \times 63 \times 63$ runs, it varies between

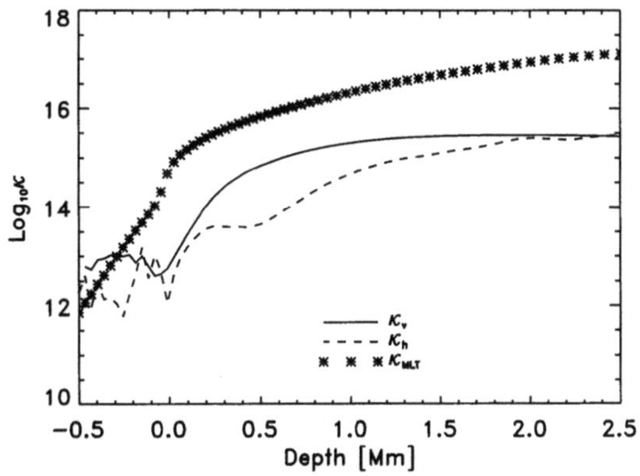

Figure 2. Depth-dependent profiles of the conductivity components for the lowest resolution $63 \times 63 \times 63$ (conductivity is expressed in cgs units).

1 and 15, with an average in the convective region of approximately 5. The average vertical to horizontal conductivity ratio varied between 2.3 and 5 between different simulation runs and resolutions. We believe these differences are largely caused by "realization" fluctuations due to the finite duration of the convection simulation, although we cannot rule out the possibility that the simulation resolution also affects the numerical value of the anisotropy.

Off-diagonal components of the conductivity tensor seem to be two orders of magnitude lower than diagonal terms. Through both the radiative zone and the superadiabatic region they are negative and do not exhibit any systematic pattern. In the convective region, κ_{hv} suddenly becomes positive, while absolute values of κ_{hh} and κ_{hv} are comparable to each other and gradually increase with depth. Close to the lower boundary, κ_{hv} becomes negative again and fluctuates; this could be a boundary effect. The measured off-diagonal components, while physically insignificant, appear to be a measure of the noise background in the least-squares calculations.

5. Conclusions

In contrast with MLT we find a smaller, anisotropic conductivity in the upper part of the solar convection zone. Since the simulation extends below the superadiabatic region it seems likely that this anisotropy persists into the deep convection zone although higher spatial dynamic range calculations may be needed to confirm this. The smaller, anisotropic, thermal diffusion observed here weakens the conclusion that photospheric heat flow perturbations (e.g. from sunspots) are coupled to the entire convection zone over long (10^5 year) timescales.

Evidently the MLT assumption of isotropic convective velocities is inadequate. This conclusion is consistent with our understanding that convection is driven by cool, rapidly descending, downflows which penetrate the ambient fluid in the vertical direction. Moreover, since the convective velocity is dominated by these plunging downflows, diffusive entropy transport over transverse length scales larger than the scale of the downdraft is impeded. This diminishes the average conductivity from the MLT values.

The numerical experiments imply a more complex picture for heat transport in the convection zone than the MLT. While a quantitative application of these results to solar observations has yet to be completed, some qualitative conclusions are apparent. It is possible for perturbations to the local heat flow (i.e. due to magnetic fields) to generate large scale thermal signatures in the solar convection zone or the photosphere. For example, the thermal pertubations due to fields near the base of the convection zone (Kuhn, 1996) or due to photospheric sunspots are not necessarily homogenized and mixed throughout the solar convection zone. It is interesting that observations of "hot" magnetic latitudes (Kuhn, 1996 and references therein) are consistent with these numerical results. In the future better constraints or direct measurements of sunspot bright rings (cf. Fowler et al., 1983) may provide additional insight into the curious heat transport properties of the solar convection zone. While the diffusion model presented here appears to capture the necessary physics while providing a tool for extending numerical insight in small-scale simulations to global solar problems, we look forward to sensitive photometric measurements which may directly test these conclusions.

References

Bevington, P. R. and Robinson, D. K.: 1992, *Data Reduction and Error Analysis for the Physical Sciences*, McGraw-Hill, New York.

Bohm-Vitense, E.: 1958, *Z. Astrophys.* **46**, 108.

Castaing, B. *et al.*: 1989, *J. Fluid Mech.* **204**, 1.

Cox, J. P. and Giuli, R. T.: 1968, *Principles of Stellar Structure*, Gordon and Breach, New York.

Fowler, L. A., Foukal, P. and Duvall, T.: 1983, *Solar Phys.* **84**, 33.

Georgobiani, D., Kuhn, J. R. and Stein, R. F.: 1997, in F. P. Pijpers *et al.* (eds.), *SCORe '96: Solar Convection and Oscillations and their Relationship*, Kluwer Academic Publishers, Dordrecht.

Kerr, R. M.: 1996, *J. Fluid Mech.* **310**, 139.

Kuhn, J. R.: 1996, in T. Roca Cortes (ed.), *VI IAC Winter School: The Structure of the Sun*, Cambridge University Press, Cambridge, p. 231.

Kuhn, J. R., Libbrecht, K. G., and Dicke, R. H.: 1988, *Science* **242**, 908.

Kuhn, J.R. and Stein, R. F.: 1996, *Astrophys. J.* **463**, L117.

Landau, L. D. and Lifshitz, E. M.: 1987, *Fluid Mechanics*, Pergamon Press, London.

Nordlund, A. and Stein, R. F.: 1990, *Comp. Phys. Comm.* **59**, 119.

Nye, A., Bruning, D., and Labonte, B. J.: 1988, *Solar Phys.* **115**, 251.

Rast, M. *et al.*: 1999, 'Observations of Sunspot Bright Rings', preprint.

Siggia, E.D., Ann. Rev. Fluid Mech., 26, 137.

Spruit, H. C.: 1977, *Solar Phys.* **55**, 3.

Spruit, H. C.: 1991, *The Sun in Time*, University of Arizona Press, Tucson, p. 118.

Spruit, H. C.: 1997, *Mem. Soc. Astron. It.* **68** (2), 397.

Stein, R. F. and Nordlund, A.: 1989, *Astrophys. J.* **342**, L95.

Stein, R. F. and Nordlund, A. 1998, *Astrophys. J.* **499**, 914.

Tikhonov, A. N. and Samarskii, A. A.: 1963, *Equations of Mathematical Physics*, Pergamon Press, London.

USING PRECISE SOLAR LIMB SHAPE MEASUREMENTS TO STUDY THE SOLAR CYCLE

J.R. KUHN (kuhn@ifa.hawaii.edu)
Institute for Astronomy, 2680 Woodlawn, Honolulu, HI 96822

L. FLOYD
Interferometrics, Inc. Chantilly, VA 20151

C. FRÖHLICH
Physikalisch-Meteorologishes Observatorium, Davos, Switzerland

J.M. PAP
Dept. Physics and Astronomy, UCLA, Los Angeles, CA, 90095

Received: 22 February 2000; Accepted in final form: 18 April 2000

Abstract. Despite 20 years of total solar irradiance measurements from space, the lack of high precision spatially resolved observations limits definitive answers to even simple questions like "Are the solar irradiance changes caused solely by magnetic fields perturbing the radiative flux at the photosphere?" More subtle questions like how the aspheric structure of the sun changes with the magnetic cycle are only now beginning to be addressed with new tools like p-mode helioseismology. Solar 5-min oscillation studies have yielded precise information on the mean radial interior solar structure and some knowledge about the rotational and thermal solar asphericity. Unfortunately this progress has not been enough to generate a self-consistent theory for why the solar irradiance and luminosity vary with the magnetic cycle. We need sharper tools to describe and understand the sun's global aspheric response to its internal dynamo, and we need to be able to measure the solar cycle manifestation of the magnetic cycle on entropy transport from the interior to the photosphere in much the same way that we study the fundamentally more complex problem of magnetic flux transport from the solar interior. A space experiment called the Solar Physics Explorer for Radius, Irradiance and Shape (SPHERIS) and in particular its Astrometric and Photometric Telescope (APT) component will accomplish these goals.

1. Overview

Why does the solar luminosity vary and could it change on human timescales by enough to affect terrestrial climate? We do not yet know the answer to this question because we do not know the physical mechanisms responsible for the solar cycle with sufficient clarity to predict even the well-observed global-scale solar magnetic field evolution. Our understanding of solar luminosity and irradiance evolution rests on a foundation built from many years of study based on correlative modeling of irradiance proxies (most accurately obtained from space) and only a few direct observations with the spatial resolution and high photometric precision needed to directly assess physical solar models. While this background to the luminosity

Space Science Reviews **94:** 169–176, 2000.

problem has been crucial, there are fundamental limitations to correlative studies where proxies are used for physically important observables. Such efforts *can not* generate answers to questions like, "Could the solar irradiance vary by more than 0.1% over decades or longer (centuries to interglacial) time scales?"

The essential goal of the SPHERIS investigation is to understand luminosity variations in the sun by tracing these changes from the base of the convection zone up through the photosphere during a three year period near solar minimum. Highly precise photometry will detect magnetically induced entropy fluctuations in the deep solar convection zone from their faint thermal shadows visible at the photosphere. SPHERIS achieves these goals by observing minute changes in the solar limb radius, shape, and brightness using the solar astrometric techniques recently demonstrated with SoHO/MDI. The SPHERIS experiment will have two instrument components: 1) The Astrometric and Photometric Telescope (APT), a new instrument concept briefly outlined here, and 2) a copy of the VIRGO instrument now operating on SoHO and providing highly accurate total and spectral irradiance measurements.

2. Key Questions

In greater detail, the key questions which SPHERIS will address are:

1. SOLAR RADIUS: Does the sun's radius vary over a timescale of a year at the level of 0.2 milliarcseconds and can these changes be used to understand how and where the solar luminosity is gated or stored in the radiative or convection zone?
2. SOLAR SHAPE: Do the solar quadrupole and hexadecapole shape terms vary with the solar cycle and are higher angular order shape terms of significant amplitude? How does the time evolution of the gravitational potential affect our understanding of the convection zone?
3. LATITUDE SURFACE BRIGHTNESS DEPENDENCE: Does the approximately 1 K solar latitudinal surface brightness variation change with the solar cycle as predicted by earlier helioseismic solar asphericity determinations? How is this change driven by the solar dynamo and how does it change the total and spectral irradiance?
4. LOCAL BRIGHTNESS CHANGES: On solar rotation timescales, do transient quiet photospheric surface brightness changes predict future photospheric magnetic structure, as some models suggest based on thermal or "antithermal" shadows caused by evolving magnetic structure from near the base of the convection zone?
5. SOLAR IRRADIANCE: What is the spectral distribution of total irradiance variability? How do the transient surface brightness changes affect the total and spectral solar irradiance? To what extent are short- and long-term irradiance

changes accounted for by surface magnetic activity, versus the global changes described in (1), (3) and (4) above?

6. ENERGY AND ENTROPY TRANSPORT: How do magnetic fields near the photosphere and in sunspots and faculae perturb the sun's emergent energy flux or affect the transport of entropy near the photosphere? Astrometric and photometric data of the Sun at this spatially resolved precision have never before been obtained, but are critical for addressing the physical mechanisms for entropy transport near the top of the magnetized convection zone.

2.1. Solar Radius

- Does the Sun's radius vary over a time scale of a year at the level of 0.2. milliarcseconds?
- Can these changes be used to understand how and where the solar luminosity is gated or stored in the radiative or convection zone?

While the sun's fundamental energy source is the nuclear conversion of H to He in the core, our immediate radiative energy source is the solar photosphere. Given long photon diffusion times in the core (10^6 yr), the emergent luminosity at its outer boundary is effectively constant on any total solar irradiance (TSI) measurement timescale. Given the observed variability, it follows that the mechanisms which deliver energy to the photosphere must be variable over the course of an 11 year solar cycle. Indeed, observations of the TSI, helioseismic studies, and precise solar photometric measurements all show that the sun varies globally, as well as locally at the photosphere, during an 11-year solar cycle.

Since the central luminosity remains constant but the solar output varies, there must be an intermediate reservoir. There are several possible mechanisms for storing energy during a solar cycle (e.g. gravitational or magnetic fields) but all will perturb the equilibrium stellar structure and change the radius (e.g. Lydon and Sofia, 1995; Sofia, 1998). Thus, a sensitive determination of the solar radius and irradiance can reveal the mechanism for any long-term irradiance changes. For example, solar luminosity changes which originate deep in the convection zone or below will cause fractional radius changes that are about 5×10^{-3} of the fractional luminosity change (on timescales less than 10^5 years). This change occurs on relatively short timescales because of the fast relaxation time of the superadiabatic zone. Perturbations which originate closer to the photosphere will cause larger relative radius changes (Dearborn and Blake, 1980).

Ground-based measurements of the solar radius exist over the last 300 years (e.g. Ribes *et al.*, 1991), however the results are controversial and inconsistent. Historical data show that the sun's radius may have been larger during the Maunder Minimum, which coincided with extremely cold periods in Europe and the Atlantic regions (Ribes *et al.*, 1991). These results are also suggested by the French CERGA radius measurements which found a larger solar radius during solar minimum (e.g. Laclare *et al.*, 1996). In contrast, Ulrich and Bertello (1995) found a positive

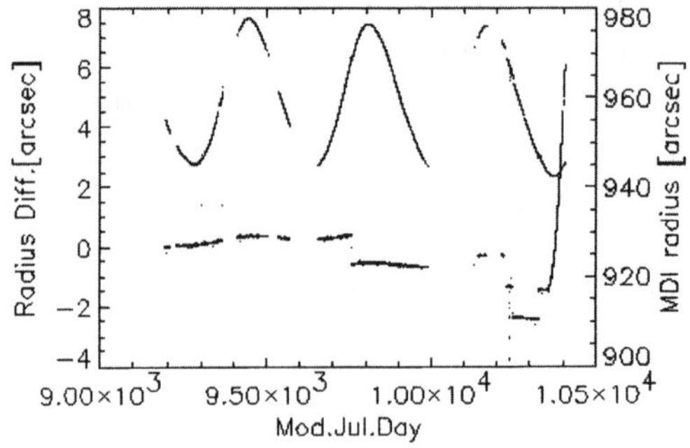

Figure 1. MDI solar radius measurements vs. modified Julian day number (JD - 2441000). The left hand scale shows the residual radius measurement (lower points) after the expected apparent solar radius (upper points and right hand scale) are subtracted.

correlation between apparent radius changes and the solar activity cycle. There are also hints of periodic solar radius variations over time scales of 1,000 days to 80 years (Gilliland, 1980), but the measurements are generally neither consistent nor conclusive (Parkinson *et al.*, 1980; Brown, 1987; Ribes *et al.*, 1991). These controversial results underscore the necessity of more sensitive efforts to measure the sun's radius. The limitation of ground-based measurements due to "seeing" from the Earth's atmosphere has also been recognized by Sofia and his collaborators in their development of the balloon borne Solar Disk Sextant (Sofia *et al.*, 1994).

The MDI experiment offered, for the first time, the promise of very accurate solar radius measurements – these could only be obtained from space. Figure 1 shows the residual MDI radius determination after correcting for the apparent variation due to the SoHO orbit. Residual instrumental errors due to a small yearly solar heating variation of the MDI front window, and larger "jumps" due to internal focus optics changes within the instrument are apparent. After correcting for these errors using an empirical thermal model for the front window, we recovered the radius measurements shown in Figure 2. The unmodeled linear residual radius change is about 0.008 arcseconds/year, but because of the MDI instrument limitations we must treat this as a systematic measurement uncertainty and an upper bound to possible cyclic or secular solar cycle variations. The MDI measurements represent the most sensitive observation of possible solar radius changes that have been obtained but they are not sensitive enough to see the solar cycle.

In contrast to MDI, APT has been specifically designed for radius (and shape) measurements and will achieve a precision at least an order of magnitude greater.

2.2. Solar Shape

– Do the solar quadrupole and hexadecapole shape terms vary with the solar cycle?

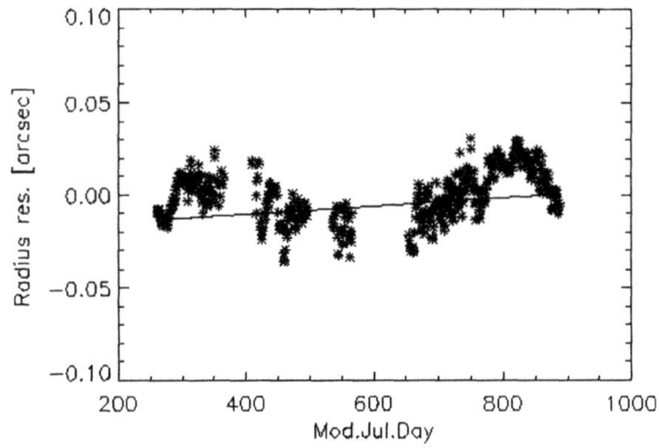

Figure 2. Derived solar radius variation after applying an empirical instrumental thermal model, and focus corrections. The residual solar radius variation (solid line) corresponds to 0.008 arcsec/yr, although it may be instrumental and non-solar.

- Are higher angular order shape terms of significant amplitude?
- How does the time evolution of the gravitational potential affect our understanding of the convection zone?

The theory of the solar limb shape is well established, although current solar shape measurements (also from MDI astrometry) are marginally inconsistent with the shape deduced from the helioseismic solar interior rotation determinations (Armstrong and Kuhn, 1999). This is an important problem since the limb shape is a sensitive function of the Sun's gravitational potential and its interior rotation. Thus, the measurement of solar cycle changes in the shape is another important tool for seeing how the solar magnetic cycle causes global changes in the solar interior stratification and dynamics. Only two high order shape experiments (Lydon and Sofia, 1996; Kuhn et al., 1998a) have been published, and they were not consistent. Since these experiments occurred at different phases of the solar cycle they are suggestive of temporal variability in the hexadecapole term. In order to reconcile the solar interior mass and rotation models with the observations, we require shape data which allow the possibility of measuring temporal solar cycle variations.

2.3. Solar Irradiance and Brightness Variations

- How do the transient surface brightness changes affect the total and spectral solar irradiance?
- To what extent are short- and long-term irradiance changes accounted for by surface magnetic activity, versus the global changes, such as radius and temperature changes?
- Does the approximately 1 K solar latitudinal surface brightness variation change with the solar cycle as predicted by earlier helioseismic solar asphericity determinations?

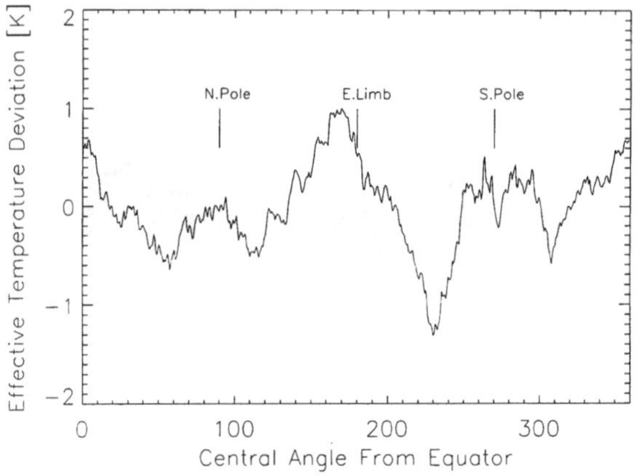

Figure 3. Latitudinal solar limb brightness variation.

- How is this brightness change driven by the solar dynamo?
- On solar rotation timescales, do transient quiet photospheric surface brightness changes predict future photospheric magnetic structure, as some models suggest based on thermal or "antithermal" shadows caused by evolving magnetic structure from near the base of the convection zone?

While considerable descriptive information exists on solar irradiance variations, we lack consistent physical models for the observations. For example, it has been shown that empirical irradiance models based solely on a "tally" of the surface manifestations of magnetic activity cannot explain all TSI changes (Fröhlich and Pap, 1989; Kuhn, 1996; Pap, 1997; Fröhlich *et al.*, 1997a; Wehrli *et al.*, 1998). Even the identification of the residual variability is a difficult problem since many mechanisms may affect the total irradiance and luminosity. These mechanisms range from photospheric temperature changes (Kuhn *et al.*, 1988), convective cells (Ribes *et al.*, 1985), large scale mixing flows (Fox and Sofia, 1994) to radius changes (Delache *et al.*, 1986; Ulrich and Bertello, 1995; Kuhn *et al.*, 1997; 1998a). Very high precision photometric and astrometric data with spatial resolution over a significant portion of the activity cycle are needed to disentangle these sources in the TSI variance.

Surface brightness measurements are often limited, not by instrumental photometric accuracy, but by the modulation of intrinsic spatial intensity gradients (e.g. limb darkening) into the brightness measurement through spatial, or registration, uncertainties in the photometric observation. To achieve high differential (or relative) photometric precision requires corresponding astrometric precision. Thus solar limb observations, which provide a fundamental fiducial when analyzed for both astrometric and photometric variations, can yield the most precise measurement of large scale solar differential surface brightness variations.

Recent precision limb brightness measurements from MDI are reproduced in Figure 3 (expressed as an effective temperature variation). These were obtained from MDI during a 10 hour period in March, 1997 when SoHO was rolled in increments of 30 degrees. These data were obtained near the minimum in the solar activity cycle and MDI magnetograms confirm that there were no sunspots and virtually no significant magnetic fields on the limb or disk. When seen with this photometric accuracy the solar limb data show dramatically that the surface brightness of the sun is not constant, and importantly, not determined *solely* by sunspots and faculae. One solar cycle earlier groundbased observations were just barely able to resolve this largescale latitudinal brightness variation using a full 4 month observing season (Kuhn *et al.*, 1985).

The temporal and spatial resolution of space experiments reveal new structural features of the solar convection zone. For example Fig 3 shows a striking dip in the temperature of the extreme south pole. This polar vortex was not visible at the north pole – presumably because it was tilted away from us at the time of these measurements.

3. The Astrometric and Photometric Telescope (APT)

The SPHERIS/APT instrument design is described in detail elsewhere (Kuhn et al. 1998b), but instrument design features that provide for highly accurate astrometry (radius sensitivity of 0.2 milliarcsec and shape coefficient accuracy of 0.1 milliarcsec) are summarized here:

1. Build a *monolithic optical structure* that maintains the primary and secondary mirrors in a rigid relative orientation using an optical and structural material with a temperature expansion coefficient of less than 100 parts per billion per degree C. *We use a simplified version of NASA's Gravity Probe B hydroxyl bonded astrometry telescope to achieve this with ULE (Ultra Low Expansion glass), Zerodur, or potentially, fused Silica.*
2. Carefully baffle radiative loads to the payload to achieve a temperature stability to the optics of better than 2 °C. *This temperature stability has been achieved in other payloads in similar orbits.*
3. Carefully control the temperature of the CCD detector to 0.1 oC and measure residual temperature changes of the detector substrate to 0.01 oC. *This allows a posteriori calibration of the radius change measurement to better than* 10^{-7}. *This detector stability has also been achieved with similar payloads.*
4. Determine the detector focus position accurately and detect focus changes during the mission. *SPHERIS/APT will periodically obtain focus calibration data by measuring small displacements of solar images formed from off-axis subapertures in the pupil plane. This "range-finder" mode will distinguish focus changes from true changes in the angular size of the solar image. For example, a detectible relative shift of 0.001 pixel between corresponding images obtained from opposite pupil masks is produced by a relative focal length*

change of one part in 10^7. *With 4 pupil masks low order optical distortion changes can also be detected.*

4. Conclusions

High photometric dynamic range requires comparable astrometric precision. A space experiment is our best hope for achieving both requirements for solar observations. A satellite experiment, comparable to SPHERIS, is needed to see entropy "shadows" or "antishadows" of magnetic changes occurring at the top of the solar interior radiative zone. Models which explain how the solar luminosity is gated by interior magnetic fields as it diffuses and convects to the photosphere will be directly tested by a solar limb photometry experiment.

References

Dearborn, D. S. P. and Blake, J. B.: 1980, *Astrophys. J.* **237**, 616.

Delache, Ph., Laclare, F. and Sadsaoud, H.: 1986, in J. Christensen-Dalsgaard and S. Frandsen (eds.), *Advance in Helio- and Asteroseismology*, IAU Press, p. 223.

Fox, P. and Sofia, S.: 1994, in J. Pap, C. Fröhlich, H. S. Hudson and S. K. Solanki (eds.), *The Sun as a Variable Star*, Cambridge University Press, Cambridge, p. 280.

Fröhlich, C. and Pap, J.: 1989, *Astron. Astrophys.* **220**, 272.

Fröhlich, C. *et al.*: 1997a, *Solar Phys.* **170**, 1.

Gililand, R. L.: 1980, *Astrophys. J.* **248**, 1144.

Kuhn, J. R.: 1996, in T. Roca Cortes (ed.), *Global Changes in the Sun, in the VI Winter School at Instituto d'Astrophysica de Canarias*, Cambridge University Press, Cambridge, p. 231.

Kuhn, J. R., Libbrecht, K. G., and Dicke, R. H.: 1985, *Astrophys. J.* **290**, 758.

Kuhn, J., Bogart, R., Bush, R., Sa, L., Scherrer, P., and Scheick, X.: 1997, in J. Provost and F.-X. Schmieder (eds.), 'Sounding Solar and Stellar Interiors', *IAU Symp.* **181**, 103.

Kuhn, J.R., Bush, R., Scheick, X., and Scherrer, P.: 1998a, *Nature* **392**, 155.

Kuhn, J. R. *et al.*: 1998b, in C. Korendyke (ed.), *SPIE '98, Missions to the Sun II*, San Diego, CA, p. 203.

Kuhn, J. R., Armstrong, J. D., Bush, R. I., and Scherrer, P.: 2000, *Nature* (in press).

Laclare, F., Delmas, C., Coin, J.P., and Irbah, A.: 1996, *Solar Phys.* **166**, 211.

Lydon, T. J. and Sofia, S.: 1995, *Astrophys. J. Suppl.* **101**, 357.

Pap, J.: 1997, in C. Castagnoli and A. Provenzale (eds.), *Past and Present Variability of the Solar-Terrestrial System: Measurement, Data Analysis and Theoretical Models*, Enrico Fermi Summer Workshop, Italian Physical Society, p. 1.

Parkinson, J. H., Morrison, L. V., and Stephenson, F. R.: 1980, *Nature* **288**, 548.

Ribes, E., Beardsley, B., Brown, T. M., Delache, Ph., Laclare, F., Kuhn, J. R., and Leister, N. V.: 1991, in C. P. Sonett, M. S. Giampapa, and M. S. Matthews (eds.), *The Sun in Time*, Arizona University Press., Tucson, p. 59.

Sofia, S.: 1998, in J. M. Pap, C. Fröhlich, and R.K. Ulrich (eds.), *Solar Electromagnetic Radiation Study for Solar Cycle 22*, Kluwer Academic Publishers, Dordrecht, Holland, p. 413.

Sofia, S., Heaps, W., and Twigg, L. W.: 1994, *Astrophys. J.* **427**, 1048.

Ulrich, R. and Bertello, E.: 1995, *Nature* **377**, 214.

Wehrli, C., Appourchaux, T., Crommelynck, D., Finsterle, W., Fröhlich, C., and Pap, J.: 1998, in F. J. Provost and F. X. Schmieder (eds.), 'Sounding Solar and Stellar Interiors Poster Volume, OCAN, Nice', *IAU Symp.* **181**, 209.

PHYSICAL CAUSES OF SOLAR VARIABILITY

Discussion Session 1b

J.R. KUHN (kuhn@ifa.hawaii.edu)
Institute for Astronomy, University of Hawaii
2680 Woodlawn Dr, Honolulu, HI, 96822, USA

M. SCHÜSSLER (schuessler@linmpi.mpg.de)
Max-Planck-Institut für Aeronomie
Max-Planck-Str. 2, 37191 Katlenburg-Lindau, Germany

Received: 8 March 2000; Accepted in final form: 19 April 2000

Abstract. This report is divided into three parts: Section 1 gives a short introduction and a summary of the topics discussed. Section 2 is a position statement by J. Kuhn on the interpretation of the irradiance measurements, while Section 3 gives a position statement by M. Schüssler discussing observations of stars that could be useful for understanding solar variability.

1. Discussion Summary

There is a considerable amount of data indicating the variability of solar magnetic activity on time scales ranging from days (formation or limb appearance of sunspots and active regions) over decades (sunspot cycle and its amplitude changes, grand minima) to millennia (cosmogenic isotopes). Satellite measurements in the last two decades have shown that solar irradiance varies together with solar magnetic activity. This applies to the passage of sunspot groups as well as to the 11-year cycle. While the (negative) contribution of sunspots to the irradiance changes is generally accepted, it is still debated whether the positive correlation between activity and irradiance on the 11-year timescale is due to small-scale surface magnetism or to large-scale variations in the convection zone, or possibly to a combination of both. Kuhn (see his position statement in Sec. 2) argued that he sees various observational and numerical results to be in conflict with the diffusive model for heat conduction in the convection zone as presented by Spruit (2000). Spruit replied that the absence of a significant global contribution is not restricted to the traditional diffusive model but also applies to the 'new paradigm' of stellar convection as being driven by strong, coherent downflows created by surface cooling.

Another topic of the discussion was the long-term variability of solar magnetic activity, in particular the origin of 'grand minima' like the Maunder minimum in the second half of the 17th century. Existing theoretical models for grand minima are based either on nonlinear effects (deterministic models, see the contribution by Weiss and Tobias, 2000) or on stochastic fluctuations interfering with the basic dynamo process (e.g., Schmitt et al., 1996; Ossendrijver and Hoyng, 1996). The presently available data seem insufficient to decide between the two possibilities. There are indications for some regularity in the occurrence of grand minima (a 205-yr signal in the ^{10}Be data, see the contribution by Beer, 2000), which would favour deterministic models; however, existing models of that kind show a behaviour that appears far more regular than the data. Stochastic models, on the other hand, can easily account for the irregularity of the data as well as for the phase-amplitude relation of the solar cycle (e.g., Ossendrijver et al., 1996) including phase jumps, one of which may have taken place during the 'Dalton minimum' around 1800.

Finally, there was some discussion about the doubling of the interplanetary magnetic field strength during the 20th century inferred from geomagnetic fluctuations by Lockwood et al. (1999). Such a variation could possibly have profound consequences for the flux of galactic cosmic rays (cf. Bazilevskaya, 2000), which, in turn, may affect cloud formation (Tinsley, 2000). The relationship of the interplanetary field to the photospheric magnetic field of the Sun is far from obvious since it is only the 'open' flux that is connected to interplanetary space. The variation of the interplanetary flux can be understood in terms of the simultaneous overall increase of solar activity since 1900, provided that part of the magnetic flux emerging in the photosphere resides there for about 4 years (Solanki, Fligge and Schüssler, in preparation).

2. On the Interpretation of Solar Variability Position Statement by J. R. Kuhn

The presentation by Spruit (2000) emphasizes how simple mixing length models lead to the conclusion that solar entropy changes are coupled to the entire convection zone and must not lead to significant solar luminosity perturbations on solar cycle timescales – here I summarize and interpret the data which contradict this theoretical *ansatz*.

2.1. Key Observational Measurements

The presentation focused on the following observational questions and results as irradiance and luminosity model discriminators:

- Resolved limb photometry measurements indicate that the solar luminosity is changing with the magnetic cycle, approximately in phase with the solar irradiance variations and with a relative amplitude of 10^{-3} (*not* 10^{-6}),
- Recent SOHO/MDI observations confirm the latitudinal surface brightness variation seen in earlier limb observations, while simultaneous magnetic field observations established that facular and sunspot magnetic fields are *not* responsible for the latitudinal effective temperature changes,

- The irradiance contribution of surface magnetic features are not simply a function of the center-to-limb angle, but are dependent on the magnetic field strength, to the extent that facular irradiance contributions can be positive *and negative*. This effect has been seen in precise photometric solar images (cf. Kuhn et al., 1999).

2.2. Interpreting Photometry, Helioseismology, and Numerical Results

Two problems with existing one-dimensional diffusive models of entropy perturbations in the solar convection zone were presented which explain their failure to account for the photometric and helioseismic observations:

1. Advective transport of magnetic flux and entropy (in a diffusive model) is highly anisotropic in the convection zone. This is apparent simply by noting that solar cycle magnetic fields are confined to the active latitudes. It has also been demonstrated in numerical simulations where entropy transport is observed to be strongly anisotropic (Kuhn, 2000).
2. The turbulent pressure is critical to understanding how entropy transport perturbs the mean solar stratification. Models which do not include this (e.g. Goldreich et al., 1991) cannot consistently describe the solar irradiance fluctuations, while accounting for helioseismic frequency changes. It was argued that numerical studies also establish this point.

Precise photometry (hopefully from space), continued helioseismological solar cycle studies and numerical convection experiments are important for extending the understanding of convection zone transport properties beyond simple mixing length theories. Without these tools we cannot proceed beyond the present inconsistent theoretical models of irradiance and p-mode frequency fluctuations.

3. What Can We Learn from Observing Other Stars? Position Statement by M. Schüssler

For the Sun we have reliable data of total irradiance variations for about two 11-year cycles. We know that the magnetic activity varies on longer time scales, so what does this imply for the variability of total irradiance, UV flux, and cosmic rays with their potential relevance for the climate? Do we have to expect changes in the variation amplitudes and secular trends over longer times? Is the present state of the Sun typical? Can we identify total irradiance variations with luminosity variations?

One possibility to obtain information to address these questions is to observe other stars that are similar to the Sun, namely cool stars with outer convection zones. The most widely used indicator of magnetic activity is the emission in the cores of the H and K lines of ionized calcium, which is well correlated with the amount of magnetic flux in the stellar atmosphere. More than 100 stars have been regularly monitored for more than three decades now and an extensive data base has been accumulated (Baliunas et al., 1995). Key results are:

1. Young, rapidly-rotating stars show a high level of activity, which varies strongly and irregularly. Older, slowly-rotating stars show solar-like activity levels and activity cycles with periods between two and over 20 years. Some stars show little variability and very low activity; these could be stars going through a 'grand minimum'. At least one star has been observed which seems to move from cyclic activity into a non-cycling state.
2. Extremely active stars (close binaries and very young stars) can show brightness variations of more than 10% caused by huge 'starspots' on their rotating surfaces. The relation between brightness and magnetic activity depends on the activity level (Radick et al., 1998): stars more active than the Sun are darker when more active (spot darkening wins), while stars of solar activity level or below become brighter at activity maximum (facular brightening wins). For the Sun, both effects nearly cancel; the ratio between facular area and spot area decreases with the overall activity level (Foukal, 1994; see also Radick, 1994), so that sunspot darkening would probably become dominant should the average activity level of the Sun double.
3. The cyclic photometric variability of the Sun is a factor 2–4 smaller than the variability of stars with a comparable level of magnetic activity (Radick et al., 1998). The Sun has an average level of activity, but a much smaller photometric variability than comparable stars. This could in principle be due to the the fact that the Earth's orbit is not far from the solar equatorial plane and that the active regions appear in low heliolatitudes (Schatten, 1993). Since the facular brightness is maximum near the solar limb, an observer over a pole of the Sun would see a stronger cyclic irradiance variation (if indeed the faculae are responsible for the brightening of the Sun during activity maximum). The measurement of cyclic solar irradiance variations from a polar vantage point is necessary in order to decide this issue. This is a key observation with regard to the question whether the irradiance variations of the present Sun are unusually small among the stars.
4. For stars with cycles, the activity level is anticorrelated with the cycle length (Baliunas and Soon, 1995). A similar kind of relationship is also indicated for the Sun (Ossendrijver and Hoyng, 1996).

As a consequence, we cannot necessarily count on the Sun staying as well-behaved as it seems to have been in the last two decades. Unless the reality of the inclination effect (see point 3 above) has been confirmed, there remains the possibility that the present solar variability is unusually weak and could possibly increase to higher 'normal' levels in the future (or has done so in the past). The amplitude of long-term secular trends in the irradiance is likewise uncertain. Even a reversal in the correlation between activity and irradiance cannot be ruled out should the Sun become much more active than today. In summary, we can say that solar-type stars seem to be capable to a richer variety of behaviour than we would imagine from the rather limited available record of systematic solar observations. So the Sun must be kept under permanent surveillance while, at the same time, it is important to continue the monitoring of cool stars in order to extend our basis of stellar data.

References

Baliunas, S., et al.: 1995, *ApJS* **438**, 269.
Baliunas, S., and Soon, W.: 1995, *ApJ* **450**, 896.
Bazilevskaya, G.: 2000, *Space Sci. Rev.,* this volume.
Beer, J.: 2000, *Space Sci. Rev.,* this volume.
Foukal, P.: 1994, *Science* **264**, 238.
Goldreich, P., Murray, N., Willette, G., and Kumar, P.: 1991, *ApJ* **370**, 752.
Kuhn, J. R., Lin, H., and Coulter, R.: 1999, *Adv. Space Res.* **24**, (2)185.
Kuhn, J. R.: 2000, *Space Sci. Rev.,* this volume.
Lockwood, M., Stamper, R., and Wild, M. N.: 1999, *Nature* **399**, 437.
Ossendrijver, A. J. H., and Hoyng, P.: 1996, *A&A* **313**, 959.
Ossendrijver, A. J. H., Hoyng, P., and Schmitt, D.: 1996, *A&A* **313**, 938.
Radick, R. R.: 1994, Science, 266, 1072
Radick, R. R., Lockwood, G. W., Skiff, B. A., and Baliunas, S.: 1998, *ApJS* **118**, 239.
Schatten, K. H.: 1993, *J. Geophys. Res.* **98**, 18,907.
Schmitt, D., Schüssler, M., and Ferriz-Mas, A.: 1996, *A&A* **311**, L1.
Spruit, H.: 2000, *Space Sci. Rev.,* this volume.
Tinsley, B. A.: 2000, *Space Sci. Rev.,* this volume.
Weiss, N. O., and Tobias, S. M.: 2000, *Space Sci. Rev.,* this volume.

II: . . . INFLUENCES . . .

THE INFLUENCE OF TOTAL SOLAR IRRADIANCE ON CLIMATE

U. CUBASCH and R. VOSS
Max-Planck-Institut für Meteorologie, Hamburg, Germany

Received: 8 December 1999; Accepted in final form: 7 March 2000

Abstract. To estimate the effect of the solar variability on the climate, two estimates of the solar intensity variations during the last three centuries have been used as forcing in numerical simulations. The model employed to carry out the experiments was the same coupled global ocean-atmosphere model used in a number of studies to assess the effect of the anthropogenic greenhouse gases on climate. The near surface temperature and the tropospheric temperature distribution shows a clear response to the variability of the solar input. Even the thermohaline circulation reacts on the large amplitudes in the forcing. In the stratosphere, the response pattern is similar as in the observations, however, the 11-year cycle found in the forcing data does not excite an appreciable response. This might be due to the missing parameterisation of the increase in the UV-radiation at the solar cycle maximum and the connected increase of the stratospheric ozone concentration.

1. Introduction

The interest in the role of the Sun in climate change has been revived by the concern and uncertainty surrounding the question about the anthropogenic influence on the climate. The focus of these studies is more towards Earth's surface and its response.

Because the global change is expected to have an influence on a century rather than on a decadal time scale, evidence has been sought and accumulated for some level of a significant effect of solar variability on this time scale (e.g., Wigley and Kelly, 1990; Friis-Christensen and Lassen, 1991; Reid, 1991; Hoyt and Schatten, 1993; Lean *et al.*, 1995a; Lean, 1997; Crowley and Kim, 1993, 1996; Cubasch *et al.*, 1997).

Research interest has also been heightened by new estimates from models of potential changes in solar radiation of 0.24 – 0.30% on centennial time scales (Hoyt and Schatten, 1993; Lean *et al.*, 1995a; Lean, 1997), which lead to a variability of the solar radiation absorbed by the surface/troposphere system of about 0.5 to 0.75 W/m^2 (Cubasch *et al.*, 1997). This number should be compared to the best available estimate of the radiative forcing by increase of anthropogenic greenhouse gases from pre-industrial (1880s) to present-day concentrations of 2.4 W/m^2 (IPCC, 1996).

Calculations with energy balance models (Wigley and Raper, 1990; Reid, 1991; Crowley and Kim, 1996) and 3-dimensional coupled ocean-atmosphere models (Wetherald and Manabe, 1975; Rind and Overpeck, 1993; Cubasch *et al.*, 1997) suggest that such relatively small changes could still cause surface temperature changes on the order of several tenths of a degree centigrade.

One of the largest effects of solar variability on the climate is found in the observations in the higher atmosphere by van Loon and Labitzke (1998, 1999). Their results, based on time series starting from 1958 and the NCEP-reanalysis data, show a consistent response pattern in all months of the year in the geopotential heights in the lower stratosphere (16 to 30 km) south of about 50° N, which to rises and falls with the 11-year sunspot cycle. The highest correlations move with the seasons meridionally.

In the present paper, after discussing the model and the experimental set-up (section 2), the simulations, which have been made with an estimate of the solar input (Lean *et al.*, 1995a) (subsequently referred to as LEAN), are compared to the studies published by Cubasch *et al.* (1997), where the estimate of Hoyt and Schatten (1993) (subsequently referred to as H&S) had been used (section 3). This analysis focuses on the near surface temperature (section 3.1) with special emphasis on historical times (section 3.1.1) and the recent 100 years (section 3.1.2). The impact of the solar variability on the ocean is investigated (section 3.2). The stratospheric results from the simulations are compared to the observations of van Loon and Labitzke (1998, 1999) in section 3.3.

2. Model and Experiment Design

In this paper experiments, which use the Lean *et al.* (1995a) radiative forcing are analysed and compared to the ones discussed by Cubasch *et al.* (1997) and Hegerl *et al.* (1997). All simulations were carried out using the ECHAM3/LSG-version (Voss *et al.*, 1998) of the ocean-atmosphere general circulation model (O-AGCM). This model has a horizontal resolution of approximately 5.6° (T21), 19 levels in the atmosphere, and 11 levels in the ocean. The model includes the diurnal cycle and comprehensive physical parameterisations (Roeckner *et al.*, 1992; Maier-Reimer *et al.*, 1993, DKRZ, 1993).

The atmospheric component of the model uses hybrid levels for the vertical discretization, i. e. near the surface the levels are terrain following sigma-levels, while in the higher atmosphere p-levels are used. A function describes the transition between sigma- and pressure-levels. The top level is located at 10 hPa, the next one at 30 hPa and the following one at 50 hPa. The radiative transfer model treats the solar radiation in four discrete bands (DKRZ, 1993), given in Table 1.

The spectral bands are too coarse to resolve shifts in the spectral bands connected to the changes in the solar radiation. The interaction of the solar radiation with the stratospheric ozone cannot be calculated interactively, i. e. the ozone concentration is prescribed.

The experiments discussed in the present study consist of a total of five simulations shown in Table II:

TABLE I
Spectral intervals and absorbers of the radiative transfer model

Absorber	O_3	H_2O	H_2O	H_2O, CO_2
Spectral Interval (μm)	0.215 – 0.685	0.685 – 0.891	0.891 – 1.273	1.273 – 3.580

TABLE II
The experiments

Experiment	H&S 1, 2	LEAN 1, 2	H&S + CO_2	CTRL
Solar Radiance	H&S	LEAN	H&S	1365 W/m^2
CO_2	constant	constant	observed	constant
Simulations	2	2	1	1
Simulated Period	1700-1992	1700-1993	1880-1992	1000 years

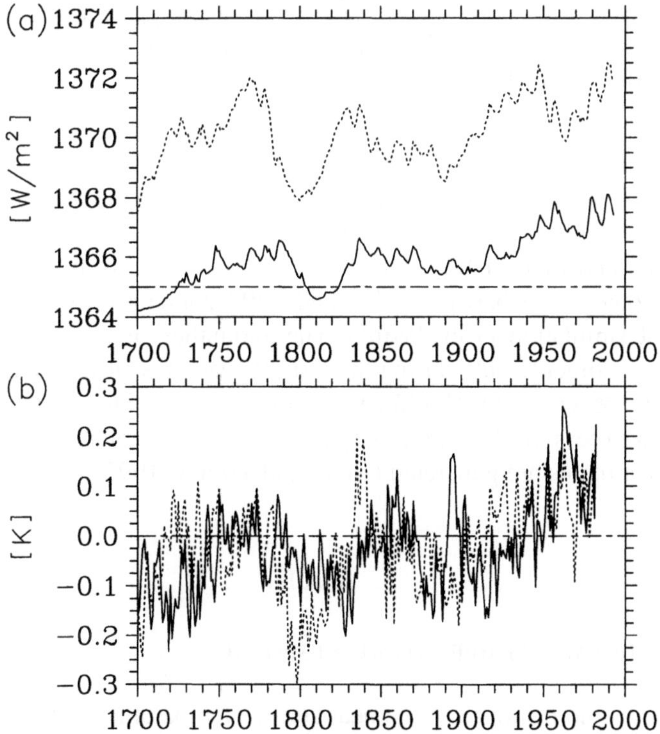

Figure 1. a) The solar irradiance after Hoyt and Schatten (1993) (dotted, H&S) and after Lean *et al.* (1995a)(solid line, LEAN), the dashed line indicates the fixed solar irradiance for the control simulation, b) the global mean near-surface temperature response to the solar variability averaged over the two simulations with H&S forcing and with the LEAN forcing, respectively.

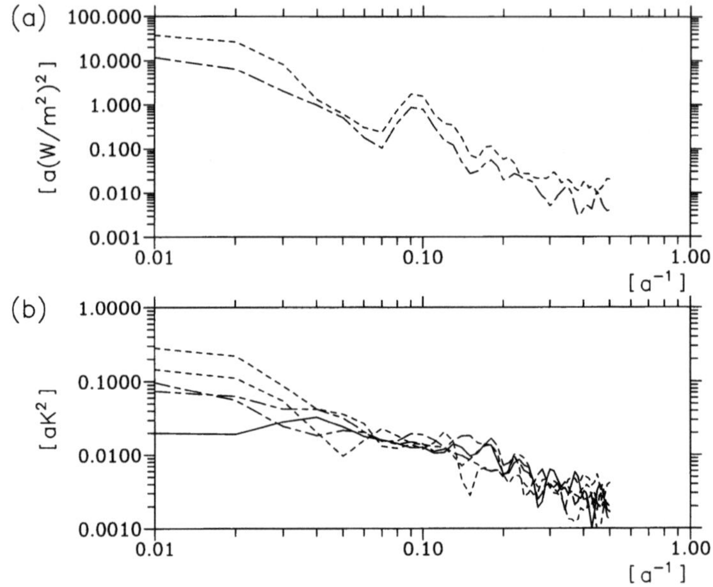

Figure 2. a) The spectrum of the solar irradiance after Hoyt and Schatten (1993) and after Lean *et al.* (1995b). b) The spectrum of the global mean near surface temperature change for the four solar experiments and the control simulation. The spectra are calculated by means of the autocovariance function using a Hanning-window with a maximum lag of 100 years.

Figure 1a shows the estimates of the solar variability since 1700 by Hoyt and Schatten (1993) as well as from Lean *et al.* (1995a). Interestingly, not even the absolute values agree due to a calibration problem of the satellite measuring devices (Fröhlich and Lean, 1998). Therefore, both curves have been normalised so that their average value between 1700 and 1992 coincides with the value used in the control simulation (1365 W/m^2). The anomalies of both estimates of the solar irradiance coincide only during the last 20 years, because here they are fixed to the satellite observations (Fröhlich and Lean, 1998). During the other years phase shifts in excess of 20 years emerge, which can be explained by the different methods the estimates are generated (Kelly and Wigley, 1992).

3. Results

3.1. Near-Surface Temperature Response

Figure 1b shows the response of the near surface temperature to the varying solar forcing. The curves show the average over the respective 2 simulations. As mentioned in Cubasch *et al.* (1997), the surface temperature follows the results of H&S quite closely with a steep drop in temperature around 1800 and a continuous rise during the recent 100 years. The response to the LEAN forcing is not quite as

TABLE III

The correlation between the two solar intensity variations (S), the observations (T Mann) and the response in the experiments (T LEAN, T H&S) averaged over the two realisations for the same forcing, and averaged over all realisations (T H&S + LEAN), for the northern hemisphere during the period 1700-1900.

	T Mann	T H&S	T LEAN	T H&S + LEAN	S H&S	S LEAN
T Mann	1.00	0.40	0.27	0.40	0.42	0.11*
T H&S		1.00	0.46	0.92	0.87	0.69
T LEAN			1.00	0.76	0.23	0.65
T H&S+LEAN				1.00	0.73	0.78
S H&S					1.00	0.37
S LEAN						1.00

*indicates a non significant correlation at the 95% level. The time series have been smoothed by a 15 year running mean.

strong around 1800 (their forcing has a smaller amplitude), but has the warming trend during the last 100 years as well. Generally, the response appears in the longer periods, while the 11 year cycle, which is clearly visible in the forcing data, cannot be found. A spectral analysis endorses this statement (Figure 2). It is interesting to note that the solar input enhances those long periods, which are under-represented in the control simulation (Stouffer *et al.*, 1999). This might be an indication that some variable external forcing is needed to improve the model performance in these spectral bands.

3.1.1. *During Historical Times*

During the years 1700 to 1900, which were chosen, because during this time-period the anthropogenic impact on climate appears to be negligible, the different simulations have been compared to the temperature reconstructions for the northern hemisphere provided by Mann *et al.* (1998). All the four simulations (as well as the observations) represent one possible realisation of the near surface temperature Voss and Cubasch (2000). If one assumes that these 4 experiments are just random samples (Monte Carlo approach, Cubasch *et al.* (1994)), and that they have a Gaussian distribution, they can be averaged to obtain the best representation of the observation. The correlation matrix (Table III) shows that the correlation with the observations of the average over all experiments is as good as the average over the H&S experiments, and better than the individual H&S experiments. The mean of the experiments with one forcing is hardly a better representation of the mean of the experiments with the other forcing than of the observations. The averaged experiments correlate generally rather well with their forcing, while the two forcings, LEAN and H&S, correlate only to 37%.

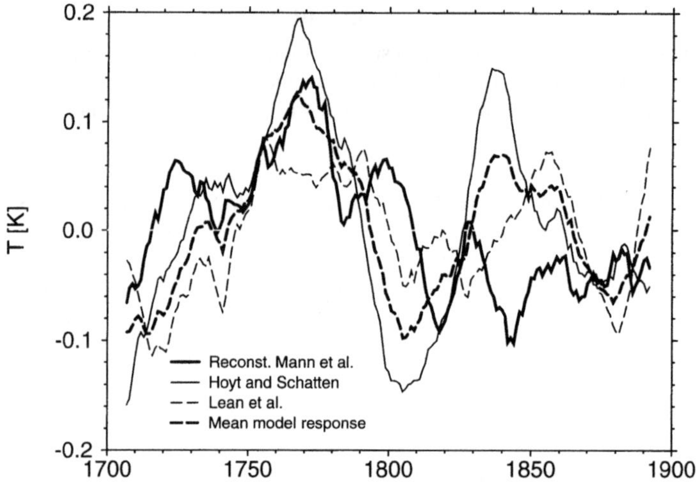

Figure 3. The reconstructed temperature change for the northern hemisphere (bold solid) from 1700 to 1900 after Mann *et al.* (1998), the average of the H&S (thin solid) and the LEAN (thin dashed) experiments separately, and the average over all four experiments (bold dashed). The long-term mean from each time series is subtracted and a 15-year running mean is applied.

The average of the two H&S-experiments (Figure 3) show that the warming starting from 1700, peaking at around 1770 and cooling until about 1820 is well simulated with an exaggeration of the maximum at 1770, while the LEAN-experiments show a similar behaviour, but with a lower amplitude. As a consequence, the mean of all experiments represents the temperature reconstruction quite well. During the second half of the 19th century the reconstruction displays a cooling, while all model experiment simulate a warming. It appears that during the second half of the 19th century the amplitude of the solar forcing variation is too small to trigger an unequivocal response in the model.

3.1.2. *During the Last 100 Years*

The global mean temperature change during the last 100 years can be found in Figure 4. During this period a linear trend can be fitted, which amounts to 0.19 K for the H&S experiments and 0.17 K for the LEAN experiments. This is well below the observed trend seen in the observations of 0.55 K (Jones and Briffa, 1992; Jones, 1994; Parker *et al.*, 1994). A set of 3 experiments (Paeth *et al.*, 1999), where the greenhouse gases where prescribed according to observations, but without the influence of the Sun, simulate a linear trend of 0.43 K. An experiment with greenhouse gases increase and solar irradiance according to H&S shows a linear trend of 0.60 K. It therefore gives the most realistic estimate. Its slight overestimation of the warming might be caused by the missing representation of the aerosol effect in these simulations (Cubasch *et al.*, 1996).

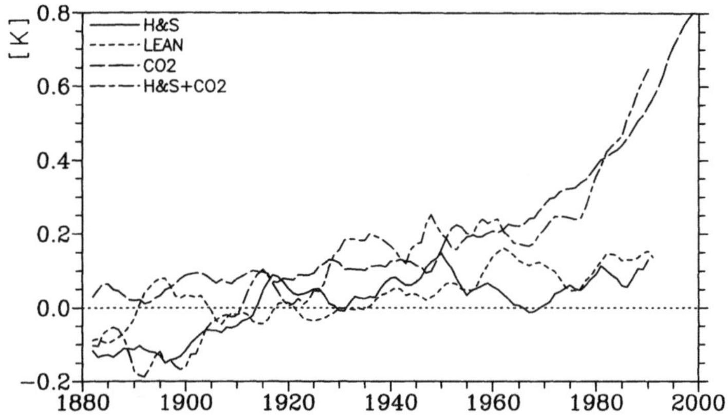

Figure 4. The simulated globally averaged near-surface temperature change from 1880 to 1992 for the solar experiments (H&S, LEAN), as well as for an ensemble of greenhouse gas simulations (CO_2, after Paeth *et al.*, 1999) and for a greenhouse gas plus solar forcing experiment (H&S+CO_2, after Voss and Cubasch, 2000). A 5-year running mean filter is applied.

3.2. Oceanic Response

Figure 5 shows the maximum of the meridional stream function in the North Atlantic (THC). In the H&S experiments the THC responds with a delay of about 30 years to the large drop during the 18th century with an increase, which is in accordance to the findings of Manabe *et al.* (1991) who find a similar response if they decrease the CO_2-concentration in the atmosphere (i.e. they decrease the atmospheric temperature). During the last 100 years the model responds to the warming due to the solar input with an decrease of the overturning intensity. This is similar to the response of most O-AGCMs (Rahmstorf, 1999) to an enhancement of the greenhouse gas concentration. This response is not found in the simulations with LEAN forcing. One can only speculate that the amplitude of the LEAN forcing is too weak to excite any appreciable response. A similar problem emerges if one analyses the lag correlation between solar forcing and THC-response (Figure 5b). In the H&S experiment a high correlation ($> 75\%$) can be found with a lag of around 25–30 years, while neither appreciable nor significant correlations can be identified in the LEAN experiments.

The signal of the temperature penetrates deeply into the Ocean giving signals of 0.01 K at 1000 m depth for the global mean (Figure 6). This signal is caused by the long periodic fluctuations (50+ years). A response on the 11-year timescale, as identified by White *et al.* (1997) cannot be found in the experiment (Voss and Cubasch, 2000).

3.3. Stratospheric Response

A large effect of solar variability on the climate has been described in the observational study of the higher atmosphere by van Loon and Labitzke (1998, 1999).

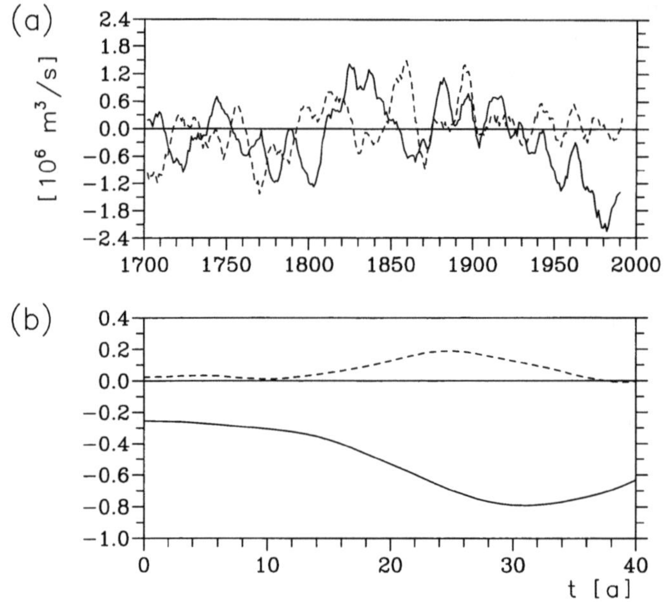

Figure 5. a) The maximum of the meridional stream function in the North Atlantic averaged over the two experiments with H&S forcing and with the LEAN forcing, respectively; b) the lag correlation between the solar forcing and the maximum of the meridional stream function. The time series are smoothed with a 5-year running mean (after Voss and Cubasch, 2000).

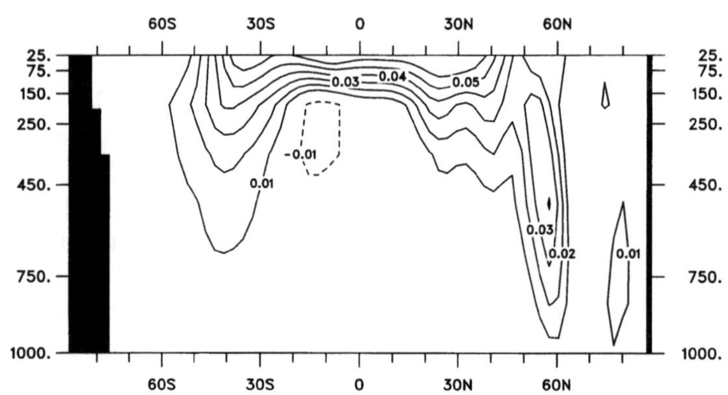

Figure 6. The regression between the solar intensity and the temperature change in the global ocean, zonally averaged. Units: $[K/(W/m^2)]$

Their results, based on the reanalysis time series by NCEP (Kalnay *et al.*, 1996) show a consistent response pattern in all months of the year in the geopotential heights in the lower stratosphere (16 to 30 km) south of about 50° N, which rises and falls with the 11-year sunspot cycle. The highest correlations moves with the seasons meridionally from one summer hemisphere to the other (Labitzke and van Loon, 1997; van Loon and Labitzke 1998, 1999).

The model simulations are compared with these results. As was done with the observations, the modelled 30 hPa height and temperature fields have been correlated with the solar variability. Due to the much longer record length, the solar data contain more frequencies than the 11-year cycles present in Labitzke's and van Loons analysis (see section 3.1). In order to separate the effect of the different periods present in the forcing, the data have been analysed unfiltered and after the application of a low pass (30-year running mean) and a bandpass filter (8 – 15 years). The filters have a rectangular characteristic.

3.3.1. *30 hPa Height Field*

The spatial distribution of the correlations between the observed mean 30 hPa height field and the solar variability in an annual average (Figure 7 of van Loon and Labitzke (2000) shows a belt of high correlation that encircles either hemisphere in a well defined latitude zone ranging from around 5° to 50° in each hemisphere. In the model simulation these belts can be found in the unfiltered data as well (Figure 7a). The low pass filtered data show that most of the correlation can be seen as response to the 80-year Gleissberg cycle, which has a high correlation in a latitudinal band ranging from 50° N to 50° S (Figure 7b), thus covering the whole tropical belt. These correlations can be related to the response of the troposphere, which influences the 30 hPa height field via the thermal structure throughout the atmosphere. The response to the 11 year cycle is not structured as the observed response (Figure 7c).

The correlations of the observed zonally averaged 30-hPa height field through the year (Figure 4 (van Loon and Labitzke, 2000)) as two-month running mean values shows that the largest correlations move inter-seasonally from lower latitudes in winter to higher latitudes in summer. The inter- seasonal meridional movement is larger in the Southern Hemisphere. This may be due to the fact that the earth is in perihelion in the southern summer and aphelion in the southern winter (Labitzke and van Loon, 1997). In the model simulation a similar behaviour can be found (Figure 8).

3.3.2. *30 hPa Temperature*

The correlation of the solar intensity with the observed zonal mean 30 hPa temperature field for the boreal summer is shown in Figure 8 of (van Loon and Labitzke, 2000). The correlation is largest in the northern (summer) hemisphere in a belt between 10° N and 60° N as well in a small belt in the southern hemisphere at around 15° S. In the simulation this signal cannot be found (Figure 7 d, e, f). Only in the bandpass filtered data the solar irradiance is correlated to the temperature (Figure 7e), however, with the opposite sign than in the observations. In the southern hemisphere summer, inverse to the northern hemisphere summer, a belt with positive correlation between solar forcing and observed 30 hPa temperature can be found in a belt between 10° S and 60° S, as well as a small region stretching from

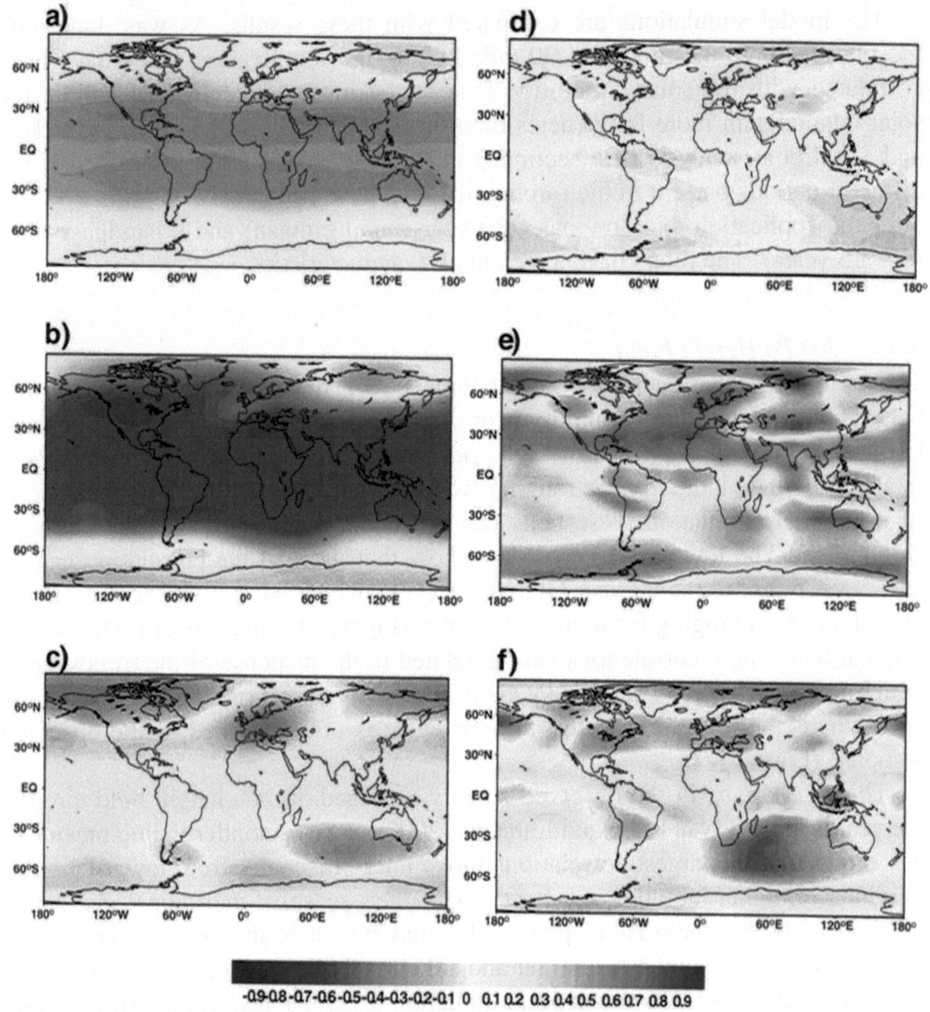

Figure 7. The correlation between the solar irradiance and simulated the 30 hPa height field (left column, annual mean): a) unfiltered, b) low pass filtered (30-year running mean), c) band pass filtered (8 – 15 years), as well as the correlation between the solar irradiance and simulated the 30 hPa temperature field during June/July (right column), d) unfiltered, e) low pass filtered (30-year running mean), f) band pass filtered (8 – 15 years)

the Caribbean to Saudi Arabia. In the simulations, however, no comparable signal can be found in this season either.

4. Discussion

The model results show clearly a response to the long periods of solar variability, but an equivocal response to the 11-year cycle, even though this has been found in

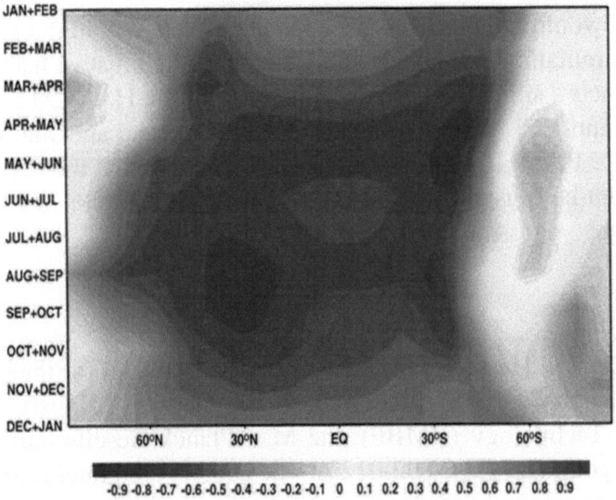

Figure 8. The correlation between the solar irradiance and the zonally and over 2 month averaged simulated 30 hPa height field.

oceanic data (White *et al.*, 1997), or in studies with a simplified model (Drijfhout *et al.*, 1999) could not be identified in the simulations.

The model has not explicitly been coded to represent the stratosphere and particularly the top layer suffers from problems connected to the upper boundary (Arpe *et al.*, 1998).

A possible explanation might be given by a number of studies, in which the effect of the interaction of ozone and the solar input has been analysed (Haigh, 1996, 1999a, b; Rind and Balachandran, 1995; Balachandran and Rind, 1995; Shindell *et al.*, 1999). They find a clear warming signal in the stratosphere due to the enhanced absorption of UV-radiation, which is increased in times of high solar intensity, by the stratospheric ozone. This effect is reinforced by the generation of ozone due to an enhanced photolysis of oxygen. These studies suggest that this effect causes a significant change in the vertical structure of the atmosphere, thus modifying not only the stratospheric, but also the tropospheric circulation.

The different response of the stratosphere to the solar forcing according to the phase of the QBO (quasi-biannual oscillation) has been described by (van Loon and Labitzke, 2000). They point out that the response is larger in the eastern phase of the QBO due to a number of positive feedbacks, but smaller in the negative phase. Since the model used in this simulation has only a weak east phase (Arpe *et al.*, 1998; Giorgetta and Bengtsson, 1999), this finding and the resulting interaction between troposphere and stratosphere (Lean and Rind, 1998) could not be verified.

Another interesting question is the response of the ocean. Since the solar variability seems to be able to influence the thermohaline circulation, the question arises, if this can be verified in historical data. The observational evidence speaks for a complete encasement of Island in ice during the Maunder Minimum, while an enhance THC would rather create the opposite effect. Within the EU-project SIDDACLICH (Simulation, Diagnosis and Detection of Climate Change), the Hadley centre performed a similar set of experiments with the HADCM2 3D model (Tett *et al.*, 1999), and UCL/ASTR with a 2d model (Bertrand and van Ypersele, 1999; Bertrand *et al.*, 1999). Both modelling groups confirm that the currently observed warming cannot be caused by solar forcing alone (Cubasch *et al.*, 1999).

Acknowledgements

The author thank K. Hasselmann, J. Lean and K. Labitzke for their scientific support. The research has been supported by the German Ministry for Education, Research and Technology (BMBF), the Max-Planck-Gesellschaft, the EC Environmental program (EV5V-CT92-0123). The authors are grateful to the staff of the DKRZ for their technical support.

References

Arpe, K., Dümenil, L., and Giorgetta, M. A. : 1998, 'Variability of the Indian monsoon in the ECHAM3 model - Sensitivity to sea surface temperature, soil moisture, and the stratospheric quasi-biennial oscillation', *J. Climate* **11**, 1,837–1,858.

Balachandran, N. K., and Rind, D.: 1995, 'Modeling the effects of UV variability and the QBO on the troposphere-stratosphere system. Part I: The middle atmosphere', *J. Climate* **8**, 2,058–2,079.

Bertrand, C., van Ypersele, J.-P., and Berger, A.: 1999, 'Volcanic and solar impacts on climate since 1700', *Climate Dynamics* **15**, 355–367.

Bertrand, C., and van Ypersele, J.-P. : 1999, 'Potential role of solar variability as an agent for climate change', *Climatic Change*, in press.

Crowley, T. J., and Kim, K.-Y.: 1993, 'Towards development of a strategy for determining the origin of decadal-centennial scale climate variability', *Quat. Sci. Rev.* **12**, 375–385.

Crowley, T. J., and Kim, K.-Y. : 1996, 'Comparison of proxy records of climate change and solar forcing', *Geophys. Res. Lett.* **23**, 359–362.

Cubasch, U., Santer, B. D., Hellbach, A., Hegerl, G. C., Höck, H., Maier-Reimer, E., Mikolajewicz, U., Stössel, A., und Voss, R. : 1994, 'Monte Carlo climate change forecasts with a global coupled ocean-atmosphere model', *Climate Dynamics* **10**, 1–19.

Cubasch, U., Hegerl, G., Hellbach, A., Höck, H., Mikolajewicz, U., Santer, B. D., and Voss, R.: 1995, 'A climate change simulation starting 1935', *Climate Dynamics* **11**, 71–84.

Cubasch, U., Hegerl, G. and Waszkewitz, J. : 1996, 'Prediction, detection and regional assessment of anthropogenic climate change', *Geophysica* **32**, 77–96.

Cubasch, U., Hegerl, G. C., Voss, R., Waszkewitz, J., and Crowley, T. C.: 1997, 'Simulation with an O-AGCM of the influence of variations of the solar constant on the global climate', *Climate Dynamics* **13**, 757–767.

Cubasch, U., Myles, A., Beniston, M., Bertrand, C., Brinkop, S., Caneill, J.-Y., Dufresne, J.-L., Fairhead, L., Filiberti, M.-A., Gregory, J., Hegerl, G., Hoffmann, G., Johns, T., Jones, G., Laurent, C., McDonald, R., Mitchell, J., Parker, D., Oberhuber, J., Poncin, C., Sausen, R., Schlese, U., Stott, P., Tett, S., leTreut, H., Ulbrich, U., Valcke, S., Voss, R., Wild, M., and van Ypersele, J.-P.: 1999, 'Simulation, diagnosis and detection of anthropogenic climate change', *in* Final report of project ENV4- CT-0102, EU-DG XII, Brussels, Belgium.

DKRZ: 1993, 'The ECHAM3 atmospheric general circulation model', *in* Tech. Rep. No. 6, DKRZ, Bundesstr 55, Hamburg, Germany.

Drijfhout, S. S., Haarsma, R. J., Opsteegh, J. D., and Selten, F. M.: 1999, 'Solar-induced versus internal variability in a coupled climate model', *Geophys. Res. Lett.* **26** 205–208.

Friis-Christensen, E., and Lassen, K. : 1991, 'Length of the solar cycle: An indicator of solar activity closely associated with climate', *Science* **254**, 698–700.

Fröhlich, C., and Lean, J. : 1998, 'The sun's total irradiance cycles, Trends and related climate change uncertainties since 1976', *Geophys. Res. Lett.* **25**, 4,377–4,380.

Giorgetta, M. A. and Bengtsson, L.: 1999, 'The potential role of the quasi-biennial oscillation in the stratosphere- troposphere exchange as found in water vapour in general circulation model experiments', *J. Geophys. Res.* **104**, 6,003–6,019.

Haigh, J. D.: 1996, 'The impact of solar variability on climate', *Science* **272**, 981–984.

Haigh, J. D.: 1999a, 'Modelling the impact of solar variability on climate', *Journ. of Atmos. and Solar-Terr. Phys.* **61**, 63–72.

Haigh, J. D.: 1999b, 'A GCM study of climate change in response to the 11-year solar cycle', *Q.J.R.M.S.* **125**, 871–892.

Hegerl, G. C., Hasselmann, K., Cubasch, U., Mitchell, J. F. B., Roeckner, E., Voss, R., und Waszkewitz, J.: 1997, 'Multi- fingerprint detection and attribution analysis of greenhouse gas, greenhouse gas-plus-aerosol and solar forced climate change', *Climate Dynamics* **13**, 613–634.

Hoyt, D. V., and Schatten, K. H.: 1993, 'A discussion of plausible solar irradiance variations, 1700-1992', *J. Geophys. Res.* **98**, 18,895–18,906.

IPCC: 1996, *in* J. T. Houghton, L. G. Meira Filho, B. A. Callander, N. Harris, A. Kattenberg and K. Maskell (eds.), *'Climate Change, 1995: The science of climate change'*, Cambridge University Press, Cambridge.

Jones, P. D. and Briffa, K. R. : 1992, 'Global surface air temperature variations during the twentieth century: Part 1, spatial, temporal and seasonal details', *The Holocene* **2**, 165–179.

Jones, P. D.: 1994, 'Hemispheric surface air temperature variations: A reanalysis and an update to 1993', *J. Climate* **7**, 1,794–1,802.

Kalnay, E., Kanamitsu, M., Kistler, R., Collins, W., Deaven, D., Gandin, L., Iredell, M., Saha, S., White, G., Woollen, J., Zhu, Y., Chelliah, M.,Ebisusaki, W., Higgins, W., Janowiak, J., Mo, K. C., Ropelewski, C., Wang, J., Leetma, A.,Reynolds, R., Jenne, R., and Joseph, D.: 1996, 'The NCEP/NCAR 40-year re-analysis project', *Bull. Am. Met. Soc.* **77**, 437–471.

Kelly, P. M. and Wigley, T. M. L.: 1992, 'Solar cycle length, greenhouse forcing and global climate', *Nature* **360**, 328–330.

Labitzke, K., and van Loon, H.: 1997, 'The signal of the 11-year sunspot cycle in the upper troposphere-lower stratosphere', *Space Sci. Rev.* **80**, 393–410.

van Loon, H., and Labitzke, K.: 1998, 'The global range of the stratospheric decadal wave. Part I: Its association with the sunspot cycle in summer and in the annual mean, and with the troposphere', *J. Climate* **11**, 1529–1537.

van Loon, H., and Labitzke, K.: 1999, 'The signal of the 11-year solar cycle in the global stratosphere', *Journ. of Atmos. and Solar-Terr. Physics* **61**, 53–61.

van Loon, H., and Labitzke, K. : 1999, 'The influence of the 11- year solar cycle on the stratosphere below 30 km: a review', *Space Sci. Rev.*, this volume.

Lean, J., Beer, J., and Bradley, R.: 1995a, 'Reconstruction of solar irradiance since 1610: Implications for climate change'., *Geophys. Res. Lett.* **22**, 3,195–3,198.

Lean, J., White, O. R., and Skumanich, A.: 1995b, 'On the ultraviolet spectral irradiance during the Maunder Minimum', *Glob. Biochem. Cycles* **9**, 171–172.

Lean, J.: 1997, 'The sun's variable radiation and its relevance to earth', *Annu. Rev. Astron. Astrophys.* **35**, 33–67.

Lean, J., and Rind, D.: 1998, 'Climate forcing by changing solar radiation', *J. Clim.* **11**, 3,069–3,093.

Maier-Reimer, E., Mikolajewicz, U., and Hasselmann, K.: 1993, 'Mean circulation of the Hamburg LSG OGCN and its sensitivity to the thermohaline surface forcing', *J. Phys. Oceano.* **23**, 731–751.

Manabe, S., Stouffer, R. J., Spelmann, M. J., and Bryan K.: 1991, 'Transient responses of a coupled ocean-atmosphere model to gradual changes of atmospheric CO_2: I Annual mean response', *J. Climate* **4**, 785–818.

Mann, M. E., Bradley, R. S., and Hughes, M. K.: 1998, 'Global scale temperature patterns and climate forcing over the past six centuries', *Nature* **392**, 779–787.

Paeth, H., Hense, A., Glowienka-Hense, R., Voss, R., and Cubasch, U.: 1999, 'The North Atlantic Oscillation as an indicator for greenhouse-gas induced climate change', *Climate Dynamics* **15**, 953–960.

Parker, D. E., Jones, P. D., Folland, C. K., and Bevan, A.: 1994, 'Interdecadal changes of surface temperature since the late nineteenth century', *J. Geophys. Res.* **99**, 14,373–14,399.

Rahmstorf, S.: 1999, 'Shifting seas in the greenhouse', *Nature* **399**, 523–524.

Reid, G. C.: 1991, 'Solar irradiance variations and the global sea surface temperature record', *J. Geophys. Res.* **96**, 2,835–2,844.

Rind, D., and Overpeck, J.: 1993, 'Hypothesized causes of decade-to-century scale climate variability: Climate model results', *Quat. Sci. Rev.* **12**, 357–374.

Rind, D. and N. K. Balachandran: 1995, 'Modeling the effects of UV-variability and the QBO on the troposphere-stratosphere system. Part II: The troposphere', *J. Climate* **8**, 2,080–2,095.

Roeckner, E., Arpe, K., Bengtsson, L., Brinkop, S., Dömenil, L., Esch, M., Kirk, E., Lunkeit, F., Ponater, M., Rockel, B., Sausen, R., Schlese, U., Schubert, S., and Windelband, M.: 1992, 'Simulation of the present-day climate with the ECHAM model: Impact of model physics and resolution'. Report No. 93, Max-Planck-Institut für Meteorologie, Bundesstr. 55, Hamburg, Germany.

Shindell, D. T., Rind, D., Balachandran, N., Lean, J., and Lonergan, P.: 1999, 'Solar cycle variability, ozone and climate', *Science* **284**, 305–308.

Spelman, M. J., and Manabe, S.: 1984, 'Influence of oceanic heat transport upon the sensitivity of a model climate', *J. Geophys. Res.* **89**, 571–586.

Stouffer, R. J., Manabe, S., and Vinnikov, K. Y.: 1994, 'Model assessment of the role of natural variability in recent global warming', *Nature* **367**, 634–636.

Stouffer, R. J., Hegerl, G. C., and Tett, S. F.: 1999, 'A comparison of surface air temperature variability found in three 1000-year coupled ocean-atmosphere model integrations', *J. Climate* **13**, 513–537.

Tett, S. F. B., Stott, P. A., Allen, M. R., Ingram, W. J., and Mitchell, J. F. B.: 1999, 'Causes of twentieth-century temperature change near the Earth's surface', *Nature* **399**, 569–572.

Voss, R., Sausen, R., and Cubasch, U.: 1998, 'Periodically synchronously coupled integrations with the atmosphere-ocean general circulation model ECHAM3/LSG', *Climate Dynamics* **14**, 249–266.

Voss, R., and Cubasch, U.: 2000, 'Climate response to variations of the solar irradiance since 1700 in coupled OAGCM simulations', *J. Atmos. and Solar-Terr. Phys.*, submitted.

Wetherald, R. T., and Manabe, S.: 1975, 'The effects of changing the solar constant on the climate of a general circulation model', *J. Atmos. Sci.* **32**, 2,044–2,059.

White, W. B., Lean, J., Cayan, D. R., and Dettinger, M. D.: 1997, 'Response of global upper ocean temperature to changing solar irradiance', *J. Geophys. Res.* **102**, 3,255–3,266.

Wigley, T. M. L., and Kelly, P. M.: 1990, 'Holocene climate change, 14C wiggles and verification in solar irradiance', *Phil. Trans. Roy. Soc.* (London) **A330**, 547–560.

Wigley, T. M. L., and Raper, S. C. B.: 1990, 'Climatic change due to solar irradiance changes', *Geophys. Res. Lett.* **17**, 2,169–2,172.

THE EFFECT OF SOLAR UV IRRADIANCE VARIATIONS ON THE EARTH'S ATMOSPHERE

ALICE LARKIN, JOANNA D. HAIGH and SAMY DJAVIDNIA
Blackett Laboratory
Imperial College of Science, Technology and Medicine, London, UK

Received: 14 January 2000; Accepted in final form: 9 March 2000

Abstract. The response of the lower and middle atmosphere to variations in solar irradiance typical of those observed to take place over the 11-year activity cycle has been investigated. The effects on radiative heating rates of changing total solar irradiance, solar spectral irradiance and two different assumptions concerning stratospheric ozone have been studied with a radiative transfer code. The response in the stratosphere depends on the changes specified in the ozone distribution which is not well known. A general circulation model (GCM) of the atmosphere up to 0.1 mbar (about 65 km) has been used to study the impacts of these changes on the thermodynamical structure. The results in the troposphere are very similar to those reported by Haigh (1999) using a quite different GCM. In the middle atmosphere the model is able to reproduce quite well the observed seasonal evolution of temperature and wind anomalies. Calculations of radiative forcing due to solar variation are presented. These show that the thermal infrared component of the forcing, due to warming of the stratosphere, is important but suggest a near balance between the longwave and shortwave effects of the increased ozone so that ozone change may not be important for net radiative forcing. However, the structure of the ozone change does affect the detailed temperature response and the spectral composition of the radiation entering the troposphere.

1. Introduction

There have been many studies with GCMs to investigate the impact on climate of changes in total solar irradiance (TSI) and these generally show very small impact for realistic variations in TSI. Middle atmosphere models have been employed (Kodera *et al.*, 1991; Rind and Balachandran, 1995) to study the effects of increases in solar UV. These suggest that interaction between vertically propagating planetary-scale waves and the zonal mean flow may have an impact which propagates downwards from the middle atmosphere to the troposphere in winter. However, because of unrealistically large changes in UV, these results cannot be directly compared with observations and they do not show any significant effect in the summer hemisphere.

Haigh (1996, 1999) and Shindell *et al.* (1999) have carried out experiments in which realistic changes in solar spectrum, together with some estimates of changes in ozone, have been imposed in GCMs. In these experiments the sea surface temperatures were fixed so that no response at the surface due to the direct absorption of extra solar irradiance was possible but the response of the atmosphere could be

Space Science Reviews **94:** 199–214, 2000.

studied. Haigh's experiments, which were carried out in perpetual January mode in a GCM with only three levels in the stratosphere, showed most significant effect in the summer hemisphere with a vertical banding structure in winds and temperatures in mid-latitudes which was found to be associated with a broadening of the tropical Hadley cells. Increases in 30 mbar geopotential height were found to be of similar structure to those seen in the data analyses of van Loon and Labitzke (2000) but only about one-fifth of the magnitude, depending on the ozone changes used. The Shindell et al model extended into the middle atmosphere and the simulations were run with a full seasonal cycle. Results were only presented for the winter hemisphere and the 30 mbar geopotential height changes, found to be due to wave-mean-flow interactions, were again similar in structure but somewhat smaller than those of van Loon and Labitzke. The mechanisms studied by Haigh (1999) and Shindell *et al.* (1999) look promising but further work is needed to understand the observed response, especially in the summer hemisphere.

Before some new GCM experiments are described (in Section 3) we present results showing the effect of changes in solar irradiance on radiative heating rates. These provide the most direct means for transmitting variations in solar activity to the atmosphere and are the basis upon which the GCM experiments are carried out.

2. Radiative Heating Rates

The solar spectrum can be approximated by that of a black body at about 5800 K, thus it peaks at 730 nm, with only 5% of the incoming energy at wavelengths, λ, less than 356 nm and 5% at $\lambda > 2230$ nm. About 47% of the solar radiation incident at the top of the atmosphere passes through the atmosphere to the Earth's surface and about 30% is reflected back to space. The absorption of the other 23% as it passes through the atmosphere depends on the spectral properties, and the distribution, of the atmospheric constituents. If a species absorbs strongly over a particular spectral range, and is present at sufficiently high concentrations, then the radiation in that spectral region will be preferentially absorbed and may not reach the lower atmosphere, or the Earth's surface. In the lower and middle atmosphere most of the absorbed energy ultimately becomes thermal energy, either through collisional deexcitation of molecules raised to an excited energy state, or through exothermic chemical recombination of species produced by photodissociation of source molecules. Thus the atmosphere is heated by the absorption of solar radiation at altitudes which depend on wavelength. It follows that the response of atmospheric solar heating rates to variations in solar input will depend on the spectrum of these variations.

In order to assess the magnitude and distribution of changes in heating rate, calculations have been carried out using the radiative transfer code of Edwards and Slingo (1996) in the same version as is used in the GCM discussed in Section 3.

In this scheme solar heating rates are calculated using a two-stream model with six spectral bands.

The left hand column of Figure 1 shows diurnally averaged solar heating rates as a function of latitude and altitude for a January situation with the Sun at (11-year cycle) minimum activity. The frame at the top shows the heating due to absorption of radiation in the 220 – 320 nm range (subsequently referred to as UV), the centre frame that between 320 and 690 nm ("VIS") and the bottom frame that due to the $\lambda > 690$ nm region ("NIR"). The ultraviolet radiation is absorbed by oxygen and ozone in the middle atmosphere causing a peak in heating rates of around 12 $\mathrm{K\,d^{-1}}$ near the summer stratopause. The visible radiation is absorbed by ozone and is mostly absorbed in the lower stratosphere with heating rates peaking at just over 1 $\mathrm{K\,d^{-1}}$. The near infrared radiation is the main contributor to solar heating in the troposphere where it is absorbed by water vapour resulting in heating rates of 0.3 $\mathrm{K\,d^{-1}}$ in the 200–300 mbar region rising to about 1 $\mathrm{K\,d^{-1}}$ in equatorial regions near the surface.

The other four columns in Figure 1 show the difference in solar heating rate between periods of minimum and maximum solar activity (over the 11-year cycle) making different assumptions concerning the nature of the imposed change. In column 2 the total solar irradiance (TSI) is increased by 0.08%, based on 17 years of satellite measurements (Fröhlich and Lean, 1998), with the same fractional increase applied across the spectrum. The change in heating rates is, not surprisingly, a fairly uniform increase of the same proportion in each wavelength region so that the largest increases are only about 0.01 $\mathrm{K\,d^{-1}}$ at the summer stratopause. In column 3 the same increase in total solar irradiance is applied but this time spectrally distributed as suggested by the results of Lean and Rind (1998), for $\lambda < 399$ nm, and Solanki and Unruh (1998) at longer wavelengths. These show largest fractional changes in the ultraviolet of around 10% near 200 nm, 4% near 300 nm reducing to about 0.1% in the visible region. The changes in the near infrared region are much smaller. The imposed increases enhance the additional UV heating by a factor of 8 – 10 and VIS by about a factor 2. The additional IR heating is reduced by a factor of 2 – 3 because the increase in irradiance is less than that applied uniformly across the spectrum in column 2.

Radiation at $\lambda < 242$ nm causes the photodissociation of oxygen molecules, and hence the formation of stratospheric ozone through the combination of the resultant oxygen atoms with other O_2 molecules. Thus enhanced levels of solar UV radiation tend to increase O_3 concentrations which further modifies (at longer wavelengths) the transfer and absorption of radiation. Columns 4 and 5 in Figure 1 show the solar heating due to changes not only in top-of-atmosphere solar spectral irradiance but also the ozone distribution. Column 4 uses the ozone changes predicted by a 2D photochemical- transport model (Haigh, 1994) and column 5 those deduced from SBUV/TOMS (The Solar Backscattered Ultraviolet and Total Ozone Mapping Spectrometer instruments on the Nimbus 7 satellite) data (McCormack and Hood, 1996). The structure of these solar-induced ozone changes is very different

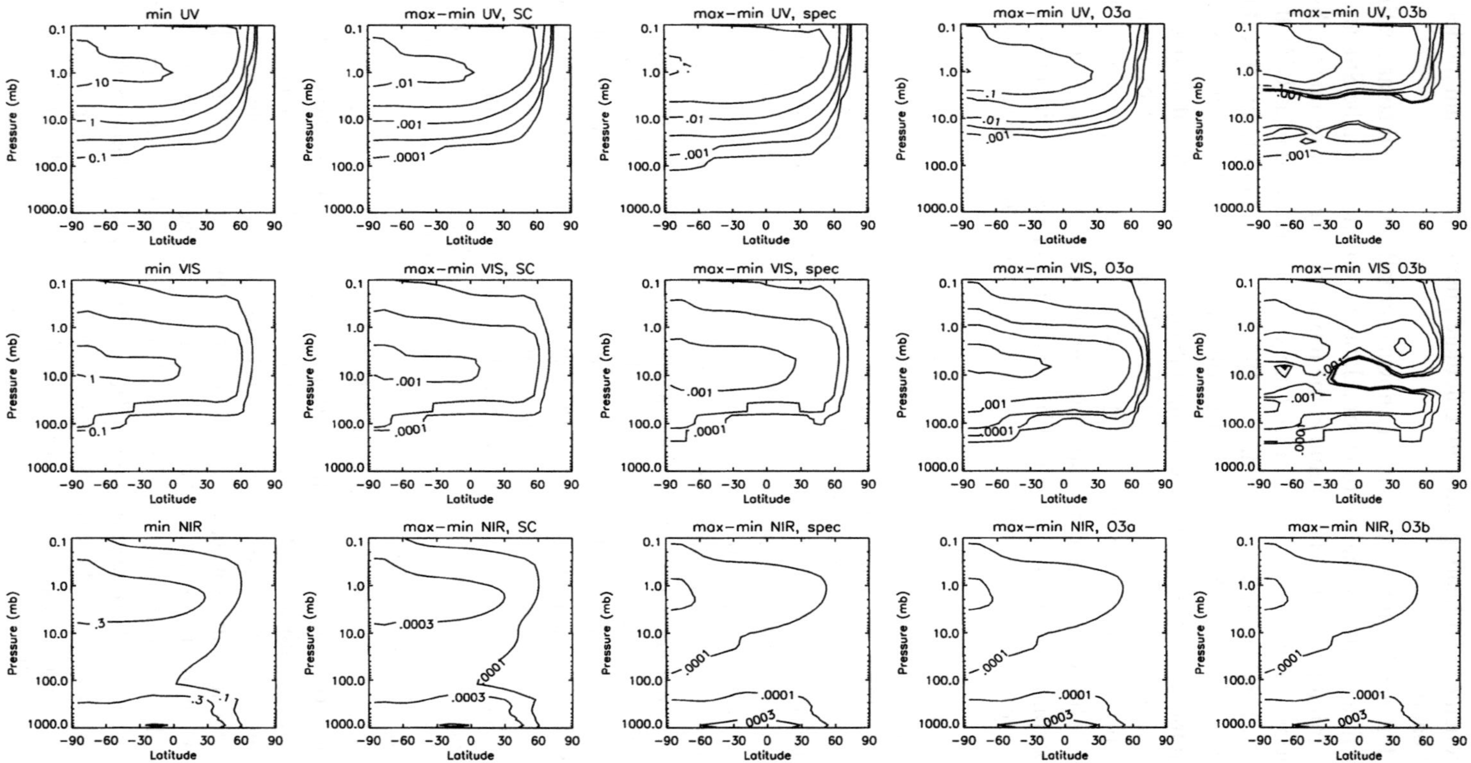

Figure 1. Solar heating rates (K d^{-1}) Top row 220 – 320 nm, middle row 320 – 690 nm, bottom row 690 – 10000 nm. First column control (solar minimum) values. Column 2 – 5 show difference between solar max. and solar min. assuming: Column 2: 0.08% in TSI uniform across spectrum. Column 3: 0.08% in TSI with spectral distribution. Column 4: As col. 3 plus ozone increase from 2D model. Column 5: As col. 3 plus ozone increase from SBUV/TOMS. Contours are shown at 1×10^n and 3×10^n with the range of n varying between frames.

with the observational data showing a minimum in the middle stratosphere which is not seen in the model results. These differences have yet to be explained as there are shortcomings in both the photochemical model and the data analysis, and it is not clear if either is "correct". The peak additional UV heating now exceeds 0.5 $K\,d^{-1}$ with the vertical structure strongly reflecting that in the imposed ozone change field. A similar, but smaller magnitude, response is also seen in the VIS but the IR is unaffected by O_3 changes.

These results show that simulations of atmospheric response to changes in solar irradiance need to properly take account of the spectral composition of the change and also the ozone response. Otherwise they will tend to underestimate solar heating in the stratosphere and overestimate the direct solar heating of the troposphere as well as the irradiance at the surface.

3. GCM Studies

In this paper we describe some experiments with the UK Meteorological Office Unified Model (UM, Cullen, 1993). This model is entirely different to the UGAMP (UK Universities' Global Atmospheric Modelling Programme) model used by Haigh (1996, 1999). For example, the radiative transfer scheme implemented in the UM is the Edwards and Slingo (1996) two-stream code with 6 spectral bands for the shortwave (i.e. for solar radiation) and 8 in the longwave (thermal infrared radiation) whereas the UGAMP model included the Morcrette (1991) emissivity code with 2 shortwave bands and 8 (different) longwave bands. Even more fundamentally the numerical formulation of the dynamical equations is different in the two models — the UGAMP model uses a spectral approach while the UM has a grid-point formulation. Haigh (1996, 1999) used a version of the UGAMP model with 19 vertical levels extending to 10 mbar and a horizontal resolution of T42 (approx. 2.75° × 2.75°), all experiments were carried out in perpetual January mode. The work described here used a version of the UM which has 58 vertical levels extending up to 0.1 mbar and a horizontal resolution of 2.5° latitude by 3.75° longitude and the full seasonal cycle was included. Both the UGAMP and UM model runs were carried out with fixed (seasonally varying for the UM) sea surface temperatures so that a full climate response to the changes in solar irradiance was not possible but the studies could focus on the internal impact of changes in UV radiation on the atmosphere.

Two experiments have been carried out with the UM. The first experiment (labelled I) consisted of a control run of 10 years in solar minimum conditions and a solar maximum run of 7 years in which the solar irradiance was enhanced according to the Lean and Rind (1998) and Solanki and Unruh (1998) spectra, as used in section 2, and ozone increased, according to the 2D model results, again as described in section 2. Longer integrations were not possible due to the decommissioning of the computer. The second experiment (II, carried out on a different

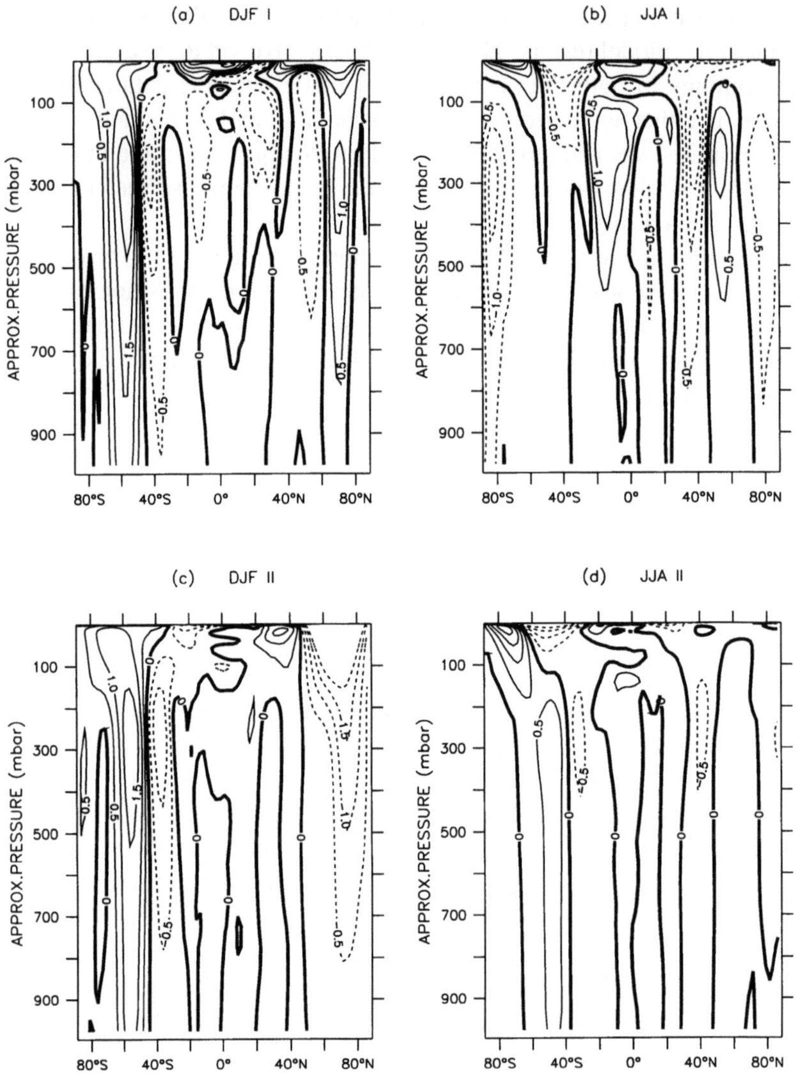

Figure 2. Difference between solar max. and solar min. in zonal mean zonal wind (m s^{-1}). Top row: Experiment I, bottom row: Experiment II. Left column: DJF, right column: JJA.

machine) consisted of a 20 year control (solar minimum) run and a 20 year solar maximum run in which the solar spectrum was increased as before and the ozone distribution change based on the SBUV data analysis, as described in section 2.

Figure 2 shows the difference between solar minimum and solar maximum in zonal mean zonal wind in DJF (December-January-February mean) and JJA (June-July-August mean) averaged over all years for each of experiments I and II. The plots are presented with a linear pressure scale as the vertical axis in order to focus on the troposphere and also to compare with the results of Haigh (1999). No clear

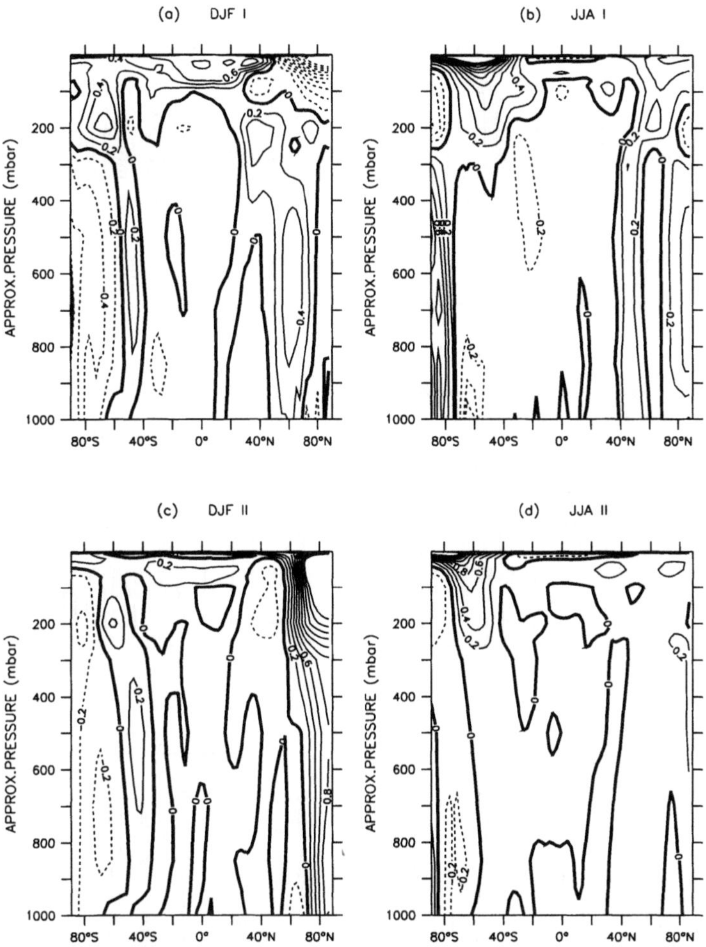

Figure 3. As Figure 2 but for zonal mean temperature (K).

pattern of response is seen in the winter hemispheres, and the statistical significance of the changes here is anyway low. In the summer hemispheres, however, (i.e. the S.H. in DJF and N.H. in JJA) a similar response can be seen in all four panels, the main feature of which is a banding structure in mid-latitudes with a negative anomaly near 40° latitude sandwiched between two positive anomalies. All cases also show easterly anomalies in the summer hemisphere low latitude lower stratosphere. This structure is statistically significant at the 95% level in three of the four cases (not JJA II which shows the same pattern but with smaller magnitude) and, more significantly, is exactly the same pattern as seen in the S. Hemisphere January in the UGAMP model experiments of Haigh (1999). [NB the bands in the two models do not occur at exactly the same positions in absolute latitude but do in terms of position relative to the sub-tropical jet which is somewhat further south in the southern hemisphere in January in the UGAMP model]. That

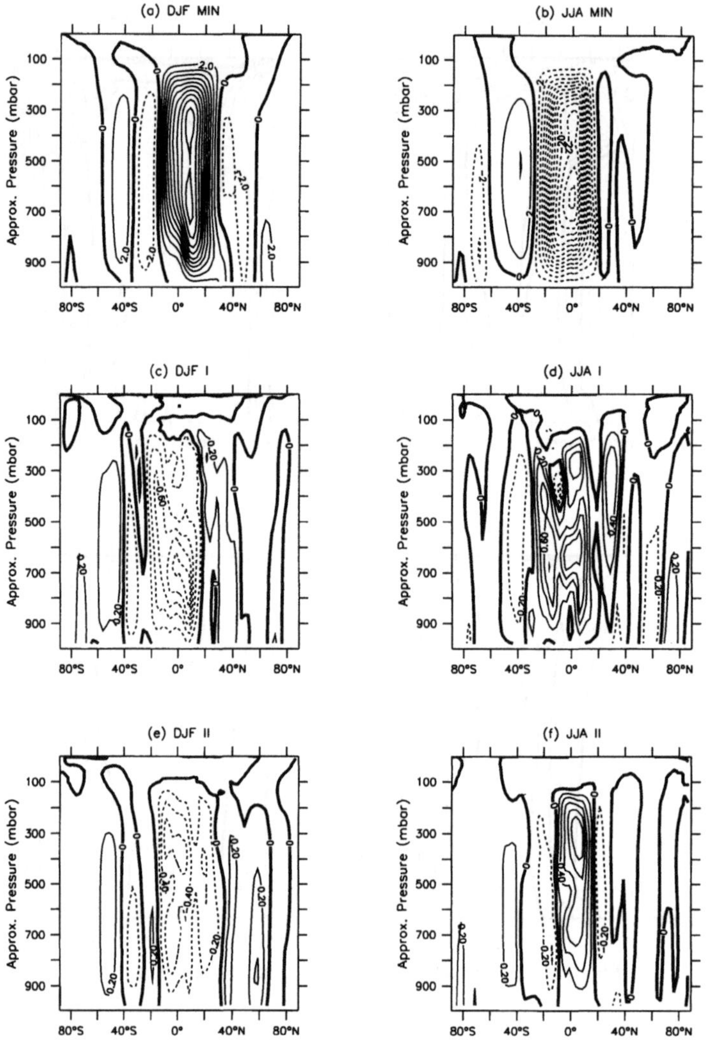

Figure 4. Mean meridional circulation. Top row: control run. Second and third rows: difference between solar min. and solar max. in Experiment I and II respectively. Left column: DJF, right column: JJA.

the same result should be found in several experiments with two quite different models strongly suggests that this mode of response in the summer hemisphere mid-latitudes might be expected to occur in the real atmosphere. The changes in zonal mean temperature are shown in Figure 3. The bands of warming in summer mid-latitudes are again consistent with those found by Haigh (1999).

Haigh (1996, 1999) also found that the anomalies in zonal mean zonal wind and temperature were associated with changes in the mean meridional circulation (MMC). In particular the tropical Hadley cells were observed to weaken and

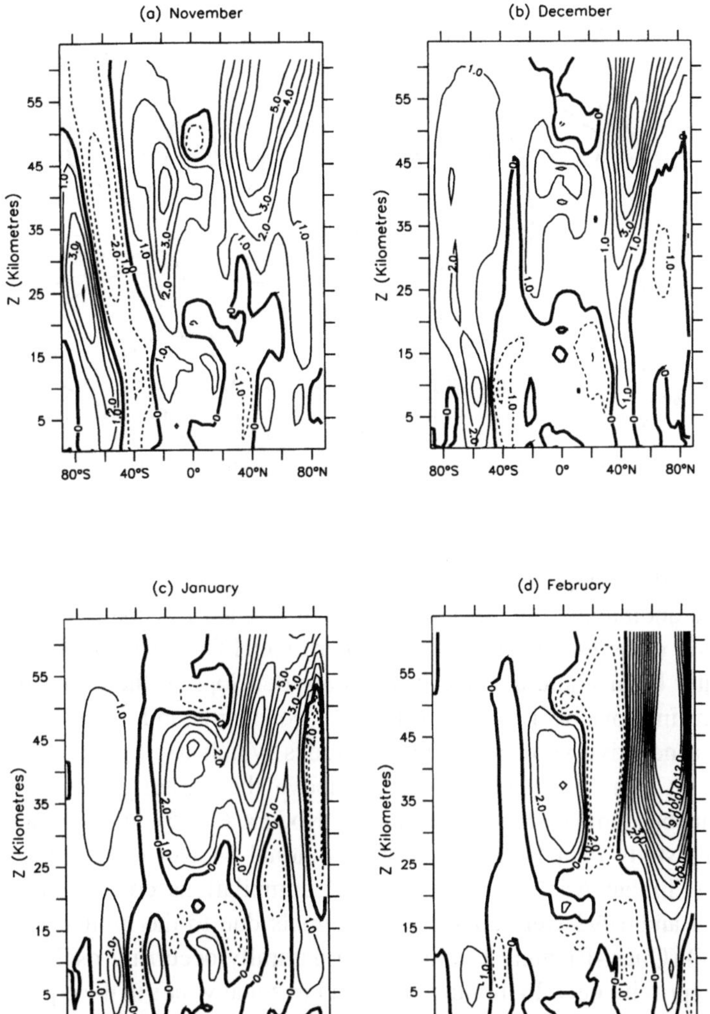

Figure 5. Difference between solar min. and solar max. in zonal mean zonal wind (m s^{-1}) from Experiment I for 4 months in N. Hemis. winter.

broaden and the summer mid-latitude Ferrel cell to move poleward for solar maximum conditions. Figure 4 shows the MMC at solar minimum from the UM for DJF and JJA together with the differences at solar maximum seen in experiments I and II. In all cases the dominant (winter) Hadley cell has weakened and broadened. In DJF in the southern hemisphere the summer Hadley cell and Ferrel cell have moved polewards. In JJA in the northern hemisphere the summer Hadley cell and Ferrel cell are very weak and no pattern of change can be distinguished. As expected there is no clear response in winter mid-high latitudes.

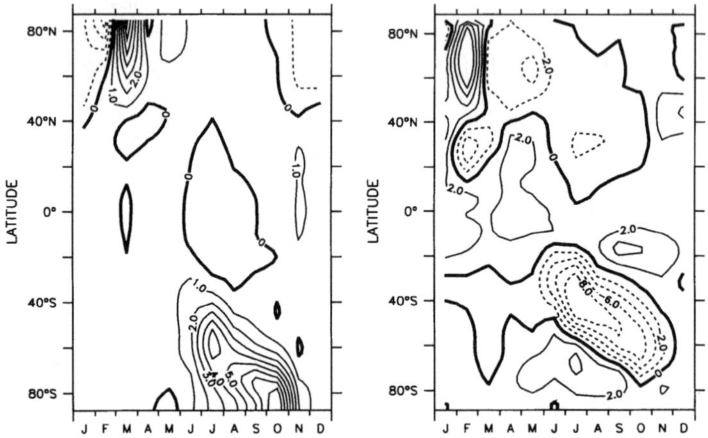

Figure 6. Difference between solar min. and solar max. throughout the year at 5 mbar in left: zonal mean temperature (K), right: zonal mean zonal wind (m s^{-1}).

In the middle atmosphere it is instructive to investigate the seasonal development of the anomalies. Figure 5 shows the difference in zonal mean zonal wind as a function of height and latitude in experiment I from November to February. In the winter hemisphere a westerly anomaly develops in the mid-latitude upper stratosphere in November which propagates polewards and downwards to be replaced by a negative anomaly in mid-latitudes which itself subsequently moves polewards by March (not shown) to lie above the westerly anomaly. This behaviour can be compared to that seen by Kodera (1995) in observational data. In Kodera's study data from the NMC (National Meteorological Center, USA) and CAC (Climate Analysis Center, USA) compilations covering the years between 1979 and 1992 were analysed for signals due to the influence of the Sun, of the QBO (Quasi-Biennial Oscillation, a naturally-occurring oscillation, with a period of approximately 26 months, observed in zonal winds in the equatorial lower stratosphere) and of volcanic eruptions. The development of the zonal wind anomaly in the model compares well with Kodera's calculations of the solar influence although the model development lags behind the observations by about one month by the end of the winter and in the model the magnitudes of the anomalies are about half those seen in the data.

The development throughout the year of the anomalies in zonal mean temperature and zonal mean zonal wind at 5 mbar can be seen in Figure 6. Largest warming is shown in late winter at mid-high latitudes in both hemispheres while the zonal wind shows positive anomalies throughout the year in the tropics and negative for most of the year in the mid-latitudes of both hemispheres. The latitudinal distribution of these changes is very similar to that deduced by Hood *et al.* (1993) to be of solar influence in an NMC data analysis.

The model shows increases in 30 mbar geopotential height (Z_{30}) in low to mid-latitudes but does not satisfactorily reproduce the changes seen by van Loon and

Labitzke (this volume, vLL). In DJF increases, of up to 20 m, are predicted in low to mid-latitudes, i.e. in the same sense but only about one-third the magnitude of the vLL results. At high latitudes the response is highly variable so that no significant signal is found. This differs from the results of Shindell *et al.* (1999), who found significant reduction in Z_{30} near the winter pole in DJF, similar to the vLL results. They ascribed this to an interaction between vertically propagating waves and anomalies in the mean temperature distribution which were permitted by the presence of a middle atmosphere in the GCM and of a quasi-interactive ozone distribution. The GCM used in the present work also has a middle atmosphere, and in addition has significantly higher horizontal and vertical resolution, so the reason for the difference in dynamical response is not clear. Certainly longer runs are needed, at least in the present work, to allow detection of a signal in the, naturally highly variable, winter polar region.

In JJA the response is small (< 10 m) in low to mid-latitudes and Shindell *et al.* (1999) also failed to simulate the vLL JJA response. This lack of agreement may be due inadequacies in the model simulations, such as the use of fixed sea surface temperatures or unrealistic representation of ozone changes. It is also possible that the vLL Z_{30} signals include the effects of other factors such as El Niño, volcanic eruptions and other natural variability difficult to disentangle in the forty-year data series. Further data analysis as well as GCM simulations are needed to answer some of these questions.

4. Radiative Forcing

A well-established approach to estimating the climate impact of a change in the Earth's radiation balance is to calculate radiative forcing (e.g. Shine, 2000). This is the induced radiative imbalance at the tropopause taking into account the radiative adjustment of the stratosphere but assuming that the conditions of the troposphere are unchanged (because the adjustment time of the stratosphere is much faster). The globally-averaged increase in downward solar irradiance at the tropopause from minimum to maximum of the 11-year solar activity cycle is approximately 0.26 W m^{-2}. About 30% of this is reflected back to space by clouds and high-albedo surfaces resulting in a direct radiative forcing of 0.18 W m^{-2}. Additionally there will be an increase in downward thermal infrared flux due to heating of the stratosphere and also radiative forcing due to ozone changes. However, the sign of the ozone effect is not well established. Haigh (1994) and Myhre *et al.* (1998) found that ozone increases *reduced* the solar radiative forcing at maximum relative to minumum points on the 11-year cycle, by about 0.1 W m^{-2} and 0.02 W m^{-2} respectively, and Wuebbles *et al.* (1998) computed a reduction of 0.13 W m^{-2} due to ozone increases since the Maunder Minimum. Hansen *et al.* (1997), however, showed an *increased* forcing of about 0.05 W m^{-2} from minimum to maximum of a solar cycle due to ozone increases and lower stratospheric warming. This disparity represents the different approaches used: Hansen *et al.* (1997) estimated the ozone effect by assuming that all the ozone change was in the 10 – 150 mbar region and extrapolating from ozone sensitivity studies carried out with a low resolution GCM. Haigh (1994), Wuebbles *et al.* (1998) and Myhre *et al.* (1998) calculated the ozone response using 2D chemical-transport models in which temperature changes are estimated using the fixed dynamical heating approximation. The negative radiative forcing values probably correspond to ozone change profiles which peak at higher altitudes, and thus have less impact on lower stratospheric temperature and long-wave radiative forcing, although the different methods for calculating temperature change may also be important.

Here we calculate the radiative forcing using the solar irradiance and ozone changes as specified in the GCM experiments described in section 3 and the stratospheric temperature changes calculated in these experiments. Figure 7a shows the shortwave radiative forcing, at solar maximum relative to solar minimum, in January. The dash-dot line, showing the response to changing the solar spectrum, has lower values than the solid line, showing the effect of just changing the solar constant, because the UV radiation, which is greater in the former, is preferentially absorbed in the middle atmosphere. The other two curves, at consistently lower values, show that the inclusion of additional ozone reduces the downward shortwave irradiance considerably because of further absorption at higher levels. Experiment I has greater impact because the total ozone column change was somewhat larger. Figure 7b shows the change in longwave forcing (i.e. downward thermal irradiance at the tropopause) for the two GCM runs with and without stratospheric temperature effects. Without these the longwave forcing (due to extra irradiance in the O_3 9.6 μm band) is only about 25 – 50% of the value when temperature effects are included, in which case there is significant positive forcing, of $0.07-0.1$ W m^{-2} in low latitudes with Experiment I showing larger values due to a larger stratospheric warming. Larger forcings, due to the warmer lower stratosphere, are seen near the summer pole. The very large change seen at winter high latitudes in Experiment II reflects a warming produced by the GCM that is not statistically significant. The total (SW + LW) forcing, including stratospheric temperature adjustment, is shown in Figure 7c. At low latitudes the values for Experiment II are slightly higher, i.e. the SW component has dominated, whereas in summer high latitudes the reverse is true.

The (area-weighted) global average value is 0.23 W m^{-2} for both Experiments I and II. This suggests that the ozone increase does not have a significant effect on solar radiative forcing — because of a compensation between the shortwave and longwave components. We did not conduct a GCM experiment with an increase in irradiance but no change in ozone so we do not have a value for the resulting longwave, or net, forcing in this case. However an estimate can be made from the values shown in Table I.

In the Table the LW value (in brackets) for the no ozone change scenario is assumed to be greater than zero (because there is unlikely to be lower stratospheric

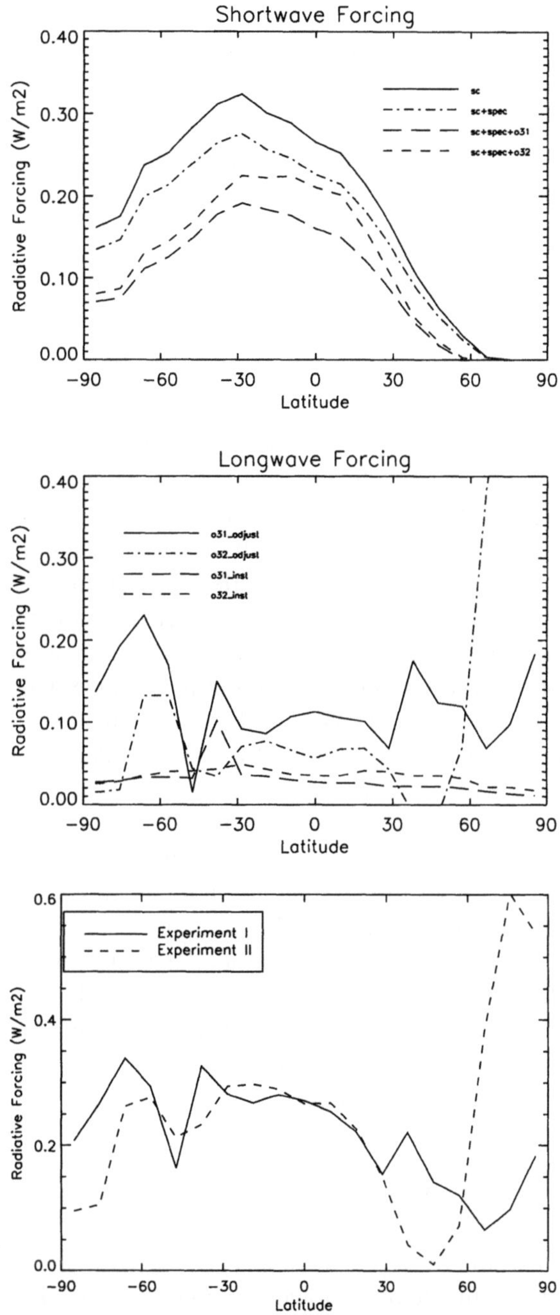

Figure 7. Solar radiative forcing (W m^{-2}) between solar min. and solar max. in January.
(a) Shortwave forcing due to change in TSI (solid line); in spectrally distributed TSI (dash-dot); spectrum plus 2D ozone (long dashes); spectrum plus SBUV ozone (short dashes).
(b) Longwave forcing due to 2D model (Experiment I) ozone and temperatures (long dashes/solid line) and SBUV (Experiment II) ozone and temperatures (short dashes/dash-dot) without/with stratospheric adjustment.
(c) Sum of shortwave and adjusted longwave forcings for Experiment I (solid line) and II (long dashes).

TABLE I

Global average radiative forcing (W m^{-2}). The figures in brackets are estimates (see text for details)

	Experiment I	Experiment II	No ozone change
SW	0.12	0.15	0.18
LW (instantaneous)	0.03	0.04	0
LW	0.11	0.08	(0-0.08)
NET = SW + LW	0.23	0.23	(0.18-0.26)

cooling) and less than 0.08 because the experiments of Haigh (1999) suggested that the LW forcing of the no ozone change case is lower than that when the ozone is increased. This gives a value for total solar radiative forcing, with no ozone change, to be in the range 0.18 – 0.26 W m^{-2}. Overall the results suggest that the inclusion of solar-induced ozone increases has very little effect on the global average solar radiative forcing, providing the LW effect due to stratospheric heating is included. However, as pointed out by Haigh (1999), the vertical and latitudinal structure of the energy absorption in the troposphere and at the surface will be very different depending on the relative SW and LW contributions to the total.

It is interesting to note that the value of 0.23 W m^{-2}derived for the amplitude of radiative forcing between minimum and maximum of the 11-year solar activity cycle is of similar magnitude to the approximately 0.3 W m^{-2} decade^{-1} deduced to be due to increases in greenhouse gas concentrations over recent years (see Shine, 2000, for further discussion of the radiative forcing context). Thus solar modulations of the overall warming trend in global average sea surface temperatures are quite likely to be observable. These cyclical modulations will not have an impact on longer term climate trends but, based on the above calculations, the increase in TSI of 2.9 W m^{-2} (Lean *et al.*, 1995) or 3.9 W m^{-2} (Hoyt and Schatten, 1998) deduced to have occurred since 1700 would suggest an increase in solar radiative forcing of 0.6 or 0.8 W m^{-2} since that date.

5. Discussion and Conclusions

A GCM has been used to investigate the impact of changing irradiance over the 11-year solar cycle on the climate of the lower and middle atmosphere. Both the spectral composition of the changing irradiance and solar-induced changes in stratospheric ozone have been taken into account. A robust pattern of response was found in the troposphere consisting of anomalies in zonal mean zonal wind in the summer hemisphere that are suggestive of a weakening and broadening of the subtropical jet. This is consistent with the imposed changes in radiative heating rates which tend to preferentially warm the summer polar stratosphere thus reducing the pole-equator temperature gradient resulting in an easterly thermal wind. The changes in zonal wind are associated with modifications to the mean meridional circulation such that the winter Hadley cell weakens and broadens and the summer hemisphere cells are moved polewards. This response is very similar to that seen in the modelling studies of Haigh (1996, 1999) despite the different model used and differences in the design of the experiments. This provides further evidence that the pattern represents a robust response; it also suggests that an improved representation of the middle atmosphere, such as included in the new runs, has little impact on the effects seen in the troposphere, at least in the summer hemisphere.

Two experiments using different assumptions concerning ozone changes were carried out. The temperature response of the middle atmosphere reflects the changes in ozone distribution. The response of the troposphere was qualitatively similar in both cases in the summer hemisphere with that of Experiment I (slightly larger ozone column changes) having somewhat larger magnitude effect.

In the middle atmosphere the development of an anomaly in zonal wind in the winter hemisphere seen by Kodera (1995) in observational data and the variations in wind and temperature seen throughout the year by Hood *et al.* (1993) in the upper stratosphere are fairly well reproduced in the model simulations. The model does not satisfactorily reproduce the changes in 30 mbar geopotential height seen by van Loon and Labitzke although it does predict a general increase in low to mid-latitudes as observed.

This work presents the first attempt to calculate radiative forcing due to changes in solar irradiance taking into account stratospheric adjustment due to temperature changes derived from the GCM runs. The thermal infrared component of the adjusted forcing is shown to be important. The global average radiative forcings of the experiments with different ozone change scenarios are similar, showing a balance between ozone-induced longwave and shortwave effects. Thus the inclusion of ozone effects may not be important in terms of calculating global average solar radiative forcing. Nevertheless, the ozone does affect the wavelength composition and the latitudinal distribution of the radiation entering the troposphere and thus the location of its absorption.

Acknowledgements

This work was supported by the UK Natural Environmental Research Council through a graduate studentship and through UGAMP.

References

Cullen, M. J. P.: 1993, 'The unified forecast/climate model', *Meteorol. Mag.* **122**, 81–94.

Edwards, J. M., and Slingo, A.: 1996, 'Studies with a flexible new radiation code. I. Choosing a configuration for a large-scale model', *Q. J. R. Meteorol. Soc.* **122**, 689–719.

Fröhlich, C., and Lean, J.: 1998, 'The Sun's total irradiance: Cycles, trends and related climate change uncertainties since 1976', *Geophys. Res. Lett.* **25**, 4377–4380.

Haigh, J. D.: 1994, 'The role of stratospheric ozone in modulating the solar radiative forcing of climate', *Nature* **370**, 544–546.

Haigh, J. D.: 1996, 'The impact of solar variability on climate', *Science* **272**, 981–984.

Haigh, J. D.: 1999, 'A GCM study of climate change in response to the 11-year solar cycle', *Q. J. R. Meteorol. Soc.* **125**, 871–892.

Hansen, J., Sato, M., and Ruedy, R.: 1997, 'Radiative forcing and climate response', *J. Geophys. Res.* **102**, 6831–6864.

Hood, L. L., Jirikowic, J. L., and McCormack, J. P.: 1993, 'Quasi- decadal variability of the stratosphere: influence of long-term solar ultraviolet variations', *J. Atmos. Sci.* **50**, 3941–3958.

Hoyt, D. V., and Schatten, K. H.: 1998, 'Group Sunspot Numbers: A new solar activity reconstruction', *Solar Phys.* **181**, 491–512.

Kodera, K.: 1995, 'On the origin and nature of the interannual variability of the winter stratospheric circulation in the northern hemisphere', *J. Geophys. Res.* **100**, 14077–14087.

Kodera, K., Chiba, M., and Shibata, K.: 1991, 'A general circulation model study of the solar and QBO modulation of the stratospheric circulation during the northern hemisphere winter' *Geophys. Res. Lett.* **18**, 1209–1212.

Lean, J., and Rind, D.: 1998, 'Climate forcing by changing solar radiation' *J. Climate* **11**, 3069–3094.

Lean, J., Beer, J., and Bradley, R. S.: 1995, 'Reconstruction of solar irradiance since 1610: Implications for climate change', *Geophys. Res. Lett.* **22**, 3195–3198.

McCormack, J. P., and Hood, L. L.: 1996, 'Apparent solar-cycle variations of upper stratospheric ozone and temperature: latitude and seasonal dependencies', *J. Geophys. Res.* **101**, 20933–20944.

Morcrette, J.-J.: 1991, 'Radiation and cloud radiative properties in the European Centre for Medium Range Weather Forecasts Forecasting System', *J. Geophys. Res.* **96**, 9121–9131.

Myhre, G., Stordal, F., Rognerud, B., and Isaksen, I. S. A.: 1998, 'Radiative forcing due to stratospheric ozone' in *Atmospheric Ozone: Proc. of the XVIII Quadrennial Ozone Symposium*, Eds. Bojkov, R. D. and Visconti, G., pp. 813–816.

Rind D., and Balachandran, N. K.: 1995, 'Modelling the Effects of UV variability and the QBO on the Troposphere- Stratosphere Systems. Part II: The troposphere', *J. Climate* **8**, 2080–2095.

Shindell, D., Rind, D., Balachandran, N., Lean, J., and Lonergan, J.: 1999, 'Solar cycle variability, ozone and climate' *Science* **284**, 305–308.

Shine, K. P.: 2000, *Space Sci. Rev.*, this volume.

Solanki, S. K., and Unruh, Y. C.: 1998, 'A model of the wavelength dependence of solar irradiance variations', *Astron. and Astrophys.* **329**, 747–753.

van Loon, H., and Labitzke, K.: 2000, *Space Sci. Rev.*, this volume.

Wuebbles, D. J., Wei, C.-F., and Patten, K. O.: 1998, 'Effects on stratospheric ozone and temperature during the Maunder Minimum', *Geophys. Res. Lett.* **25**, 523–526.

COSMIC RAYS, CLOUDS, AND CLIMATE

NIGEL MARSH and HENRIK SVENSMARK
Danish Space Research Institute,
Juliane Maries Vej 30, DK-2100 Copenhagen Ø, Denmark.

Received: 2 March 2000; Accepted in final form: 10 May 2000

Abstract. A correlation between a global average of low cloud cover and the flux of cosmic rays incident in the atmosphere has been observed during the last solar cycle. The ionising potential of Earth bound cosmic rays are modulated by the state of the heliosphere, while clouds play an important role in the Earth's radiation budget through trapping outgoing radiation and reflecting incoming radiation. If a physical link between these two features can be established, it would provide a mechanism linking solar activity and Earth's climate. Recent satellite observations have further revealed a correlation between cosmic ray flux and low cloud top temperature. The temperature of a cloud depends on the radiation properties determined by its droplet distribution. Low clouds are warm (>273 K) and therefore consist of liquid water droplets. At typical atmospheric supersaturations ($\sim 1\%$) a liquid cloud drop will only form in the presence of an aerosol, which acts as a condensation site. The droplet distribution of a cloud will then depend on the number of aerosols activated as cloud condensation nuclei (CCN) and the level of super saturation. Based on observational evidence it is argued that a mechanism to explain the cosmic ray-cloud link might be found through the role of atmospheric ionisation in aerosol production and/or growth. Observations of local aerosol increases in low cloud due to ship exhaust indicate that a small perturbation in atmospheric aerosol can have a major impact on low cloud radiative properties. Thus, a moderate influence on atmospheric aerosol distributions from cosmic ray ionisation would have a strong influence on the Earth's radiation budget. Historical evidence over the past 1000 years indicates that changes in climate have occurred in accord with variability in cosmic ray intensities. Such changes are in agreement with the sign of cloud radiative forcing associated with cosmic ray variability as estimated from satellite observations.

1. Introduction

The Sun is a variable star, which emits both electromagnetic radiation and energetic particles known as the solar wind – these are released as a plasma carrying a fingerprint of the solar magnetic field throughout inter-planetary space. Effects from the solar wind are felt at distances well beyond Neptune, possibly up to 200 AU from the Sun, in a region of space known as the Heliosphere (Fig. 1). The flux of the inter-planetary magnetic field (IMF) at 1 AU is $\sim$5 nT. Variability in solar activity affects both the radiative output of the Sun and the strength of the IMF carried by the solar wind. The IMF shields the heliosphere from galactic cosmic radiation which consists of energetic particles, mainly protons that are accelerated through stellar processes in our galaxy. Thus, solar variability modulates both the flux of incoming galactic cosmic radiation (GCR) and the amount of solar radiation received by the planets.

Space Science Reviews **94:** 215–230, 2000.

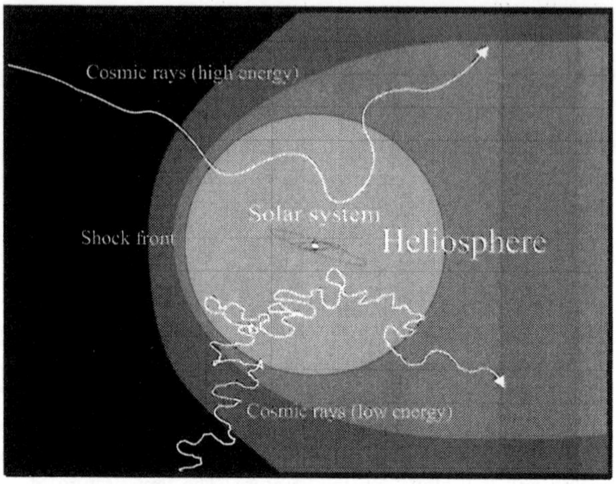

Figure 1. Schematic of Heliosphere

Satellite measurements of total solar radiation over the past 20 years have revealed that the total radiative output of the Sun varies by $\sim$0.1% (1.4 Wm^{-2}) over a solar cycle (approx. 11 years). Currently this is believed to be too small to significantly influence observed surface temperatures directly, although this may have been larger back in time (Lean et al., 1995). However, the UV region of the solar radiation spectrum is observed to vary by up to 10%. UV is absorbed by ozone and can lead to variations in Stratospheric heating. It has been suggested that variability in this heating due to changes in solar activity can have an influence on the dynamical properties of the troposphere below (Geller and Alpert, 1980; Haigh, 1996; Shindell et al., 1999). It remains to be seen how this affects surface climate once taking into account the many feedback mechanisms involved. GCRs interact with the Earth's atmosphere through nuclear collisions producing secondary particles (protons, neutrons and muons) which can penetrate deeper into the atmosphere; these undergo further collisions, which leads to a cascade of particles. The cascade ends when the nucleonic component i.e., neutrons and protons, have reached energies that are too low for further particle production. The primary energy spectrum results in a maximum production rate at about 16 km. Below 16 km, the charge particle intensity drops off and it is mainly the muons and electrons, products of very high-energy GCR collisions, which contribute to the charged particle intensity at low altitudes (Lal and Peters, 1967; Herman and Goldberg, 1978). Observations of both muons and neutrons at the Earth's surface contain information related to atmospheric ionisation at different altitudes and can reveal variability between different energies of the GCR spectrum (Fig. 2).

Cosmic rays are responsible for the ionisation in the lower atmosphere below 35 km. Atmospheric profiles of ion pair production taken during solar maximum and solar minimum can be seen in Fig. 3. The difference between max and min peaks at around 13 km and continues down to the surface.

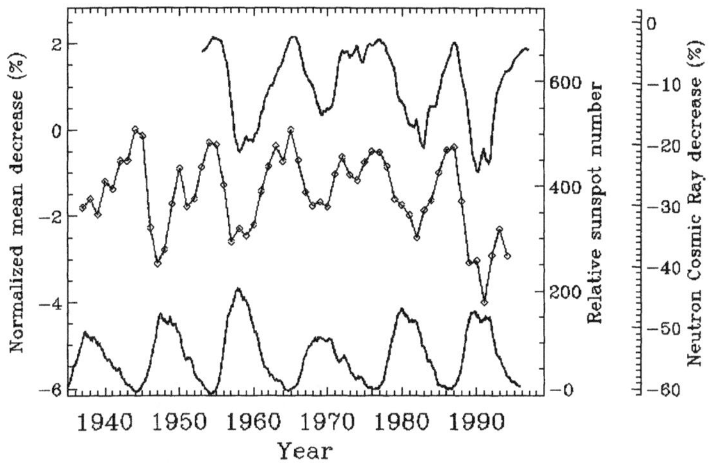

Figure 2. Temporal variability of GCR measured at the Earth surface together with the sunspot number (adapted from Svensmark, 1998). This reveals a clear solar cycle modulation of the Cosmic Ray flux. The top curve is monthly averages of GCR as measured by the CLIMAX neutron monitor (cut-off rigidity >3 GeV), Colorado (1953–1996), values have been normalised to May 1965. The middle curve is the annual mean variation in GCR captured by ionisation chambers (1934–1994), values are normalised to 1965 (adapted from Ahluwalia, 1997). The bottom curve is the relative sunspot number.

Edward P. Ney (1959) observed that atmospheric ionisation was 'the meteorological variable subject to the largest solar cycle modulation in denser layers of the atmosphere'. How this might influence climate is open to speculation, but Ney suggested a possible link might be through an influence on 'storminess'. It was not until 1997 that Svensmark and Friis-Christensen (1997) showed from observations that ionisation might be influencing global cloud cover.

2. Clouds and Cosmic Rays

Figure 4 is a composite of Satellite observations of Earth's total cloud cover from Svensmark (1998). The cloud data comprise the NIMBUS-7 CMATRIX project (Stowe et al., 1988) (triangles), and the International Satellite Cloud Climatology Project (ISCCP) (Rossow and Schiffer, 1991) (squares). Finally data from the Defence Satellite Meteorological Program (DMSP) Special Sensor Microwave/Imager (SSM/I) (diamonds) (Weng and Grody, 1994). For details see Svensmark (1998). In the figure the cloud data is compared with variation in GCR flux and the 10.7 cm radio flux from the Sun. This radio flux follows closely variations in total solar irradiance, soft X-rays, and ultraviolet radiation over this period. There are clearly differences between the variation of GCR and the radio flux with a lag of almost two years before 1987. Crucially, Earth's cloud cover follows the GCR flux more closely than the 10.7 cm flux. This observation suggests that a mechanism linking

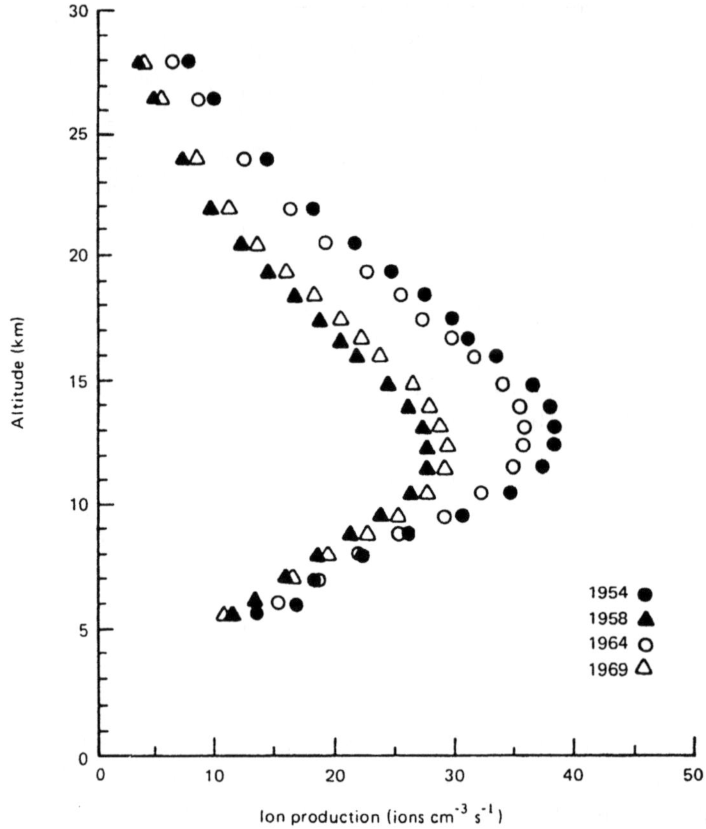

Figure 3. Variability over a solar cycle in the vertical profiles of Ion pair production over Thule, which is situated at a high magnetic latitude (adapted from Neher, 1971). Due to geomagnetic shielding, the levels of ionisation are reduced at lower latitudes, but essential features in the vertical profile are similar.

cloud cover with solar activity will involve changes in ionisation properties of the atmosphere rather than incoming solar radiation.

Clouds influence vertically integrated radiative properties of the atmosphere by cooling through reflection of incoming shortwave radiation, and heating through trapping of outgoing longwave radiation. The net radiative impact of a particular cloud is mainly dependent upon its height above the surface and its optical thickness as can be seen from Tab. I. High optically thin clouds tend to heat while low clouds, which are generally thick, tend to cool (Hartmann, 1993). With a current climatic estimate for the net forcing of the global cloud cover as a 27.7 Wm^{-2} cooling in the atmosphere, clouds clearly play an important role in the Earth's radiation budget (Hartmann, 1993; Ohring and Clapp, 1980; Ramanathan et al., 1989; Ardanuy et al., 1991). A significant solar influence on global cloud properties is potentially important for the Earth's radiation budget, but the sign and magnitude will be determined by the cloud types affected.

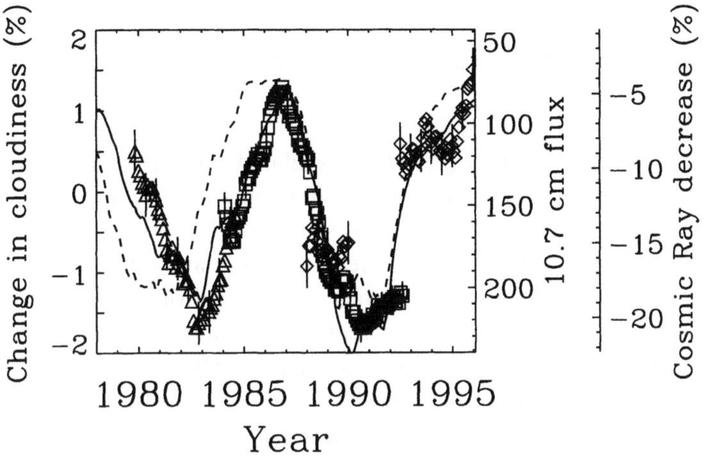

Figure 4. Composite figure showing changes in Earth's total cloud cover for four cloud data sets obtained from satellite observations. Also plotted are the cosmic rays fluxes from Climax (solid curve, normalised to May 1965), and 10.7-cm Solar flux (broken curve, in units of $10^{-22}\,Wm^{-2}Hz^{-1}$). Triangles are the Nimbus-7 data, squares are the ISCCP-C2 and ISCCP-D2 data, and diamonds are the DMSP data. All the displayed data have been smoothed using a 12-month running mean. The Nimbus-7 and the DMSP data is total cloud cover for the Southern Hemisphere over oceans, and the ISCCP data have been derived from geostationary satellites over oceans with the tropics excluded.

TABLE I

Radiative forcing from ERBE experiment (adapted from Hartmann, 1993).

	High Clouds		Middle Clouds		Low Clouds	Total
	Thin	*Thick*	*Thin*	*Thick*		
Global Cloud Fraction %	10.1	8.6	10.7	7.3	26.6	63.3
Net CRF Wm^{-2}	2.4	−7.0	1.1	−7.5	−16.7	−27.7

Recently it has been possible to observe cloud properties at different altitudes from the global dataset International Satellite Cloud Climate Project (ISCCP-D2) (Schiffer and Rossow, 1983; Schiffer and Rossow, 1985; Rossow and Schiffer, 1991; Rossow et al., 1996). These are derived from Top Of Atmosphere (TOA) radiances for the period July 1983 to September 1994. Monthly averages of infrared (IR) measurements are preferred due to their superior spatial and temporal homogeneity over visual observations that can only be detected during daylight. Cloud cover is obtained from an algorithm using the TOA IR statistics to identify the cloudiness on an equal area grid (280 km × 280 km). Cloud top temperatures (CT) and pressures (CP) are obtained from an ISCCP IR model constrained by water vapour and vertical temperature profiles retrieved from the TIROS Observed Vertical Sounder (TOVS) (Rossow et al., 1996). CT and CP are found by assuming an opaque cloud, i.e. emissivity $\epsilon = 1$, and adjusting the clouds pressure level (effectively cloud height) in the model until the reconstructed outgoing IR flux at TOA matches that observed. Based on retrieved CP, clouds are divided into Low >680 hPa (<3.2 km), Middle ~680–440 hPa (3.2–6.5 km), and High <440 hPa (>6.5 km), Fig. 5.

Clearly, GCR flux over this period is correlated with a ~2% absolute change (~7% relative change) in low cloud cover, while there is no correlation with middle and high clouds. However, at these time scales GCR ionisation is not the only mechanism affecting low clouds, there are of course many other decadal processes in the climate system which are important, and one should not expect a perfect correlation. The small differences between Low cloud cover and GCR seen in Fig. 5c are much smaller than the lag between the 10.7-cm flux and total cloud cover (Fig. 4), and are close to the satellite's resolution. What is surprising is that despite these limitations a signal of solar variability in low clouds is dominant at time scales longer than 1 year. Since low clouds tend to be optically thick they are efficient reflectors of sunlight and have a negative impact on the Earth's radiation budget (Tab. I). Based on observations made under the Earth Radiation Budget Experiment (ERBE) the average low cloud radiative forcing contributes a cooling of $-16.7\ \mathrm{W/m^2}$. Thus a ~2% absolute change in low cloud cover over a solar cycle, would give a change in net low cloud forcing of $\sim 1.2\ \mathrm{W/m^2}$.

In addition to cloud cover, the IR model also generates an additional cloud parameter, cloud top temperature (CT), at Low, Mid, and High altitudes. The spatial correlation map in Figure 6 shows how low cloud top temperatures covary with GCR flux, particularly over oceans in regions where stratocumulus and marinestratus clouds are dominant (for details see (Marsh and Svensmark, 2000). The lack of correlation at high latitudes is currently not understood, but may be related to differences in processes of cloud formation.

However, the opaque cloud assumption excludes microphysical properties and so constrains cloud variability to appear only in cloud 'model height', thus introducing an element of artificial variability into CT. Observed properties of low level maritime clouds suggests that they are not opaque (Heymsfield, 1993). Relaxing the opaque assumption allows cloud variability to manifest itself through changes in cloud optical depth, i.e., $\epsilon < 1$.

3. Clouds and Aerosols

Cloud optical depth depends on processes affecting the cloud drop size distribution, and cloud thickness which is influenced by atmospheric vertical temperature profiles. Since low clouds are warm (>273 K) they consist of liquid water drops which form on cloud condensation nuclei (CCN), and so the drop size distribution depends on the density of atmospheric aerosols activated as CCN. The abundance of CCN is determined by both the level of supersaturation and the number of aerosols present in the atmosphere able to grow into CCN. Increases in supersaturation,

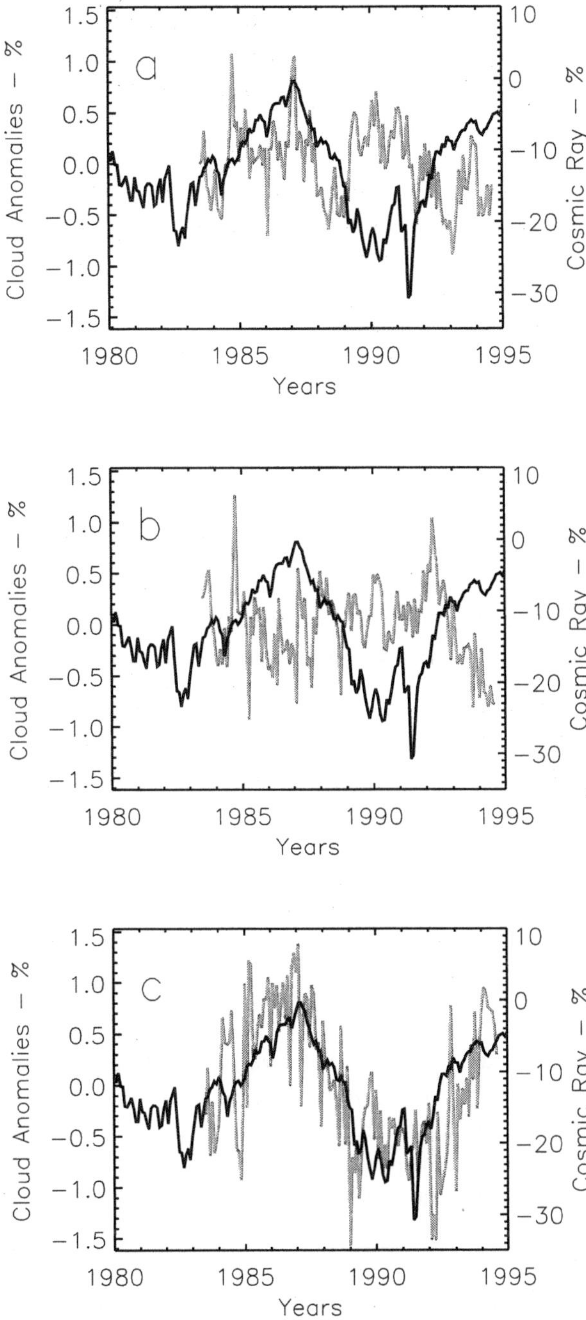

Figure 5. Monthly mean values for global anomalies of; a) high (<440 hPa), b) middle (440–680 hPa), and c) low (>680 hPa) cloud cover (gray). Galactic cosmic ray fluxes (GCR) from Climax (black, normalised to May 1965) are used as a proxy for solar variability. The global average of the temporal mean over this period for high, middle and low IR detected clouds is 13.5%, 19.9%, and 28.0% respectively.

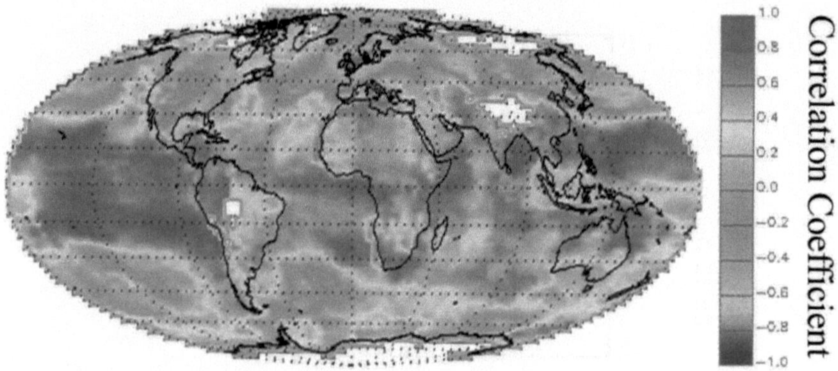

Figure 6. Global correlation map of GCR with anomalies of Low cloud top temperature (CT). The fraction of Earth possessing a correlation coefficient >0.6 is 29.6% which is significant at >99.9%.

typically between 0.1% and a few percent, activates increasingly smaller aerosols. A solar signal could enter low cloud properties through influencing atmospheric vertical temperature profiles, water vapour, or aerosol to CCN activation processes.

TOVS observations of the vertical profiles of water vapour and temperature demonstrate little correlation with GCR (Fig. 7). This suggests that solar induced variability of local thermodynamic properties in the atmosphere cannot be entirely responsible for the observed changes in low cloud properties. Since the vertical atmospheric structure depends on radiatively driven large-scale circulation, this might not be surprising given that variability in solar irradiance agrees poorly with cloud properties (Fig. 4).

4. Aerosols and Cosmic Rays

Assuming typical atmospheric water vapour saturations of up to a few percent, the abundance of CCN is determined through properties of the background aerosol size distribution (~0.1–1.0 mm). Production of aerosol can be due to many processes involving: gas-particle conversion, droplet-particle conversion, i.e., evaporation of water droplets containing dissolved matter, and bulk particles from the surface, e.g., smoke, dust, or pollen (Pruppacher and Klett, 1997). Observations of spectra in regions of low cloud formation indicate that aerosols are produced locally. In the troposphere it is believed that ionisation contributes to the gas-particle formation of ultrafine (<20 nm) aerosol, and the subsequent growth into the matured aerosol distributions which act as CCN (Turco et al., 1998; Hõrrak et al., 1998; Yu and Turco, 2000). Figure 8 shows the ionisation density profile of the atmosphere and the major ion species produced. The complexity of ions increases with decreasing altitude. It is the ions, produced within the troposphere by cosmic rays, which are potentially important for aerosol production.

a

b

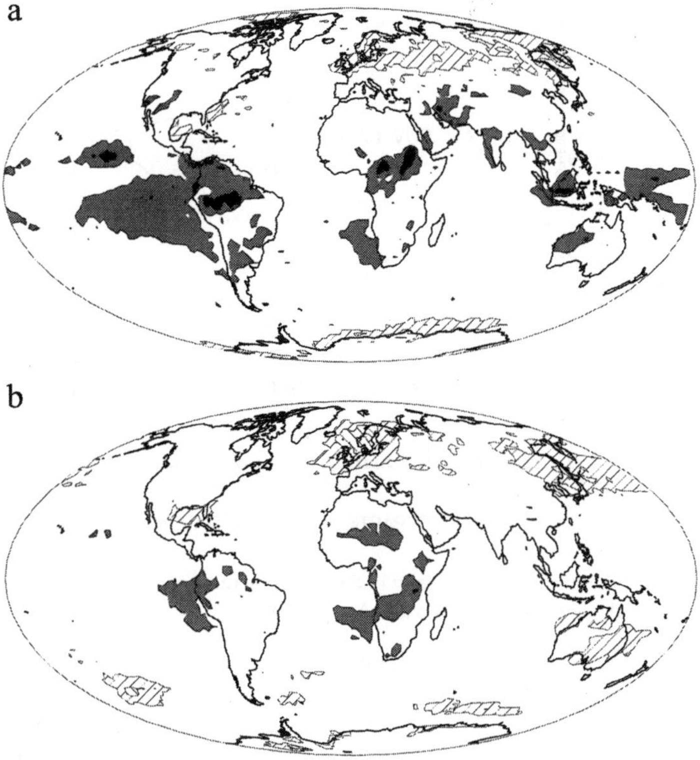

Figure 7. Global correlation maps of GCR with anomalies of a) atmospheric water vapour (1000–680 hPa), and b) atmospheric temperature at 740 hPa. The water vapour and temperature are given by TOVS observations covering the period July 1983 – September 1994 (Rossow, 1996). Anomalies and correlation coefficients are calculated as in Figs. 5 and 6 (see both figure captions). The shading corresponds to filled contours, where correlation coefficient $r > 0.8$ - black, $r > 0.6$ - grey, and $r < -0.6$ - hatched, there are no regions with $r < -0.8$. The fraction of Earth possessing a correlation coefficient > 0.6 is a) 11.7% and b) 3.4% which is not significant.

Currently it is uncertain whether variability in atmospheric ionisation due to the GCR flux could have a significant effect on either aerosol production or droplet growth. However, models have revealed that nucleation through ion-ion recombination (Turco et al., 1998) is capable of maintaining a background aerosol distribution with realistic concentrations in the troposphere, Tab. II.

This is just one possible mechanism, but it emphasises the potential importance of ionisation for troposphere aerosol production. It is not unrealistic to imagine that a systematic change in cosmic ray ionisation could lead to a change in aerosol populations acting as CCN and thus influence cloud properties.

Note that the process of aerosols acting as a source of CCN for low clouds is quite different to the source of nuclei for higher altitude ice clouds. Therefore, the influence of ionisation on aerosol processes affecting low cloud CCN need not

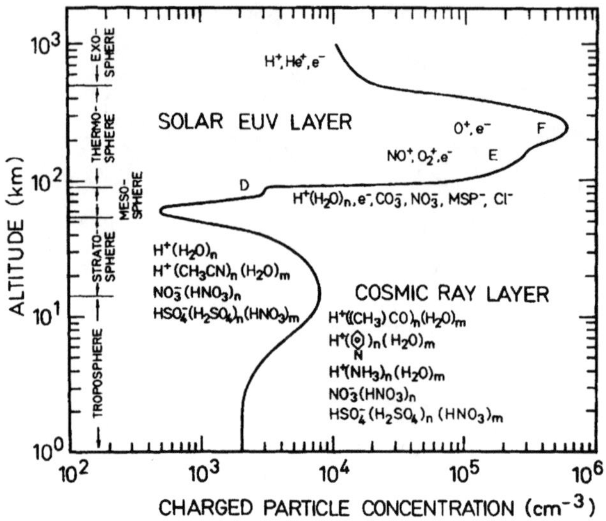

Figure 8. Vertical Profile of ionisation density in the atmosphere. Included are the major ion species at various altitudes (adapted from Viggiano and Arnold, 1995).

TABLE II
Aerosols

	Concentration - cm^{-3}
CCN in a maritime air mass at 1% supersaturation over the Pacific (Pruppacher and Klett, 1997).	~100
Background aerosol in a maritime air over the Pacific (Pruppacher and Klett, 1997).	~200
Background aerosol distribution maintained by ion-ion recombination (Turco et al., 1998).	~200–500

be important for high clouds. A detailed discussion of ice nuclei can be found in Pruppacher and Klett (1997).

5. Influence of the Aerosol Distribution on Cloud Radiative Properties

The effect of increased aerosol on cloud radiative properties can be observed directly from ship tracks. Ship exhaust emits aerosol into low cloud layers thus increasing the local background aerosol. Consequently, the cloud droplet distribution is modified which influences the radiative properties of the cloud. Two satellite images at different wavelengths can be seen in Fig. 9, where a number of ship tracks are observable. In the visible (upper image), these show up as regular

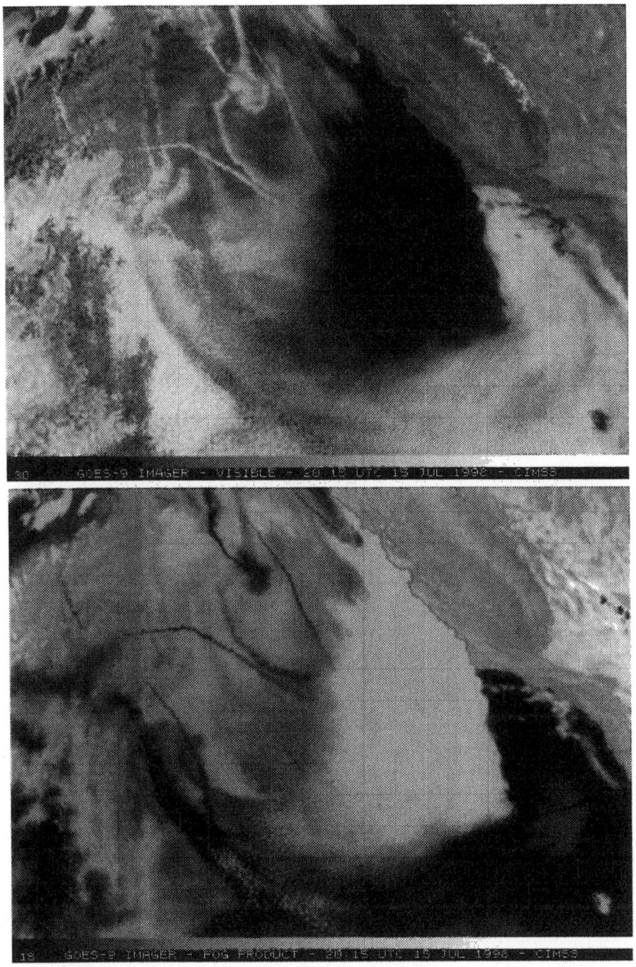

Figure 9. Images of Ship Condensation Trails Off the California Coast taken on 15 July 1998 by GOES-9. The upper image is in the visible (0.7 μm) and the lower image is the IR Fog Product (10.7 μm – 3.9 μm).

bright white lines indicating an increase in cloud albedo. The lower image is the difference between NIR and IR where the same tracks show up as dark regions. The NIR (3.2 μm) is dominated by absorption properties of the cloud while the IR (10.7 μm) is the result of the cloud emissivity. Subtracting the NIR image from the IR enhances regions of increased absorption and emission such as the dark tracks due to ship exhaust.

In-situ aircraft measurements made through shiptrack contaminated cloud give detail to the radiative changes (King et al., 1993). Figure 10 shows radiation intensities from an aircraft path through the middle of such cloud looking both in the nadir (down) and zenith (up). In the visible, the greatest influence is from backscatter

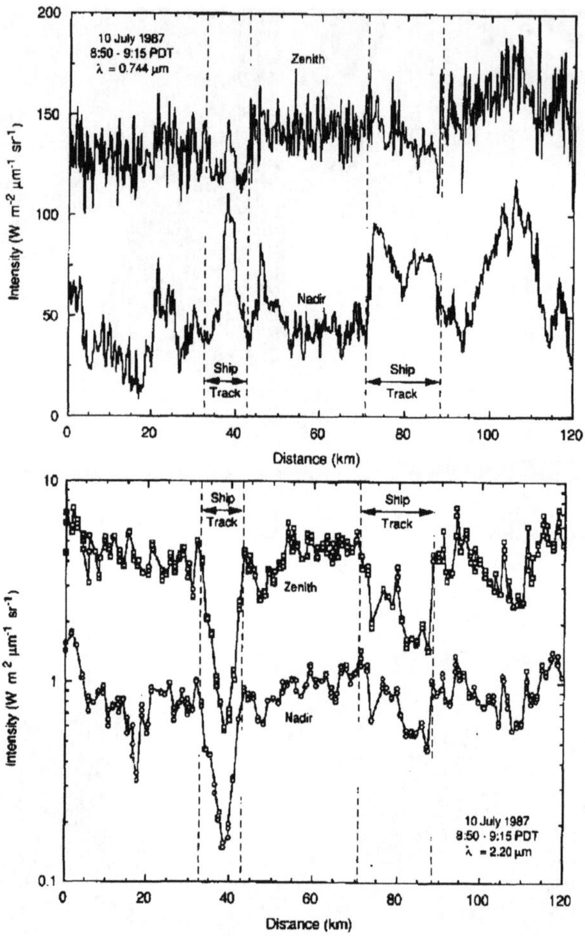

Figure 10. Visible 0.7 μm (top) and Near IR 2.2 μm (bottom) aircraft observations of ship tracks in low cloud (adapted from King et al., 1993).

below the aircraft, while the NIR reveals strong absorption both from above and below. In fact, the absorption is much stronger from above indicating the cloud is absorbing a greater flux of incoming radiation. This suggests that an increase in aerosol gives a decrease in droplet size and an increase in cloud liquid water, thus enhancing absorption of incoming radiation from above the cloud. Consequently, the cloud will heat in these regions, which explains the darkened tracks in the bottom satellite image of Fig. 9.

From the ship tracks, it is apparent that increases in aerosol leads to both an increase in albedo and an increase in absorption and therefore cloud temperature. If an increase in GCR can lead to a similar increase in aerosol, then this would explain the positive correlations between GCR and low cloud top temperatures (Fig. 7) over a large fraction of the globe.

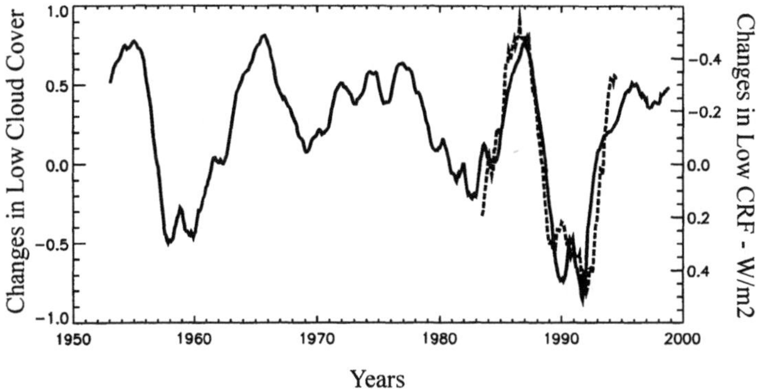

Figure 11. Reconstructed low cloud cover (solid) from galactic cosmic ray fluxes measured with a neutron counter at Huancayo (cut-off rigidity 12.91 GeV). Also shown is the 12-month smoothed global low cloud cover (dashed) from Fig. 5c. The right axis is an estimated of changes in Cloud Radiative Forcing (CRF) based on the ERBE results displayed in Tab. I. See Tab. III for estimates of long term trends in CRF from this reconstruction.

6. Estimates of Past Cloud Radiative Forcing

The radiative impact from cosmic ray influence on clouds can be extended back in time from a reconstruction assuming a linear response function (Svensmark, 1998). It was argued above, that cosmic ray ionisation in the lower troposphere affecting aerosol distributions could explain the solar variability observed in low clouds. The high-energy secondary particles penetrate the atmosphere to depths of low cloud formation, thus the Huancayo neutron counter with cut-off rigidity at 12.91 GeV is used in the reconstruction. From the following relationship:

$$\text{Reconstructed low cloud cover} = \alpha \times \text{GCR} + \beta \qquad (1)$$

a history of low cloud cover is found, where α and β are obtained from a least squares fit over the period of available ISCCP low cloud data (1983–1994). The reconstructed low cloud cover is shown in Fig. 11.

Levels of cosmic rays received at Earth are influenced by the strength of the solar magnetic field. Thus, relative changes in the solar magnetic field can be used to estimate relative changes in cloud forcing. Lockwood et al. (1999) have reconstructed a history of solar source magnetic flux for the past century which agrees well with the available satellite observations for the period 1964–1995. This allows an estimate for cloud radiative forcing to be made for the period before cosmic ray observations were available. The solar magnetic field is observed to have more than doubled over the past century while at the same time cosmic radiation measured at Huancayo has decreased by 3.5% (Tab. III). Using the reconstruction of low cloud cover in Fig. 11 and the ERBE estimates for cloud radiative forcing (Tab. I) one can infer a $\sim$1.4 $\mathrm{Wm^{-2}}$ warming attributed to low cloud cover changes over the past century (1901–1995). This is potentially important when considering that over the same period the estimated heating from increased CO_2 emissions is $\sim$1.5 $\mathrm{Wm^{-2}}$

TABLE III

Estimates of radiative heating from the reconstruction of low clouds shown in Fig. 11.

Estimated Relative Change (%) in:	Solar Max-Min 1983–1987	30-year trend 1964–1995	Century trend 1901–1995
Cosmic Ray (>12.91 GeV)	3.5	−3.5	−11.2
Solar Source Magnetic Flux		41	131
Global Low Cloud Fraction (ISCCP)	6.1	−3.5	−8.6
Net Change (Wm^{-2}) in			
ERBE Low CRF ($-16.7\,Wm^{-2}$)	−1.0	0.5	1.4

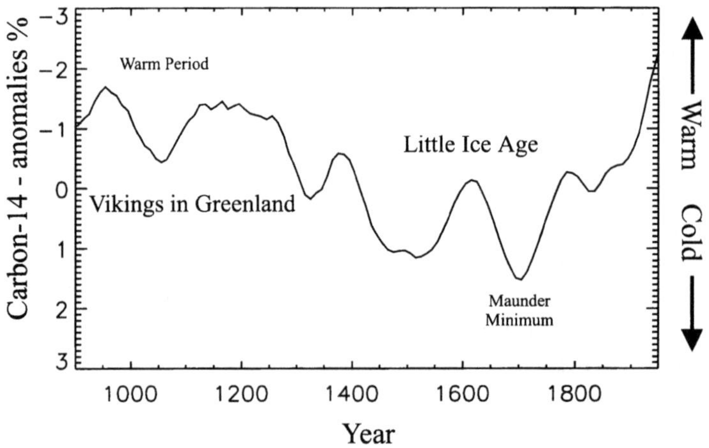

Figure 12. Changes in ^{14}C production over the past 1000 years relative to 1950. ^{14}C is produced through nuclear interactions of cosmic ray particles in the atmosphere. Cosmic rays are modulated by the solar wind; thus, variability in ^{14}C production is mainly a result of changes in solar activity. High solar activity leads to stronger shielding from the solar wind and thus a reduction in the production rate. Note the inverted axis for ^{14}C anomalies. The sharp deviation of ^{14}C during the last century is due to the burning of ^{14}C depleted fossil fuels (the Suess effect).

and changes in solar irradiance received at Earth are $0.4\,Wm^{-2}$ (Lockwood and Stamper, 1999).

A comparison between cosmic rays and climate can be extended yet further back in time, as a history of Solar activity is well known for the past 10,000 years. This is from observing changes in the production of radioisotopes in the atmosphere by galactic cosmic rays. Variations in ^{14}C production for the last millennium together with major climate epochs can be seen in Fig. 12.

Although we do not have a good estimate of the absolute global temperature record covering this period, some of the major changes in ^{14}C appear to coincide with significant climate shifts. Solar activity was particularly high from AD 1000–

1300, known as the medieval warm period, when wine was made from grapes grown in England, and the Vikings colonised Greenland. This was followed by a cold period known as the Little Ice Age, which lasted until the middle of the 19th century. During this period, the Vikings left Greenland and the River Thames regularly froze in London. The Maunder Minimum (1645–1715) was a particularly cold spell when there was a complete absence of sunspots indicating very low solar activity. This evidence suggests that in the past people experienced climate changes that have occurred in concurrence with extremes in solar variability.

7. Conclusions

Based on the ISCCP-D2 IR cloud data there is a clear correlation between solar activity and properties of low clouds in contrast to middle and high clouds. The correlation is seen in two independent parameters; a) Low cloud cover, and b) Low cloud top temperature. Observations of atmospheric parameters from TOVS do not support a solar-cloud link through tropospheric dynamics influenced by UV absorption in the stratosphere. However, since low clouds are warm and comprise of water droplets nucleated by aerosol, a link may be found through a GCR influence on the atmospheric aerosol distribution. Ship tracks suggest that low cloud radiative properties are sensitive to the background aerosol distribution such that radiative forcing due to increased aerosol leads to a warming of the cloud. If an increase in GCR can lead to similar increases in aerosol, then this would explain the positive correlation between GCR and low cloud top temperatures over a large fraction of the globe. Historical records suggest that systematic changes in global climate follow observed changes in GCR. This is consistent with low GCR corresponding to fewer low clouds and therefore a warmer climate. Crude estimates of changes in cloud radiative forcing over the past century, when the solar magnetic flux more than doubled, indicates that a GCR-cloud mechanism could have contributed $\sim$1.4 Wm^{-2} to the observed global warming. These observations provide compelling evidence to warrant further study of the effect of GCR on clouds.

References

Ardanuy, P., Stowe, L.L., Gruber, A., and Weiss, M.: 1991, 'Shortwave, longwave, and net cloud-radiative forcing as determined from Nimbus-7 observations', *J. Geophys. Res.* **96**, 18,537.

Geller, M.A., and Alpert, J.C.: 1980, 'Planetary wave coupling between the troposphere and the middle atmosphere as a possible sun–weather mechanism', *J. Atmos. Sci.* **37**, 1197.

Haigh, J.D.: 1996, 'The impact of solar variability on climate', *Science* **272**, 981.

Hartmann, D.L.: 1993, *Radiative effects of clouds on Earth's climate, in Aerosol-Cloud-Climate Interactions*, Academic Press.

Herman, J.R., and Goldberg, R.A.: 1978, *Sun, Weather, and Climate*, NASA SP 426.

Heymsfield, A.J.: 1993, *Microphysical Structures of Stratiform and Cirrus Clouds, in Aerosol-Cloud-Climate Interactions*, Academic Press.

Hõrrak, U., Salm, J., and Tammet, H.: 1998, 'Bursts of intermediate ions in atmospheric air', *J. Geophys. Res.* **103**, 13,909.

King, M.D., Radke, L.F., and Hobbs, P.V.: 1993, 'Optical properties of marine stratocumulus clouds modified by ships', *J. Geophys. Res.* **98**, 2729.

Lal, D., and Peters, B.: 1967, 'Cosmic ray produced radioactivity on the Earth', in *Encyclopaedia of Physics*, Springer-Verlag, Berlin.

Lean, J., Beer, J., and Bradley, R.: 1995, 'Reconstruction of solar irradiance since 1610: Implications for climate change', *Geophys. Res. Lett.* **22**, 3195.

Lockwood, M., Stamper, R., and Wild M.N.: 1999, 'A doubling of the Sun's coronal magnetic field during the past 100 years', *Nature* **399**, 437.

Lockwood, M. and Stamper, R.: 1999, 'Long-term drift of the coronal source magnetic flux and the total solar irradiance', *Geophys. Res. Lett.* **26**, 2461.

Marsh, N.D. and Svensmark, H.: 2000, 'Low cloud properties influenced by solar activity', *Phys. Rev. Lett.*, submitted.

Neher, H.V.: 1971, 'Cosmic-Rays at high latitudes and altitudes covering four solar maxima', *J. Geophys. Res.* **76**, 1637.

Ney, E.R.: 1959, 'Cosmic radiation and the weather', *Nature* **183**, 451.

Ohring, G., and Clapp, P.F.: 1980, 'The effect of changes in cloud amount on the net radiation at the top of the atmosphere', *J. Atmos. Sci.* **37**, 447.

Pruppacher, H.R., and Klett, J.D.: 1997, *Microphysics of Clouds and Precipitation*, Kluwer, Dordrecht.

Ramanathan, V., et al.: 1989, 'Results from the Earth's radiation budget experiment', *Science* **243**, 57.

Rossow, W.B., and Schiffer, R.A.: 1991, 'ISCCP Cloud Data Products', *Bull. Am. Met. Soc.* **72**, 2.

Rossow, W.B., Walker, A.W., Beuschel, D.E., and Roiter, M.D.: 1996, *International Satellite Cloud Climatology Project (ISCCP): Documentation of New Cloud Datasets*, World Meteorological Organization, Geneva.

Schiffer, R.A., and Rossow, W.B.: 1983 'The International Satellite Cloud Climatology Project (ISCCP): The first project of the World Climate Research Programme', *Bull. Am. Met. Soc.* **64**, 779.

Schiffer, R.A., and Rossow, W.B.: 1985, 'ISCCP Global Radiance Data Set: A New Resource for Climate Research', *Bull. Am. Met. Soc.* **66**, 1498.

Shindell, D., Rind, D., Balabhandran, N., Lean, J., and Lonergan, P.: 1999, 'Solar cycle variability, ozone, and climate', *Science* **284**, 305.

Stowe, L.L., Wellemayer, C.G., Eck, T.F., Yeh, H.Y.M., and the Nimbus-7 team: 1988, *J. Clim.* **1**, 445.

Svensmark, H., and Friis-Christensen, E.: 1997, 'Variation of Cosmic Ray Flux and Global Cloud Coverage – A Missing Link in Solar-Climate Relationships', *J. Atm. Sol. Terr. Phys.* **59**, 1225.

Svensmark, H.: 1998, 'Influence of cosmic rays on climate', *Phys. Rev. Lett.* **81**, 5027.

Turco, R.P., Zhao, J.-X., and Yu, F.: 1998, 'A new source of tropospheric aerosols: Ion-ion recombination', Geophys. Res. Lett. **25**, 635.

Viggiano, A.A., and Arnold, F.: 1995, *Ion Chemistry and Composition of the Atmosphere, in Handbook of Atmospheric Electrodynamics*, CRC Press.

Weng, F., and Grody, N.C.: 1994, *J. Geophys. Res.* **99**, 25,535.

Yu, F., and Turco, R.P.: 2000, 'Ultrafine aerosol formation via ion-mediated nucleation', *Geophys. Res. Lett.* **27**, 883.

Address for Offprints:

Nigel Marsh and Henrik Svensmark, Danish Space Research Institute, Juliane Maries Vej 30, DK-2100 Copenhagen Ø, Denmark, hsv@dsri.dk

INFLUENCE OF SOLAR WIND ON THE GLOBAL ELECTRIC CIRCUIT, AND INFERRED EFFECTS ON CLOUD MICROPHYSICS, TEMPERATURE, AND DYNAMICS IN THE TROPOSPHERE

BRIAN A. TINSLEY
University of Texas at Dallas

Received: 12 November 1999; Accepted in final form: 14 June 2000

Abstract. There are at least three independent ways in which the solar wind modulates the flow of current density (Jz) in the global electric circuit. These are (A) changes in the galactic cosmic ray energy spectrum, (B) changes in the precipitation of relativistic electrons from the magnetosphere, and (C) changes in the ionospheric potential distribution in the polar caps due to magnetosphere-ionosphere coupling. The current density J_z flows between the ionosphere and the surface, and as it passes through conductivity gradients it generates space charge concentrations dependent on J_z and the conductivity gradient. The gradients are large at the surfaces of clouds and space charge concentrations of order 1000 to 10,000 elementary charges per cm^3 can be generated at cloud tops. The charge transfers to droplets, many of which are evaporating at the cloud-clear air interface. The charge remains on the residual evaporation nuclei with a lifetime against leakage of order 1000 sec, and for a longer period the nuclei also retain coatings of sulfate and organic compounds adsorbed by the droplet while in the cloud.

The charged evaporation nuclei become well mixed with more droplets in many types of clouds with penetrative mixing. The processes of entrainment and evaporation are also efficient for these clouds. The collection of such nuclei by nearby droplets is greatly increased by the electrical attraction between the charge on the particle and the image charge that it creates on the droplet. This process is called electroscavenging. Because the charge on the evaporation nuclei is derived from the original space charge, it depends on J_z, giving a rate of electroscavenging responsive to the solar wind inputs.

There may be a number of ways in which the electroscavenging has consequences for weather and climate. One possibility is enhanced production of ice. The charged evaporation nuclei have been found to be good ice forming nuclei because of their coatings, and so in supercooled clouds droplet freezing can occur by contact ice nucleation, as the evaporation nuclei are electroscavenged. Although quantitative models for the all the cloud microphysical processes that may be involved have not yet been produced, we show that for many clouds, especially those with broad droplet size distributions, relatively high droplet concentrations, and cloud top temperatures just below freezing, this process is likely to dominate over other primary ice nucleation processes. In these cases there are likely to be effects on cloud albedo and rates of sedimentation of ice, and these will depend on J_z.

For an increase in ice production in thin clouds such as altocumulus or stratocumulus the main effect is a decrease in albedo to incoming solar radiation, and in opacity to outgoing longwave radiation. At low latitudes the surface and troposphere heat, and at high latitudes in winter they cool. The change in meridional temperature gradient affects the rate of cyclogenesis, and the amplitude of planetary waves. For storm clouds, as in winter cyclones, the effect of increased ice formation is mainly to increase the rate of glaciation of lower level clouds by the seeder-feeder process. The increase in precipitation efficiency increases the rate of transfer of latent heat between the air mass and the surface. In most cyclones this is likely to result in intensification, producing changes in the vorticity area index as observed. Cyclone intensification also increases the amplitude of planetary waves, and shifts storm tracks, as observed.

Space Science Reviews **94:** 231–258, 2000.

In this paper we first describe the production of space charge and the way in which it may influence the rate of ice nucleation. Then we review theory and observations of the solar wind modulation of J_Z, and the correlated changes in atmospheric temperature and dynamics in the troposphere. The correlations are present for each input, (A, B, and C), and the detailed patterns of responses provide support for the inferred electrical effects on the physics of clouds, affecting precipitation, temperature and dynamics.

1. Introduction

The relative variations of solar wind parameters such as velocity and magnetic field strength in the neighborhood of the earth are large (from 1 to 10 times the mean values) although the energy flux is very small. There are at least three processes in which variations in solar wind parameters can be coupled into the stratosphere and troposphere, as indicated in Figure 1, and each of these modulate the flow of ionosphere-earth current density (J_z) in the global electric circuit. The coupling processes are (A) changes in the galactic cosmic ray (GCR) flux at energies below about 10 GeV; (B) changes in the precipitation of relativistic electrons from the magnetosphere; and (C) changes in the ionospheric potential distribution in the polar caps due to magnetosphere-ionosphere coupling.

Over the past 30 years changes in the temperature and dynamics in the troposphere have been found to correlate separately with each of these three inputs, as summarized by Tinsley (1996), and as we will review in Section 4. The agreement in onset and duration on the day-to-day timescale between the solar wind modulated inputs (A, B, and C) and the individual tropospheric responses (Roberts, Wilcox and Mansurov/Page effects) rules out mechanisms for the responses involving solar UV or total irradiance (see Tinsley and Deen, 1991; Tinsley and Heelis, 1993; Tinsley *et al.*, 1994; Kirkland *et al.*, 1996). These correlations prompted a search for a mechanism that would connect the common feature of all of them, i. e. the response of J_z to the inputs, with a process that could release free energy in the atmosphere. The amplification of the input by a process of this type is necessary because of the very small amounts of energy implicit in the J_z changes.

In Section 2 we consider the effect of J_z on production of space charge in the atmosphere, particularly at the interfaces between cloudy and clear air, and the transfer of this charge onto droplets and its retention by their residues after evaporation (evaporation nuclei). In Section 3 we consider the effect of this charge on the collision rate of such charged evaporation nuclei with droplets (electroscavenging; Tinsley *et al.*, 2000). Because the charge on the evaporation nuclei is derived from the space charge, the rate of electroscavenging depends on J_z. One possible mechanism connecting changes in weather and climate with J_z is indicated in the box in Figure 1. This is the enhancement of ice production in clouds by the electroscavenged evaporation nuclei acting as ice forming nuclei, and then

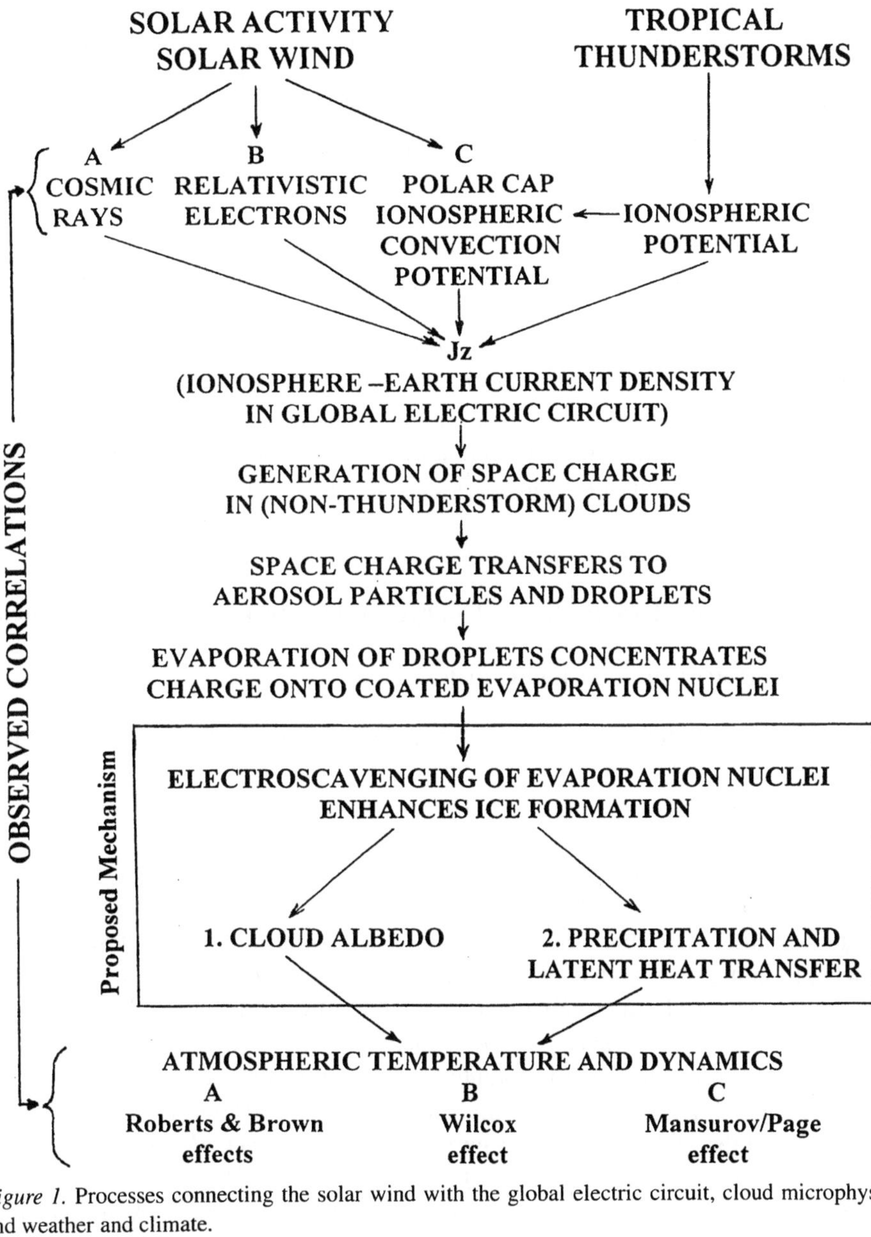

Figure 1. Processes connecting the solar wind with the global electric circuit, cloud microphysics, and weather and climate.

consequent changes in cloud albedo, precipitation, and latent heat release, affecting tropospheric temperature and dynamics.

This is only one of what could be a number of cloud microphysical processes affected by charge distributions in the atmosphere responding to solar wind inputs. For example, it has been suggested that ionization produced by cosmic rays increases the production of condensation nuclei at upper tropospheric levels, and so affects the concentration and size distributions of cloud particles (Svensmark and Friis-Christensen, 1997). This mechanism is consistent with changes in low-latitude cloud cover on the solar cycle time scale, but does not seem capable of explaining correlations of atmospheric temperature and dynamics with the inputs B and C.

In Section 4 we consider the solar wind modulation of the inputs A, B, and C, and the observed correlations of both J_z and of atmospheric temperature and dynamics with these inputs. Data on J_z variations are sparse and noisy, and the correlations of J_z with the solar wind parameters are in most cases less significant than the correlations of the meteorological parameters with the solar wind parameters. This can be understood by comparing the quality of available data; the meteorological databases are much more extensive, with averaging over many thousands of square kilometers, as compared with intermittent point measurements of atmospheric electricity in the midst of considerable natural electrical noise at the surface.

2. Global Electric Circuit, Space Charge, and Charged Evaporation Nuclei

2.1. Production of Space Charge

Figure 2a is a schematic diagram of the global electric circuit. The most important generator in the circuit, and that responsible for maintaining the ionospheric potential, V_i , of about 250 kV positive with respect to the earth's surface, is the totality of global thunderstorms. These are concentrated in the tropics over land masses, and inject a time varying current of about 1000 A into the ionosphere at any one time. Except within the polar caps, the global ionosphere is effectively an equipotential. Also, the conductivity in the mesosphere and upper stratosphere is so high, compared to that of the troposphere and lower stratosphere, that the ionospheric potential effectively extends down to mid-stratospheric levels.

The situation is like a spherical shell capacitor with a leakage current between the plates. The ionosphere-earth potential difference gives rise to a vertical current density J_z, which in clear weather can be measured in the range 1–4 pA m^{-2}. A recent review of the global electric circuit is given by Bering *et al.* (1998), and earlier comprehensive reviews by Israël (1973) and NAS (1986). The global integral of J_z constitutes the return current and is equal to the charging current. At any location on the globe J_z is given by the product of the overhead ionospheric potential V_i and the resistance of the vertical earth-ionosphere column:

$$J_z = V_i \int dz/\sigma(z) \qquad (1)$$

where z is altitude and $\sigma(z)$ is the conductivity, which increases strongly with altitude in clear air, except within the layer of mixed air in the lowest one or two km over land.

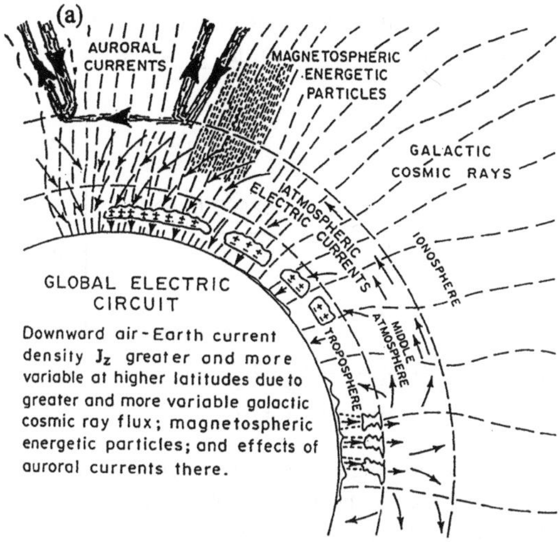

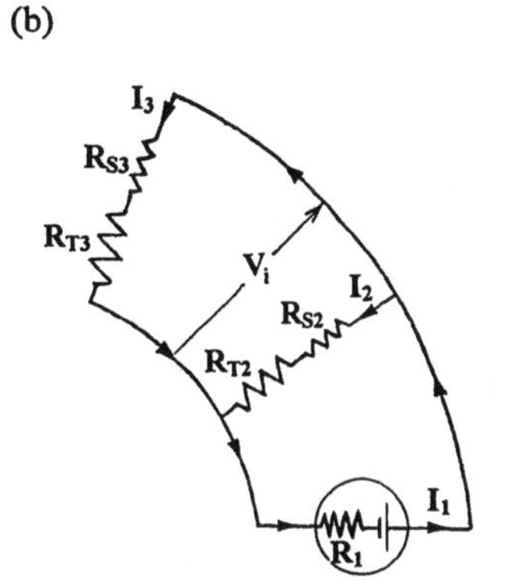

Figure 2. (a) Illustration of meridional section of the global electric circuit, with about 25 times vertical exaggeration. The arrows represent the flow of current density for which the main generator is tropical thunderstorms. The ionosphere-earth current density J_z of 1–4 pA m^{-2} is driven by the ionosphere-earth potential difference V_i of about 250 kV acting across the vertical column resistance at each location. (b) A simplified equivalent circuit with the only horizontal variation due to magnetic latitude. The ionosphere-earth return path is divided into a low latitude branch (with relatively small variations due to solar wind inputs) and a high latitude branch (with variations in the inputs at least several times greater).

The solar wind modulated inputs can affect J_z by either changing $\sigma(z)$ in the troposphere and/or lower stratosphere, or by changing V_i, especially in the polar regions. Both V_i and $\sigma(z)$ have systematic and irregular natural variations, and the small scale chaotic variations of $\sigma(z)$ in the mixing layer make it difficult to measure at low altitude sites on land the systematic variations of J_z due to the solar wind inputs. It is possible also that the thunderstorm charging current and V_i are affected by the small (3–6%) changes in the cosmic ray flux in the equatorial regions due to solar wind variations. In Figure 2b we show an equivalent circuit for the global circuit, with the only horizontal variation due to magnetic latitude, and with the return path divided into one representing the high latitude regions, and another representing the low latitude regions. We divide the vertical column resistance integrated over either part into tropospheric (R_T) and stratospheric (R_S) components. The high latitude path is subject to solar wind influences at least several times stronger than the low latitude path. The latitude gradient in the modulation of vertical column conductivity redistributes J_z between the high and low latitude return paths. The cosmic ray flux changes modulate $\sigma(z)$ between about 2 km and 60 km altitude. The relativistic electron flux and associated bremsstrahlung appear to affect J_z by modulating $\sigma(z)$ between about 20 km and 60 km during episodes of high stratospheric resistivity R_S. The polar cap potentials have large ($\pm$50 kV)

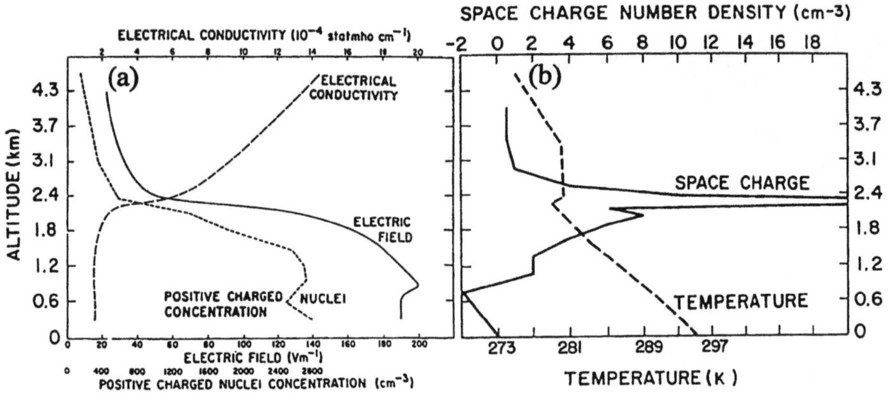

Figure 3. (a) Measurements from an aircraft made on 20 August, 1953, between 1311 to 1416 EST by Sagalyn and Faucher (1954) modified from Sagalyn and Burke (1985); showing (a) the conductivity gradient at the top of the mixing layer and corresponding changes in electric field and the concentration of positively charged nuclei, and (b) the space charge density and the temperature.

deviations from the otherwise general ionospheric equipotential that prevails up to about 50° geomagnetic latitude. We will discuss these effects further in Section 4. For now we will focus on the production of space charge $\rho(z)$, that is due to the flow of current density J_z through regions of varying conductivity in the troposphere.

Figure 3 shows measurements of tropospheric conductivity, electric field, concentration of positively charged aerosol particles (nuclei), temperature, and calculated space charge, from the review of Sagalyn and Burke (1985), based on aircraft observations by Sagalyn and Faucher (1954). The strong gradient in electrical conductivity at 2.2 km altitude is due to the top of the mixing layer being at this level. The conductivity below the top of the mixing layer is decreased by the attachment of the high mobility air ions to more massive and therefore lower mobility aerosol particles brought up from the surface. To maintain the constant vertical current density J_z in the presence of the decreasing conductivity at lower levels, space charge $\rho(z)$ accumulates to provide an increasing vertical electric field E_z, as seen in Figure 3. The amount of space charge is given by Gauss's equation; $\rho(z)/\varepsilon_0 = \nabla \cdot \mathbf{E}$. Horizontal stratification and constant J_z can be assumed to apply, so that $\nabla \cdot \mathbf{E} = dE_z/dz$ and $E_z = J_z/\sigma(z)$ so that:

$$\rho(z) = \varepsilon_0 J_z (d/dz)(1/\sigma(z)) \tag{2}$$

As can be seen in Figure 3b, there is a strong concentration of space charge in the conductivity gradient at the top of the mixing layer, but there is also significant space charge in the levels below.

When clouds are present in the atmosphere they create very large conductivity gradients. Within clouds and within layers of haze in which the relative humidity is approaching 100% there is a large decrease in conductivity compared to that of

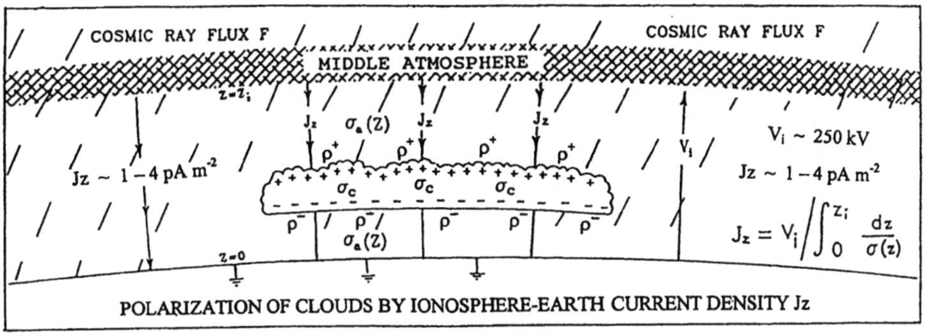

Figure 4. The accumulation of space charge at the conductivity gradients at the tops and bottoms of clouds. If the conductivity σ decreases by a factor of 10 from outside to inside the cloud at 5 km in say 10 meters, then $\rho \sim 10^9 e\mathrm{m}^{-3}$ for $J_z = 2\,\mathrm{pA\,m}^{-2}$. The layer of charge density represents a surface charge of $10^{10} e\ \mathrm{m}^{-2}$ that is provided by J_z in a time $\sim 10^3$ s, or 16 minutes. Mixing redistributes the charge density, but the distribution still scales as J_z.

clear air at the same altitude (Anderson and Trent, 1966; Reiter, 1992, pages 193–200; Dolezalek, 1963). In their comprehensive review Pruppacher and Klett (1997, pages 798–802) compare observations and theory and report that within clouds the conductivity is reduced by a factor of between 3 and 40. Rust and Moore (1974) found that the conductivity of cloudy air was 10% of that of clear air at the same altitude. So layers of space charge will develop at the tops and bottoms of clouds (positive and negative space charge respectively). Figure 4 illustrates the situation for a stratus type cloud (Tinsley, 1996; MacGorman and Rust, 1998).

To estimate an approximate charge density at, say, the top of a cloud such as in Figure 4 we can assume that σ decreases by a factor of 10 from outside the cloud at 5 km, where $\sigma \sim 10^{-13}\Omega^{-1}\ \mathrm{m}^{-1}$ (Gringel *et al.*, 1986) to inside where $\sigma \sim 10^{-14}\Omega^{-1}\ \mathrm{m}^{-1}$. If the width of the interface is 1 m, then $(d/dz)(1/\sigma) \sim 9 \times 10^{13}\Omega$. Taking $J_z \sim 2 \times 10^{-12}\ \mathrm{Am}^{-2}$ then $\rho = 1.6 \times 10^{-9}\ \mathrm{Cm}^{-3}$. In terms of the elementary charge e, and CGS units $\rho = 10^4 e\ \mathrm{cm}^{-3}$. If the width of the interface is 10 m, then $\rho = 10^3 e\ \mathrm{cm}^{-3}$. In either case the layer of charge density represents a surface charge $S = 1.6 \times 10^{-9}\ \mathrm{Cm}^{-2}$, that is provided by J_z in a time $\sim 10^3$ s, or 16 minutes.

In the more general case with shear flow and turbulence at the interface, the charge density will be distributed on parcels mixed outward and inward, but the overall distribution will still scale as J_z. In general, the space charge distribution throughout the troposphere will scale as J_z, and space charge carried up into convective clouds, and the screening layers generated around the clouds in response, will all scale as J_z. The above situation is applicable to conditions before clouds have developed significant amounts of ice or precipitation. When that happens, powerful charge separation processes begin, producing much larger values of ρ.

2.2. Transfer of Space Charge to Droplets

The dominant source of ionization in the troposphere and up to about 60 km altitude is trails of pairs of electrons and positive ions created by GCR particles, except near land surfaces. Radiation from land includes radon gas and radioactive dust particles and gamma radiation from soil and rocks. The radioactive dust and gas can be mixed upward by convection, and is the dominant source of electron-ion pairs in the lowest one or two km over land (Hoppel *et al.*, 1986, Figure 11.1). The electrons immediately attach to air molecules, and the resulting negative ions, together with the positive ions (light ions) react within milliseconds with other molecules, including the formation of cluster ions with molecules of water and other vapors. In the presence of aerosol particles the light and cluster ions attach to the particles, which can grow to haze particles as the relative humidity approaches 100%.

The space charge in the air near clouds is the difference between the concentration of positive and negative ions, and within the mixing layer (also known as the planetary boundary layer) most of the positive and negative ions will be in the form of charged aerosol particles (nuclei). At the interface between cloudy and clear air many of the charged aerosol particles will become attached to droplets. The presence of electric fields associated with the development of space charge layers at the cloudy-clear air interfaces allows droplets to accumulate many elementary charges of the same sign. Processes listed by Beard and Ochs (1986) for charge accumulation by droplets in clouds (prior to rain stage) include diffusion charging, drift charging, and selective ion charging. The charging of droplets to high levels due to scavenging of aerosol particles with charge of predominantly one sign (in space charge) is an additional charging process that is a consequence of electroscavenging, as discussed by Tinsley *et al.* (2000). This depends on a hitherto neglected effect of image charges on the rate of collection of charged particles by droplets. In electroscavenging, as in drift charging and selective ion charging, the distribution of droplet charge with droplet size depends approximately on the square of the droplet radius.

Observations of charges on droplets in and near the boundaries of clouds are generally consistent with these charging processes, especially for stratus and stratocumulus clouds, with typically several hundred positive elementary charges found on droplets of radius 25–40 μm near cloud tops (Reiter, 1992, section 4.2.2.2; Pruppacher and Klett, 1997, section 18.4; MacGorman and Rust, 1998, chapter 2). A layer of negative charge is often found at cloud base.

The processes we have described so far, i. e., the changes in space charge and droplet charge that follow J_z and that are present through the troposphere all the way down to the surface, are well established. When averaged over areas of order 10^3 km^2 the amplitude of the electrical changes ($\sim$10%) induced by solar wind variability becomes significant compared to the levels of natural variability. The amplitude is relatively large compared to the $\sim$0.1% of total solar irradiance changes, and the ~1% of solar UV changes. Taken with their occurrence throughout the troposphere, where they may influence clouds and therefore the flow of latent heat and of visible and infrared radiation, the variations in cloud charging become a realistic prospect for a link between the solar wind and changes in weather and climate.

In the next sections we will discuss the production of charged evaporation nuclei and the effects of electroscavenging of them by droplets. We will limit discussion to only one of a number of possible mechanisms for electrical effects on clouds and on weather and climate, which is the one that appears qualitatively consistent with the observed correlations in Figure 1. This mechanism is a consequence of the evaporation of charged droplets in weakly charged clouds, followed by electroscavenging of the charged residual nuclei. It entails an increase in the rate of contact ice nucleation and changes in the albedo and sedimentation rates of ice from those clouds that are supercooled (droplets at temperatures below freezing). Other possibilities include space charge and electroscavenging processes affecting not only ice nucleation, but also ice aggregation and crystal growth, particularly in cirrus clouds. Effects on coalescence efficiencies and other processes described in Chapter 18 of Pruppacher and Klett (1997) are likely to be important in thunderstorms, where there are very large electric fields and droplet charges. Whether the thunderstorm charges and fields respond to solar activity depends on the unresolved issues as to whether or not thunderstorm electrification depends on initial cloud charges and fields and on changes in atmospheric conductivity related to cosmic ray flux changes. Other possibilities concern the electroscavenging of aerosol particles changing the chemical and physical structure and concentration of what will become condensation nuclei for further cloud formation. This may affect drop size distributions and thereby droplet coalescence processes, and precipitation and radiative properties of later clouds.

2.3. Evaporation Nuclei

As pointed out by Beard (1992) mixing processes at the cloud-to-clear air interface will result in evaporation of many of the charged droplets. Little charge is lost from droplets by the evaporation process (Robertson, 1969), and also the aerosol particles and adsorbed vapors that the droplet has previously scavenged are retained. The result is a population of highly charged evaporation nuclei, coated at least temporarily with materials such as sulfate and organic compounds. The electrical charge per unit mass increases during evaporation by a factor of $(r_0/r)^3$ and the electrical energy per unit mass increases by a factor of $(r_0/r)^4$, where r_0 is the original radius and r is the radius after evaporation. (The mass decreases as the cube of the radius, and the electrical potential energy of the charge on the spherical droplet surface increases as r^{-1}). The increase in electrical potential energy comes from thermal energy of escaping water molecules. So as the droplet shrinks from say 25 μm to 0.25 μm then the electrical energy per unit mass increases by a factor of 10^8. The high charge/mass ratio means that movement of the particle by electrical forces can dominate over movement due to other forces, with just a few tens of elementary charges on the evaporation nucleus (Tinsley *et al.*, 2000). Also, the high energy/mass ratio means that at the moment of contact with droplets there is energy available to overcome the energy barrier to ice nucleation.

The lifetime for loss of charge from these 'evaporation nuclei' is of order 10^3 s at 5 km altitude, provided the relative humidity remains near 100% (Tinsley *et al.*, 2000) and would be about a factor of ten less if the relative humidity dropped to zero. Droplets may evaporate on mixing with air of less than 100% humidity, or in downdrafts (relative humidity only slightly below 100%) or in a combination of these processes. The evaporation nuclei have surface properties that enhance ice formation on them, evidently due to the soluble sulfate compounds and insoluble organic compounds that were previously scavenged from the air by the droplet (Garrett, 1978; Rosinski and Morgan, 1991; Beard, 1992; Rosinski, 1995). The ice nucleating activity of the evaporation nuclei was found to decrease after the temperature was cycled through higher values, or when the humidity was cycled through lower values. The situation of interest in the context of electroscavenging is when cloudy air is at temperatures between about −5 to −20°C , and when continued penetrative mixing occurs, so that the air parcels containing the charged evaporation nuclei are mixed back into contact with cloudy air parcels containing supercooled droplets. Then the electric charge on the evaporation nuclei (some of which may be covered by ice by that time) greatly enhances the rate at which they are electroscavenged by the supercooled droplets, and the effect is to considerably enhance the overall rate of ice formation by contact ice nucleation. Electroscavenging is numerically modeled by Tinsley *et al.* (2000) and its consequences for ice formation in clouds is discussed by Tinsley (2000).

3. Electroscavenging and Possible Atmospheric Consequences

3.1. Estimates of Relative Importance of Contact vs. Other Ice Nucleation Processes

Observations of clouds made by Hobbs and Rangno (1985) and Rangno and Hobbs (1991), and which are consistent with similar observations by others, showed that in about half the clouds studied the ice particle concentrations were at least 50 times higher than could be explained by the measurements of ice nucleus concentrations, and sometimes 10^4 or 10^5 times larger. The measurements were of activity of ice forming nuclei in standard ice nucleus counters, which count particles active in the deposition and sorption- freezing modes, but not in the contact mode. The study included stratiform (altostratus, stratocumulus and stratus) clouds, as well as cumuliform clouds. It was found that the ice particle concentrations were a strong function of the broadness of the droplet size distribution. Hobbs and Rangno (1985) suggested that contact nucleation might be responsible for the freezing of the larger droplets, with subsequent augmentation by ice multiplication mechanisms such as splinter production during riming, and crystal fragmentation. However, it was argued by others that the observed ice particle concentrations were also very much higher than expected from modeling of contact nucleation rates. Beard (1992) suggested that electrically enhanced scavenging could be one way of resolve the discrepancy, especially for scavenging of evaporation nuclei. The electroscavenging calculations made by Tinsley *et al.* (2000) amount to a detailed investigation of that suggestion.

It was shown by Tinsley (2000) that the effects of electroscavenging leading to contact ice nucleation in clouds depends strongly on the broadness of the droplet size distribution. Calculations of electroscavenging and contact nucleation for clouds that had a broad or bimodal size distribution and high droplet concentration (i. e., high liquid water content) lead to estimated rates of contact ice nucleation that were comparable to, or exceeded the rates of deposition nucleation. Both sets of nucleating particles were considered to be part of the same aerosol population. For the clouds with measured droplet size distributions given by Hobbs and Rangno (1985) Figures 9 and 17, and by Pruppacher and Klett (1997) Figures 2.8(c), 2.9, 2.11(c), and 2.15(a), and for cloud top temperatures near $-15°C$, the estimated ratio of contact nucleation rates to deposition nucleation rates ranged from 1 up to 10. The clouds all had broad droplet size distributions, and were maritime cumuliform clouds, except for Figure 17 which was an average of a number of maritime stratocumulus clouds and Figure 2.8(c) which was an average of four cumuli embedded in stratiform clouds in Montana.

Contact nucleation is likely to be the dominant nucleation process giving rise to ice precipitation from clouds for values of the above ratio not only unity, but of considerably less than unity. The reason is that while ice particles of any size can grow by vapor deposition, the frozen droplets of radii of order 10 μm from contact nucleation start larger and add mass faster than those particles of radii of order 1 μm from deposition nucleation. In addition, the larger particles have much greater fall speeds and rates of growth due to riming and collision-coalescence (Braham, 1986). They are also more likely to give rise to secondary ice production.

3.2. Uncertainties in Absolute Rates of Contact Nucleation

To resolve the puzzle posed by the observations of Hobbs and Rangno (1985) it is necessary to show not only that contact nucleation rates are greater than deposition rates, but in addition that the absolute rates of contact nucleation are high enough to account quantitatively for the observed production of ice, or at least that the uncertainties in modeling ice production can allow the rates to be sufficiently high. The results discussed above were obtained using measurements on ice forming nuclei in both contact and deposition modes made by Cooper (1980) in equipment that did not entail any electrical charge on the contacting particle. The measurements showed that for temperatures $-20°C$ to $-15°C$ the fraction of ice forming nuclei that cause freezing on contact with a supercooled droplet was an order of magnitude higher than the fraction that undergo freezing in other modes, and at $-10°C$ and warmer the fraction was 30 to 100 times larger. These results are supported by theory and laboratory work (Pruppacher and Klett, 1997).

In the case of electroscavenging leading to contact ice nucleation, this enhancement of activity may be even greater. The image charge forces increase approximately inversely as the square of the separation between the aerosol particle and the droplet surface, and as the square of the particle charge. The evaporation nucleus is accelerated as it approaches the droplet, and the increased speed at impact increases the area of the interface zone with high energy which has an important role in overcoming the nucleation energy barrier (see Fukuta, 1975a, b). In addition to such direct mechanical effects, it seems that direct electrical effects may also be important. (Electrical enhancement of ice nucleation probability has been termed electrofreezing). Numerical simulations of molecular dynamics were carried out by Borzsák and Cummings (1997) and showed that a combination of electric fields with shear motions in bulk supercooled water samples would induce freezing, whereas shear motion by itself would not. Also, Gavish *et al.* (1992) reported on the ice nucleating capability of polar and non-polar crystals that were otherwise closely matched, with neither having a structural match to ice. The polar crystals were much more efficient nucleators than were the non-polar crystals. The ice nucleation was observed to occur at submicroscopic cracks in the crystals, which suggests that it was the microscale electric fields (in this case due to electrostatic charges that appear on opposite walls of cracks in polar crystals) that were the significant physical ordering agent promoting ice nucleation. In the case of electroscavenging, there is a strong electric field present between the charged particle and its image on the water surface at the moment and point of contact. Thus the absolute rates of contact nucleation may be increased by such electrofreezing processes.

In addition, the observations of Rosinski and Morgan (1991) showed that the ice nucleating activity of the coated evaporation nuclei was found to decrease after the temperature was cycled through higher values, or when the humidity was cycled through lower values. Thus there is uncertainty as to how high the nucleating activity might be in the 10^3 s immediately after formation when electroscavenging is most likely to take place. The basic enhancement of contact ice nucleation rates due to electroscavenging, together with the above uncertainties, appear to be great enough to accommodate the high ice formation rates observed by Hobbs and Rangno (1985), especially when further amplified by the uncertainties of secondary ice production. Thus, the estimates and the uncertainties discussed above make it plausible that contact ice nucleation, responsive to J_z and its variations, is the dominant mechanism for production of ice particles large enough for precipitation, for many clouds that are undergoing mixing and have broad droplet size distributions.

3.3. Consequences for Atmospheric Temperature and Dynamics

The production of ice particles large enough to precipitate or sediment from a cloud has two important effects for atmospheric temperature and dynamics, as noted in Figure 1. The first effect is for relatively thin clouds, such as stratocumulus, and is simply that the flux of sedimenting ice particles (virga or fall streaks) removes the ice to lower and warmer levels, where it evaporates. Because the lifetime for the cloud can be several hours, an enhanced rate of sedimentation and dissipation of the cloud can reduce the effective cloud cover. Also, sedimentation prevents the cloud from evaporating in place, and perhaps reforming in that air mass at a later time. The glaciation and dissipation of the cloud decreases the cloud albedo to incoming solar radiation, and decreases the opacity to outgoing infrared. The net effect on tropospheric and surface temperature depends on the cloud height, whether day or night, and whether high or low latitude. The change in meridional temperature profile changes the meridional pressure gradient (Dickinson, 1975) and can affect rates of cyclogenesis and the amplitude of planetary waves.

The second effect of changes in the sedimentation of ice particles occurs in relatively thick clouds. For example, in winter cyclones the sedimentation of ice from the tops of storm clouds will increase the precipitation efficiency by the 'seeder feeder' process, as discussed by Rutledge and Hobbs (1983). They found that in typical conditions the flux of ice crystals from higher levels produced roughly a doubling of precipitation rates in warm frontal rainbands. Typical precipitation efficiencies (ratio of liquid plus solid water removed from a cloud to that initially contained in it) are a few tens of percent or less, so there is considerable scope for increase the amount of precipitation from a cloud. This reduces the amount of water in the cloud to re-evaporate with entrainment or downdrafts. This decrease in latent heat transfer increases the temperature of the air mass containing the cloud, compared to the case without precipitation enhancement. Effectively, an increase in amount of rainfall results in a warmer air mass remaining. It has been shown from numerical models by Pauley and Smith (1988), Zimmerman *et al.* (1989), and Mallet *et al.* (1999), and also analytically by van Delden (1989) that the release of heat at mid levels on a mid-latitude winter cyclone will intensify it. This intensification involves the conversion of available kinetic energy from the shear in the winter circulation into eddy kinetic energy.

This second set of processes completes the connection between Forbush decreases of the GCR flux and the Roberts effect in Figure 1; i.e., a decrease in the strength of winter cyclones, as measured by the vorticity area index (VAI), responding to a decrease in J_z. We will discuss the observational data in Section 4.2.1. This set of processes also applies to the Wilcox effect, which is a similar change in the VAI, associated with a similar change in J_z, but in this case associated with a decrease in the flux of precipitating relativistic electrons (Section 4.3).

The numerical models of storm development (Pauley and Smith, 1988; Zimmerman *et al.*, 1989; Mallet *et al.*, 1999) also show that the storm intensification (or weakening) affects the storm track. Cumulative effects over a winter may affect the meridional transport of heat and momentum enough to significantly affect the amplitude of planetary waves, which would be an explanation for the Brown effect noted in Figure 1. The Brown effect is a shift in the mean latitude of storm tracks in the eastern north Atlantic between sunspot maximum (lower GCR flux and lower J_z) and sunspot minimum (higher GCR flux and higher J_z). The longitude where the effect of J_z variations on winter storms would be expected to be greatest is where higher geomagnetic latitudes are found at a given geographic latitude, and where a cyclogenesis region exists east of a continent. The north Atlantic is a particularly favorable region. Cold air masses in winter encountering the relatively warm Gulf Stream in the western north Atlantic, so that a shift in storm track latitude becomes apparent as the storm moves across to the eastern North Atlantic.

4. Solar Wind Modulation of J_z and Correlations of Atmospheric Temperature and Dynamics

4.1. Inputs to Global Circuit and Quality of Data on Responses

In this section we will consider the nature of the solar wind modulation of the three inputs (A) the galactic cosmic ray flux; (B) the precipitation of relativistic electrons from the magnetosphere; and (C) the ionospheric potential variations in the polar cap regions. In each case we will discuss the theoretically expected variations of J_z with the solar wind inputs, and the available observations of responses of J_z (or its proxy E_z) and the observed correlations of meteorological parameters.

The J_z or E_z observations represent intermittent point measurements from only a few observatories, and appear noisy compared to the meteorological measurements, which in most cases represent averages over many thousands of square kilometers. We include the measurements of J_z variations to show that they are not inconsistent with the theoretical expectations, nor with the role of J_z as a link between solar wind variations and the changes in atmospheric parameters. An earlier discussion of these correlations was given by Tinsley (1994, 1996).

4.2. Galactic Cosmic Ray Flux Changes and the Redistribution of J_z with Latitude

The solar wind modulates the flux of lower energy (below about 10 GeV) GCR as they propagate in towards the Earth from the outer solar system, and encounter changing magnetic field strengths and directions in the outwardly streaming plasma. The modulation occurs on all time scales. Forbush decreases of tens of percent can occur in a few hours as coronal mass ejections travel outwards past the earth. The GCR flux typically recovers after these on the time scale of about a week. Longer term changes of comparable magnitude occur on the quasi-decadal sunspot cycle, and the quasi- bidecadal solar magnetic cycle. There are variations on the century time scale that have been recorded in carbon-14 and beryllium-10 isotope ratios in tree rings and ice cores.

Theoretical analyses of the response of the global electric circuit to changes in the GCR flux have been made with complex numerical simulations (Roble and Hays, 1979; Tzur *et al.*, 1983; Sapkota and Varshneya, 1990). The solar wind attenuates lower energy fluxes more than higher energy fluxes, and the geomagnetic field confines the lower energy fluxes (less than about 5 GeV) to higher geomagnetic latitudes. The solar wind modulation of the GCR flux into the atmosphere, and therefore of the atmospheric column conductivity, occurs with several times higher amplitude at higher latitudes than at lower latitudes. Considering the simple two return path equivalent circuit in Figure 2b, the internal impedance of the low latitude current source R_1 is calculated to be much greater than the resistance of the two return paths in parallel, so that the current I_1 is thought to be essentially independent of the load. So during Forbush decreases or at sunspot maximum when the high latitude cosmic ray flux and high latitude column conductivity are decreased, I_3 and J_z at high latitudes decrease; while V_i and I_2 and J_z at low latitudes increase, while I_1 can remain constant. (In other words, part of the load increases at high latitudes while the total current through the load remains constant, causing a redistribution in latitude of J_z). The two return path equivalent circuit is further discussed by Tinsley (1996, 1994) and is consistent with the results of the complex numerical simulations. The situation of I_1 being independent of the load of the return path does not mean that it is constant in time: there are regular annual and diurnal variations in universal time and irregular day-to-day variations of I_1 and V_i due to varying thunderstorm activity. In addition, if the internal impedance and the current output of these sources are directly affected by the relatively small (3–6%) changes in cosmic ray flux at low latitudes (which accompanies the several times larger changes at high latitudes), then it is possible that there are changes in I_1 and V_i due to solar wind changes.

4.2.1. *Short term GCR changes and responses in the troposphere*

Figure 5 shows superposed epoch analyses keyed to the onset times of Forbush decreases. Figures 5a and 5b are from Tinsley and Deen (1991), and are for Forbush decreases between 1953 and 1985. Figure 5a is for a set of larger events (averaging 5% decrease) and Figure 5b is for smaller events (averaging 3% decrease). A comparison is made of time variations of the neutron monitor count rates at Climax and Mt. Washington (representing the cosmic ray flux at energies above about 3 GeV/nucleon) with the geomagnetic Ap index (representing geomagnetic storm activity) and the VAI (representing the strength of cyclones). The VAI or vorticity area index, was defined by Roberts and Olson (1973) and is the area (in units of 10^5 square kilometers) covered by values of absolute vorticity above a threshold of 20×10^{-5} s^{-1} plus the area for vorticity above 24×10^{-5} s^{-1}. The area covered is for the northern hemisphere, poleward of 20° latitude, in the cold season November through March. Not all data sets are complete. Figure 5c is from Märcz

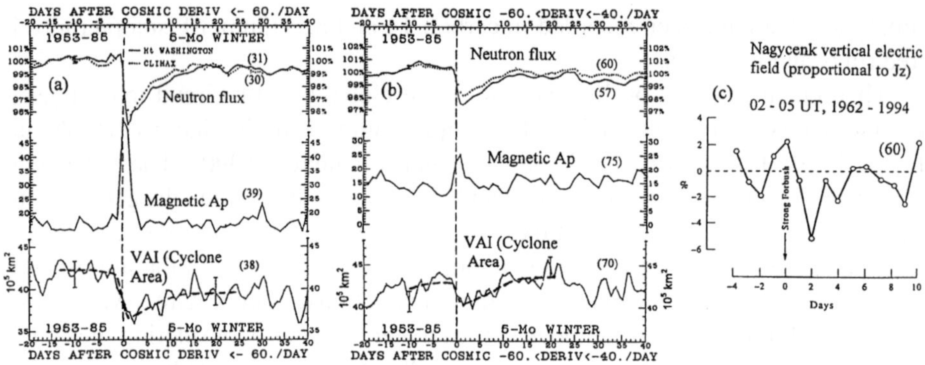

Figure 5. Superposed epoch analysis of (a) changes in neutron monitor count rate, magnetic Ap index, and the vorticity area index for large Forbush decreases 1953–85; (b) as for (a) but for smaller Forbush decreases; (c) changes in the vertical electric field , representing J_z changes, at Nagycenk, Hungary for Forbush decreases in relatively calm weather, 1962–94.

(1997) and is a superposed epoch analysis of the changes in vertical electric field E_z at the Nagycenk observatory in Hungary. This is in the high latitude branch for J_z variations, and a reduction in J_z is expected. Just above the surface E_z is a proxy for J_z, because the conductivity near the surface is dominated by surface radioactivity, and is independent of the cosmic ray flux. The E_z data analyzed was for 60 events with less disturbed meteorological conditions between 1962 and 1994. Even with the averaging in better weather conditions, there is a large amount of electrical noise due to meteorological disturbance evident in the figure, but a reduction in E_z of a few percent during the Forbush decreases is evident and is consistent with the theoretical expectations.

The signal/noise for the VAI data in Figures 5a and 5b is also not great. However, the dashed lines, drawn to smooth the noise, are consistent with a reduction which correlates with the cosmic ray flux change, and is of larger amplitude for the larger Forbush decreases relative to its amplitude for the smaller ones, as required for a causal relationship. The analysis was taken further in Tinsley and Deen (1991) and the VAI response to Forbush decreases was present in both halves of the data, when the data was separated for the first and second halves of the 33 year period; when it was separated for the first and second halves of the November-March intervals; and when it was separated for east and west phases of the quasi-biennial oscillation of equatorial stratospheric winds. The latter separation constitutes nearly alternate yearly November-March intervals. Thus the VAI response to Forbush decreases is well distributed in the data, and cannot be attributed to a few large noise excursions. An analysis with respect to latitude zones and longitude sectors showed the stronger responses in the zone 40°N–60°N and in longitude sectors over oceans compared to those over continental regions.

4.2.2. *Quasi-decadal and Quasi-bidecadal GCR changes and responses in the troposphere*

The quasi-decadal (solar cycle) variations in the galactic cosmic ray flux above about 3 GeV/nucleon can be inferred as a function of latitude from neutron monitor count rates. At geomagnetic latitudes above about 50° they are 15% to 25%, while for low and equatorial latitudes, where the geomagnetic field allows particles with rigidity only above about 13 GV (energy of 13 GeV for protons) the variation is 3% to 6%. These and larger amplitude variations (30% – 40%), that have been measured for energies above about 100 MeV/nucleon from balloons in the polar stratospheres, are reviewed by Bazilevskaya (2000). There is a quasi-bidecadal (Hale) cycle also evident in the cosmic ray variations, consistent with the theory of drift motions in the heliosphere as its magnetic field polarity varies on the Hale cycle. There are many correlations of weather and climate parameters with solar variations on these timescales. Reviews are given by Herman and Goldberg (1978), NAS (1982, 1994), Tinsley (1996), and Hoyt and Schatten (1997). A striking correlation extending over the last 300 years in Greenland ice core data is given by Stuiver *et al.* (1995). A correlation involving low latitude cloud cover is discussed by Svensmark and Friis-Christensen (1997) and Marsch and Svensmark (2000). There are wintertime changes in cyclone frequency and the related storm track latitude shift in the north Atlantic (the Brown effect), and in pressure gradients in the south Atlantic (Tinsley, 1996) that correlate with the quasidecadal solar cycle. These correlations are consistent with the same processes (atmospheric electricity affecting clouds and cyclone intensity) that we discussed in Section 3 as explanations of the VAI changes.

4.3. Relativistic Electron Precipitation and Responses in the Stratosphere and Troposphere

Figure 6 is from Tinsley *et al.* (1994) and in Figure 6a we compare the solar wind speed, the relativistic electron flux, and the Climax neutron monitor count rate, for solar wind magnetic sector boundary crossings in 1982–88. (For a list of such crossings, and an extension of these analyses with other spacecraft and more recent data, see Kirkland *et al.*, 1996). The solar wind speed shows a well-defined drop at the crossings, which are associated with the extension of the streamer belt and which usually separate regions of high speed streams. The relativistic electron flux given here is for the geomagnetic latitude of 67°, corresponding to geosynchronous orbit. The phase delay of about two days from the solar wind velocity variations to the relativistic electron flux variations depends on geomagnetic latitude and electron energy (Kirkland, 1996; Li *et al.*, 1997). Figure 6b shows the reduction of the VAI that correlates with the relativistic electron flux change, and Figure 6c shows similar changes in the tropospheric electric field E_z that is a proxy for proportional J_z changes. The neutron monitor count rate shows no systematic change that could give rise to the Roberts effect (see also Laštovicka, 1987). Thus we

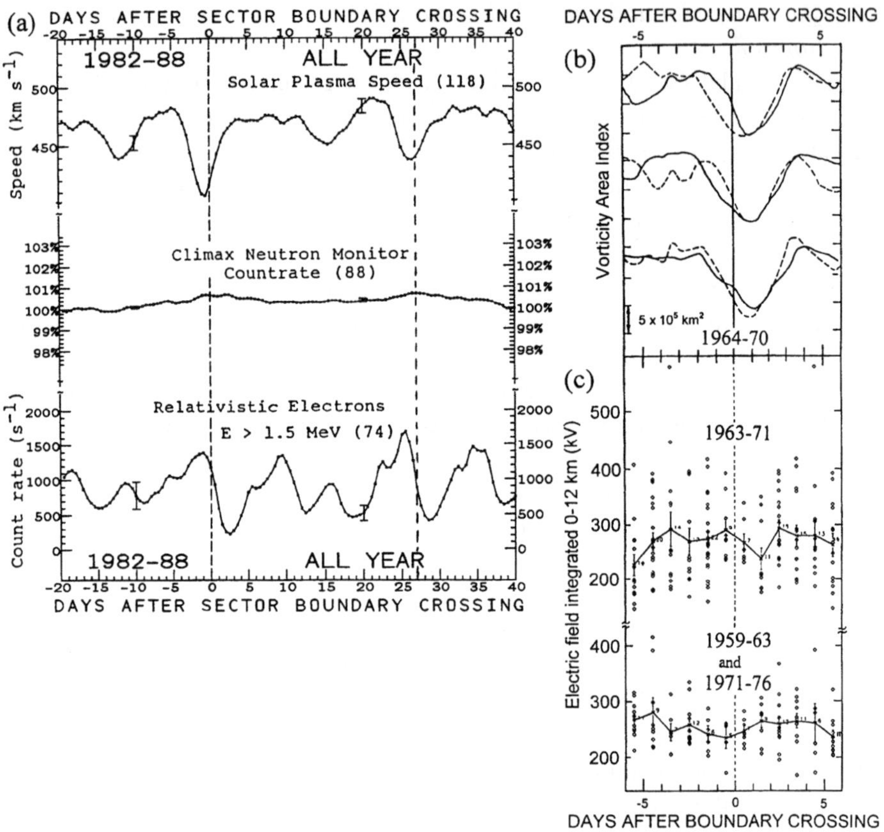

Figure 6. Superposed epoch analysis of effects at solar wind magnetic sector boundary crossings. (a) Upper trace shows 15% decrease in solar wind velocity; middle trace shows no systematic change in cosmic ray flux; lower trace shows 80% reduction in relativistic electron flux. (b) VAI variations 1964–70 showing response distributed through the data set with upper traces for 24 +/- crossings (solid) and 30 -/+ crossings (dashed); middle traces for 32 early winter crossings (solid) and 22 late winter crossings (dashed); lower traces for 26 crossings 1964–66 (solid) and 28 crossings 1967–70 (dashed). (c) Changes in tropospheric vertical electrical field with upper trace indicating 10–20% decrease for 1963–71; lower trace with no observable decrease when the 1964–70 period with high levels of stratospheric volcanic H_2SO_4 is excluded.

are left with the precipitation of the relativistic electrons as a candidate for the forcing agent for the J_z changes and the responses of the VAI that are shown in Figure 6c and Figure 6b. Calculated ion production rates for relativistic electron precipitation in the sub-auroral region are given by Fram *et al.* (1997). The direct ion production by such electrons is present only above about 50 km; however the X-ray Bremsstrahlung produces ionization down to about 20 km. For the higher flux rates the ion production rate exceeds that due to GCR down to about 30 km.

The VAI response is known as the Wilcox effect, and was detected for crossings between 1964 and the early 1970s (Wilcox *et al.*, 1973), but then the effect declined relative to the noise level. The inability to demonstrate its presence after 1974 prompted claims that the phenomenon was a chance coincidence of statistical fluctuations. (Williams and Gerety, 1978; Shapiro, 1979; Williams, 1979). However, Tinsley *et al.* (1994) pointed out that the 1964–70 period, when the Wilcox effect was relatively strong, was the time when the mixing ratio of H_2SO_4 in the stratosphere was increased by more than an order of magnitude. This was due to the eruptions of the volcano Agung in 1963 and subsequent eruptions of some other volcanoes. The H_2SO_4 was present in the form of liquid aerosol particles of H_2SO_4 / H_2O and as H_2SO_4 vapor, that only slowly decayed in the late 1960s (Castleman *et al.*, 1974). The presence of the H_2SO_4 in the stratosphere increased the resistivity there (by promoting attachment and recombination of ions), and it seems likely that resistivity was increased to the extent that the stratospheric column resistance became a significant component of the vertical column resistance between the earth and the ionosphere. The increase of the stratospheric column resistance (shown as R_S in Figure 2b) and its modulation by Bremsstrahlung from relativistic electron precipitation, may be the reason why at these times the J_z and VAI variations in the mid-high latitudes (where relativistic electron precipitation occurs) follow the relativistic electron flux. The lower trace in Figure 6c shows that for times when the H_2SO_4 mixing ratio averaged much lower values the dip in J_z on days 1 and 2 after the crossing was weak or absent.

To further test the mechanism proposed above, the variations of the VAI at boundary crossings before, during, and after the period 1983–86 were examined (Tinsley *et al.*, 1994). This period corresponded to the El Chicon eruption, which again greatly increased stratospheric H_2SO_4 mixing ratios. A VAI variation with a well-defined dip reaching a minimum on day 2 was again found, and for some years before 1983 and after 1986 it was weak or absent. A further opportunity to make a similar test occurred with the Pinatubo eruption and enhancement of stratospheric H_2SO_4 in 1992–94 (Kirkland *et al.*, 1996), and a similar reduction of the VAI for day $+2$ relative to day -2 was found, not present in 1987–91.

It might be asked whether at times of high fluxes of strongly modulated relativistic electrons (as found when the solar cycle is in its descending and minimum phase when the solar wind contains stable high speed streams) and relatively low concentrations of stratospheric H_2SO_4, as in 1972–75, there will be a detectable Wilcox effect. A very sensitive technique was applied to the time intervals October-March 1972–74 and January 1975 by Larsen and Kelley (1977). They evaluated the VAI from the North American Fine Mesh Meteorological grid data, and also evaluated it from the 12 hr and 24 hr forecasts of the Fine Mesh data. They then determined the correlation coefficient between the actual VAI and the forecast VAI. This technique cancels out much of the day-to-day variability in the VAI that is not due to external forcing. Their results are shown in Figure 7. Since North America is at mid-high geomagnetic latitudes and the meteorological observations were closely spaced and of high quality, this was an optimum application of the technique. Their result of a clear demonstration of a VAI decrease following sector boundary crossings is not

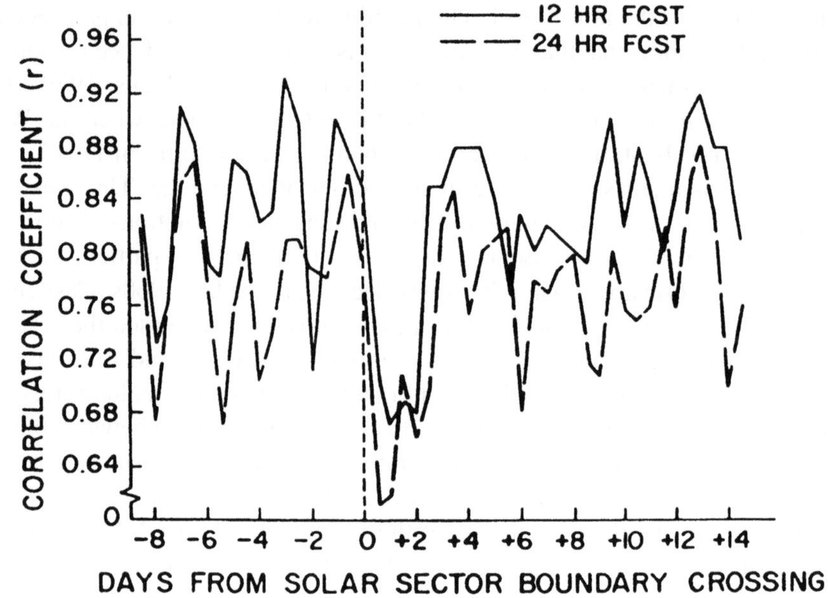

Figure 7. Superposed epoch analysis keyed to 47 sector boundary crossings October - March 1972–74 and January 1975, of correlation coefficients between the forecasted 500 mb VAI and the actual VAI. From Larsen and Kelley (1977).

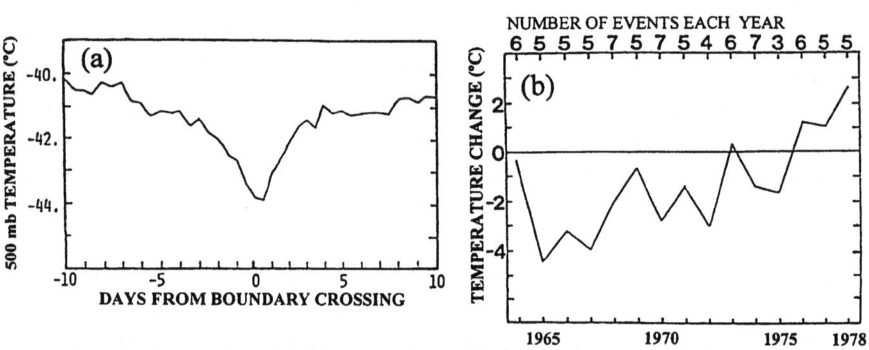

Figure 8. (a) Superposed epoch analysis keyed to sector boundaries for temperatures at 500 mbar in the Arctic for winters 1965–67; (b) decline of the amplitude of the effect as the H_2SO_4 from Agung and Helka and other eruptions gradually cleared from the stratosphere. From Misumi (1983).

inconsistent with the marginal Wilcox effect for moderately low H_2SO_4 mixing ratios in 1971–74 found by Tinsley *et al.* (1994, Figure 2). The latter result is for the less sensitive direct VAI determination, made on lower resolution grid data, averaging the VAI for the whole of the northern hemisphere.

Another tropospheric response that is closely related to the Wilcox effect is the reduction in 500 mbar temperatures in winter at high latitudes, following sector boundary crossings. Figure 8 is from Misumi (1983) and shows a temperature reduction that minimized on day 1 after the crossing (Figure 8a) and that was as much as 4°C in 1965 (Figure 8b). The effect decayed through the late 1960s and early 1970s, reflecting the clearing out of H_2SO_4 from the stratosphere. Fig 8b shows a considerably reduced amplitude for the period 1972–75 of the Larsen and Kelley (1977) data. The analysis of Misumi (1983) was made as a function of geographic latitude and the maximum reduction was at the geographic pole. This is not entirely consistent with the modulation of stratospheric conductivity and of J_z by relativistic electrons, since these are particles precipitating from trapped orbits in the radiation belts, and show minima in fluxes over the magnetic poles. However, the magnetic poles are $\sim$15° from the geographic poles, and there is direct evidence of E_z, and therefore of J_z reductions at the magnetic poles at boundary crossings, from the observations in 1974 at Vostok, Antarctica, by Park (1976). A similar result was found by Frank-Kamenetsky *et al.* (1999) for 1979–80 crossings in Vostok E_z observations. While the mixing ratio of H_2SO_4 was not high in 1979–80, it is possible that an equivalent effect occurs with the particles of polar stratospheric clouds. The magnetic poles are in a region of open magnetic field lines connected to the solar wind, and the results of Švestka *et al.* (1976) showed reductions in solar wind relativistic electrons (Jovian or solar) before and at the times of crossings, and increases in solar wind protons afterwards (see also Wilcox, 1979). Further studies of changes at boundary crossings of solar wind particle fluxes and of polar cap E_z responses are needed.

4.4. Polar Cap Ionospheric Potential Distributions and Responses in the Troposphere

The ionosphere from equatorial latitudes to about 50° geomagnetic latitude is essentially an equipotential according to models (e. g. Hays and Roble, 1979) and as found from satellite measurements. But at higher latitudes there are superimposed potential distributions, due to solar wind - magnetosphere interactions that couple into the northern and southern polar cap ionospheres. Relatively large potentials are superimposed on the low latitude value, and these vary strongly with the solar wind velocity and magnetic field orientation. The current systems were illustrated schematically in Figure 2, with the dusk side on the left and the dawn side on the right, with dawn-dusk potential differences averaging about 80 kV (Richmond, 1986; Roble and Tzur, 1986).

The value of J_z in the stratosphere and troposphere below the potential pattern (e. g., at the South Pole) is proportional to the overheard ionospheric potential V_i. The potential V_i is the sum of the low latitude ionospheric potential (with its various time variations) and the superimposed convection potential for that geomagnetic latitude and geomagnetic local time. Very little correction to V_i is needed for downward mapping, as the scale of the pattern is large (3000 km between the dawn maximum and the dusk minimum) compared to the vertical distance of less than 100 km to the surface. To verify that E_z (and by proxy J_z) near the surface

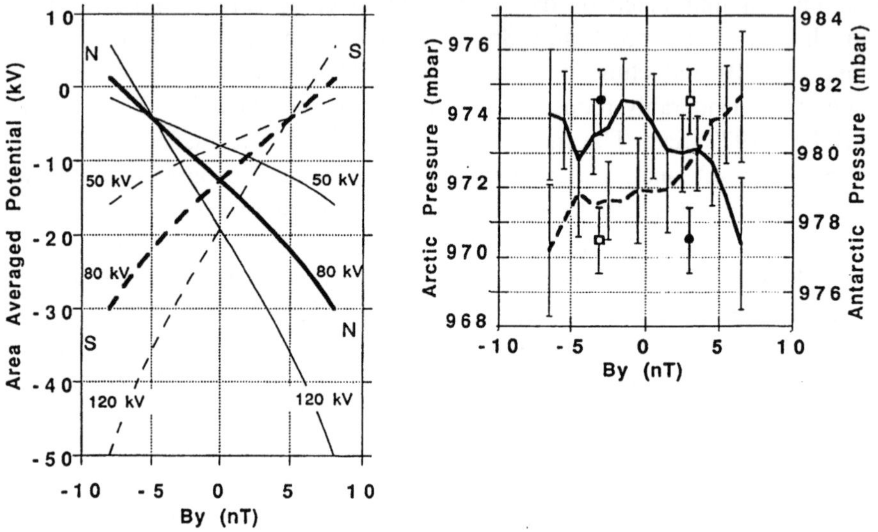

Figure 9. (a) The averages of the potential due to magnetosphere-ionosphere coupling for the regions within circles of radius 10° of latitude from each geomagnetic pole. The abscissa is the solar wind B_y component in nT. The solid lines are for the north polar cap for three values of the dawn-dusk potential difference, determined by B_z and the solar wind velocity. The dashed lines are for the southern polar cap. (b) The variation of surface pressure with B_y at four separate polar stations. The solid line is for Thule (north) and the dashed line is for McMurdo (south) after Page (1989). The solid circles are for Mould Bay (north) and Dumond d'Urville (south) after Mansurov *et al.* (1974). The pressure changes show the same variations as the averaged ionospheric potential and therefore averaged J_z variations, i. e., they are of opposite sign in the northern and southern polar caps as a function of B_y. From Tinsley and Heelis (1993).

respond to changes in the overhead V_i, an analysis of measurements of E_z from the South Pole between 1982 and 1986 was made by Tinsley *et al.* (1998). The correlations found demonstrated the expected response to V_i, in the presence of local meteorologically induced electrical noise, and irregular variations in the low latitude thunderstorm generators, and differences between the real ionosphere and the model.

It is possible to calculate averages over large areas of the superimposed polar cap potential. Figure 9 is from Tinsley and Heelis (1993) and Figure 9a shows plots of averages of the potential over the regions within circles of radius 10° of latitude from each geomagnetic pole. The abscissa is the solar wind B_y component in nT. The solid lines are for the northern polar cap for three values of the dawn-dusk potential difference (determined by B_z and the solar wind velocity) and the dashed lines are for the southern polar cap. If values of J_z were averaged in a similar way, they would be expected to show similar variations, and in particular, the opposite changes with B_y in the southern polar cap as compared with the northern polar cap.

The Mansurov/Page effect listed in Figure 1 consists of variations of surface pressure of about 4 mbar that correlate with the inferred J_z variations. Mansurov *et al.* (1974) analyzed 1964 observations of surface pressure at Mould Bay and Dumond d' Urville, which are two stations about 10° from the north and south geomagnetic poles respectively. On average the pressure varied by about 4 mbar between periods corresponding to B_y negative and positive. These pressure variations are represented in Figure 9b by the solid and open circles respectively. A consistent effect was found by Page (1989) who analyzed daily average surface pressures at Thule, Greenland, and McMurdo, Antarctica, for B_x values in 1 nT intervals for the period 1964–1974. These stations are also about 10° from the geomagnetic poles. The geometry of the solar wind magnetic field allows the transformation of B_x to B_y values, and in Figure 9b we show the Thule variation as a solid line, and the McMurdo variation as a dashed line. The pressure variations again show the opposite changes in the southern polar cap as compared with the north. Overall, the pressure variations of Figure 9b represent four independent stations, and apply to two different time periods, and show the same relationship to B_y as the potential variations of Figure 9a, which are a proxy for the J_z variations in the same area.

Thus the Mansurov/Page effect supports the proposal that J_z variations affect cloud physics, and in this case the changes in cloud cover and latent heat exchange presumably affect the dynamics of the polar cap atmosphere, and hence the surface pressure. There is a need to examine polar cap geopotential height and temperature changes at mid- tropospheric levels, where the effects are likely to be strongest (at these altitudes at lower latitudes the amplitude of temperature and pressure changes correlating with J_z are considerably larger than at the surface).

5. Discussion and Conclusions

Of the three inputs into the global electric circuit that are modulated by the solar wind, two of them do not involve cosmic rays or other particles that are energetic enough to penetrate to cloud levels and directly affect clouds. But all three inputs cause J_z variations that extend all the way down to the surface. So we have concluded that in all three cases it is most likely that J_z variations are causing changes in cloud microphysics that then produce correlated changes in weather and climate.

The J_z variations cause changes in space charge at the boundaries of clouds, and there may be several ways in which this affects cloud microphysics. New numerical models show that electrical effects on scavenging of aerosol particles are not negligible for non-thunderstorm clouds. One effect of electroscavenging involves an increase in ice production with increasing J_z , where mixing and evaporation are occurring at cloud boundaries. Electroscavenging leading to contact ice nucleation requires supercooled clouds with a broad droplet size distribution and relatively high droplet concentration. These are more typical of maritime clouds than continental clouds. Greater ice production will result in earlier glaciation and dissipation of clouds. For thin clouds the ice particles sediment as fall streaks and evaporate at lower levels. This reduces both the albedo to incoming short wave solar radiation, and the opacity to outgoing longwave radiation, with climatic effects depending on cloud latitude and cloud height and whether day or night.

In winter cyclones the ice produced from middle and upper level clouds or the tops of deep clouds sediments into dense cloud masses at lower levels. This increases the precipitation efficiency by the seeder-feeder process. This increase in precipitation in turn increases the latent heat transfer between the air mass and the surface. This has been shown to intensify winter cyclones, increase the vorticity area index, and cause the storm tracks to shift. Thus observed changes in the vorticity area index and in the amplitude of planetary waves and in regional climate can be accounted for.

There is much research needed to narrow the uncertainties in the various processes connecting the solar wind to weather and climate. Measurements from high quality mountain and polar ice cap sites are needed to monitor and quantify the changes in the amount and large-scale distribution of current density in the global electric circuit. Continuous measurements from several such sites have long been needed just to provide reasonably low-noise data for evaluating the variability of the global electric circuit. With precise daily values of J_z representing regional or zonal mean averages one would expect to see direct correlations between the VAI and J_z, irrespective of whether the J_z variations were due to external or internal forcing. (Atmospheric dynamical responses to internal changes in the global circuit would of course be of considerable interest in their own right.) There is a need for a quantitative model, utilizing the theory of dusty plasmas, of effects of high mixing ratios of H_2SO_4 and relativistic electron precipitation and Bremsstrahlung X-rays on stratospheric conductivity.

Measurements in clouds of ice particle concentrations and simultaneous droplet size distributions, ice nucleus concentrations, droplet and aerosol particle charge distributions, and associated current density, electric fields, and conductivity are a high priority. Much insight can also be gained from laboratory experiments on ice nucleation.

Acknowledgements

This work has been supported by NSF grants ATM 95 04828 and ATM 99 03424. The treatment of cloud microphysics has benefited from many stimulating discussions with K. V. Beard.

References

Abbas, M. A., and Latham, J.: 1969, 'The electrofreezing of supercooled water droplets', *J. Meteorol. Soc. Japan* **47**, 650–74.

Anderson, R. V., and Trent, E. M.: 1966, 'Evaluation of the use of atmospheric-electricity recordings in fog forecasting', *Naval Research Lab.* Report No. **6424**, 20 pages.

Baker, D. N., McPherron, R. L., Cayton, T. E., and Klebesdal, R. W.: 1990, 'Linear prediction filter analysis of relativistic electron properties at 6.6 RE', *J. Geophys. Res.* **95**, 15133–15140.

Bazilevskaya, G. A.: 2000, 'Observations of Variability in Cosmic Rays', *Space Sci. Rev.*, this volume.

Beard, K. V.: 1992, 'Ice initiation in warm-base convective clouds: An assessment of microphysical mechanisms', *Atmosph. Res.* **28**, 125–152.

Beard, K. V., and Ochs, H. T.: 1986, 'Charging mechanisms in clouds and thunderstorms', *The Earth's Electrical Environment*, National Academy Press, Washington, D.C., pp. 114–130.

Bering, E. A., III, Few, A. A., and Benbrook, J. R.: 1998, 'The global electric circuit', *Physics Today* **51**, 24–30.

Borszák, I. B., and Cummings, P.: 1997, 'Electrofreezing of water in molecular dynamics simulation accelerated by oscillatory shear', *Phys. Rev.* **E 56**, R6279–R6282.

Braham, R. B.: 1986, 'Coalescence-freezing precipitation mechanism', *10th Conf. Planned and Inadverdent Weather Modification*, Am. Meteorol. Soc., Boston, 142–145.

Castleman, A. W., Jr., Munkelwitz, H. R., Manowitz, B.: 1974, 'Isotopic studies of the sulfur component of the stratospheric aerosol layer', *Tellus* **26**, 222–234.

Cooper, W. A.: 1980, 'A method of detecting contact ice nuclei using filter samples', *8th Intern. Conf. Cloud Phys.*, Clermont-Ferand, France, 15-19 July, 1980, 665–668.

Dickinson, R. E.: 1975, 'Solar variability and the lower atmosphere', *Bull. Am. Meterol. Soc.* **56**, 1240–1248.

Dolezalek, H.: 1963, 'The atmospheric fog effect', *Rev. Geophys.* **1**, 231–282.

Fischer, H. J., and Mühleisen, R.: 1980, 'The ionospheric potential and the solar magnetic sector boundary crossings', Rep. Astron. Inst., Univ. Tübingen, Ravensburg, Germany.

Fram, R. A., Winningham, J. D., Sharber, J. R., Link, R., Crowley, G., Gaines, E. E., Chenette, D. L., Anderson, B. J., and Potemera, T. A.: 1997, 'The diffuse aurora: A significant source of ionization in the middle atmosphere' *J. Geophys. Res.* **102**, 28,203–28,214.

Frank-Kamenetsky, A. V., Burns, G. B., Troshichev, O. A., Papitashvili, V. O., Bering, E. A., and French, W. J. R.: 1999, 'The geoelectric field at Vostok, Antarctica: it's relation to the interplanetary magnetic field and the cross polar cap potential difference', *J. Atmos. Solar Terr. Phys.* **61**, 1347–1356.

Fukuta, N.: 1975a, 'A study of a mechanism for contact ice nucleation' *J. Atmos. Sci.* **32**, 1597–1603.

Fukuta, N.: 1975b, 'Comments on 'A possible mechanism for contact ice nucleation', *J. Atmos. Sci.* **32**, 2371–2373.

Garrett, W. D.: 1978, 'The impact of organic material on cloud and fog processes', *Pageoph.* **116**, 316–334.

Gavish, M., Wang, J.-L., Eisenstein, M., Lahav, M., Lieserowitz, L.: 1992, 'The role of crystal polarity in alpha-amino acid crystals for induced nucleation of ice', *Science* **256**, 815–818.

Gringel, W., Rosen, J. M., and Hoffman, D. J.: 1986, 'Electrical structure from 0 to 30 kilometers', *The Earth's Electrical Environment*, NAS Press, Washington, D.C., pp. 166–182.

Hays, P. B., and Roble, R. G.: 1979, 'A quasi-static model of global atmospheric electricity, I. The lower atmosphere', *J. Geophys. Res.* **84**, 3291–3305.

Herman, J. R., and Goldberg, R. A.: 1978, 'Sun, Weather, and Climate', NASA , SP-426, Washington, D. C.

Hobbs, P. V., and Rangno, A. L.: 1985, 'Ice particle concentrations in clouds', *J. Atmos. Sci.* **42**, 2523–2549.

Hoppel, W. A., Anderson, R. V., and Willet, J. C.: 1986, 'Atmospheric electricity in the planetary boundary layer', *The Earth's Electrical Environment*, NAS Press, Washington, D.C., pp. 195–205.

Hoyt, D. V., and Schatten, K. H.: 1997, 'The Role of the Sun in Climate Change', Oxford University Press, Oxford.

Israël, H.: 1973, Atmospheric Electricity, vol. II, translated from German, Israel Program for Scientific Translations, Jerusalem.

Kirkland, M. W.: 1996, 'Further Evidence for Solar Wind Forcing of Tropospheric Dynamics via the Global Electric Circuit', Ph. D. Dissertation, University of Texas at Dallas.

Kirkland, M. W., Tinsley, B. A., and Hoeksema, J. T.: 1996, 'Are stratospheric aerosols the missing link between tropospheric vorticity and earth transits of the heliospheric current sheet?' *J. Geophys. Res.* **101**, 29,689–29,699.

Larsen, M. F., and Kelley, M. C.: 1977, 'A study of an observed and forecasted meteorological index and its relation to the interplanetary magnetic field', *Geophys. Res. Lett.* **4**, 337–340.

Laštovicka, J., 1987, 'Influence of the IMF sector boundaries on cosmic rays and tropospheric vorticity', *Stud. Geophys. Geod.* **31**, 213–218.

Li, X., Baker, D. N., Temerin, M., Larson, D., Lin, R. P., Reeves, G. D., Looper, M., Kanekal, S. G., and Mewaldt, R. A.: 1997, 'Are energetic electrons in the solar wind the source of the outer radiation belt?' *Geophys. Res. Lett.* **24**, 923–926.

MacGorman, D. R., and Rust, W. D.: 1998, The Electrical Nature of Storms, Oxford University Press, Oxford.

Mallet, I., Cammas, J.-P, Mascart, P., and Bechtold, P.: 1999, 'Effects of cloud diabatic heating on the early development of the FASTEX IOP17 cyclone', *Q. J. Roy. Meteorol. Soc.* **125**, 3439–3467. .

Mansurov, S. M., Mansurova, L. G., Mansurov, G. S., Mikhenvich, V. V., and Visotsky, A. M.: 1974, 'North-south asymmetry of geomagnetic and tropospheric events', *J. Atmos. Terr. Phys.* **36**, 1957–1962.

Märcz, F.: 1997, 'Short term changes in atmospheric electricity associated with Forbush decreases', *J. Atmos. Solar-Terr. Phys.* **59**, 975–982.

Marsh, N. and Svensmark, H.: 2000, *Space Sci. Rev.*, this volume.

Misumi, Y.: 1983, 'The tropospheric response to the passage of solar sector boundaries', *J. Meteorol. Soc. Japan* **61**, 686–694.

NAS: 1982, 'Solar Variability, Weather and Climate', Geophys. Res. Board, Nat. Acad. Press, Washington, D. C.

NAS: 1986, *The Earth's Electrical Environment*, Geophys. Res. Board, Nat. Acad. Press, Washington, D. C.

NAS: 1994, 'Solar Influences on Global Change', Geophys. Res. Board, Nat. Acad. Press, Washington, D. C.

Page, D. E.: 1989, 'The interplanetary magnetic field and sea level polar pressure', *in* S. K. Avery and B. A. Tinsley (eds.) *Workshop on Mechanisms for Tropospheric Effects of Solar Variabilty and the Quasi-Bienniel Oscillation*, Univ. of Colorado, Boulder, pp. 227–234.

Park, C. G.: 1976, 'Solar magnetic sector effects on the vertical atmospheric electric field at Vostok, Antarctica', *Geophys. Res. Lett.* **3**, 475–478.

Pauley, P. M., and Smith, P. J.: 1988, 'Direct and indirect effects of latent heat release on a synoptic wave system', *Mon. Weather Rev.* **116**, 1209–1235.

Pruppacher, H. R., and Klett, J. D.: 1997, 'Microphysics of Clouds and Precipitation', 2nd rev. ed., Kluwer, Dordrecht.

Rangno, A. L., and Hobbs, P. V.: 1991, 'Ice particle concentrations and precipitation development in small polar maritime cumuliform clouds', *Q. J. Roy. Meteorol. Soc.* **117**, 207–241.

Reiter, R.: 1992, *Phenomena in Atmospheric and Environmental Electricity*, Elseiver, Amsterdam.

Richmond, A. D.: 1986, 'Upper-atmosphere electric field sources', *The Earth's Electrical Environment*, National Academy Press, Washington, D.C., pp. 195–205.

Roberts, W. O., and Olson, R. H.: 1973, 'Geomagnetic storms and wintertime 300 mb trough development in the North Pacific–North America area', *J. Atmos. Sci.* **30**, 135–140.

Robertson, J. A.: 1969, 'Interactions Between a Highly Charged Aerosol Droplet and the Surrounding Gas', Ph. D. Thesis, University of Illinois.

Roble, R. G., and Hays, P. B.: 1979, 'A quasi-static model of global atmospheric electricity, II. Electrical coupling between the upper and the lower atmosphere', *J. Geophys. Res.* **84**, 7247–7256.

Roble, R. G., and Tzur, I.: 1986, 'The global atmospheric-electrical circuit', *The Earth's Electrical Environment*, National Academy Press, Washington, D.C., pp. 206–231.

Rosinski, J.: 1995, 'Cloud condensation nuclei as a real source of ice forming nuclei in continental and marine air masses', *Atmosp. Res.* **38**, 351–359.

Rosinski, J., and Morgan, G.: 1991, 'Cloud condensation nuclei as a source of ice-forming nuclei in clouds', *J. Aerosol. Sci.* **22**, 123–133.

Rust, W. D., and Moore, C. B.: 1974, 'Electrical conditions near the bases of thunderclouds over New Mexico', *Q. J. Roy. Meteorol. Soc.* **100**, 450–468.

Rutledge, S. A., and Hobbs, P. V.: 1983, 'The mesoscale and microscale structure and organization of clouds and precipitation in midlatitude cyclones, VIII: A model for the feeder-seeder process in warm frontal rainbands', *J. Atmos. Sci.* **40**, 1185–1206.

Sagalyn, R. C., and Faucher, G. A: 1954, 'Aircraft investigation of the large ion content and conductivity of the atmosphere and their relation to meteorological factors', *J. Atmos. Terr. Phys.* **5**, 253–272.

Sagalyn, R. C., and Burke, H. K.: 1985, 'Atmospheric Electricity', *in* A. S. Jursa (ed.), *Handbook of Geophysics and the Space Environment*, Air Force Geophysics Lab., Bedford, MA.

Sapkota, B. K., and Varshneya, N. C.: 1990, 'On the global atmospheric electrical circuit', *J. Atmos. Terr. Phys.* **52**, 1–20.

Shapiro, R.: 1979, 'An examination of a certain proposed sun- weather correlation', *J. Atmos. Sci.* **36**, 1105–1116.

Stuiver, M., Grootes, P. M., and Braziunas, T. F.: 1995, 'The GISP2 δ^{18}O climate record of the past 16,500 years and the role of the sun, ocean, and volcanoes', *Quarternary Res.* **44**, 341–354.

Švestka, Z., Fritzova-Svestkovă, L., Nolte, J. T., Dodson-Prince, H. W., and Hedeman, E. R.: 1976, 'Low energy particle events associated with sector boundaries', *Sol. Phys.* **50**, 491–500.

Svensmark, H., and Friis-Christensen, E.: 1997, 'Variation of cosmic ray flux and global cloud coverage - a missing link in solar climate relations', *J. Atmos. Solar Terr. Phys.* **59**, 1225–1232.

Tinsley, B. A.: 1994, 'Solar wind mechanism suggested for weather and climate change', *Eos*, Trans. Am. Geophys. Un. **75**, 369–374.

Tinsley, B. A.: 1996, 'Correlations of atmospheric dynamics with solar wind induced air-earth current density into cloud tops', *J. Geophys. Res.* **101**, 29,701–29,714.

Tinsley, B. A.: 2000, 'Electroscavenging and contact nucleation in clouds with broad droplet size distributions', *Proceedings of the 13th Int. Conf. on Clouds and Precipitation, ICCP*, Reno, Nevada, 14–18 August, 2000, in press.

Tinsley, B. A., and Deen, G. W.: 1991, 'Apparent tropospheric response to MeV-GeV particle flux variations: A connection via electrofreezing of supercooled water in high level clouds?' *J. Geophys. Res.* **96**, 22283–22296.

Tinsley, B. A., and Heelis, R. A.: 1993, 'Correlation of atmospheric dynamics with solar activity: Evidence for a connection via the solar wind, atmospheric electricity, and cloud microphysics', *J. Geophys. Res.* **98**, 10375–10384.

Tinsley, B. A., Hoeksema, J. T., and Baker, D. N.: 1994, 'Stratospheric volcanic aerosols and changes in air-earth current density at solar wind magnetic sector boundaries as conditions for the Wilcox tropospheric vorticity effect', *J. Geophys. Res.* **99**, 16,805–16,813.

Tinsley, B. A., Liu, W., Rohrbaugh, R. P., and Kirkland, M.: 1998, 'South pole electric field responses to overhead ionospheric convection', *J. Geophys. Res.* **103**, 26,137–26146.

Tinsley, B. A., Rohrbaugh, R. P., Hei, M., and Beard, K. V.: 2000, 'Effects of image charges on the scavenging of aerosol particles by cloud droplets, and on droplet charging and possible ice nucleation processes', *J. Atmos. Sci.* **57**, 2118–2134.

Tzur, I., Roble, R. G., Zhuang, H. C., and Reid, G. C.: 1983, 'The response of the earth's global electric circuit to a solar proton event', *in* B. M. McCormac (ed.), *Weather and Climate Responses to Solar Variations*, Colo. Assoc. Univ. Press, Boulder, pp. 427–435.

van Delden, A.: 1989, 'On the deepening and filling of balanced cyclones by diabatic heating', *Meteor. Atmos. Phys.* **41**, 127.

Wilcox, J. M.: 1979, 'Tropospheric circulation and interplanetary magnetic sector boundaries followed by MeV proton streams', *Nature* **278**, 840–841.

Wilcox, J. M., Scherrer, P. H., Svalgaard, L., Roberts, W. O., and Olson, R. H.: 1973, 'Solar magnetic structure: Influence on stratospheric circulation', *Science* **180**, 185–186.

Williams, R. G.: 1979, 'Reply (to Wilcox and Scherrer)', *Nature* **280**, 846.

Williams , R. G., and Gerety, E. J.: 1978, 'Does the troposphere respond to day-to-day changes in solar magnetic field?', *Nature* **275**, 200–201.

Zimmerman, J. E., Smith, P. J., and D. R. Smith, 1989, 'The role of latent heat release in the evolution of a weak extratropical cyclone', *Mon. Wea. Rev.* **117**, 1039–1057.

THE INFLUENCE OF THE 11-YEAR SOLAR CYCLE ON THE STRATOSPHERE BELOW 30 KM: A REVIEW

H. VAN LOON (vanloon@ncar.ucar.edu)
NCAR, P.O. Box 3000, Boulder, CO 80307, USA.
NCAR is sponsored by the National Science Foundation.

K. LABITZKE (labitzke@strat01.met.fu-berlin.de)
Meteorologisches Institut FUB, C.-H. Beckerweg 6–10, Berlin 12165, Germany.

Received: 12 July 1999; Accepted in final form: 1 December 1999

Abstract. The NCEP/NCAR re-analyses of the global data as high as 10hPa have made it possible to examine the influence of the 11-year sunspot cycle on the lower stratosphere over the entire globe. Previously, the signal of the solar cycle had been detected in the temperatures and heights of the stratosphere at 30hPa and below on the Northern Hemisphere by means of a data set from the Freie Universität Berlin. The global re-analyses show that the signal exists on the Southern Hemisphere too, and that it is almost a mirror image of that on the Northern Hemisphere. The largest temperature correlations with the solar cycle move from one summer hemisphere to the other, and the largest height correlations move poleward within each hemisphere from winter to summer.

The correlations are weakest over the whole globe in the northern winter. If, however, one divides the data into the winters when the equatorial Quasi-Biennial Oscillation was easterly or westerly, the arctic correlations become positive and large in the west years, but insignificantly small over the rest of the earth. The correlations in the east years are negative in the Arctic but positive in the subtropics and tropics on both hemispheres.

The difference between the east and west years in January-February can be ascribed to the fact that the dominant stratospheric teleconnection and the solar influence work in the same direction in the east years but oppose each other in the west years.

1. Introduction

Data from the stratosphere became comparatively abundant below 25km on the Northern Hemisphere after the International Geophysical Year (1957/1958). The situation elsewhere is still not so favorable as in the lower stratosphere on the Northern Hemisphere for an investigation of the influence of the 11-year sunspot cycle (hereafter abbreviated as SSC). It was therefore natural to concentrate on the Northern Hemisphere when we began the work described in the following, and we were lucky to have at our disposal the data set produced in the Stratospheric Research Group at the Freie Universität Berlin (FUB) and initiated by Richard Scherhag in 1957.

The 30-hPa level was the primary level in our studies, since it was the highest level with a continuous series of daily analyses. Our first map with correlations

Space Science Reviews **94:** 259–278, 2000.

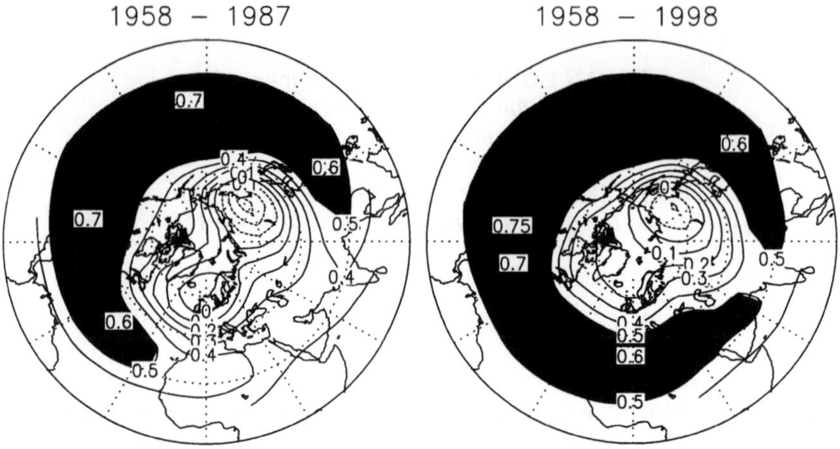

Figure 1. Correlations between the 30-hPa annual mean heights and the 10.7 cm solar flux (11-year solar cycle): (a) 1958–1987, (b) 1958–1998. Shaded for emphasis. Data from the Freie Universität Berlin (van Loon and Labitzke, 1990), updated.

between 30 years of annual mean 30-hPa heights and the SSC, with data from 1958 to 1987, is reproduced in Fig. 1a. The pattern of correlation is characteristic of the solar signal, with the best correlations in the regions between about 20°N and 45°N. Since then, another 11 years have become available — a whole solar cycle — and the results of the correlations between the SSC and 41 years of 30-hPa data appear in Fig. 1b. In contrast to the many apparent solar relationships which have vanished when more data were added, the addition of one more SSC strengthens the early pattern and adds to its statistical significance.

We have over the years monitored the annual mean 30-hPa heights at a point near Hawaii (30°N 150°W) where the correlations consistently have exceeded 0.7; that is, where half the interannual variance of the heights is associated with the solar cycle. The time series of the heights at this point, Fig. 2, brings out two important aspects: first, an independent sample (the last solar cycle) fits the solar curve perfectly; and second, the amplitude of the signal appears to be dependent on the amplitude of the individual 11-year cycle.

Further encouragement came from the inclusion of the Southern Hemisphere in the analysis (van Loon and Labitzke, 1998, 1999) since, as seen below, a clear solar signal — almost a mirror image of that on the Northern Hemisphere — emerged south of the equator.

In the following, we shall paint a fuller picture of the effect on the stratosphere of the SSC with data which now cover nearly four cycles and the whole globe below the 10-hPa level. As measure of the solar activity in the 11-year periods we have used the 10.7 cm solar flux. No causal relationship is implied by the use of the solar flux; it is meant strictly to be an index.

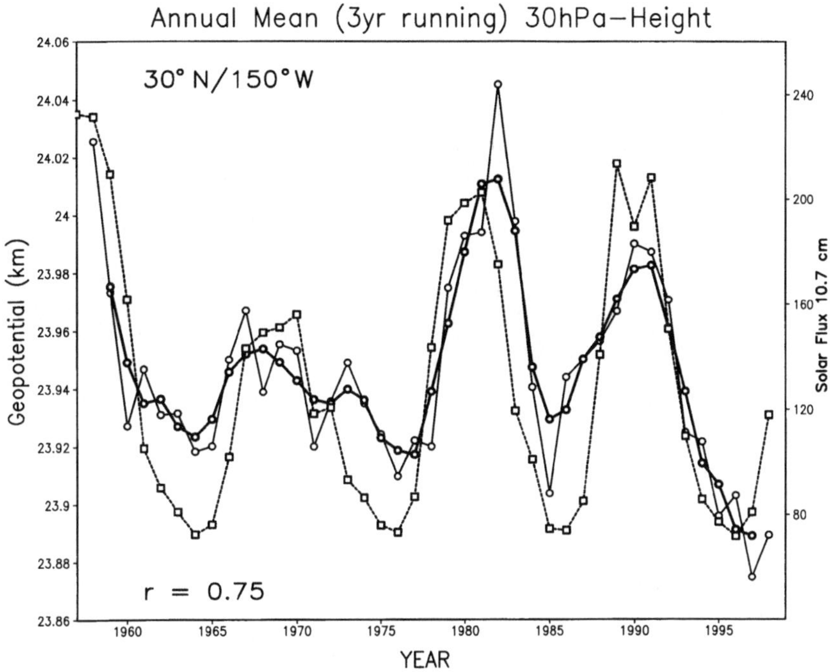

Figure 2. Time series of the 10.7 cm solar flux (dashed line) and of the annual mean 30hPa-heights (thin solid line) and their three-year running means (heavy solid line); at 30°N 150°W. In geopotential kilometer. Data from the Freie Universität Berlin (Labitzke and van Loon, 1995), updated.

2. Geopotential Heights

All *global* data below the 10-hPa level (about 30 km) have lately been used in a re-analysis of the state of the atmosphere since the International Geophysical Year. The agencies responsible for this analysis are the US National Centers for Environmental Prediction (NCEP) and the National Center for Atmospheric Research (NCAR). The correlation values obtained for the Northern Hemisphere with this data set (Fig. 3) are close to those in Fig. 1b which were calculated with the FUB data for the same period. The analysis methods differ between the NCEP/NCAR and the FUB series: the former uses a mathematical model and the latter is made by experienced analysts, but this does not greatly affect the annual and monthly averages used in the calculation of the correlation coefficients.

The correlation maximum on the Southern Hemisphere also encircles the hemisphere but it lies somewhat closer to the equator and appears to be weaker than the one on the Northern Hemisphere. According to the correlations in Fig. 3, the Arctic and Antarctic show no evidence of a solar connection; but the interannual variability is large in the polar regions and tends to obscure a solar signal. The next figure (Fig. 4) demonstrates that the influence of the SSC nevertheless can be found

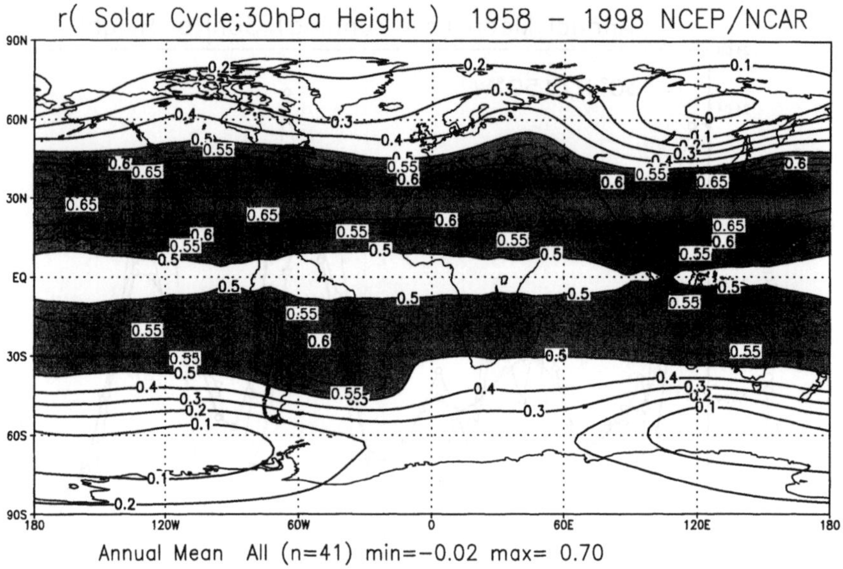

Figure 3. Correlations between the 10.7 cm solar flux and annual mean 30-hPa heights, 1958–1998. Shaded for emphasis. Data from the NCEP-NCAR re-analyses, which is also so in those figures where no source is shown (van Loon and Labitzke, 1998), updated.

in high latitudes too: this figure has on top the correlations between the SSC and the zonally averaged, annual mean geopotential heights between 300 hPa and 10 hPa (about 8 to 30 km). A downward extension and attenuation of the two correlation maxima in Fig. 3, as well as the unimportant values at high latitudes, are evident. The middle section of Fig. 4 contains the standard deviations of the annual mean heights. As defined by the standard deviation, the interannual variability is large in both polar regions and on the equator, and it increases with increasing height all over the earth.

A correlation coefficient is determined by a fraction in which the numerator consists of the covariance of the two quantities which are being correlated, and the denominator is the product of their standard deviations [$r = \mathrm{COV}_{x,y}/(\mathrm{SD}_x\mathrm{SD}_y)$]: the interannual variability. A large interannual variability — that is, a large denominator — lowers the value of the correlation coefficient. The high correlations in Figs. 1, 3, and 4 therefore do not necessarily show where the solar influence is strong, but rather where the interannual variability is low. The covariances at the bottom of Fig. 4 are as high in the polar regions as in lower latitudes, and thus indicate that the SSC affects these regions too, although the signal is obscured by interannual noise.

There is a relative correlation minimum in the stratosphere on the equator (see also Fig. 3). The interannual variability is comparatively large in the equatorial regions because of the Quasi-Biennial Oscillation, Fig. 4 middle, but there is no

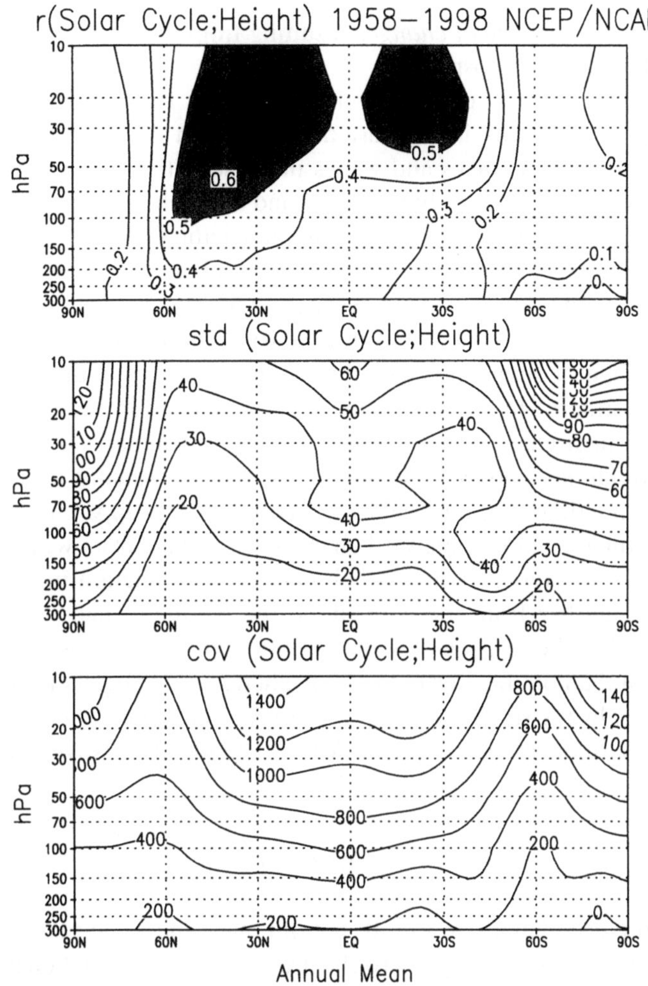

Figure 4. Vertical meridional sections of (top) correlations between the zonally averaged, annual mean heights and the 10.7 cm solar flux; (middle) the standard deviations of the heights; and (bottom) the covariances of the heights with the solar flux.

minimum there in the covariances (Fig. 4 bottom) which are almost the same between 40°N and 40°S.

The covariances are also large in the both polar regions in winter (not shown), which excludes a direct influence of the sun there. Experiments with the general circulation model at the Goddard Institute for Space Studies by Shindell et al. (1999) demonstrate a mechanism by which interaction in the upper stratosphere between ozone and ultraviolet radiation from the sun induces circulation changes which amplify the effect of observed, small changes in solar irradiance over the 11 year solar cycle. What we observe in the lower half of the stratosphere are thus principally *indirect, dynamical changes* resulting from the interaction between UV and ozone in the upper stratosphere.

The annual mean pattern of the correlations in Fig. 1b is repeated with varying intensity in the correlations in the various seasons on the Northern Hemisphere (Fig. 5). The solar signal waxes and wanes in phase with the elevation of the sun, being statistically significant — both locally and as a field — over nearly all of the hemisphere in summer, but in winter only in the Pacific Ocean.

The signal in the Southern Hemisphere is weakest in the southern summer (Fig. 6), that is, at the time when the weakest signal is observed on the Northern Hemisphere. The weak global signal in January–February is associated with the influence of the Quasi-Biennial Oscillation on the dynamics of the unstable winter vortex on the Northern Hemisphere, and will be discussed in Section 4.

The interseasonal changes in the solar signal are summarized in Fig. 7 which shows the correlations between the solar flux and the zonally averaged 30-hPa heights as two-month overlapping values. It is immediately apparent that the largest correlation coefficients move poleward on each hemisphere from winter to summer, and that they are weakest in January–February. The correlation maximum moves across latitudes farther from the equator on the Northern than on the Southern Hemisphere; and in the respective winters it lies appreciably closer to the equator on the Southern than on the Northern Hemisphere. On the equator the correlation coefficients are always smaller than in the tropical and subtropical latitudes, and they are weakest in the months January–March and October–November.

3. Temperatures

The temperature correlations in Fig. 8, bottom, have the same pattern for the annual values as the heights in Fig. 3. To illustrate the interseasonal change (top and middle in Fig. 8) we have chosen four months centered on the solstices, and find that although there is a correlation maximum on each hemisphere in both seasons, the largest correlations follow the sun from one summer hemisphere to the other.

The use of zonal averages brings out clearly the summer-to-summer movement of the correlation maximum in the stratospheric temperature: the correlations in Fig. 9 have been made with the same four-month averages as in Fig. 8, and they show the movement of the major peak from middle northern latitudes in the northern summer to the middle latitudes on the Southern Hemisphere in the southern summer. The minor peak in the tropics moves from one winter hemisphere to the other.

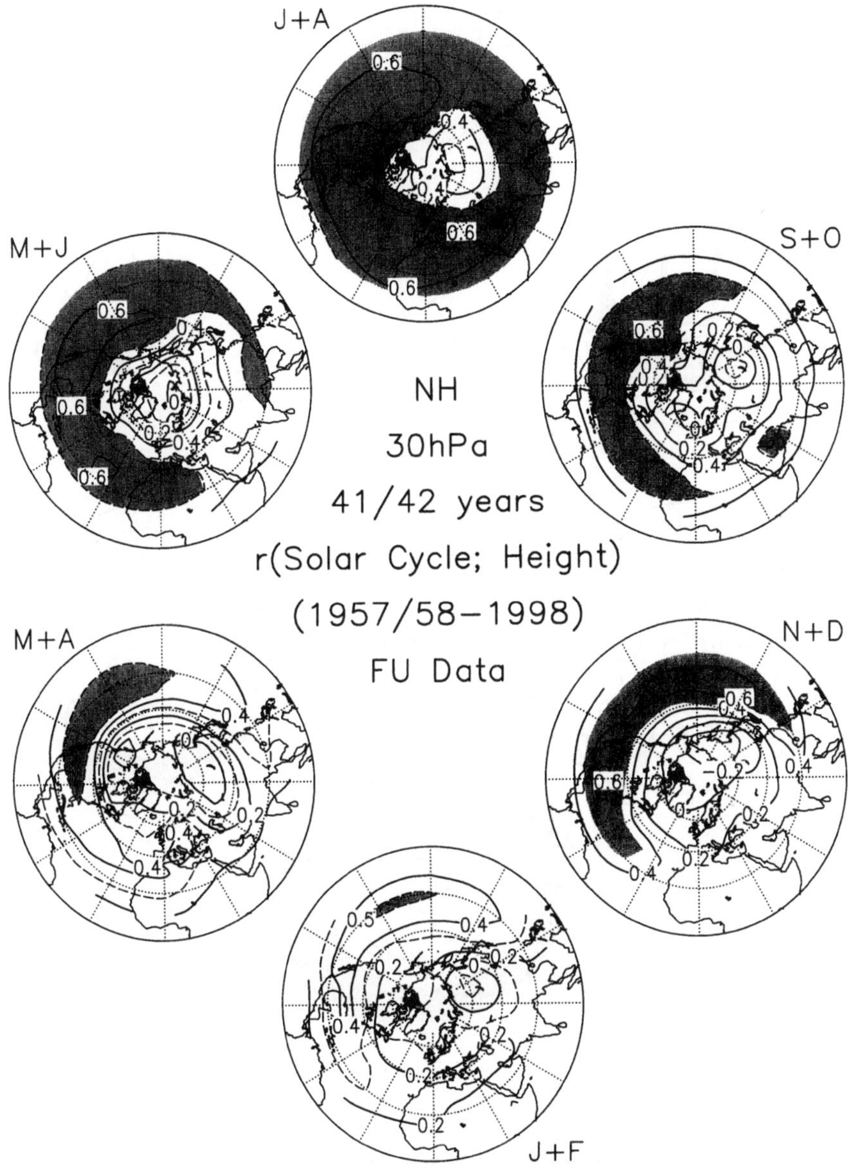

Figure 5. The correlations between two-month averages of 30-hPa heights on the Northern Hemisphere and the solar flux, 1958–1998. Shaded for emphasis. Data from the Freie Universität Berlin (Labitzke and van Loon, 1999).

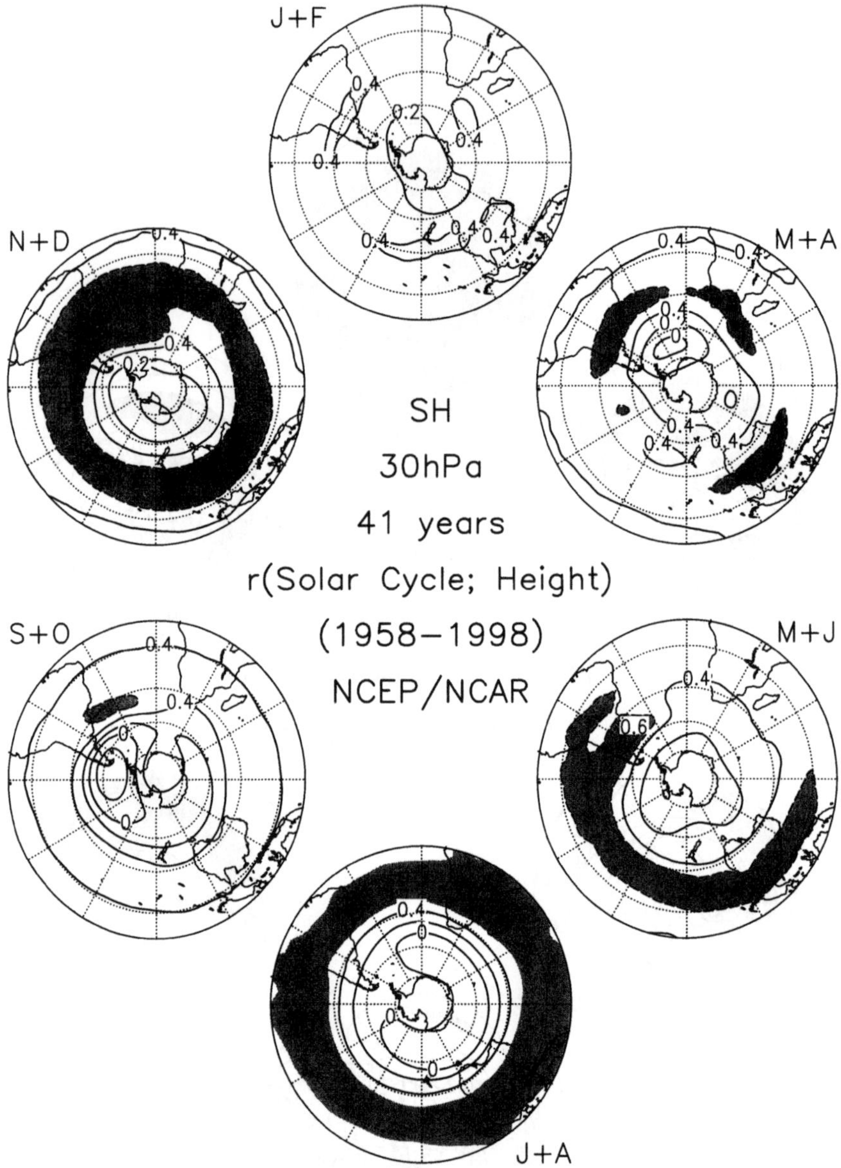

Figure 6. The same as Fig. 5, but for the Southern Hemisphere.

4. The QBO and the solar signal in the northern winter.

The incentive to study the influence of the 11-year sunspot cycle in the stratosphere came from the discovery that the Quasi-Biennial Oscillation in the equatorial stratospheric winds modulates the solar signal (Labitzke, 1987; see also Labitzke and van Loon, 1997 and 2000; and van Loon and Labitzke, 1994). The

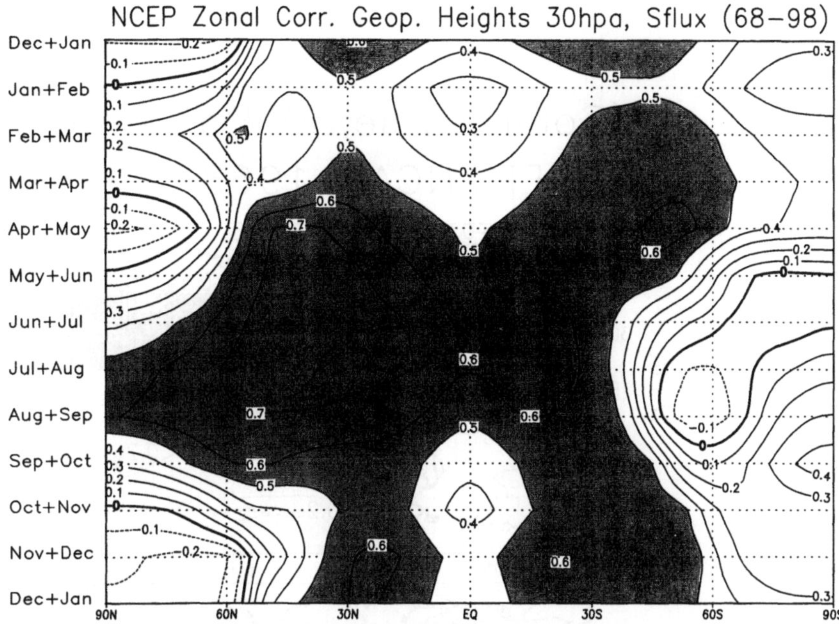

Figure 7. Two-month running correlations between the 30-hPa heights and the 10.7 cm solar flux, 1968–1998. Shaded for emphasis (van Loon and Labitzke, 1999).

solar signal is weak or non-existing on both hemispheres in the *northern* winter if one uses a full series of stratospheric heights for correlation with the SSC (Figs. 5,6,7, and 10 top), but if the data are grouped into years when the QBO in the lower stratosphere (about 45 hPa) was westerly and years when it was easterly, the solar correlations become positive and large in the Arctic in the west years and negative in the east years (Fig. 10 middle and bottom). Beyond the Arctic, and as far as the Antarctic, the correlation coefficients approach zero in the west years (Fig. 10 bottom), whereas they are positive and large in the subtropics and tropics in the east years (Fig. 10 middle). The pattern is thus the same in the east years in winter as that obtained with undivided series in the other seasons (see, e.g, van Loon and Labitzke, 1994; and Labitzke and van Loon, 1997).

Recently Salby and Callaghan (1999) have confirmed that the influence of the SSC on the arctic winter temperatures and the polar night vortex in the stratosphere “rests on a connection to the quasi-biennial oscillation (QBO) of equatorial wind”. They also showed that the QBO itself is modulated by the solar cycle.

According to the correlations in the east and west years in Fig. 10, the geostrophic west wind in the polar night jet stream north of 30°N tends to be substantially different in the west and east years of the QBO: in solar maxima it is usually strong (cold vortex) in the east years and weak (warm vortex) in the west years, and conversely in SSC minima. The dominant teleconnection pattern in the stratosphere of the Northern Hemisphere (Shea et al., 1992) is a strong inverse relationship —

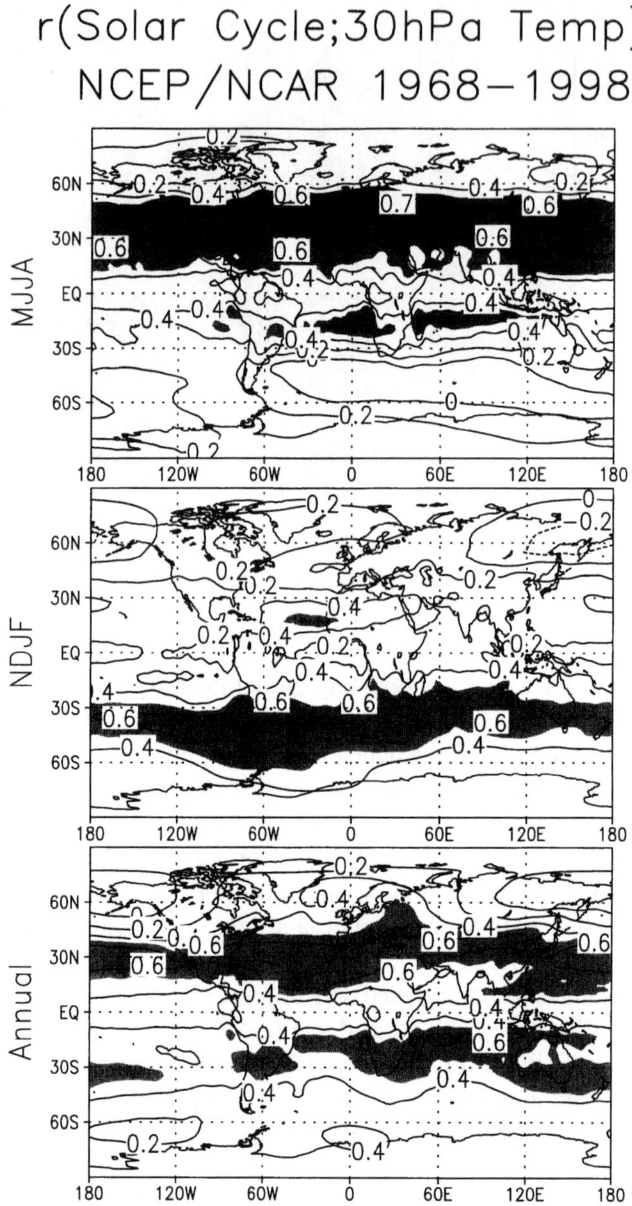

Figure 8. Correlations between the 30-hPa temperature and the 10.7 cm solar flux, 1968–1998: (top) for May-June-July-August; (middle) for November-December-January-February and (bottom) for the year. Shaded for emphasis.

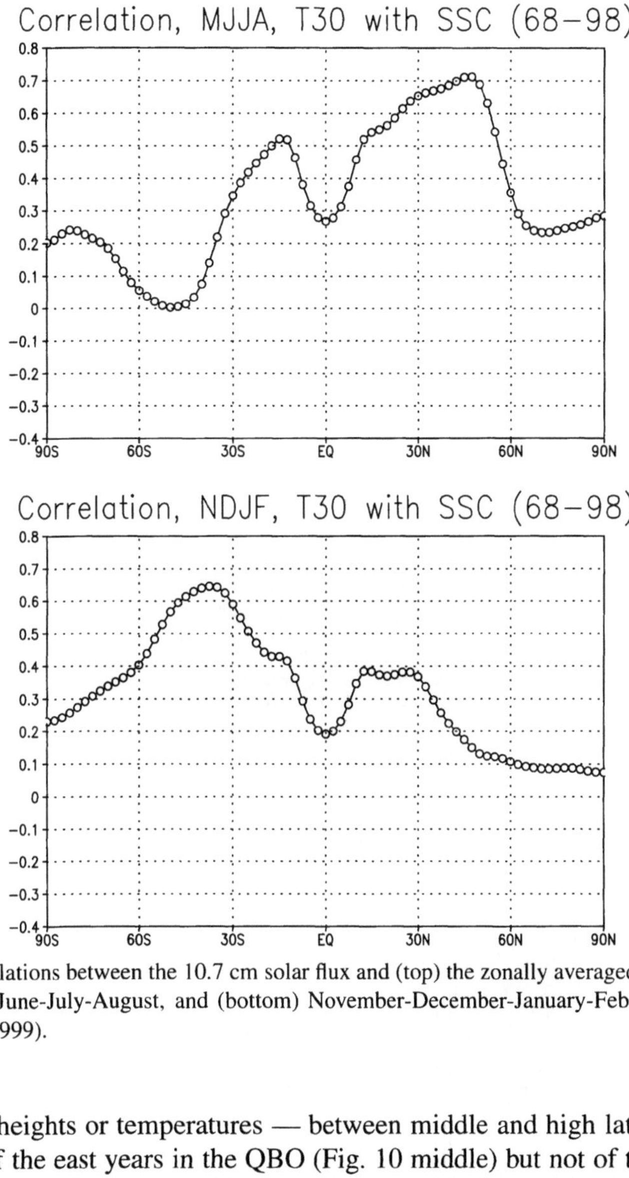

Figure 9. Correlations between the 10.7 cm solar flux and (top) the zonally averaged 30-hPa temperatures in May-June-July-August, and (bottom) November-December-January-February (van Loon and Labitzke, 1999).

a seesaw of heights or temperatures — between middle and high latitudes; that is a property of the east years in the QBO (Fig. 10 middle) but not of the west years (Fig. 10 bottom). The large positive correlations with the SSC in the Arctic in the west years (Fig. 10 bottom) is associated with the fact that the polar vortex in nearly all the 40 years of observation has broken down and warmed in those solar maxima which occurred in west years.

In July-August, Fig. 11, dividing the data according to the phase of the QBO makes little difference; there is then no visible interaction between the SSC and the QBO with respect to the *south polar* winter vortex. This vortex is cold and stable and no major midwinter breakdowns, such as those which occur in more than a third of the winters in the north polar vortex, have yet been observed below 10 hPa.

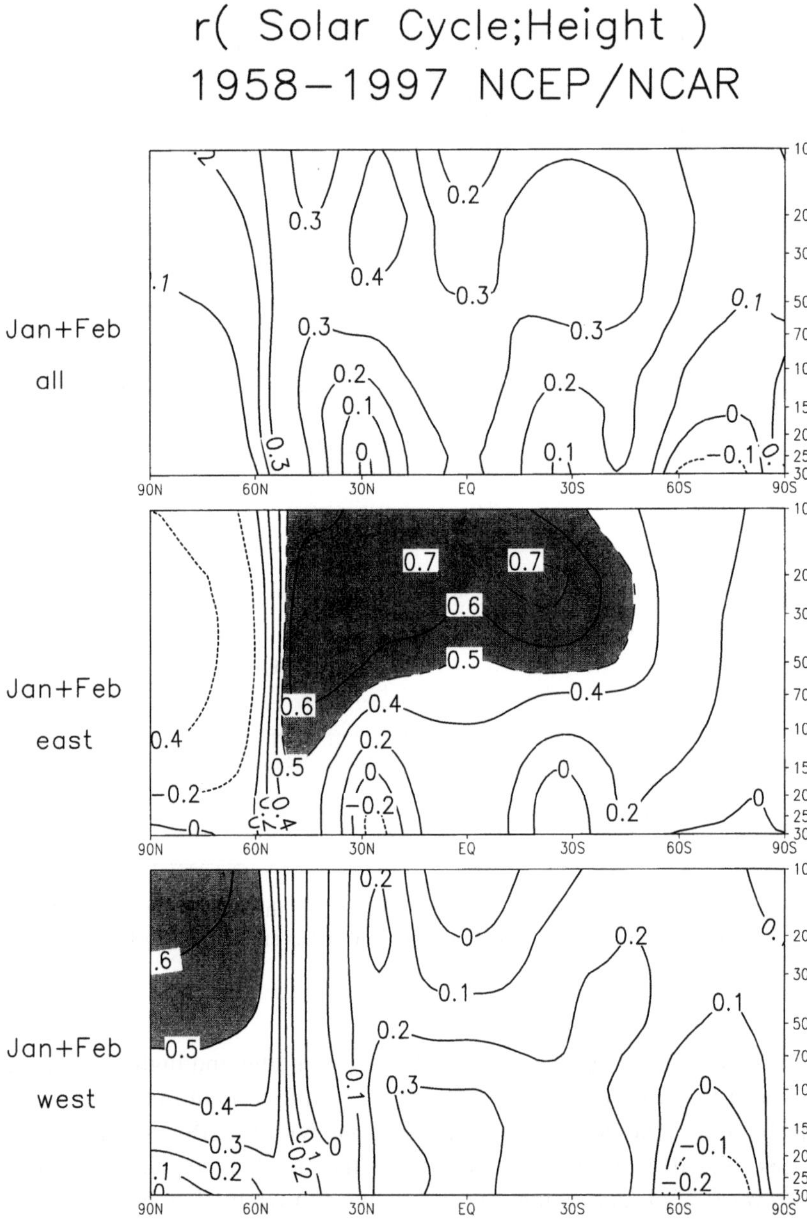

Figure 10. Correlations between the 10.7 cm solar flux and (top) the zonally averaged 30-hPa heights in January-February, all years 1958–1997; (middle) the same for the east years in the equatorial Quasi-Biennial Oscillation; and (bottom) the same for the west years in the QBO. Shaded for emphasis (Labitzke and van Loon, 2000).

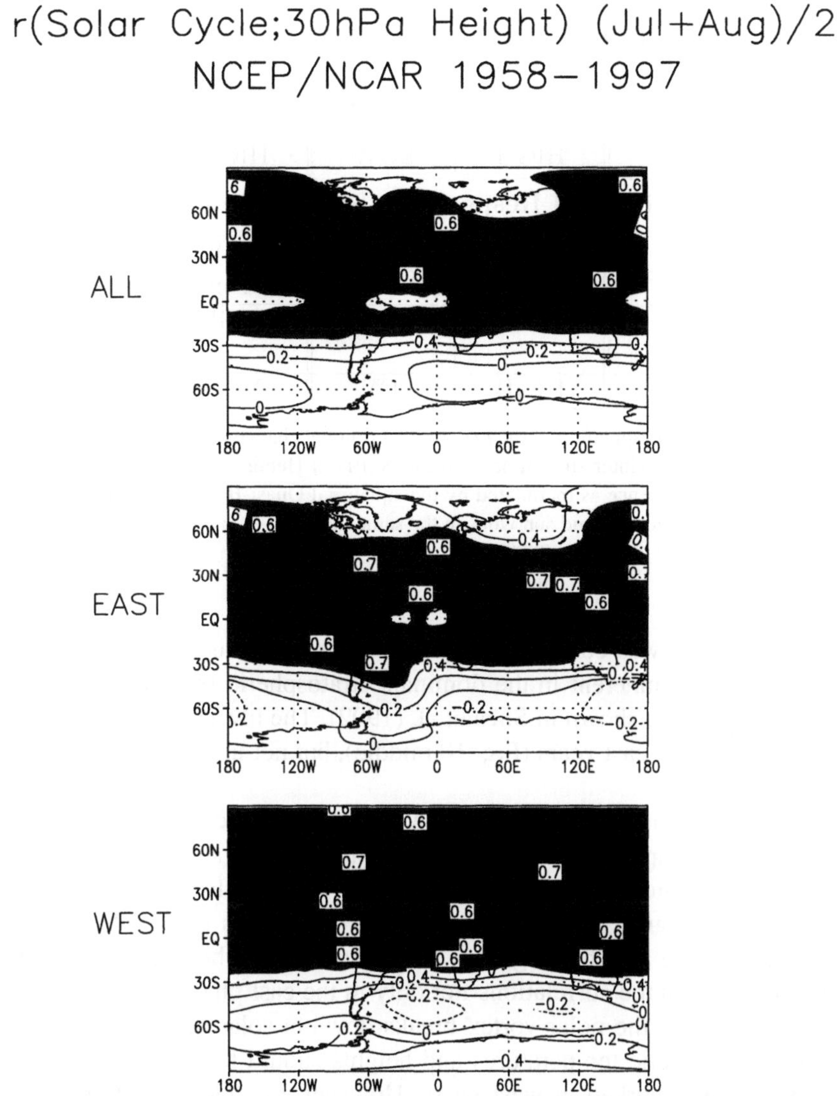

Figure 11. Correlations between the 10.7 cm solar flux and (top) the 30-hPa heights in July-August, all years 1958-1997; (middle) the same for the east years in the equatorial Quasi-Biennial Oscillation; and (bottom) the same for the west years in the QBO. Shaded for emphasis. (Labitzke and van Loon, 2000).

Since the east phase in the northern winter yields solar correlations in the tropics and subtropics which are similar to those made with full data series in other seasons (Fig. 10, middle), the reason for the weak global signal of the SSC in the northern winter (Fig. 10 top) must bc sought in the west phase of the QBO:

Fritz and Soules (1970) observed that stratospheric temperature changes in high northern latitudes are accompanied by changes in the opposite sense as far south

Expected TELECONNECTIONS from Solar Influence				
	solar max.	solar min.	solar max.	solar min.
Arctic	LOW	HIGH	LOW	HIGH
Extra–Arctic	HIGH	LOW	HIGH	LOW

Observed Response to Solar Influence				
	EAST QBO		WEST QBO	
	solar max.	solar min.	solar max.	solar min.
Arctic heights and temperatures	LOW	HIGH	HIGH	LOW
Extra–Arctic	HIGH	LOW	No clear response	

Figure 12. Schematic representation of the *expected* meridional changes to follow from the influence of the solar cycle in the winter stratosphere on the Northern Hemisphere (top); and the *observed* response to the solar influence as modulated by the equatorial Quasi-Biennial-Oscillation (bottom). In the east years teleconnection and sun work in the same direction — in the west years they oppose each other. (Labitzke and van Loon, 2000).

as Antarctica. This was confirmed for the Northern Hemisphere in van Loon et al. (1975), and it is inherent in the dominant stratospheric teleconnections on the Northern Hemisphere shown in Shea et al. (1992). The inverse teleconnection pattern is obvious in the east years (Fig. 10 middle), but not in the west years (Fig. 10 bottom).

Midwinter warmings with a breakdown of the arctic cyclonic vortex in *east years* of the QBO happen mainly in minima of the SSC, as suggested by the correlations in Fig. 10 middle. The accompanying cooling and lowering of the heights at lower latitudes agree with the teleconnection. At peaks in the SSC the vortex is intense and stable and the heights at lower latitudes are high; the lower latitude response is found in the correlations not only in the east phase in winter but in all other seasons. Since the polar vortex breaks down at solar activity *peaks* in west years, the stratospheric temperatures and heights should fall beyond the Arctic if the teleconnection worked in west years. However, the teleconnection pattern is counterbalanced by the solar effect, which in the peaks in the SSC creates higher heights and temperatures in temperate and low latitudes. Therefore, we end up with little or no correlation with the SSC beyond the Arctic (Fig. 10 top).

This description is schematically illustrated by the table in Fig. 12. The top part shows the state of the geopotential heights in the stratosphere *expected* from the solar influence, *if* the teleconnection functioned undisturbed by the QBO: in the latitudes beyond the Arctic the heights are higher at solar maximum than minimum, and conversely for the polar regions. The influence of the QBO is seen in the lower part of Fig. 12: in the east phase the solar effect is the same as in the undisturbed conditions in the upper part. In the west phase and solar maximum the lower-latitude heights should be high if the SSC were the only influence; but the polar

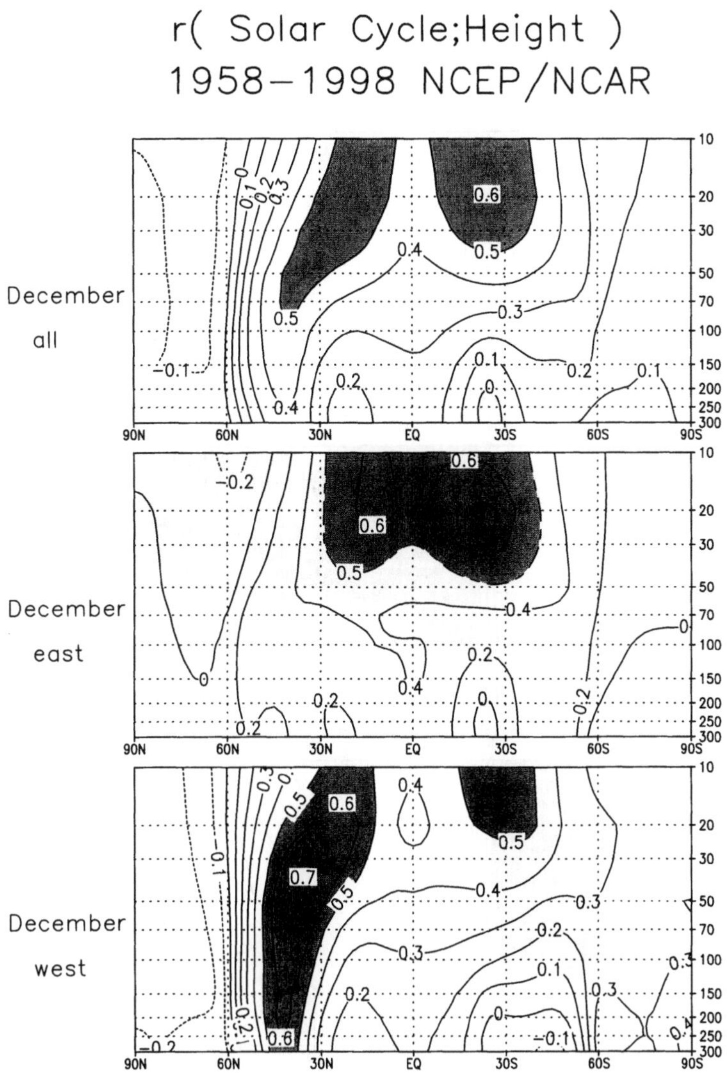

Figure 13. Correlations between the 10.7 cm solar flux and the zonally averaged heights between 300 hPa and 10 hPa: (top) all Decembers; (middle) Decembers in the east years of the QBO; and (bottom) Decembers in the west years. 1958–1998. Shaded for emphasis (Labitzke and van Loon, 2000).

vortex breaks down at solar peaks in this phase and the heights rise in the Arctic. The teleconnection pattern which is associated with rising heights in the Arctic yields lower heights at lower latitudes; but that cancels the teleconnection pattern associated with a solar maximum. The vortex warmings and breakdowns in QBO west phase thus cancel the effect of the SSC at its peak by lowering the heights (and temperatures) at lower latitudes and raising them in the Arctic.

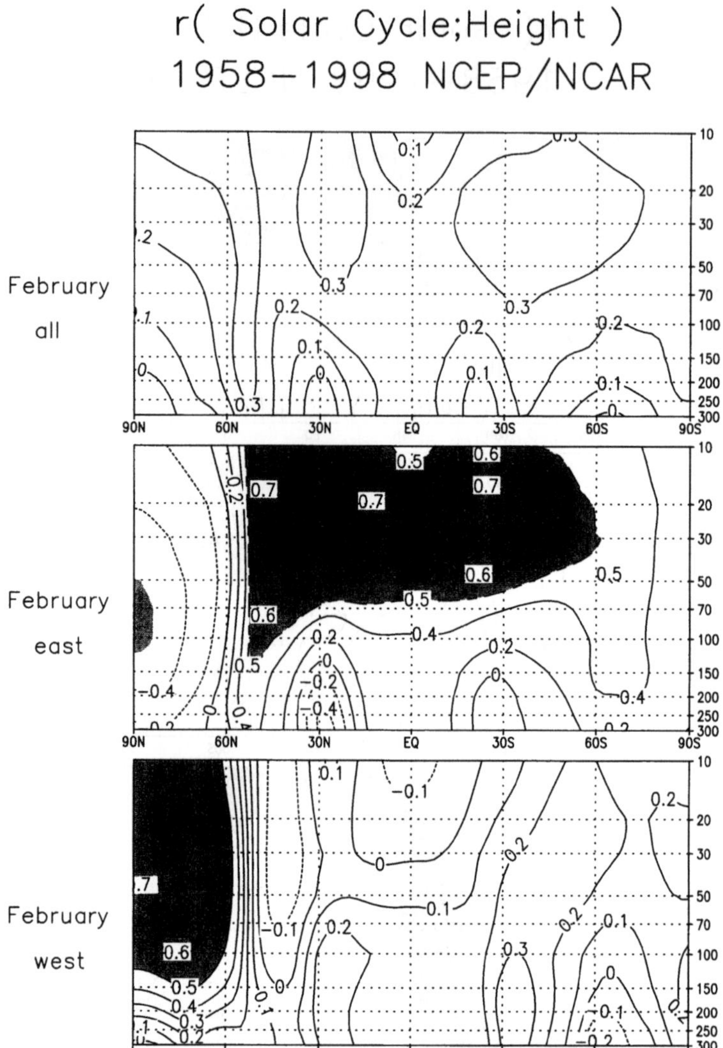

Figure 14. The same as Fig. 13, but for February (Labitzke and van Loon, 2000).

The question is: why does the north polar vortex in the *west* years and solar peaks tend to be warmer and weaker than in the east years and solar peaks? Qualitatively, one can understand what happens in the east years, because the already strong stratospheric Aleutian high in the east years is enhanced in solar peaks, whereby the condition is favorable for lower polar heights and temperatures according to the dominant teleconnection pattern. But whatever the reason is for the vortex behavior at solar peaks in the west years, it is clear that *the solar maxima disturb the stratosphere.*

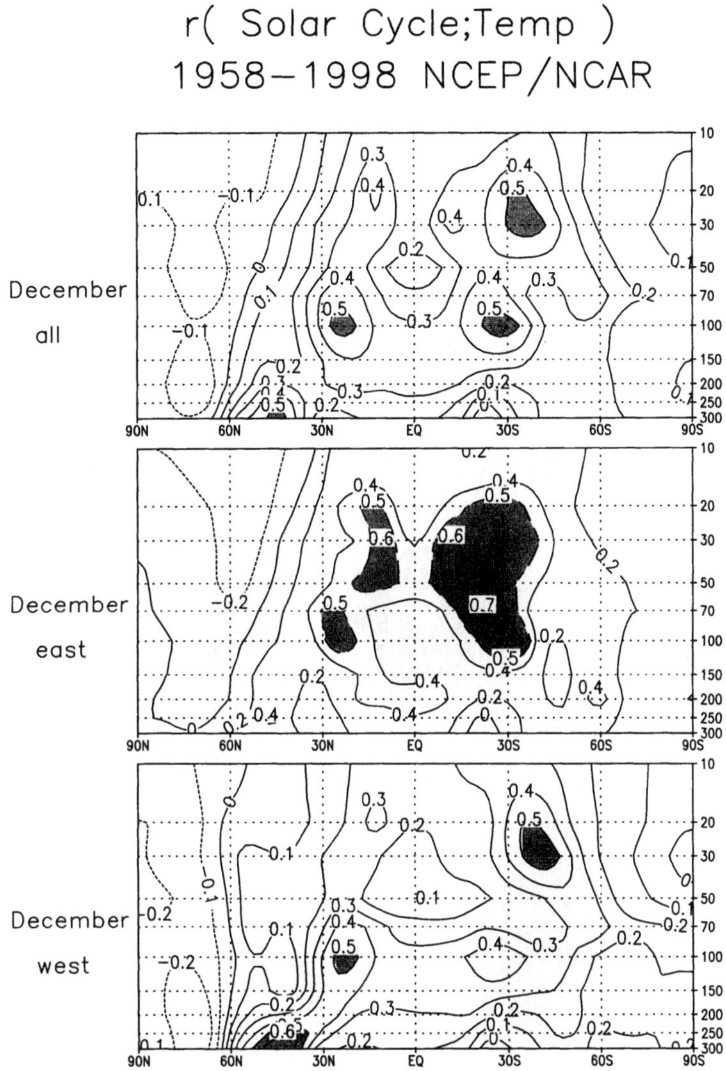

Figure 15. The same as Fig. 13, but for temperature (Labitzke and van Loon, 2000).

The QBO effect is not the same in all of winter, as illustrated by means of the correlations in the vertical, meridional sections in Figs. 13 and 14. In December (Fig. 13 top) one recognizes the established pattern of enhanced heights at lower latitudes in the stratosphere, see Fig. 4 top. Above 50 hPa the pattern is about the same in the east and west years of the QBO, but in the west years the large correlations extend downward beyond the 300-hPa level. There is therefore a tendency in early winter for stronger westerlies north of 45°N, and thus a stronger vertical wave guide in the west than in the east years. The downward extension of high correlations are a reflection of the intensification of the stratospheric Aleutian high

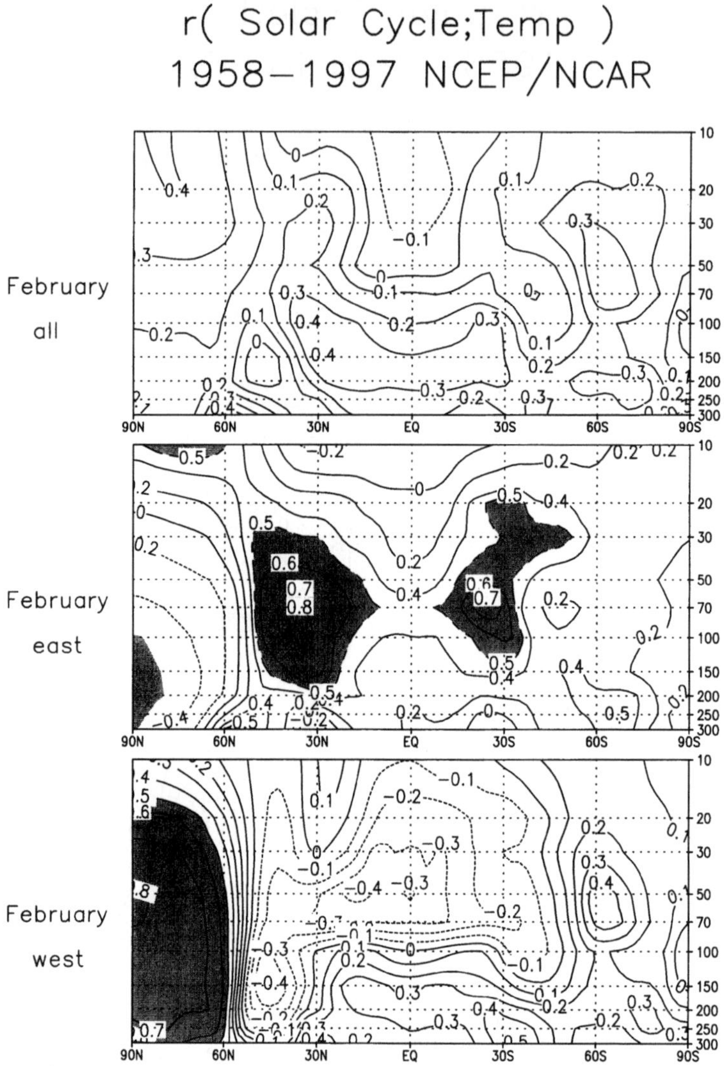

Figure 16. The same as Fig. 14, but for temperature (Labitzke and van Loon, 2000).

in solar maximum in the west phase which is described in van Loon and Labitzke (1994, Fig. 10) as a precursor to a major midwinter warming in the west phase of the QBO.

In February (Fig. 14, top) the correlations are negligible when the full time series is used, but increase when the data are divided according to the QBO: in the east phase (Fig. 14, middle) large positive correlations cover the region above the tropopause from 55°N to 55°S, whereas the correlations in the north polar region are negative. In the west phase the arctic correlations are large and positive (Fig. 14, bottom), but they are negligible over the rest of the globe.

The reaction of the Arctic stratosphere to the modulation of the influence of the sunspot cycle by the QBO is thus principally a phenomenon of the latter half of the winter.

The temperature correlations with the SSC in the northern winter (Figs. 15 and 16) are similar to the height correlations in Figs. 13 and 14, but the largest values are at levels lower than those of the height correlations, which is consistent with the hydrostatic relationship. Although the correlations are not extraordinarily high in the region of the lower latitude tropopause region, it is worth noting the difference in February (Fig. 16) between the east and west years there: in the east years (Fig. 16 middle) the correlation coefficients are positive and in the west years (Fig. 16 bottom) negative in the layers at and above the lower-latitude tropopause. One might interpret this difference to mean that at solar maximum the tropopause region is lifted in the west phase (cooling), which would fit the positive correlations at 300 and 200hPa, where an upward motion in increased convection would warm the upper troposphere through release of latent heat.

In contrast, the positive correlations in the tropopause region at lower latitudes in the east years would indicate a downward, or decreased upward motion in the tropical stratosphere. The opposite correlations in the Arctic are then a sign of increased downward motion in the west years, which would be consistent with the warmings in the west years at peaks in the solar cycle. In the east years the negative Arctic correlations would be indicative of a weakening of the downward motion which normally works against the seasonal radiative cooling.

5. Conclusion

- Examination of the global re-analyses of the stratosphere as high as the 10-hPa level reveals a signal of the 11-year sunspot cycle on both hemispheres.
- The signal is strongest in the northern summer. In the northern winter the unstable dynamics of the cyclonic vortex on the Northern Hemisphere counteracts the solar influence in the west years of the Quasi-Biennial Oscillation in the equatorial winds.
- The highest correlations between the solar cycle and the stratospheric heights move poleward from winter to summer in both hemispheres. The highest correlations between the solar cycle and the temperatures in the stratosphere move from one summer hemisphere to the other.
- Experiments with atmospheric models indicate that the high correlations with the solar cycle in the lower stratosphere are an indirect, dynamic result of interaction in the upper atmosphere between ozone and ultraviolet radiation from the Sun.

References

Fritz, S. and Soules, S. D.: 1970, 'Large-scale Temperature Changes in the Stratosphere Observed from Nimbus III', *J. Atmospheric Sci.* **27**, 651–661.

Labitzke, K.: 1987, 'Sunspots, the QBO, and the Stratospheric Temperature in the North Polar Region', *Geophys. Res. Lett.* **10** 535–537.

Labitzke, K. and van Loon, H.: 2000: 'The QBO Effect on the Global Stratosphere in Northern Winter', *J. Atmospheric Solar-Terrest. Phys.* **62**, submitted.

Labitzke, K. G. and van Loon, H.: 1999, *The Stratosphere: Phenomena, History, and Relevance*, Springer-Verlag, Berlin, 190 pp.

Labitzke, K. and van Loon, H.: 1997, 'The Signal of the 11-Year Sunspot Cycle in the Upper Troposphere–Lower Stratosphere', *Space. Sci. Rev.* **80**, 393–410.

Labitzke, K. and van Loon, H.: 1995, 'Connection between the Troposphere and Stratosphere on a Decadal Scale', *Tellus* **47A**, 275–286.

Salby, M. and Callaghan, P.: 1999, 'Connection between the Solar Cycle and the QBO: The Missing Link', *J. Clim.* **12**, in press

Shea, D.J., van Loon, H., and Labitzke, K., 1992, 'Point Correlations of Geopotential Height and Temperature at 30 mb and between 500 mb and 30 mb', *NCAR/TN-368+STR*, 6 pp. plus 285 pp. maps.

Shindell, D., Rind, D., Balachandran, N., Lean, J., and Lonergan, J.: 1999, 'Solar Cycle Variability, Ozone, and Climate', *Science* **284**, 305–308.

van Loon, H. and Labitzke, K.: 1990, 'Association between the 11-Year Solar Cycle and the Atmosphere. Part IV: The Stratosphere, Not Grouped by the Phase of the QBO', *J. Clim.* **3**, 827–837.

van Loon, H. and Labitzke, K.: 1994, 'The 10–12 Year Atmospheric Oscillation', *Met. Z., N.F.*, **3**, 259–266.

van Loon, H. and Labitzke, K.: 1998, 'The Global Range of the Stratospheric Decadal Wave', *J. Clim.* **11**, 1529–1537.

van Loon, H. and Labitzke, K.: 1999, 'The Signal of the 11-Year Solar Cycle in the Global Stratosphere', *J. Atmospheric Solar-Terrest. Phys.* **61**, 53–61.

van Loon, H., Madden, R. A., and Jenne, R. L.: 1975, 'Oscillations in the Winter Stratosphere', *Mon. Wea. Rev.* **103**, 154–162.

Address for Offprints: K. Labitzke, Meteorologisches Institut FUB, C.-H. Beckerweg 6–10, Berlin 12165, Germany. E-mail: labitzke@strat01.met.fu-berlin.de

AMPLIFICATION OF THE INFLUENCE OF SOLAR FLUX VARIATIONS ON THE WINTER STRATOSPHERE BY PLANETARY WAVES

N.F. ARNOLD (nfa1@ion.le.ac.uk) and
T.R. ROBINSON (txr@ion.le.ac.uk)
Department of Physics and Astronomy, University of Leicester, University Road, Leicester, LE1 7RH, U.K.

Received: 14 September 1999; Accepted in final form: 20 December 1999

Abstract. Recent observational evidence has suggested that variations in solar activity may affect winter stratospheric polar ozone and temperature levels. The paucity of direct sunlight available during this season points strongly to a dynamical mechanism. We have carried out several large ensemble experiments within the middle atmosphere and the coupled middle atmosphere and lower thermosphere to simulate the radiative/dynamical coupling via planetary waves for a range of solar fluxes. In the former case, the model response in the winter stratosphere was linear and of the order of the summer stratopause forcing, whilst in the latter, the level of correlation in the winter stratosphere remained high, but was diluted over a wider volume. The inclusion of the upper atmosphere enhanced the winter polar stratospheric response by a factor of three.

1. Introduction

Attempts to establish a physical mechanism whereby solar variability influences climate must overcome several obstacles. Firstly, the solar constant has been observed to depart from a constant value by only a few tenths of a percent in the past thirty years (Pap and Fröhlich, 1999). Secondly, the natural internal variability of the atmosphere is considerable at all temporal and spatial scales. There is a distinct possibility that detection of a terrestrial response to this forcing is impossible (James, 1993). However, numerical simulations provide an opportunity for carrying out mechanistic studies where the natural variability of the atmospheric system is kept to a minimum (Arnold and Robinson, 1998) and/or the solar forcing is artificially raised (Kodera et al., 1991).

Variations in solar output are not uniformly distributed across the electromagnetic spectrum (see for example Brasseur and Solomon, 1986). Between 200–300 nm, the variability rises to a few per cent (Hood, 1999). This increases to a factor of two at 100 nm and in the soft X-ray part of the spectrum, it can rise by a factor of 100. For this reason, the thermosphere, where the bulk of the high energy photons gets absorbed, undergoes considerable changes in response to solar activity (Richmond, 1983).

Space Science Reviews **94:** 279–286, 2000.

Whilst the upper mesosphere and the thermosphere contain a very small fraction of the total mass, energy and momentum, it has been recognised that gravity waves in this region are an important driving force for the extra-tropical middle atmosphere circulation via 'downward control' (Haynes et al., 1991). These waves also contribute to the strength of the equatorial Quasi-Biennial Oscillation (Mayr et al., 1997a; Mayr et al., 1997b). Mesospheric gravity wave fluxes have been reported to be anti-correlated with solar activity levels (Gavrilov, 1997). In addition, the 11-year solar cycle has been observed to modulate the influence of the QBO on the winter Northern Hemisphere stratospheric circulation (Labitzke and van Loon, 1988). Planetary wave activity is considerably more pronounced during solar maximum / western phase QBO and to a lesser extent during solar minimum / eastern phase. This has resulted in significant solar activity induced extrema in high latitude total ozone concentrations (Bjarnason et al., 1993) and temperatures and geopotentials (Labitzke and van Loon, 1993).

Efforts to reproduce these observations have largely focussed on planetary wave - mean flow mechanisms within the middle atmosphere. Kodera et al. (1991) obtained qualitatively reasonable atmospheric responses to solar flux variations. Haigh (1994) obtained a stronger signal by allowing feedback with ozone. As a precursor to considering the contribution from gravity wave momentum deposition in a physically realistic fashion, it is necessary to extend the vertical boundary up to the lower thermosphere. In so doing, a region of the atmosphere is introduced that is strongly influenced by direct solar activity variations. Arnold and Robinson, (1998) found that enhanced stratospheric forcing can be obtained by considering the planetary wave coupling between the middle and the upper atmospheres. In the present paper, we compare the relative significance and amplitude of the stratospheric response to changes in solar flux levels in a middle atmosphere model and also in the extended version that incorporates the upper mesosphere and lower thermosphere.

2. The Influence of Solar Variability in a Middle Atmosphere Model

There are a number of pathways for varying solar energy fluxes to affect the atmosphere and initiate complex interactions that rapidly disguise simple cause and effect. For this reason, comprehensive General Circulation Models can suffer from too many degrees of freedom, whereby a relatively weak signal is readily overwhelmed by internal climate 'noise'. Therefore, it is of considerable interest to establish a solid foundation of relatively unambiguous results that help to develop a conceptual framework for future studies by using a relatively simple model that is still able to reproduce the interactions between planetary waves and the mean circulation. To achieve this, an ensemble of experiments was carried out with a version of the UK Met. Office Stratosphere Mesosphere Model (Austin and Butchart, 1992). This model has been used extensively over a number of years

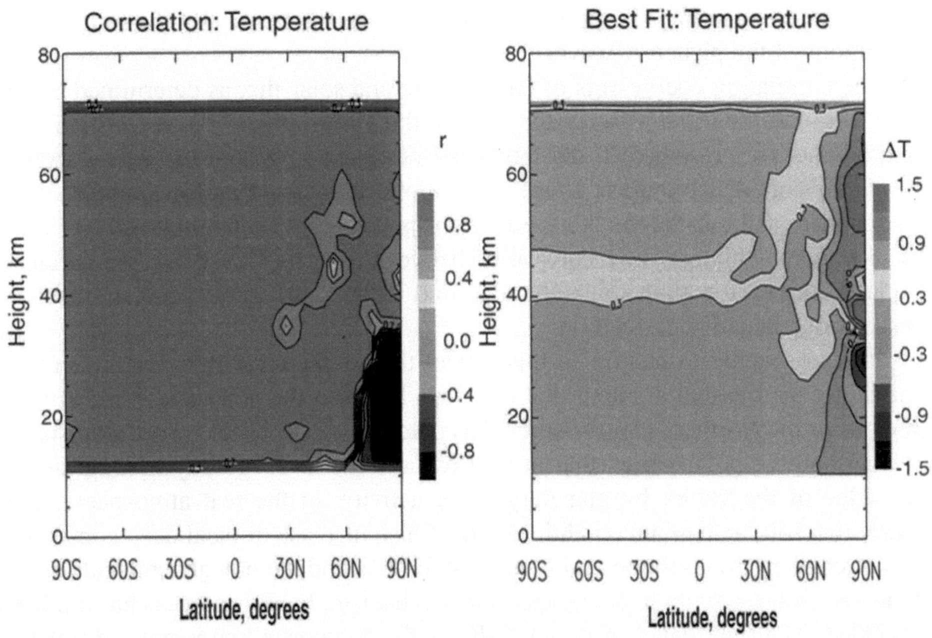

Figure 1. Left panel: Correlation coefficient of temperature against increasing solar flux levels using 16 10-day zonal means between 16–25th February 1993 after 75 days integration. *Right panel:* Best fit temperature difference between solar minimum and solar maximum using the above simulations.

for the investigation of the interplay between dynamics, radiation and chemistry (see for example O'Neill and Pope, 1988 and Austin et al., 1992).

Realistic amplitudes and phases of vertically propagating planetary waves during the winter months were provided at a lower boundary of 314 mb by assimilated Stratospheric Sounding Unit data (Miller et al., 1980). In the current experiments, observed ozone climatological values were employed. Also, the infrared heating and cooling rates were calculated every three hours at eight equally spaced longitudes.

This enabled the contribution of non-axisymmetric radiative transfer during the displacement of the gravest planetary wave modes during warming events to be taken into account. For the purposes of this experiment, the default ultraviolet heat flux was taken to be the solar minimum level and increased values taken from Brasseur and Solomon (1986) were selected for solar maximum for each of the bands specified by Strobel (1978).

To assess the relative contribution of internal variability within the model in contrast to that due to the solar fluxe changes, sixteen simulations were carried out, starting with lower boundary forcing data appropriate for 1st December, 1992. Equally spaced solar flux values ranged from solar minimum to solar maximum. The experiments were integrated in time for three months to allow the small changes in the radiative heating the opportunity to influence systematically the propagation conditions of the planetary waves.

The correlation coefficients of temperature and solar flux as determined by the F10.7 cm radio emission proxy for each of the 10-day zonally averaged latitudes and heights are presented in the left panel of figure 1. Where the values exceed ± 0.5, the confidence level is greater than 95 %. It is clear that for the bulk of the mesosphere and outside the Northern Hemisphere polar winter stratosphere, there is an extremely high positive correlation between solar flux levels and temperature. Apart from a narrow transition zone at around 70 °N, much of the polar stratosphere has a strong negative correlation.

It is reasonable to deduce in this model that as the solar flux levels increase, the larger the thermal gradient that develops between the polar and extra-tropical regions of the Northern Hemisphere. This gradient enhances the upper atmosphere jets around the polar vortex that in turn reduces the level of eddy mixing around the edge of the vortex by planetary wave activity. In the real atmosphere, it is likely that solar maximum conditions would also increase tropical deep convection and strengthen tropospheric baroclinic instabilities and thereby generate additional planetary wave activity to counteract this mechanism, but this process has not been included. The right panel of figure 1 shows the composite temperature difference between solar minimum and solar maximum obtained from a best fit of the sixteen model runs. The summer hemisphere and tropical Northern Hemisphere exhibited a response to the forcing that can be explained in terms of simple radiative heating, whereas the winter polar region requires a dynamical explanation, given the low levels of direct insolation available in this area. The maximum temperature drop at the North Pole was around 1 K, which is considerably less than that reported by Labitzke and van Loon (1993), but greater than that achieved in the summer hemisphere by direct insolation. Similarly, mesospheric temperatures have increased in a complementary fashion. Therefore, in a controlled environment, planetary wave amplification of the solar forcing within the winter middle atmosphere appears to take place at low amplitude, but in an extremely well correlated fashion.

3. Thermospheric Influences on the Middle Atmosphere

As the altitude of the upper boundary of the model is raised into the lower thermosphere, the number of processes with a potentially significant role in modifying the response of the middle atmosphere to solar variability also increases. Not least because the relative impact of flux changes grows with height as relatively more high energy photons are absorbed. Above 70 km, gravity wave deposition and tidal interactions with planetary waves and the mean circulation will have a large effect on the transport of energy and momentum. Charged particle interactions with the neutral atmosphere also become important, especially at high latitudes. Whilst a full consideration of all of these coupling processes is beyond even the most sophis-

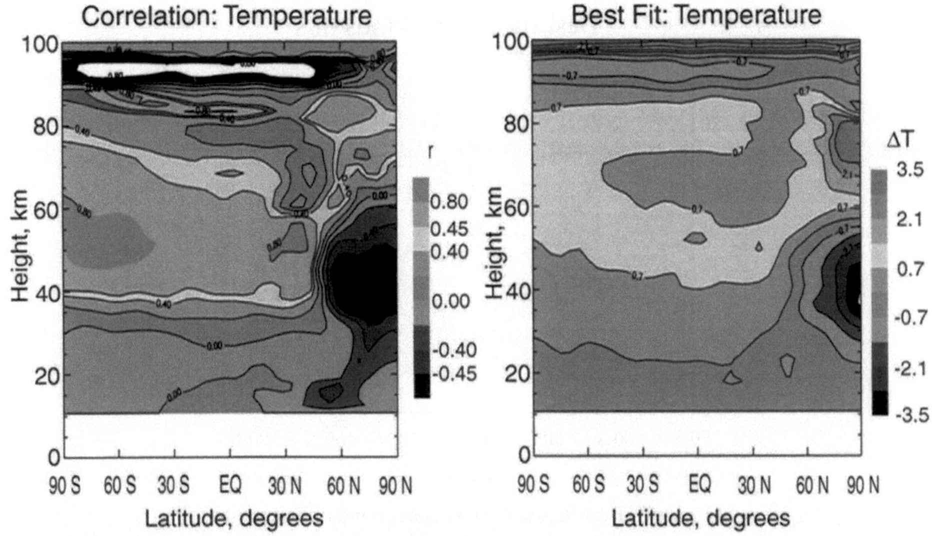

Figure 2. Left panel: Correlation coefficient of temperature against increasing solar flux levels using 16 10-day zonal means between 16-25th February 1993 after 75 days integration. *Right panel:* Best fit temperature difference between solar minimum and solar maximum using the above simulations.

ticated models available at present, we are able to assess the interaction of planetary waves with the mean circulation up to and including the lower thermosphere. The details of the model above 70 km have been laid out in (Arnold and Robinson, 1998); its principal features being the linear relaxation of the model temperatures to the empirical MSIS-90E model (Hedin, 1983; Hedin, 1991). The modifications to the model below 70 km, as described in the section above, were retained.

For this ensemble of simulations, the same lower boundary conditions and range of solar flux values were used. Above 100 km, there is a very strong correlation between atmospheric temperature and solar forcing at all latitudes. This is mostly due to the increasing influence of the empirical temperatures provided by MSIS, where this relationship is well known.

Below 90 km, correlation coefficients rarely match the values found in the previous section, indicating that the interactions between solar-induced heating and dynamical processes have become more complex (figure 2). However, it is important to note that there remain substantial regions where the statistical significance is high. The region in black is negatively correlated above the 95 % confidence threshold. The summer upper stratosphere / lower mesosphere still exhibits the expected heating in line with increased solar flux levels, as does the winter mesosphere, albeit to a lesser extent. What is most interesting is the fact that the extent of the winter stratospheric region with the high negative correlation has grown considerably compared with the simpler model, perhaps as the effect is becoming diluted by the increased complexity of the radiative/dynamical interactions involved. There has been an effective trebling of the amplitude of the

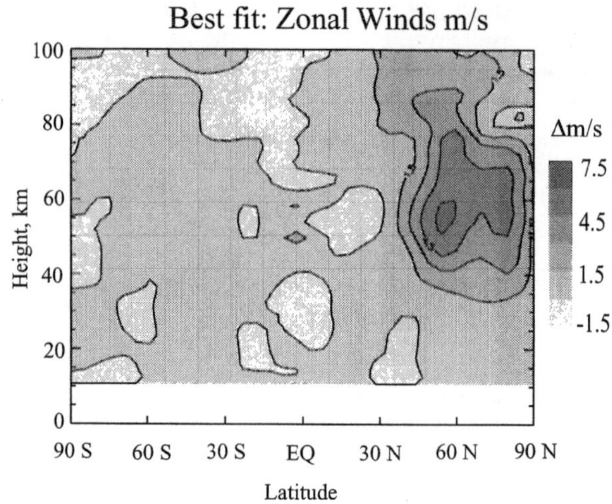

Figure 3. Best fit zonal mean zonal wind differences.

temperature response in the winter hemisphere (figure2b) in both the stratosphere and the mesosphere compared with the previous model where the thermosphere was absent.

There are also very good correlations within the upper air jets in both hemispheres particularly between 30–60 °(not shown). However, the amplitude deviation of the Southern Hemisphere zonal mean zonal winds between solar minimum and solar maximum is too weak to be apparent in the best fit data given in figure 3. At solar maximum, the winter jet is up to 10 ms^{-1} stronger than for minimum conditions and this strengthening is extended polewards even though the significance is not as great.

4. Discussion and Conclusions

Numerical simulations of the atmosphere allow sensitivity experiments to be performed under carefully controlled conditions. Provided that the noise generated within the model remains within acceptable levels, it is then possible to identify relatively weak signals that are often swamped by the considerable natural variability of the atmosphere proper. The first set of experiments using a middle atmosphere model indicated that there was an unambiguous response of the winter stratosphere to solar forcing through feedback between dynamics and the radiative changes. However, the signal was so small that it would be nearly impossible to detect experimentally. Certainly, it could not explain the findings of Labitzke and van Loon (1993).

Consideration of additional feedback mechanisms, such as ozone chemistry is required (Haigh, 1994; Balachandran and Rind, 1995). Also, consideration of the interaction between planetary waves and the upper mesosphere and the thermosphere circulations appears to be necessary. By their nature, the behaviour of non-linear processes tends to make the determination of causality more difficult to realise and it should not be surprising that simple correlations between solar flux variations and stratospheric parameters often become unclear. However, this in no way detracts from the importance of the contribution of external solar forcing to the stratospheric circulation.

Experiments by Kodera et al. (1991) indicated that the equatorial Quasi-Biennial Oscillation plays an important role in moderating the extra-tropical planetary wave activity that brings about the solar modulation of the stratospheric circulation. Holton (1994) has suggested that the atmosphere switches between a relatively small number of states involving processes such as the QBO. Therefore it is conceivable that during solar extrema, the natural variability of the atmosphere and hence the planetary wave activity is controlled such that the solar modulation is observable. Trials with more sophisticated gravity wave schemes, such as that developed by Hines (1991) and a realistic QBO will be needed to test this hypothesis further.

Acknowledgements

The authors are particularly grateful to Drs. J. Austin and N. Butchart for providing the original Met. Office Stratosphere Mesosphere model and Al Hedin for both MSIS-90E and the Horizontal Wind Model. N.F. Arnold was supported by a grant from the Particle Physics and Astronomy Research Council of the United Kingdom.

References

Arnold, N. F. and T. R. Robinson: 1998, 'Solar cycle changes to planetary wave propagation and their influence on the middle atmosphere circulation', *Ann. Geophysicae* **16**, 69–76.

Austin, J. and N. Butchart: 1992, 'A three-dimensional modeling study of the influence of planetary wave dynamics on polar ozone photochemistry', *J. Geophys. Res.* **97**, 10,165–10,186.

Austin, J., N. Butchart and K. P. Shine: 1992, 'Possibility of an Arctic ozone hole in a doubled-CO_2 climate', *Nature* **360**, 221–225.

Balachandran, N. K., D. Rind: 1995, 'Modeling the effects of UV variability and the QBO on the troposphere-stratosphere system. Part I: the middle atmosphere', *J. Climate* **8**, 2058–2079.

Bjarnason, G. G., Ö. E. Rögnavaldsson, T. I. Sigfússon, T. Jakobsson, and B. Thorkelsson: 1993, 'Total ozone variations at Reyjavik since 1957', *J. Geophys. Res.* **98**, 23,059–23,077.

Brasseur, G. and S. Solomon: 1986, 'Aeronomy of the middle atmosphere', *Reidel*, Dordrecht, 452.

Gavrilov, N. M.: 1997, 'Climatology and hydrodynamic sources of internal gravity waves in the middle and upper atmosphere'. In: K. Hamilton (ed.): Gravity wave processes, Vol. 50 of NATO ASI Series I: Global Environmental Change, Springer-Verlag, Berlin, 45–62.

Haigh, J. D.: 1994, 'The role of stratospheric ozone in modulating the solar radiative forcing of climate', *Nature* **370**, 540–546.

Haynes, P. H., C. J. Marks, M. E. McIntyre, T. G. Shepherd, and K. P. Shine: 1991, 'On the downward control of extratropical diabatic circulations by eddy-induced mean zonal forces', *J. Atmos. Sci.* **48**, 651–679.

Hedin, A. E.: 1983, 'A revised thermospheric model based on mass spectrometer and incoherent scatter data: MSISE', *J. Geophys. Res.* **88**, 10170–10188.

Hedin, A. E.: 1991, 'Neutral atmosphere empirical model from the surface to the lower exosphere: MSISE', *J. Geophys. Res.* **96**, 1159–1172.

Hines, C. O.: 1991, 'The saturation of gravity waves in the middle atmosphere. Part II: development of Doppler Spread theory', *J. Atmos. Sci.* **48**, 1360–1369, 1991.

Holton, J. R.: 1994, 'The quasi-biennial oscillation in the Earth's atmosphere and its links to longer period variability', In: E. Nesme-Ribes (ed.): *The solar engine and its influence on terrestrial atmosphere and climate*, Vol. 25 of *NATO ASI Series 1: Global Envronmental Change*. Springer-Verlag, Berlin, pp. 259–273.

Hood, L. L.: 1999, 'Effects of short-term solar UV variability on the stratosphere', *J. Atmos. Solar-Terr. Phys.* **61**, 45–51.

James, I. N.: 1993, 'Introduction to circulating atmospheres', *Cambridge University Press* Pp 446.

Kodera, K., M. Chiba, and K. Shibata: 1991, 'A general circulation model study of the solar and QBO modulation of the stratospheric circulation during the northern hemisphere winter', *Geophys. Res. Lett.* **18**, 1209-1212.

Labitzke, K. and H. van Loon: 1988, 'Association between the 11-year solar cycle, the QBO and the atmosphere. Part 1: the troposphere and stratosphere in the Northern Hemisphere in winter', *J. Atmos. Terr. Phys.* **50**, 197-206.

Labitzke, K. and H. van Loon: 1993, 'Some recent studies of probable connections between solar and atmospheric variability', *Ann. Geophysicae.* **11**, 1084-1094.

Lean, J., J. Beer and R. Bradley: 1995, 'Reconstruction of solar irradiance since 1610: implications for climate change', *J. Atmos. Solar-Terr. Phys.* **61**, 25-36.

Mayr, H. G., J. G. Mengel, C. O. Hines, K. L. Chan, N. F. Arnold, et al.: 1997, 'The gravity wave Doppler Spread Theory applied in a numerical spectral model of the middle atmosphere, Part I: formulation and global scale seasonal variations', *J. Geophys. Res.* **102**, 26,077–26,092.

Mayr, H. G., J. G. Mengel, C. O. Hines, K. L. Chan, N. F. Arnold, et al.: 1997, 'The gravity wave Doppler Spread Theory applied in a numerical spectral model of the middle atmosphere, Part II: equaotial oscillations', *J. Geophys. Res.* **102**, 26,093–26,106.

Miller D. E., J. L. Brownscombe, G. P. Carruthers, D. R. Pick, and K. H. Stewart: 1980, 'Operational temperature sounding of the stratosphere', *Phil. Trans. R. Soc. Lond.* **A292**, 65–71.

O'Neill, A. and Pope V. D.: 1988, 'Simulations of linear adn non-linear disturbances in the stratosphere', *Quart. J. R. Met. Soc.* **114**, 1063–1110.

Pap, J. M. and C. Fröhlich: 1999, 'Total solar irradiance variations', *J. Atmos. Solar-Terr. Phys.* **61**, 15–24.

Richmond, A. D.: 1983, 'Thermospheric dynamics and electrodynamics', In: R. L. Carovillano and J. M. Forbes (eds.): *in Solar Terrestrial Physics - principles and theoretical foundations*. Reidel, Dordrecht, pp. 523–607.

Shine, K. P.: 1987, 'The middle atmosphere in the absence of dynamical heat fluxes', *J.Roy. Met. Soc.* **113**, 603–633.

Strobel, D. F.: 1978, 'Parameterization of the atmospheric heating rate from 15 to 120 km due to O_2 and O_3 absorption of solar radiation', *J. Geophys. Res.* **83**, 6225–6230.

Address for Offprints: N.F. Arnold, Department of Physics and Astronomy, University of Leicester, University Road, Leicester, LE1 7RH, U.K., nfa1@ion.le.ac.uk

THE IMPACT OF SOLAR FORCING ON THE VARIABILITY IN A COUPLED CLIMATE MODEL

R.J. HAARSMA (haarsma@knmi.nl), S.S. DRIJFHOUT,
J.D. OPSTEEGH, F.M. SELTEN
Royal Netherlands Meteorological Institute, P.O. Box 201, 3730 AE De Bilt, The Netherlands

Received: 28 October 1999; Accepted in final form: 20 December 1999

Abstract. The impact of variations in solar irradiance on the variability of climate is still a topic of debate. Herein we assess the response of a coupled General Circulation Model (GCM) of intermediate complexity to an estimate of the solar variability since 1700 and to a series of idealized sinusoidal solar forcings. On the continental to global scale and averaged over periods longer than 30 years, the solar-induced variability dominates internal variability in the annual global mean surface air temperature. Locally and on the regional scale, the internal variability dominates. The dominant patterns of natural variability and explained variance are not affected by a variable solar forcing, the spectra however are sensitive. The control run shows a preferred decadal time scale of 18 year in a sea surface temperature mode associated with the North Atlantic Oscillation. The preferred decadal time scale disappears for a variable solar forcing. This is caused by small changes in oceanic circulation resulting in subsurface oceanic modes with modified structure and time scale.

1. Introduction

Satellite observations have revealed an 11-yr cycle in solar irradiance with an amplitude of about 0.1 % (Lean and Rind, 1998). It has been hypothesized that longer term variations in solar irradiance are associated with larger amplitudes in the order of 0.3% (1–2 Wm^{-2}) (Hoyt and Schatten, 1993, hereafter HS; Lean et al., 1995; Lean and Rind, 1998). These changes could have significantly contributed to the observed climate changes. Until recently, the main competitor to the recent rise in global mean air temperature for the enhanced greenhouse effect was thought to be natural climate variability associated with large-scale atmosphere-ocean interactions (Stouffer et al., 1994; Tett et al., 1997). At present, this is challenged by the newly forwarded ideas about the impact of solar variability. In this study we aim to assess the relative roles for climate change caused by internal ocean-atmosphere interactions versus climate change associated with solar variability for a whole range of different amplitudes and frequencies.

So far, most of the research about the impact of solar variability on climate has been concentrated on the direct response patterns of atmospheric variables like temperature and wind (Hoffert et al., 1988; Reid, 1991; White et al., 1998). The possible effect of solar variability on the patterns and the spectra of the dominant modes of variability has received relatively little attention. A dominant mode of

Space Science Reviews **94:** 287–294, 2000.

natural variability on the North Atlantic ocean, affecting the climate of Western Europe, is the North Atlantic Oscillation (NAO) (Hurrell, 1995), which is characterized by the pressure difference between Island and Lissabon. This pressure difference displays marked variability on interannual and decadal time scales. It is argued that the decadal variability of the NAO is affected by the ocean circulation (Sutton and Allen, 1997; Rodwell et al., 1999). During the last three decades the NAO displays a marked upward trend. At present it is not clear whether this upward trend is part of the natural variability of the system or whether it is induced by changing external climatic forcings like greenhouse forcing and/or solar variability. A second goal of this study is to investigate the response of the model's equivalent of the NAO to different forcings.

2. Climate Model and Experimental Setup

This study has been performed with ECBILT (Opsteegh et al., 1997), which is a coupled atmosphere-ocean-sea ice model of intermediate complexity. The horizontal resolution is about 500 km. The vertical resolution is 3 levels in the atmosphere and 12 in the ocean. It contains simple parameterisations, including a full hydrological cycle, for the diabatic processes. The cloud coverage is prescribed. The model is driven by annual varying solar insulation. The model is realistic in the sense that it contains the minimum amount of physics that is necessary to simulate the mid-latitude planetary and synoptic-scale circulations in the atmosphere as well as its variability on various time-scales (Opsteegh et al., 1997). The model is two orders of magnitude faster than state-of-the-art climate models, which makes it feasible to perform long simulations and to do numerous sensitivity experiments.

The model has been integrated towards a quasi-stationary steady state (Opsteegh et al., 1997). From this state the model has been integrated for 1000 years in which the solar forcing has been kept constant. This will be called the control run. The effect of variations in solar irradiance on the climate has been investigated by comparing the results from the control run with runs in which the solar constant is varied in period from 11–160 yrs and in amplitude from 0.5–6 Wm^{-2} during the integration time. In addition 4 ensemble experiments with an estimate of the the variations in solar irradiance since 1700 (HS) have been made. The different initial conditions for these ensemble integrations are obtained from states of the control run separated 50 years from each other.

3. Results

3.1. Surface Air Temperature (SAT)

Due to the quasi-geostrophic dynamics the variability in the tropics is largely underestimated in the model. Therefore, in the subsequent analyses of the annual mean global mean SAT (GM-SAT), we have excluded the latitude band between 22.5°S and 22.5°N. In the extratropics the spatial pattern and the amplitude of the model variability compares favourably with the observations, although averaged over the area poleward of 22.5°S and 22.5°N, the standard deviation of SAT in the model is about half as large as in the observations (Drijfhout et al., 1999). The response in GM-SAT is well correlated with the solar forcing. The ratio between internal and solar induced variability in GM-SAT is comparable to GCM's (Drijfhout et al., 1999). Figure 1 shows that the response patterns of SAT on the solar forcing depend on the amplitude of the forcing. Stronger forcing results in a more thermodynamic response and less influence of dynamical processes.

The variance spectra of GM-SAT for different modulations of the solar irradiance are shown in Fig. 2. Likewise as in the observations (Jones, 1994), the spectrum of GM-SAT is red, denoting an increase in temperature variance with decreasing frequency (Fig. 2a). For the HS reconstruction of the solar forcing the variability of GM-SAT at the dominant forcing periods of 50 and 80 years is typically five times larger compared to the control run (Fig. 2c). Even for an amplitude in the solar irradiance of 0.5 Wm^{-2} and a period of 80 years the response is significantly larger than the backround variability (Fig. 2b). At centennial time scales a solar irradiance amplitude of 0.5 Wm^{-2} is substantially less than reconstructed by HS and Lean et al. (1995). When for this amplitude in the solar irradiance the period is reduced to shorter periods of 11 and 22 years the response is no longer significant from the backround internal variability.

The role of internal variability becomes more important when the domain under consideration is smaller. For instance, considering the temperature variability in Western Europe (40°N – 55°N, 10°W – 10°E), only with a solar irradiance amplitude of 6 Wm^{-2} is the response at the 80-year period clearly significant. When the solar irradiance amplitude is 0.5 Wm^{-2} the response at this period is indistinguishable from the backround internal variability. Locally the internal variability always dominates over the solar induced variability.

3.2. Modulation of the North Atlantic Oscillation

The patterns obtained from a singular value decomposition (SVD) analysis of North Atlantic sea surface temperature (SST) and 800 geopotential height (Φ 800), shown in Fig. 3, reveal for the control run a mode of variability with resembles the observed patterns of the North Atlantic Oscillation (NAO) (Selten et al., 1999). The spectrum of SST is red and displays a significant peak at period of about 18 years. Variations in solar irradiance do not change the patterns in SST and Φ 800. The spectra belonging to these patterns however appear to be sensitive. Fig. 4 shows the SST variance spectra for different variations in solar irradiance. The significant peak in SST at 18 years in the control run is only marginally significant for an amplitude of 0.5 Wm^{-2} in the solar irradiance and a period of 11 years whereas it

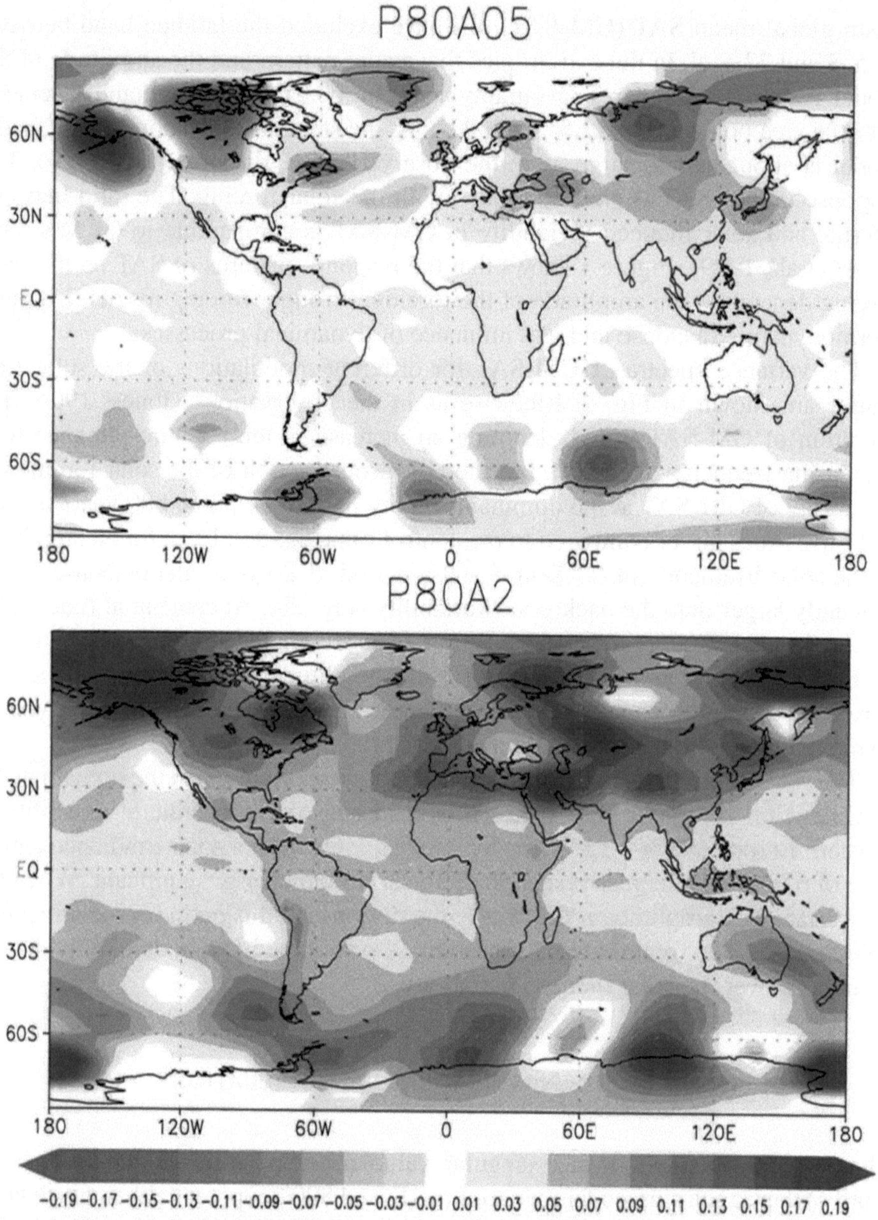

Figure 1. Response in SAT (K) for a solar irradiance with a period of 80-yrs and amplitudes of 0.5 Wm^{-2} (a) and 2 Wm^{-2} (b). Regions below the 95% confidence interval are white.

disappears completely when the period is increased to 22 years. Also for stronger forcing amplitudes and larger periods the peak disappears.

In Selten et al. (1999) it is shown that the dominant period in the SST spectra originates from a mode with a preferred time scale of about 18 years in the subsur-

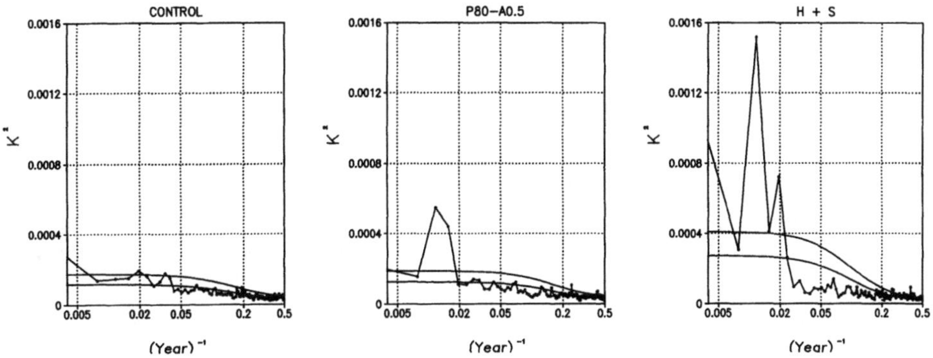

Figure 2. Variance spectra of GM-SAT for different modulations of the solar irradiance: constant (a),amplitude of 0.5 Wm^{-2} and a period of 11 yrs (b), estimated changes since year 1700 (c).

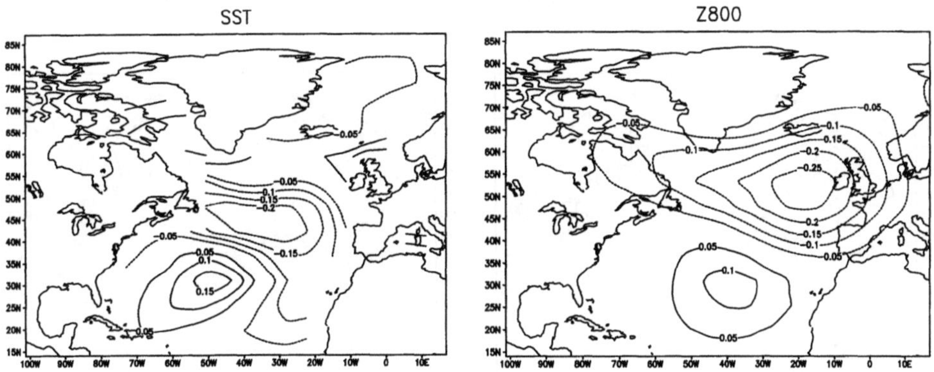

Figure 3. Patterns of the second SVD mode of wintermean SST and geopotential height at 800 hPa.

face temperatures. An optimal autocorrelation functions (OAF) analysis (Selten et al., 1999) of the subsurface temperature variability reveals that with a variable solar forcing the subsurface mode still exist but that pattern and timescale are modified. The time scale is significantly shortened to 8 years (Fig. 6) and the pattern is slightly changed (Fig. 5), projecting less well onto the SST pattern of the SVD mode associated with the NAO. This change in the structure of the subsurface pattern is responsible for the disappearence of the peak in the SST spectra. The cause of these changes in the structure and time scale of the subsurface mode is a subject of ongoing research, but is likely to be associated with changes in convective activity that result from small rectifications due to periodical solar forcing.

The insensitiveness of the patterns of the NAO mode to variations in solar forcing is in agreement with the ideas put forward by Palmer (1999). Using the Lorenz model as a metaphor for our climate he argues that a small external forcing will not change the dominant patterns of variability but only the probability density functions (pdf's). Recent analysis of NCEP by Corti et al. (1999) showes that the dominant patterns of the NAO haven't changed in contrast to the pdf's, which do reveal significant changes. The analysis of Corti et al. (1999) does not answer the

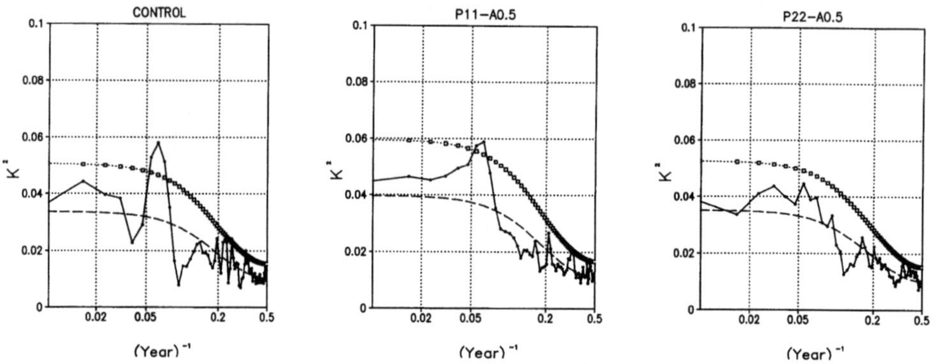

Figure 4. Variance spectrum of North Atlantic SST belonging to the second SVD of wintermean SST and geopotential height at 800 hPa for different modulations of the solar irradiance: constant (a), amplitude of 0.5 Wm^{-2} and a period of 11 yrs (b), amplitude of 0.5 Wm^{-2} and a period of 22 yrs (c).

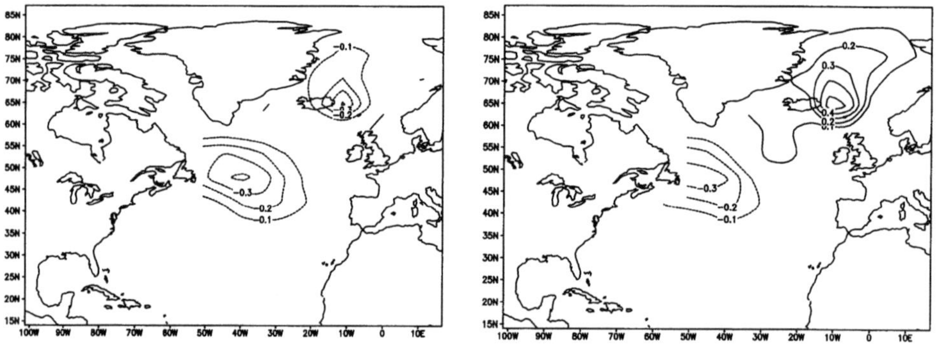

Figure 5. Subsurface temperature patterns associated with the decadal mode for the control run (a) and for a periodical solar forcing (b).

question whether these changes in the pdf's are due to internal variability or are caused by a change in an external force. Our results suggest that even a weak change in the external forcing can significantly modify the pdf's.

4. Conclusions and Discussion

The results we have presented here indicate that, averaged over periods of at least 30 years, the solar-induced variability dominates the internal variability in the annual mean GM-SAT. In contrast, locally and on the regional scale the internal variability dominates the solar-induced variability for periods up to at least 160 years.

The HS reconstruction implies a rise of 2.3 Wm^{-2} in solar irradiance over the last 30 years, which can be converted to a change in radiative forcing of $0.7/4 * 2.3 = 0.4\,Wm^{-2}$ where 0.7 accounts for reflection out to space of a portion of

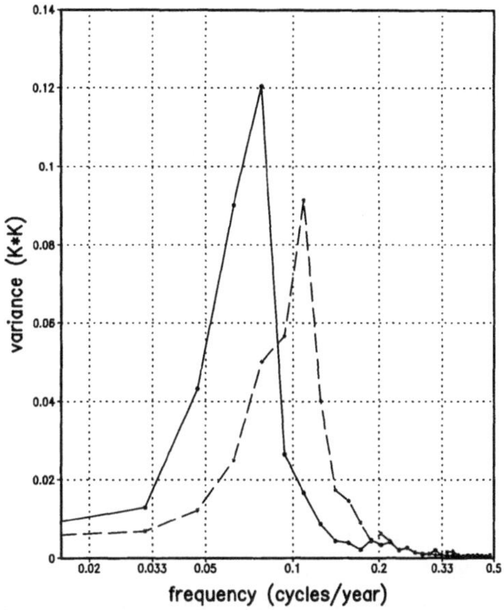

Figure 6. Spectra of the subsurface mode for the control run (solid) and for a periodical solar forcing (dashed).

the incident solar energy and 4 is the spherical average (Lean and Rind, 1998).The response in our model to such a rise in radiative forcing is in the order of 0.1°C, which is estimated to be about 50% of the response of a state-of-the-art GCM. This implies that the solar forcing could explain about 50% of the temperature rise over the last 30 years (0.4°C) when the HS reconstruction is to be believed. It should be noted however, that in general the HS reconstruction is a rather generous estimate of the solar variability over the last 30 years (compare Lean et al., 1995; Stevens and North, 1996) and should be interpreted more as an upper bound.

The patterns of natural variability like the NAO are not affected by variations in solar forcing. The spectra are however sensitive. A preferred decadal time scale of about 18 years in the control run disappears for a variable solar forcing. This is caused by changes in the structure and time scale of the subsurface ocean mode associated with the decadal peak in the NAO spectrum. As a result, the pdf of the NAO mode has changed due to solar forcing. Our results are an indication that the recent upward trend in the NAO index is not necessarily caused by the internal variability of the system but may have an external cause. The insensitiveness of the NAO patterns is in agreement with the results of Palmer (1999) and Corti et al. (1999).

Acknowledgements

R.J. Haarsma was supported by the Scale INTeracion EXperiments (SINTEX) project of the European Commission, contract number ENV4-CT98-0714. F.M. Selten was supported by a grant of the Netherlands National Research Programme on Global Air POllution and Climate Change.

References

Corti, S., Molteni, F., and Palmer, T. N.: 1999, 'Signature of Recent Climate Change in Frequencies of Natural Atmospheric Circulation Regimes', *Nature* **398**, 799–802.

Drijfhout, S. S., Haarsma, R. J., Opsteegh, J. D., and Selten, F. M.: 1999, 'Solar-Induced Versus Internal Variability in a Coupled Climate Model', *Geophys. Res. Lett.* **26**, 205–208.

Hoffert, M. I., Frei, A., and Narayanan, V. K.: 1988, 'Application of Solar Max ACRIM Data to Analysis of Solar-Driven Climatic Variability on Earth', *Clim. Change* **13**, 267–285.

Hoyt, D. V. and Schatten, K. H.: 1993, 'A Discussion of Plausible Solar Irradiance Variations, 1700–1992', *J. Geophys. Res.* **98**, 18895–18906.

Hurrell, J. W.: 1995, 'Decadal Trends in the North Atlantic Oscillation: Regional Temperatures and Precipitation', *Science* **269**, 676–679.

Jones, P. D.: 1994, 'Hemispheric Surface Air Temperature Variations: a Reanalysis and an Update to 1993', *J. Clim.* **7**, 1794–1802.

Lean, J. and Rind, D.: 1998, 'Climate Forcing by Changing Solar Radiation', *J. Climate* **11**, 3069–3094.

Lean, J., Beer, J., and Bradley, R.: 1995, 'Reconstruction of Solar Irradiance Since 1610: Implications for Climate Change', *Geophys. Res. Lett.* **22**, 3195–3198.

Opsteegh, J. D., Haarsma, R. J., and Selten, F. M.: 1997, 'An Atmospheric Climate Model of Intermediate Complexity: a Suitable Alternative to Mixed Boundary Conditions in Ocean Models', *Tellus*, submitted.

Palmer, T. N.: 1999, 'A Nonlinear Dynamical Perspective on Climate Prediction', *J. Climate* **12**, 575–591.

Reid, G. C.: 1991, 'Solar Total Irradiance Variations and the Global Sea Surface Temperature Record', *J. Geophys. Res.* **96**, 2835–2844.

Rodwell, M. J., Rowell, D. P., and Folland, C. K.: 1999, 'Oceanic Forcing of the Wintertime North Atlantic and European Climate', *Nature* **398**, 320–323.

Selten, F. M., Haarsma, R. J., and Opsteegh, J. D.: 1999, 'On the Mechanism of the North Atlantic Decadal Variability', *J. Climate* **12**, 1956–1973.

Stevens, M. J. and North, G.: 1996, 'Detection of the Climate Response to the Solar Cycle', *J. Atmospheric Sci.* **53**, 2594–2608.

Stouffer, R. J., Manabe, S., and Vinnikov, K. Y.: 1994, 'Model Assesment of the Role of Natural Variability in Recent Global Warming', *Nature* **367**, 634–636.

Sutton, R. T. and Allen, M. R.: 1997, 'Decadal Predictability in Gulf Stream Sea Surface Temperature', *Nature* **388**, 563–567.

Tett, S. F. B., Johns, T. C., and Mitchell, J. F. B.: 1997, 'Global and Regional Variability in a Coupled AOGCM', *Clim. Dyn.* **13**, 303–323.

White, W. B., Cayan, D. R., and Lean, J.: 1998, 'Global Upper Ocean Heat Storage to Radiative Forcing from Changing Solar Irradiance and Increasing Greenhouse Gas/Aerosol Concentrations', *J. Geophys. Res.* **103**, 21,355–21,366.

Address for Offprints: R.J. Haarsma, Royal Netherlands Meteorological Institute,P.O. Box 201, 3730 AE De Bilt, The Netherlands, haarsma@knmi.nl

SOLAR VARIABILITY AND THE SEARCH FOR CORRESPONDING CLIMATE SIGNALS

WERNER MENDE (mende@bbaw.de) and RITA STELLMACHER
Freie Universität Berlin

Received: 4 February 2000; Accepted in final form: 24 July 2000

Abstract. Instrumental and paleodata from the last centuries are investigated to get circumstantial evidence for external influences on the Eartht's climate machine. Such influences could be of extraterrestrial and/or anthropogenic origin. Anthropogenic influences are separated from solar on superdecadal time scales and on a hemispheric level using a non-linear regression model. The function to be explained is the northern hemispheric temperature. The model contains two forcing components explicitly: A parameterized anthropogenic component, which describes the aggregated effect of greenhouse gases, aerosols and other anthropogenic climate impacts. A solar component, which describes the solar variability history. The solution of the regression model allows, under certain assumptions, a functional separation of the variability components and provides an estimation of their relative contributions to global warming during the last 140 years.

1. Introduction

A global warming with a linear trend of 0.6 K during the last 140 years has been observed. An often addressed question is, which contribution to this warming is produced by human activity or by geophysical processes (such as changes of ice cover, eruption of volcanoes etc.), which by the climate machine itself and which, last but not least by extraterrestrial variability sources such as solar forcing (for the sun-climate-connection see e.g. the excellent survey of Hoyt and Schatten, 1997)?

We will confine our approach to this question on variability from decadal to secular time scales and on a hemispheric resp. global aggregation level. This high aggregation level is more an advantage than a drawback as many variability sources are acting on a global scale. Moreover, the forcing functions are often more simple and stable on global scales than on smaller ones.

The present investigation uses among others the observed surface temperature on both hemispheres from 1856 to 1998, represented by monthly means from Jones *et al.* (1998). In this time window orbital forcing effects can be neglected. Most of the variance of these series is due to the seasonal cycle: 99.6% in the northern and 98.6% in the southern hemisphere. Therefore, we use temperature anomalies and apply spectral lowpass filterings with different cut-off frequencies, whereby almost all seasonal variance is removed. The remaining variability is only a tiny part (about 1%) of the original series. These small effects can only be detected by adequate filterings.

Space Science Reviews **94:** 295–306, 2000.

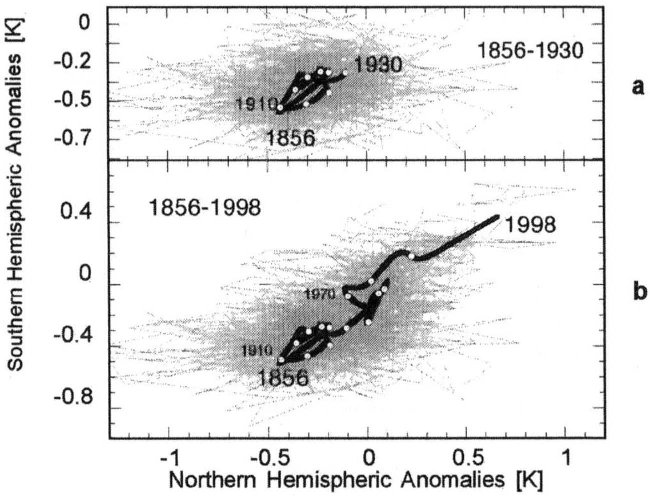

Figure 1. Variability of hemispheric temperature anomalies for (a) the period 1856–1930 and (b) the full period up to 1998. The grey lines follow the phase between monthly mean southern and northern hemispheric temperature anomalies. The bold black trajectory represents a 10-year lowpass filtering. After 1930 the global temperature escapes from the range of earlier historical variability (Fig. 1b). (Original data from Jones *et al.*, 1998).

2. Hemispheric Temperatures

First we consider the interrelationship of hemispheric temperatures on different filter time scales to demonstrate the existence of global warming and the similarity of temperature evolution on both hemispheres.

Figure 1 shows phase plots of the hemispheric temperature anomalies for the period 1856–1998. From 1856 to 1930 the bold black trajectory exhibits a quasi-stationary movement around the centre of its variability region (Fig. 1a). From 1935 to 1998 the trajectory shows in contrast to this an unstable behaviour, escaping the range of its earlier historical variability region (Fig. 1b). This run-away phenomenon demonstrates that global warming cannot be understood simply as an episodic fluctuation inside a stationary variability range. It expresses the secular instationarity of the temperature evolution at least in our time window. The black trajectory is now outside the 3-σ level (SH) resp. the 2-σ level (NH) of the filter-related deviation.

In the following we provide qualitative arguments for the existence of variability sources external to the climate machine in the investigated time window (Figs. 2–5).

The 6-year filtered temperature anomalies (Fig. 2) reveal a trend of clear warming of both hemispheres. Strong exchanges of heat between the hemispheres (movement from the lower right to the upper left or backwards) are rare events. However, a relative high synchronization of both hemispheres is obvious. Sometimes the

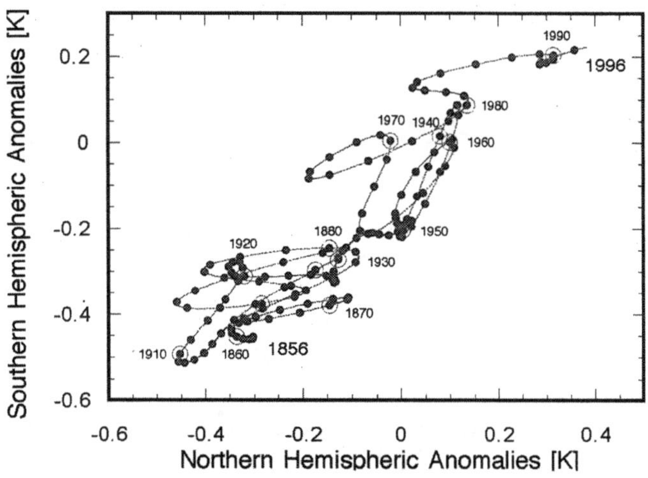

Figure 2. Phase plot of the northern against the southern hemispheric temperatures (6-year lowpass filtered) for the period 1856–1996. The trajectory exhibits sometimes common warming and cooling periods approximately along the same path and reveals the global synchronization of the two hemispheres. (Original data from Jones *et al.*, 1998).

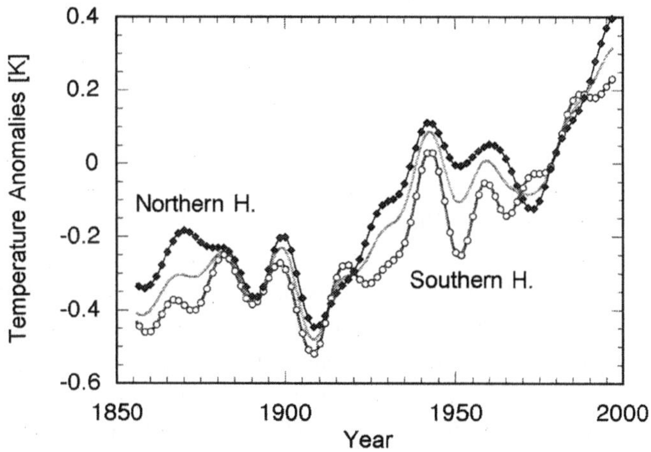

Figure 3. 13-year spectral lowpass filtering of the global and hemispheric temperature anomalies (1856–1996): The parallelism and the synchronous behaviour of the curves dominate all other variability features. (Original data from Jones *et al.*, 1998).

trajectory shows a parallel warming and cooling of the two hemispheres nearly along the same path returning to almost the same point. Such a quasi-stationary behaviour could in principle be generated by the coupled atmosphere-ocean-system under a constant driving force. However, the remarkable temperature synchronization between the two hemispheres indicates more likely a common response to slowly changing external drivers.

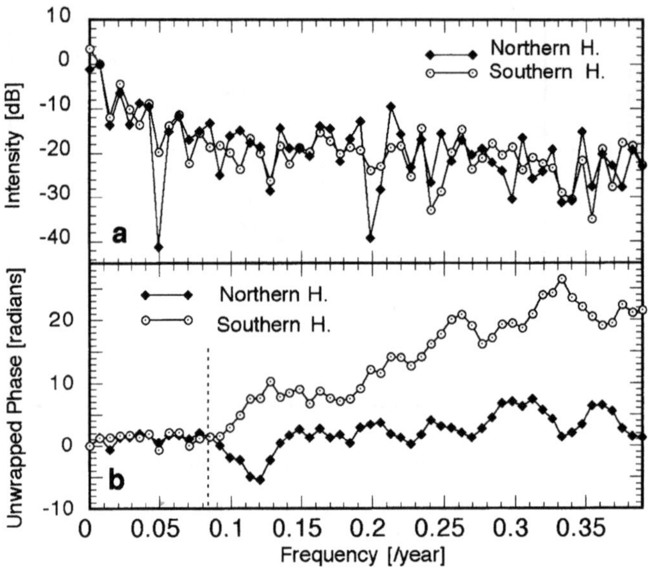

Figure 4. Spectral representation of hemispheric temperature anomalies (1856–1996): (a) Low frequency part of the Fourier power spectra: For periods longer than 4 years the two hemispheres show remarkable similarities. (b) The phase spectra reveal high in-phase correlation for periods longer than 12 years and a anticorrelation in a band between 3 and 12 years. (Original data from Jones *et al.*, 1998).

The monthly temperature anomalies of the two hemispheres and of the globe in a 13-year spectral lowpass filtering from 1856 to 1996 are shown in Fig. 3. In this filtering effects of single strong eruptions of volcanoes and short-term changes in circulation patterns like El Niño are widely removed and the high degree of parallelism in the temperature evolution of the two hemispheres becomes obvious. The filterings reveal two conspicuous flanks of relatively steep temperature increase: one between 1920 and 1940 and the other from 1975 to the present, which both represent global phenomena. Also the temperature decrease between 1940 and 1975 is global and is difficult to understand as an anthropogenic effect.

The 13-year lowpass filtered northern hemispheric temperature curve from Fig. 3 is used in Sect. 4 as the function to be described by our regression model.

Another way to demonstrate the parallelism in the temperature evolution of the two hemispheres is to investigate the Fourier power and phase spectra. The long-wave parts (>4 years) in the two spectra (Fig. 4a) show remarkable similarities indicating the presence of the same spectral variability pattern in both hemispheres. It seems to us unlikely that on the very different hemispheres (different in respect of heat capacity etc.) such similar frequency pattern would be generated by the coupled atmosphere-ocean-system alone. This is another argument for the presence of variability sources external to the climate machine.

For periods longer than about 12 years the phase spectra (Fig. 4b) are dominated by an in-phase variability of both hemispheres. For periods between 3 and 12 years, however, the phase relations are mostly in anticorrelation. Thus, the phase spectra show a divide of variability patterns in a subdecadal and a superdecadal regime in both hemispheres which points to different origins.

What are the geophysical mechanisms which could generate and maintain a global in-phase behaviour of the temperature anomalies of the two hemispheres? The global ocean conveyor belt may redistribute heat, but may barely upwell heat to the surface in all oceans simultaneously. Again, the prevailing in-phase synchronization of both hemispheres is a strong argument for an external forcing of the climate machine. In principle, this could be caused by anthropogenic influences and/or by extraterrestrial (e.g. solar) forcing. In conclusion, we need an external driver to explain the variability of the observed temperature anomalies during the last 140 years. The material presented in this section reinforces the existence of such an external driver on time scales >10 years.

3. Proxies for Surface Temperatures and for Solar Variability

The lack of reliable information about historical solar irradiance variations (total and spectral) is certainly responsible for many uncertainties in seeking sun-climate-connections. For example continuous observations of total solar irradiance (TSI) outside the atmosphere by radiometers on board of satellites have been made only since Nov. 1978 (e.g. Hoyt, 1993; Fröhlich, 2000). Sufficient long-term stability is even now not yet guaranteed.

For secular time scales we depend on reconstructions of the solar variability by solar proxy data. Cosmogenic radionuclides, e.g. ^{10}Be from ice cores, provide information on solar variability (Beer, 2000, this volume and Beer *et al.*, 1994a, b, 1996, 2000). The production of this nuclide is modulated by magnetic fields carried with the solar wind and is related to solar variability. Because the ^{10}Be-signal is correlated with the TSI during the solar Schwabe cycle time scale it is not unreasonable to assume that a similar correlation does exist also on longer time scales. The ^{10}Be data, extracted from Greenland ice cores covering the period 1423 to 1985, support this assumption and can be used to estimate the secular solar variation: The variation of the ^{10}Be concentration since the Maunder Minimum (1645–1715) up to present is approximately four times larger than during solar cycle 22 (Beer *et al.*, 1994a). Under the assumption that the relation is linear and can be expressed as an equivalent irradiance variation, this change would correspond to an increase in TSI since the Maunder minimum of about 0.4%. This suggests that slow (>10 years) solar forcing changes can generate much higher surface temperature changes than Schwabe cycle induced variations and could play an important role for the long-term evolution of the Earth's climate (Beer *et al.*, 2000).

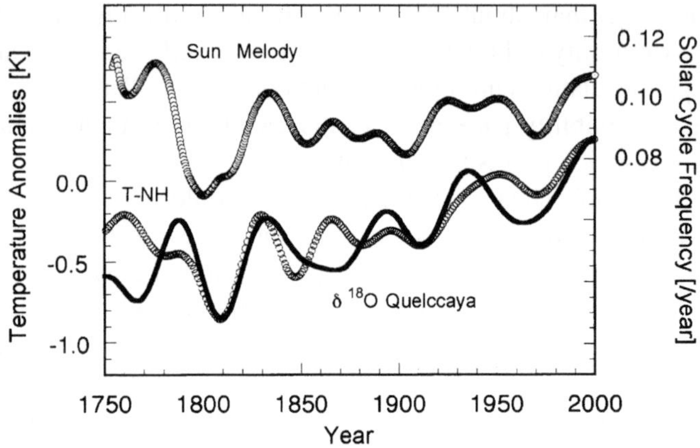

Figure 5. The northern hemispheric temperature record and the Quelccaya δ^{18}O-record in an about 30-year lowpass filtering are compared with the Sun melody in the period 1749–2000. The Sun melody is reflected in both temperature records. Data for the NH composite are from Jones *et al.* (1998) and Groveman and Landsberg (1979); data for Quelccaya from Thompson *et al.* (1994).

For a comparison of solar proxies with surface temperatures beyond the instrumental based hemispheric record only temperature proxies are available. Such a proxy is e.g. the yearly δ^{18}O-record from the Quelccaya glacier (Thompson *et al.*, 1994).

Figure 5 shows the most recent part of the Quelccaya record together with the northern hemispheric temperature composite with data from Groveman and Landsberg (1979) and Jones (1994) about 30-year lowpass filtered (most appropriate for this comparison). The local δ^{18}O-record scaled to the global temperature anomalies yields $\Delta T = 5.5 + 0.328\delta^{18}\mathrm{O}$. This glacier is at an extreme altitude and is strongly influenced by the Pacific; it is very far from the sites the NH-composite is based on. Both temperature curves show remarkable similarities and a good concordance with the Sun melody (defined in Sect. 4 and depicted in the upper curve of Fig. 5): e.g. the deep depression around 1800, followed by a peak around 1830. (The existence of this peak is supported, e.g. by a map from Bradley (2000, page 397, Fig. 4.3.)). Also, the depressions after 1900 and around 1970 coincide quite well. The mentioned conspicuous flanks (Sect. 2, cf. Fig. 3) beginning 1920 and 1975 are also in parallel. All three curves show the temperature decrease between 1940 and 1975. These similarities provide empirical evidence for a connection between the Sun melody and the hemispheric temperatures on longer than decadal time scales. The concordance between the Sun melody and the temperature curves makes the Sun a possible candidate for a lowfrequent variability source. The physical mechanisms that cause the conspicuous parallelism are not discussed here.

4. Separation of the Anthropogenic From the Solar Forcing Component on Longer Than Decadal Time Scales

Using a non-linear regression model we separate the anthropogenic influences on global climate from other variability sources on decadal/secular time scales. The function to be described ("explained") in our regression model is the 13-year filtered northern hemispheric temperature record (cf. Fig. 3). We express this function by a combination of a parameterized growth function for the anthropogenic effect and a proxy function for the influence of solar variability (Mende and Stellmacher, 1994, 1997; Beer *et al.*, 2000). The separation is based on the different growth behaviour of the two explicitly modelled forcing components. Our separation is based on a number of assumptions, which are obvious simplifications, but they are needed to render the problem tractable. The most important are:

- With respect to their amplitudes, the anthropogenic and solar components are the two most influential variability sources under the applied filtering. All other variability amplitudes are assumed to be smaller in the variability band and the considered time window.
- Global and hemispheric aggregation leads to physically meaningful and stable quantities.
- All acting feedback couplings do not essentially modify the composition of the regression model and are constant in our time window.
- The solar driving force is a linear function of the frequency modulation function of the solar cycle (Sun melody).
- The anthropogenic components can be modelled by hyperbolas with approximately the same exponent and can be aggregated into a single hyperbolic compartment.
- The Sun melody contains no significant hyperbolic component in our time window and the hyperbolic component is purely anthropogenic.
- The temperature can be expressed as a linear superposition of the model terms.

In favour of the first, most critical assumption, we accumulated some circumstantial evidence in the preceding sections. The found concordance between the hyperbolically detrended temperature curve and the Sun melody (cf. Fig. 7b) supports this approach too.

In the following we discuss the two main forcing components of the model: proxies for solar and anthropogenic influence.

Solar proxy: A spectral band of 6–24 years of the monthly sunspot number record comprises the period lengths of solar cycles since 1749. From this band a frequency modulation function has been computed by complex demodulation using the Hilbert transform. The resulting instantaneous frequency function is then 13-year lowpass filtered in order to remove strong changes of phase velocity during solar cycles (Fig. 6a) and to enable comparison with similarly filtered temperature records. We call this filtered frequency modulation function Sun melody (Mende *et al.*, 1993).This method exploits the complete frequency information of the band-

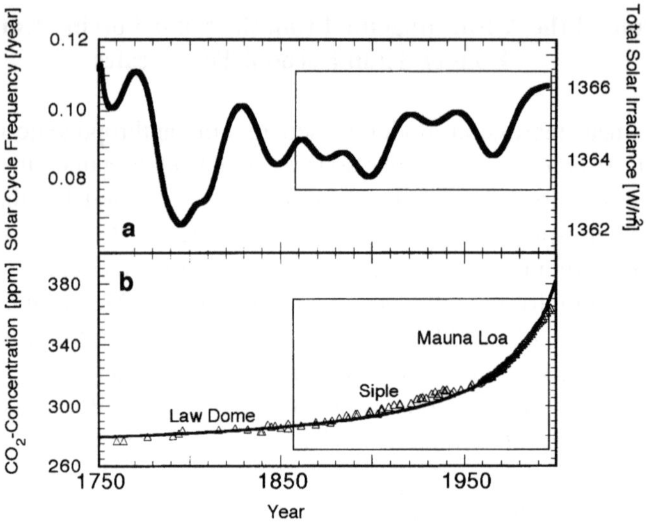

Figure 6. The main climate forcing components used in our regression model: (a) The Sun Melody (1749–1996), which is the frequency modulation function of the solar Schwabe cycle. The scaling of the Sun melody in TSI-equivalent on the right side is derived from the regression results under the assumption of a global climate sensitivity of 0.9 K/(W/m^2) for solar forcing. It is only a theoretical amount of TSI, which would generate the attributed temperature variation corresponding to this sensitivity. The absolute calibration is the average of the VIRGO data from 1996 to 1999 (Fröhlich, 2000). (b) Hyperbolic growth of CO_2 concentrations from pre-industrial times up to 1997. Data since 1958 from Keeling and Whorf (1998), before from Neftel *et al.* (1985, 1994). The parts in the inset windows are used for comparisons with the temperature in the framework of our regression model.

filtered record; it uses more than just the local information contained near the extremes as done in other solar proxy reconstructions (see e.g. Hoyt and Schatten, 1997). The solar influence on climate (TSI plus possibly other solar effects) and more specifically on the Earth's temperature is assumed to be linearly proportional to the Sun melody. Until now we do not know in detail the contribution of TSI or other solar influences to changes to the observed surface temperature variation. In the sun-climate influence many other variability aspects, like UV-spectral irradiance changes and others are probably involved, but this cannot be discussed here.

Anthropogenic proxy: The anthropogenic influence is taken as the aggregated effect of aerosols and greenhouse gases together with all other climate relevant man-made effects like albedo changes, land use etc.. As a proxy for this component we use measurements of greenhouse gases in the atmosphere and from ice cores. For the period from 1958 to present, the Mauna Loa CO_2-record is used (Keeling and Whorf, 1998). After seasonal detrending by spectral filtering a hyperbola is fitted to these data in order to estimate the exponent of the hyperbola. The hyperbolic term has the form $c + A/(t - t_p)^m$ (with $c = 268.3, A = 4431.1, t_p = 2031.1, m = 1.06$). Note that a hyperbolic model fits the data better than any exponential model. This hyperbola also describes reasonably well the Law Dome and the Siple ice core CO_2 data from Antarctica (Neftel *et al.*, 1985, 1994, Fig. 6b), and is in good agreement with the commonly used estimation of the pre-industrial CO_2 level of 280 ppmv. The use of a hyperbola for other anthropogenic influences is justified by the fact that many global industrial indicators as well as paleo-records of human activity (CO_2, methane etc.) show a typical hyperbolic growth behaviour. This hyperbola serves as a simplified – but not unrealistic – model for the global trend of anthropogenic climate influence. For prediction purposes it has to be substituted by an appropriate sigmoid (hyperlogistic) growth model. We use in the regression model only the structural form of the hyperbola and the exponent. As transfer function between CO_2 concentration and temperature a logarithmic relation is used (Wigley, 1987).

The regression model: The minimal parametric non-linear regression model for the long-term trend of the hemispheric temperature anomalies $T(t)$ has the following form:

$$T(t) = a_0 + a_1 \,\mathrm{Hyp}(t - t_1) + a_2 \,\mathrm{Sun}(t - t_2) + \mathrm{Resid}(t)$$

with $\mathrm{Hyp}(t)$ a monotonic and hyperbolically growing component as a proxy for anthropogenic influences and $\mathrm{Sun}(t)$ a non-monotonic component derived from the Sun melody as a proxy for the solar influence. The residual component $\mathrm{Resid}(t)$ represents all not explicitly modelled variability sources.

The model parameters are optimised under minimisation of a quadratic distance function using the Levenberg-Marquardt-algorithm. This guarantees that the two main components are optimised simultaneously and not sequentially. The included time lags (t_1, t_2) require for a stable convergence accurate spectral interpolation during each optimisation step. Convergence of the optimisation procedure has proven to be stable and there were no other deep minima of the distance function in an acceptable region of the parameter space.

5. Results of the Regression Model and Discussion

The solution of the regression model provides a scaling of the forcing components in surface temperature. It determines the contributions of temperature variation that can be attributed to the solar and to the anthropogenic variability component (Fig. 7). The agreement between the hyperbolically detrended temperature anomalies and the Sun melody (both curves 13-year filtered) is quite good ($r = 0.87$) (Fig. 7b). The solution yields for the time lags $t_1 = 0.6$, $t_2 = 10.1$ years.

For the time window 1856–1996 we identified in the northern hemispheric temperatures a filter trend of 0.73 K, of which 0.38 K are due to the anthropogenic and 0.37 K to the solar influence. The small residual of 0.02 K is a consequence of our model design. The non-monotonic filter trend is strongly dependent on the boundaries of the considered time window, the linear trend even more so. In terms

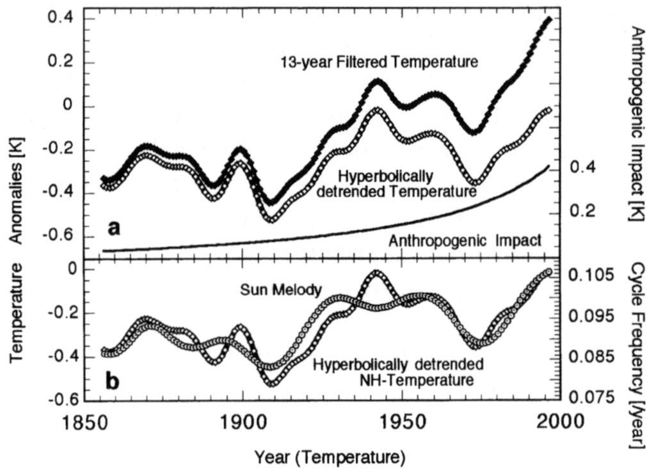

Figure 7. The upper panel (a) shows the separation of the anthropogenic forcing component from the northern hemispheric temperature anomalies (1856–1996). The area between the two upper curves represents the evolution of the anthropogenic influence on temperature. It is equal to the hyperbola at the bottom of this panel. In the lower panel (b) the Sun melody (scaled in temperature) is compared with the hyperbolically detrended northern hemispheric temperature anomalies.

of variance, the hyperbolically detrended temperature curve contains 41% of the original filtered temperature curve. 31% can be attributed to the solar component and 10% to the residual component. Giving all these numbers without the corresponding precise time window is misleading. For example, the solar influence filter trend can become zero (Fig. 7b, e. g. between 1890 and 1975) and even negative (between 1870 and 1975).

The solution of the non-linear regression model provides a connection between northern hemispheric temperature anomalies and the solar cycle frequency. The corresponding phase plot is shown in Fig. 8. In our time window the linear relation between the shifted Sun melody and the temperature anomalies is $\Delta T = -1.90 + 17.87f$, with ΔT temperature change [K] and f [year^{-1}] frequency of the Sun melody.

A similar, albeit slightly lower, correlation between solar forcing and surface temperature results for the Southern Hemisphere. This demonstrates the global character of the lowfrequent temperature variation and points again to anthropogenic effects and solar variability as external driving forces.

6. Conclusions

We could accumulate some evidence for the existence of two main forcings external to the climate machine on superdecadal time scales: the hyperbolically increas-

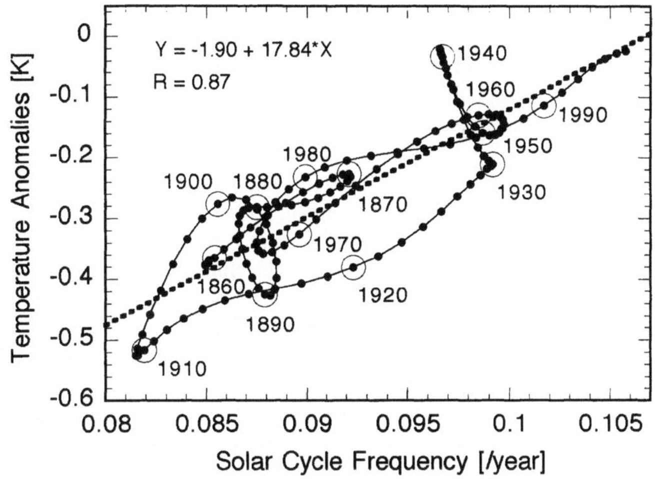

Figure 8. Connection between surface temperature and solar cycle frequency (1856–1996): The trajectory gives a relation between solar cycle frequency and northern hemispheric surface temperature variation during the last 140 years. It is based on the scaling of cycle frequency in temperature by the regression model.

ing anthropogenic and the presently also increasing solar influence on the Earth's climate.

The separation of an anthropogenic component from a solar forcing component in the framework of a non-linear regression model gives a relative consistent picture of the interrelationship of these two forcings on a hemispheric level and for the time window 1856–1996. This picture suggests that solar forcing plays a significant role in climate change, even now while the anthropogenic climate influence has reached its highest amount in the history of the Earth. Presently both influences are driving in the same direction toward a further global warming. According to our results the solar variability was apparently a major variability source in the past (illustrated here back to 1750) and will probably remain a significant one for climate change in the future.

Acknowledgements

This work was supported in part by the BMBF (07 KFT 55A7, 07VKV01 161/38 until 1997). We are indebted to Jürg Beer (EAWAG, Zürich) and to Oran R. White (HAO/NCAR, Boulder) for support and for the fruitful collaboration.

References

Beer, J., Baumgartner, St., Dittrich-Hannen, B., Hauenstein, J., Kubik, P., Lukasczyk, Ch., Mende, W., Stellmacher, R., Suter, M.: 1994a, 'Solar variability traced by cosmogenic isotopes', *in* J. M. Pap, C. Fröhlich, H. S. Hudson, S. K. Solanki (eds.), *The Sun as a variable star* , Cambridge Univ. Press, 291–300.

Beer, J., Joos. F., Lukascyk, Ch., Mende, W., Rodriguez, J., Siegenthaler, U., Stellmacher, R.: 1994b, '^{10}Be as an indicator of solar variability and climate', *in* E. Nesme-Ribes (ed.), *The solar engine and its influence on terrestrial atmosphere and climate*, NATO ASI Series I **25**, 221–233.

Beer, J., Mende, W., Stellmacher, R., White, O. R.: 1996, 'Intercomparison of proxies for past solar variability', *in* Ph. D. Jones, R. S. Bradley, J. Jouzel (eds.), *Climate variations and forcing mechanisms of the last 2000 years*, NATO ASI Series I **41**, pp. 501–517.

Beer, J., Mende, W., Stellmacher, R.: 2000, 'The role of the sun in climate forcing', *Quarternary Science Reviews* **19**, 401–415.

Beer, J.: 2000, 'Long-term indirect indices of solar variability', *Space Sci. Rev.*, this volume.

Bradley, R. S.: 2000, 'Past global changes and their significance for the future', *Quarternary Science Review* **19**, 391–402.

Fröhlich, C.: 2000, 'Observation of irradiance variations', *Space Sci. Rev.*, this volume.

Groveman, B. S., and Landsberg, H. E.: 1979, 'Reconstruction of northern hemispheric temperature 1579–1880', Publication No. 79-181, University of Maryland, 46 pp.

Hoyt, D. V.: 1993, 'A discussion of plausible solar irradiance variations 1700-1992', *J. Geophys. Res.* **98**, A11, 18,895–18,906.

Hoyt, D. V., Schatten, K. H.: 1997, 'The role of the Sun in climate change', Oxford University Press, New York, Oxford, 279 pp.

Jones, P.D.: 1994, 'Hemispheric surface air temperature variations, A reanalysis and an update to 1993', *J. Climate* **7**, 1794–1802.

Jones, P. D., Parker, D. E., Osborn, T. J., Briffa, K. R.: 1998, 'Global and hemispheric temperature anomalies - land and marine instrumental records', in *Trends: A Compendium of Data on Global Change*, Carbon Dioxide Information Analysis Centre, Oak Ridge National Laboratory, Oak Ridge, Tenn., U.S.A.

Keeling C. D., Whorf, T. P.: 1998, 'Atmospheric CO_2-records from sites in the SIO air sampling network', in *Trends: A Compendium of Data on Global Change*, Carbon Dioxide Information Analysis Center, Oak Ridge National Laboratory, Oak Ridge, Tenn., U.S.A.

Mende, W., Stellmacher, R., Olberg, M.: 1993, 'Relationships between relative sunspot numbers and air temperature', *Meteorol. Zeitschrift N.F.* **2**, 121–126.

Mende, W., and Stellmacher, R.: 1994, 'Solar radiative forcing und Klimaentwicklung', *PIK Reports* **1**, 189–192.

Mende, W., and Stellmacher, R.: 1997, 'Analyse der säkularen Klimavariabilität mit Hilfe von Methoden der Signalanalyse unter Einbeziehung von Satellitenirradianzmessungen und ^{10}Beryllium', *Annalen der Meteorologie* **34**, 171–172.

Neftel, A., Moor, E., Oeschger, H., Stauffer, B.: 1985, 'Evidence from polar ice cores for the increase in atmospheric CO_2 in the past two centuries', *Nature* **315**, 45–47.

Neftel, A., Friedly, H., Moor, E., Lötscher, H., Oeschger, H., Siegenthaler, U., Stauffer, B.: 1994, 'Historical CO_2-record from Siple Station ice core', *in* Boden, T. A., Kaiser, D. P., Sepansky, R. J. and Stoss, F. W. (eds.), *Trend '93: A Compendium of Data on Global Change*, Carbon Dioxide Information Analysis Center, Oak Ridge National Laboratory, Oak Ridge, Tenn., U.S.A., 11–14.

Thompson, L. G., Davis, M. E., Mosley-Thompson, E.: 1994, 'Glacial records of global climate, A 1500-year tropical ice core record of climate', *Human Ecology* **22**, 83–95.

Wigley, T. M. L.: 1987, 'Relative contribution of different trace gases to the greenhouse effect' *Climate Monitor* **16**, 14–29.

. . . CLIMATE

CLIMATE OBSERVATIONS - THE INSTRUMENTAL RECORD

D. E. PARKER, T. A. BASNETT, S. J. BROWN, M. GORDON, E. B. HORTON and N. A. RAYNER
Hadley Centre, The Met. Office, United Kingdom

Received: 12 October 1999; Accepted in final form: 7 February 2000

Abstract. A survey is given of the available instrumental data for monitoring and analysis of climatic variations. We focus on temperature measurements, both over land and ocean, at the surface and aloft.

Over land, the older observations were subject to exposure changes which may not have been fully compensated. The effects of urbanization have been largely avoided in studies of climatic change over the last 150 years. There are few records for pre-1850 outside Europe and eastern North America, and the global network shows a recent decline. Over the ocean, sea surface temperature (SST) has been measured using buckets, engine intakes, hull sensors, buoys, and satellites. Many of these data have been effectively homogenized, but new challenges arise as observing systems evolve. Available SST and marine air temperature datasets begin in the 1850s. The data are concentrated in shipping lanes especially before 1900, and very sparse during the world wars, but additional historical data are being digitized.

The radiosonde record is short (~40 years) and has major gaps over the oceans, tropics and Southern Hemisphere. Instrumental heterogeneities are beginning to be assessed and removed using physical and statistical techniques. The MSU record is complete but only began in 1979, and is not highly resolved in the vertical: major biases, mainly affecting the lower-tropospheric retrieval, have been reduced as a result of recent analyses.

Advanced interpolation or data-assimilation techniques are being applied to these data, but the results must be interpreted with care.

1. Introduction

The signals being sought in many climate variability and change detection studies are small. So it is important that the observational data be homogeneous, i.e. that changes of climate are not misrepresented as a result of changes in instrumentation, observing practice, or the local environment. Because of natural inderdecadal variability, it is important that the records used for climate change detection be long; because of spatial variations in climate anomalies, global coverage is needed. So here we review some aspects of the homogeneity, length and geographical coverage of the instrumental record of temperature. We include both land and marine surface temperatures, and temperatures in the troposphere and lower stratosphere. We also present a recently-developed interpolation technique which can be used to alleviate some of the difficulties caused by poor spatial coverage and provide an estimate of the uncertainty in areal averages of incomplete fields.

Space Science Reviews **94:** 309–320, 2000.

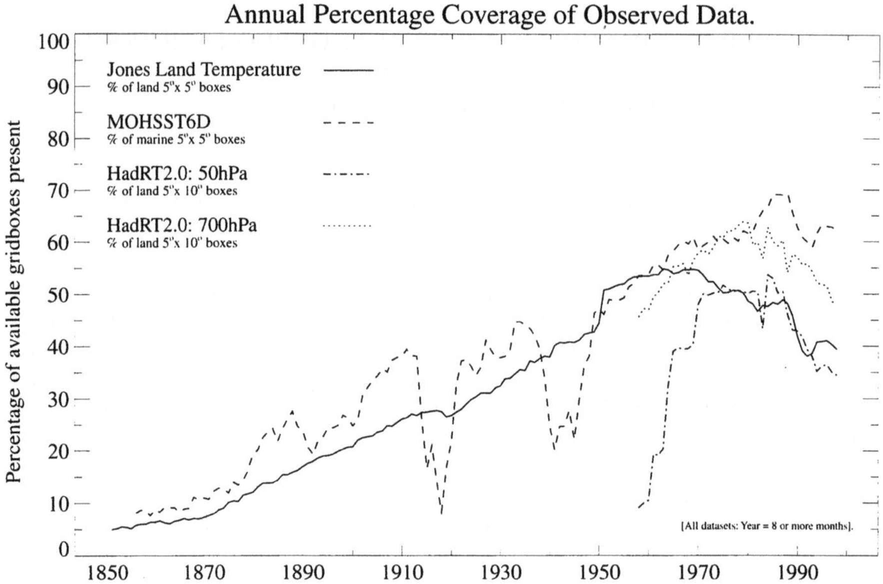

Figure 1. Annual coverage (%) of observed data. Data sources: Land air temperature, Jones *et al.* (1999); Met. Office Historical sea surface temperatures (MOHSST6), Parker *et al.* (1995); Radiosonde temperatures (HadRT), Parker *et al.* (1997).

2. Land Surface Air Temperatures

Major efforts have been made by several groups to create long, homogeneous worldwide data bases of land surface air temperature (Hansen *et al.*, 1999; Jones *et al.*, 1999; Peterson *et al.*, 1998b). A wide range of statistical methods of improving the homogeneity of station temperature and other records has been developed (Peterson *et al.*, 1998a). The impact of urban warming on global and large regional temperature trends has been shown to be minimal (Peterson *et al.*, 1999). The different techniques used for estimating global and hemispheric temperature changes yield very similar results, though the Hansen *et al.* (1999) method gives smoother fields and reduced interannual variance of the global and hemispheric anomalies (Peterson *et al.*, 1998b). Nevertheless, some biases remain. The instrumental exposures in the nineteenth century often resulted in overheating by day and excessive cooling at night. In some tropical countries the thermometers were suspended under thatched sheds, where temperatures were typically 0.2 to 0.7°C higher than in Stevenson screens. A review of these and other early exposures and their effects is given by Parker (1994). Caution is also needed when using more recent data: for example, Quayle *et al.* (1991) reported artificial warming of 0.3°C by night and cooling of 0.4°C by day when liquid in glass thermometers in wooden shelters were replaced by thermistor-based instruments in plastic shelters in the USA. The longest instrumental record is that of Central England Temper-

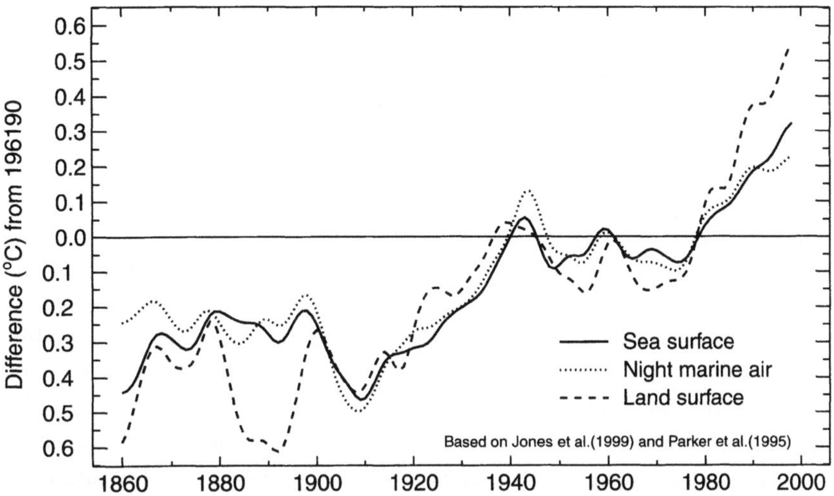

Figure 2. Smoothed annual global temperatures, 1861-1998 relative to 1961-1990 climatology. Land: Jones *et al.* (1999); sea surface temperature and night marine air temperature: Parker *et al.* (1995).

ature (CET): monthly since 1659 (Manley, 1974) and daily since 1772 (Parker *et al.*, 1992). However, the coverage of available land surface air temperature data before the mid nineteenth century is largely restricted to Europe and eastern North America. By 1910, the gaps had shrunk to the Arctic and Antarctic and the interiors of the tropical continents (Parker et al., 1994) and by the 1950s coverage was good except over Antarctica. Since then, observations have been made in Antarctica, but coverage has declined in some other areas, especially in the tropics, resulting in an overall reduction (Figure 1). The Global Climate Observing System (GCOS) Surface Network (GSN) (Peterson *et al.*, 1997) has been set up to ensure the maintenance of a network of homogeneous data for monitoring and researching into global climate, and to ensure the wide availability of the data. Key findings from the land network are that the warming since the 1970s has been greater than over the oceans (Figure 2), and night-time warming has exceeded that by day in most continental areas in the last several decades (Easterling *et al.*, 1997).

3. Sea Surface and Marine Air Temperatures and Comparison with Land

Sea surface temperatures (SSTs) and marine air temperatures (MATs) require major adjustments to ensure homogeneity. The older SSTs were adjusted by Folland and Parker (1995) to compensate for the use of uninsulated or partly-insulated buckets to sample the water. Corrections approached 1°C over the Gulf Stream and Kuroshio in winter, and 0.5°C in most of the tropics, if canvas buckets were used. However, modern SSTs measured from ships may also need adjustments

(Kent *et al.*, 1993). Careful adjustment is also required when blending satellite-based SSTs with *in situ* data (Reynolds and Smith, 1994), because the former are measures of skin rather than bulk temperature, and may be biased by atmospheric aerosol or cloud. MATs are also biased, principally by historically varying on-deck heating, so that only night-time data should be used. Adjustments are also required for changing deck-elevations. Guided by SSTs, MATs were also adjusted for non-standard observing practices in the nineteenth century and during the second world war (Parker *et al.*, 1995; Peterson *et al.*, 1998a).

Coverage of marine temperature data is limited to the main shipping lanes in the mid nineteenth century, with a steady expansion until 1910 (Parker *et al.*, 1994). There were major reductions during both world wars (Parker *et al.*, 1995 and Figure 1) followed by an expansion to most areas except the Southern Ocean and the ice-covered Arctic. These areas are now analysed with the aid of satellite data and advanced interpolation schemes (Reynolds and Smith, 1994; Rayner *et al.*, 1996; Section 5). Historical coverage is now being improved by the digitization of several million observations from ships (Parker *et al.*, 1995), including SST, MAT, pressure and wind.

Support for both marine and land surface air temperature records comes from the agreement of their global trends (Figure 2). Comparison of collocated SSTs with air temperatures from islands and coasts (Parker *et al.*, 1995) suggests that the nineteenth century biases were small on a global average and that the difference in global anomalies in 1880-1895 (Figure 2) arose from atmospheric circulation anomalies and/or sparse sampling. The relative warmth on land in 1900-1930 is likely to have been at least partly the result of the exposures used in the tropics (Section 2). Enhanced westerly winds at midlatitudes in the northern winter may also have raised air temperatures on windward coasts more than collocated SST. The relative warmth over land since 1980 comes from the high latitude northern continents where winters have been anomalously warm owing to an increase in westerly wind flow (Wallace *et al.*, 1996) and springs have been warmed through a positive feedback through reduction in snow cover (Groisman *et al.*, 1994).

Figure 3 compares seasonal anomalies (relative to 1961-1990 climatology) of SST and night MAT for four ocean regions: the North Atlantic north of 35°N, the tropical east Pacific (20°N–20°S, east of 170°W), the Indian Ocean north of 50°S, and the South Pacific north of 50°S. Series for the latter two regions are shown in Figure 4.1 of Folland *et al.* (1999) who also address patterns of oceanic temperature variability. Here, however, the SSTs are from a new analysis, MOHSST7, in which random and sampling errors are reduced by reducing the deviations from an initial reduced-space optimum interpolation (RSOI, Kaplan *et al.*, 1998; Folland *et al.*, 2000) of the available SST data. The reductions are greater where data are sparse. In addition, the night MATs are from a new analysis, MOHMAT5, based on RSOI. In the extratropical North Atlantic, differing trends of SST and MAT may be a result of atmospheric circulation changes, with relatively lower MAT in more westerly epochs such as 1900-1930. In the east tropical Pacific and the Indian

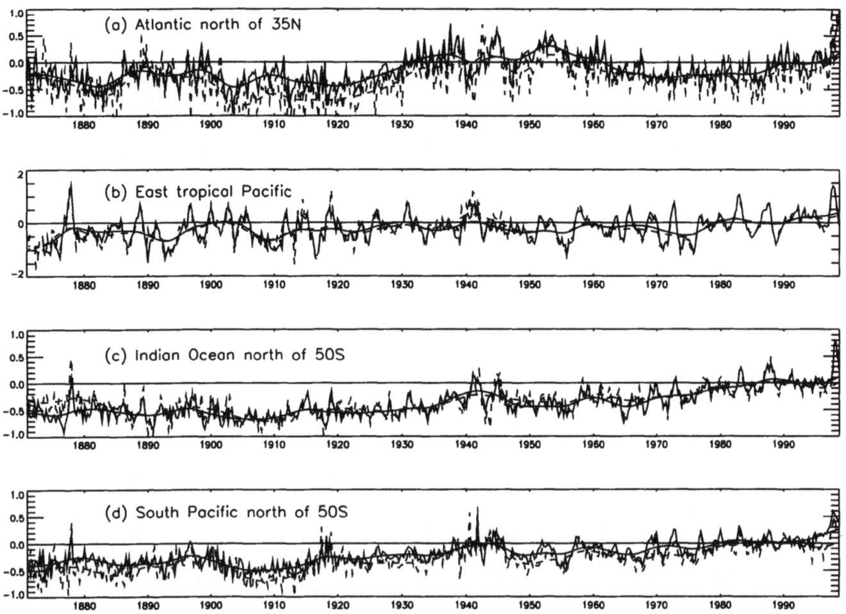

Figure 3. Seasonal anomalies (relative to 1961-1990 climatology) of SST (MOHSST7, solid) and night MAT (MOHMAT5, dashed) for four ocean regions: the North Atlantic north of 35°N, the tropical east Pacific (20°N–20°S, east of 170°W), the Indian Ocean north of 50°S, and the South Pacific north of 50°S. Data are not collocated, i.e. the coverage of SST and night MAT may differ.

Ocean, SST and MAT trends are very similar. In the South Pacific, MAT has not followed SST in showing strong warming in the 1990s. The reasons for this are still under investigation and may include atmospheric and/or oceanic circulation changes, sampling errors owing to data sparsity, and biases in the measurement of SST and/or MAT from the heterogeneous network of *in situ* measuremant platforms (Kent *et al.*, 1993). Figure 3 shows that the SST and MAT data are able to distinguish different climatic variations in different ocean basins: for example, the warming since the beginning of the 20th Century has been steady in the Indian Ocean, whereas there was a marked mid-20th Century warm peak, and a subsequent cool period, in the extratropical North Atlantic.

4. Temperatures in the Troposphere and Lower Stratosphere

The radiosonde network became worldwide in 1958 for the International Geophysical Year. However, coverage of the Southern Hemisphere remained sparse, and the oceans and parts of the tropical continents have remained poorly represented. Stratospheric coverage is sparser than that in the troposphere, and there has been a recent decline in overall coverage (Figure 1). So the GCOS Upper Air Network (GUAN) is being established, to ensure the maintenance of key stations and availability of their data (Folland *et al.*, 2000). A major task of the Comprehensive

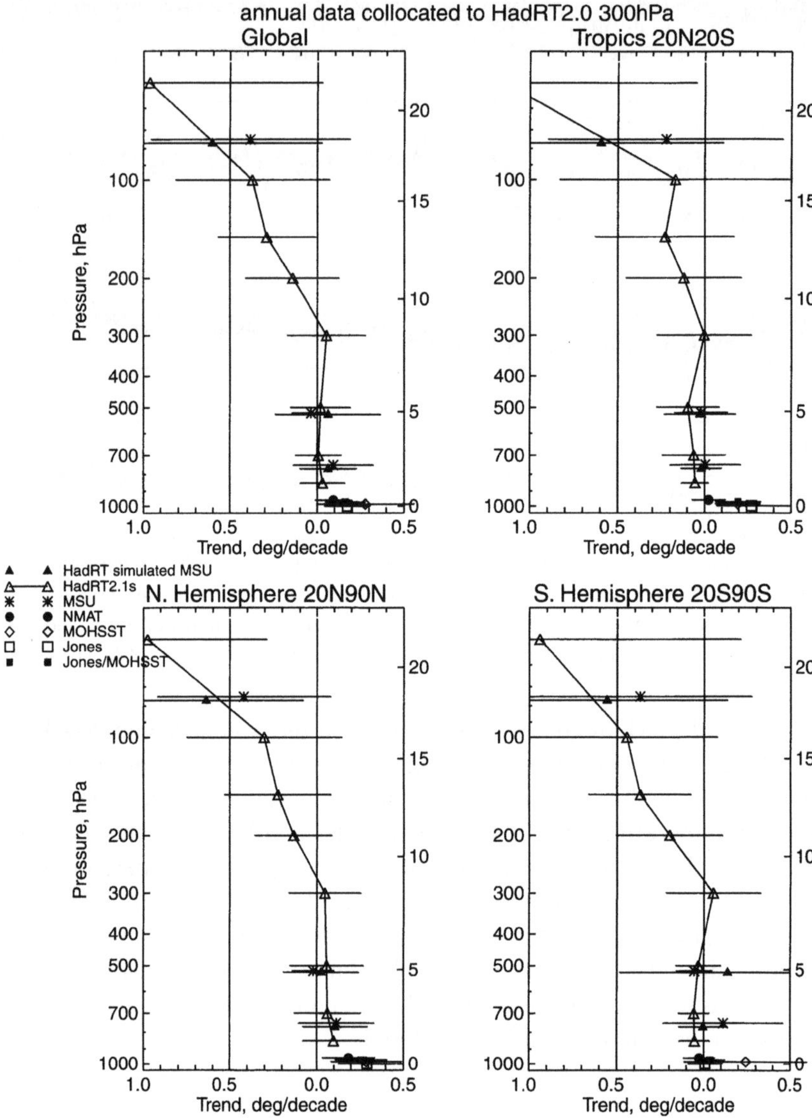

Figure 4. Profiles of temperature trends for 1979-1998: Globe, Tropics 20°N–20°S, N. Hemisphere 20° – 90°N, S. Hemisphere 20° – 90°S. Trends and their 2σ-error bars are estimated by restricted maximum likelihood (Diggle *et al.*, 1994) allowing for first order autoregression. Data sources: land surface air temperature: Jones *et al.* (1999); SST and night MAT: Parker *et al.* (1995); tropospheric temperature: HadRT2.0 (update of Parker *et al.* (1997), India omitted; no bias-adjustments), MSU2 and MSU2LT; stratospheric temperature: HadRT2.1 (India omitted: bias adjustments if stations had metadata, using MSU4 as a reference) and MSU4. All surface records are collocated to HadRT2.0 coverage at 300 hPa.

Aerological Reference Data Set (CARDS) project (Eskridge *et al.*, 1995) is the homogenization, archiving, and provision of data from GUAN and other stations.

The historical radiosonde record is affected by many changes of instrumentation and operating practices including observing time. Basic quality controls such as hydrostatic checks (Eskridge *et al.*, 1995; Parker and Cox, 1995) can remove random errors incurred during measurement, transcription or telecommunication, but systematic biases remain. Essential for the estimation and removal of these are accurate metadata describing the exact nature and timing of the instrumental and procedural changes (Gaffen, 1994; Gaffen, 1996; Parker and Cox, 1995; Parker *et al.*, 1997; Gaffen *et al.*, 1999). The remit of the CARDS Project includes the improvement of the metadata base.

Radiosonde data may be adjusted using an independent record for reference. This was done by Parker *et al.* (1997) using the retrievals made since 1979 from Microwave Sounding Units (MSU). The stratospheric "MSU4" (Spencer and Christy, 1993) and the low-mid tropospheric "MSU2R" (now known as "T2LT") retrievals (Spencer and Christy, 1992) were used to adjust monthly radiosonde temperatures since 1979. Adjustments were only applied in cases of known instrumental or procedural change at a station. Some changes in instrumentation had resulted in spurious cooling of up to 3°C in the lower stratosphere. Biases were much smaller in the troposphere. However, tropospheric MSU temperatures were shown by Wentz and Schabel (1998) to have been biased by orbital decay of the satellites. In addition, the MSU record has been affected by orbit time changes, variations in the temperature of the instruments, and the brevity of some of the overlaps between successive instruments. Improvements have now been made (J. Christy, pers. comm., 1999), but detailed validation is required. Furthermore, MSU cannot be used to adjust radiosonde data older than 1979, and although MSU has global coverage, it samples the atmosphere with much coarser vertical resolution than radiosondes (Spencer and Christy, 1992, 1993).

Radiosonde temperatures can also be adjusted using heat transfer models of the instruments, and the results verified using day-night differences (Luers and Eskridge, 1998). These models have now been developed for the majority of radiosonde types used since 1960. In some cases erroneous radiation corrections need to be removed before application of the model-based adjustments. Within the CARDS project, the models are being applied to the daily radiosonde data. As a result, a basic network of homogeneous radiosonde temperature data should soon be available for climate change detection.

Radiosonde data may also be used with other sources of data in model-based reanalyses (Section 5). However, if the data supplied to a reanalysis have biases, the output is also likely to be biased. A review of the impacts of the choice of reanalysis or radiosonde dataset and processing technique on estimates of temperature trends aloft is given by Santer *et al.* (1999). These authors conclude that biases in the National Centers for Environmental Prediction reanalysis (Kalnay *et al.*, 1996) diminish its usefulness for climate change detection studies.

Comparisons of surface, radiosonde and MSU data have shown an apparent decrease in global and tropical lower-tropospheric lapse rate during the last 20 years. Figure 4 shows profiles of temperature trends for 1979-1998 for the globe, Northern Hemisphere 20° – 90°N, Southern Hemisphere 20° – 90°S and the tropics 20°N–20°S. Reasons for a real change in lapse rate include weakening of high-latitude winter surface inversions by strengthening atmospheric circulation (Hurrell and Trenberth, 1996), and thermal forcing by stratospheric ozone loss (Hansen *et al.*, 1995). Spurious changes in lapse rate may result from biases in any of the data sets, or from the sampling errors of the sparse radiosonde network. However, longer term surface and tropospheric trends (not shown) are in closer agreement, in accord with Nicholls *et al.* (1996).

5. Advances in Interpolation Techniques

Reanalyses (e.g. Kalnay *et al.*, 1996) use all available historical data, along with a fixed state-of-the-art dynamical model's data assimilation scheme, to create a homogeneous set of analyses of the past state of the atmosphere (or ocean). Within the assimilation scheme, errant data are amended or excluded on the basis of comparisons with neighbours and a first-guess analysis. The latter is based on a previous analysis and the model's dynamics and physics. Thus, the quality control of the data incorporates real physical constraints. However, biases affecting large regions, e.g. when radiosondes are changed simultaneously in a country, may not be detected by such schemes, because the initial analyses will be biased. Furthermore, in areas with few *in situ* data the reanalysis may be affected by, for example, inadequate model physics or biased satellite data: when data quality and/or quantity improve, these biases are reduced but the overall result is a non-climatic change in the reanalysis (Santer *et al.*, 1999).

Model-independent interpolation schemes are therefore being developed. Optimum interpolation (Reynolds and Smith, 1994) uses the spatial covariance structure of the data, along with estimated random and spatially-coherent data-errors, to yield complete fields of both the parameter being analysed and the error of estimate. Optimum averaging (Kagan, 1979) uses the same theory to provide the average value of the parameter over the field, along with error bars. In reduced-space optimum interpolation (Kaplan *et al.*, 1998) and averaging, the covariance matrix is truncated to remove those eigenvectors which are regarded as noise. Reduced-space optimum smoothing (Kaplan *et al.*, 1998) incorporates a temporal dependence, e.g. Markov red noise, into the interpolation algorithm. All these techniques assume that the covariance statistics, which must be derived for some well-sampled period, remain valid throughout the period analysed. It is essential to detrend the data before applying optimum interpolation and related techniques; otherwise, there will be a bias towards climatology in data-sparse areas. It is also necessary to remove biases, e.g. by applying bucket-corrections to SST (Folland

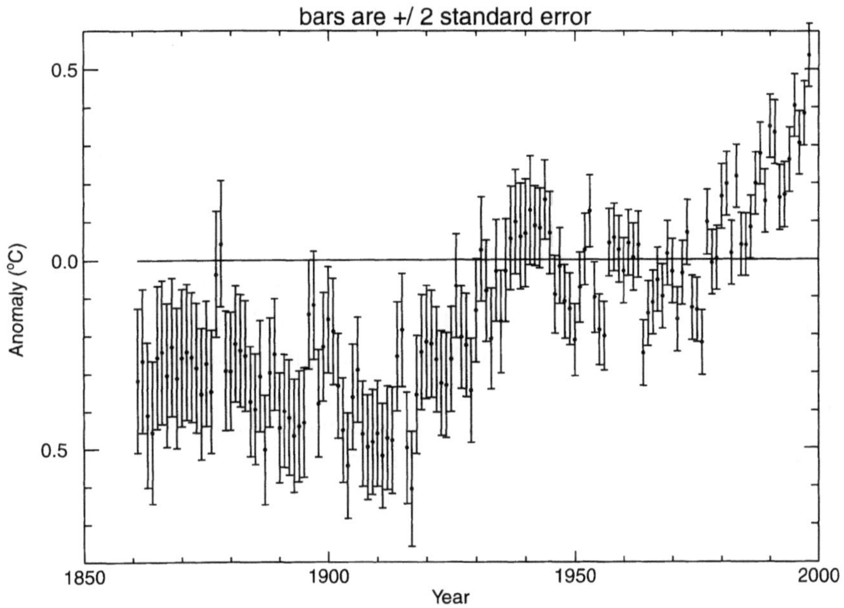

Figure 5. Annual combined land surface air temperature and SST anomalies, relative to 1961-1990 climatology, for the globe, 1861-1998. The averages and associated 2σ-error bars were calculated using reduced space optimum interpolation including an estimate of the uncertainties in the bucket-corrections to SST.

and Parker, 1995) or by adjusting satellite SSTs for their biases (Reynolds *et al.*, 1989; Reynolds, 1993; Reynolds and Smith, 1994.

In Figure 5 we show annual combined land surface air temperature and SST anomalies, relative to 1961-1990 climatology, for the globe for 1861-1998. The averages and associated 2 σ-error bars were calculated using reduced space optimum averaging. Note the steady reduction in the width of the error bars from the nineteenth century to the late twentieth century, with temporary widenings during the two world wars when data were sparser. Recently digitized data for 1911-1920 are expected to reduce the errors of estimate in future analyses for that period. There has not been a significant recent increase in the estimated errors, despite the decline in data availability over land (Figure 1) because the marine data coverage has been maintained. The error bars are similar to those obtained by Jones *et al.* (1997, 1999) who used analytic spatial correlation fits to estimate the standard errors of gridbox data and the number of independent gridbox data in the globe. Global temperature anomalies calculated by area-weighted averaging of gridbox values (as shown e.g. in Nicholls *et al.* (1996) generally lie well within the error bars in Figure 5.

6. Conclusions

The instrumental surface temperature record is marginally adequate for monitoring variations on multidecadal to century time scales. Although some biases may remain, the land-based and marine records offer mutual support, especially since the 1890s. In addition, the SST and night MAT records offer very good evidence for the diverse behaviour of different ocean basins.

With care it should be possible to remove many radiosonde temperature biases and allow a realistic assessment of interdecadal tropospheric and lower stratospheric temperature changes since the 1960s. The MSU record, though only available since 1979, is geographically complete; its lower stratospheric data are likely to be less affected by biases than its lower tropospheric retrievals.

Advanced interpolation techniques, such as model-based Reanalyses and reduced space optimum interpolation, show promise, but the results must be interpreted with care. In particular, these methods cannot remove biases from the data to which they are applied.

Acknowledgements

C. K. Folland provided helpful comments. This work was supported by the U.K. Department of the Environment, Transport and the Regions under contract PECD/7/12/37 and the Public Meteorological Service Research and Development program under contract MSG-2/99.

References

Diggle, P. J., Liang, K.-Y., and Zeger, S. L.: 1994, 'Analysis of Longitudinal Data', Clarendon Press, Oxford, Chap. 4, p. 64.

Easterling, D. R., Horton, E. B., Jones, P. D., Peterson, T. C., Karl, T. R., Parker, D. E., Salinger, M. J., Razuvayev, V. N., Plummer, N., Jamason, P., and Folland, C. K.: 1997, 'Maximum and minimum temperature trends for the globe', *Science* **277**, 364–367.

Eskridge, R. E., Alduchov, O. A. Chernykh, I. V., Zhai, P., Polansky A. C. and Doty, S. R.: 1995, 'A Comprehensive Aerological Reference Data Set (CARDS): Rough and systematic errors', *Bull. Amer. Meteorol. Soc.* **76**, 1759–1775.

Folland, C. K. and Parker, D. E.: 1995, 'Correction of instrumental biases in historical sea surface temperature data', *Quart. J. Roy. Meteorol. Soc.* **121**, 319–367.

Folland, C. K., Parker, D. E., Colman, A. W. and Washington, R. 1999: 'Large scale modes of ocean surface temperature since the late nineteenth century', *in* A. Navarra, (Ed.), *Beyond El Niño: Decadal and Interdecadal Climate Variability.*, Springer-Verlag., pp. 73–102.

Folland, C. K., Frich, P., Basnett, T. A., Rayner, N. A., Parker, D. E. and Horton, E. B.: 2000, 'Uncertainties in climate datasets: a challenge for WMO', *WMO Bulletin* **49**, 59–68.

Gaffen, D. J.: 1994, 'Temporal inhomogeneities in radiosonde temperature records', *J. Geophys. Res.* **99**, 3667–3676.

Gaffen, D. J.: 1996, 'A digitized metadata set of global upper-air station histories', NOAA Technical Memorandum ERL-ARL 211, Silver Spring, MD, pp. 38.

Gaffen, D. J., Sargent, M. A., Habermann, R. E. and Lanzante, J. R.: 1999, 'Sensitivity of tropospheric and stratospheric temperature trends to radiosonde data quality', *J. Climate*, in press.

Groisman, P., Ya., Karl, T. R., and Knight, R. W.: 1994, 'Observed impact of snow cover on the heat balance and the rise of continental spring temperatures', *Science* **263**, 198–200.

Hansen, J., Wilson, H., Sato, M., Ruedy, R., Shah, K. and Hansen, E.: 1995, 'Satellite and surface temperature data at odds?', *Climatic Change* **30**, 103–117.

Hansen, J., Ruedy, R., Glascoe, J., and Sato, M.: 1999, 'GISS analysis of surface temperature change', *J. Geophys. Res.* **104**, 30997–31022.

Hurrell, J. W., and Trenberth, K. E.: 1996, 'Satellite versus surface estimates of air temperature since 1979', *J. Climate* **9**, 2222–2232.

Jones, P. D., Osborn, T. J. and Briffa, K. R.: 1997, 'Estimating sampling errors in large-scale temperature averages', *J. Climate* **10**, 2548–2568.

Jones, P. D., New, M., Parker, D. E., Martin, S. and Rigor, I.G.: 1999, 'Surface air temperature and its changes over the past 150 years', *Rev. Geophys.* **37**, 173–199.

Kagan, R. L.: 1979, 'Averaging of meteorological fields', Gidrometeoizdat, Leningrad (now St. Petersburg). Also published in English, L. S. Gandin and T. M. Smith (Eds.), as *Atmospheric and Oceanographic Sciences Library* Vol. **19**, , Kluwer Academic Publishers, Dordrecht/Boston/London, pp. 279.

Kalnay, E., Kanamitsu, M., Kistler, R., Collins, W., Deaven, D., Gandin, L., Iredell, M., Saha, S., White, G., Woollen, J., Zhu, Y., Chelliah, M., Ebisuzaki, W., Higgins, W., Janowiak, J., Mo, K. C., Ropelewski, C., Wang, J., Leetmaa, A., Reynolds, R., Jenne, R., and Joseph, D.: 1996, 'The NCEP/NCAR 40-year reanalysis project', *Bull. Amer. Meteorol. Soc.* **77**, 437–471.

Kaplan, A., Cane, M. A., Kushnir, Y., Clement, A. C., Blumenthal, M. B., and Rajagopalan, B.: 1998, 'Analyses of global sea surface temperature 1856-1991', *J. Geophys. Res.* **103**, 18567–18589.

Kent, E. C., Taylor, P. K., Truscott, B. S., and Hopkins, J. S.: 1993, 'The accuracy of voluntary observing ships' meteorological observations - Results of the VSOP-NA', *J. Atmos. Oceanic Tech.* **10**, 591–608.

Luers, J. K. and Eskridge, R. E.: 1998, 'Use of radiosonde temperature data in climate studies', *J. Climate* **11**, 1002–1019.

Manley, G.: 1974, 'Central England Temperatures: monthly means 1659 to 1973', *Quart. J. Roy. Meteorol. Soc.* **100**, 389–405.

Nicholls, N., Gruza, G. V., Jouzel, J., Karl, T. R., Ogallo, L. A. and Parker, D. E.: 1996, 'Observed climate variability and change', *in* J. T. Houghton, L. G. Meira Filho, B. A. Callander, N. Harris, A. Kattenberg and K. Maskell (Eds.), *Climate Change 1995 - The Science of Climate Change'. The second assessment report of the Intergovernmental Panel on Climate Change: Contribution of Working Group I.*, Cambridge University Press, Cambridge, UK, pp. 133–192.

Parker, D. E.: 1994, 'Effects of changing exposure of thermometers at land stations', *Internat. J. Climatol.* **14**, 1–31.

Parker, D. E. and Cox, D. I.: 1995, 'Towards a consistent global climatological rawinsonde database', *Internat. J. Climatol.* **15**, 473–496.

Parker, D. E., Legg, T. P., and Folland, C. K.: 1992, 'A new daily Central England Temperature series, 1772-1991', *Internat. J. Climatol.* **12**, 317–342.

Parker, D. E., Jones, P. D., Folland, C. K. and Bevan, A.: 1994, 'Interdecadal changes of surface temperature since the late 19th century', *J. Geophys. Res.* **99**, 14377–14399.

Parker, D. E., Folland, C. K. and Jackson, M.: 1995, 'Marine surface temperature: observed variations and data requirements' *Climatic Change* **31**, 559–600.

Parker, D. E., Gordon, M., Cullum, D. P. N., Sexton, D. M. H., Folland, C. K., and Rayner, N.: 1997, 'A new gridded radiosonde temperature data base and recent temperature trends', *Geophys. Res. Lett.* **24**, 1499–1502.

Peterson, T. C., Daan, H., and Jones, P. D.: 1997, 'Initial selection of a GCOS Surface Network', *Bull. Amer. Meteorol. Soc.* **78**, 2145–2152.

Peterson, T. C., Easterling, D. R., Karl, T. R., Groisman, P., Nicholls, N., Plummer, N., Torok, S., Auer, I., Boehm, R., Gullett, D., Vincent, L., Heino, R., Tuomenvirta, H., Mestre, O., Szentimrey, T., Salinger, M. J., Forland, E. J., Hanssen-Bauer, I., Alexandersson, H., Jones, P. D., and Parker, D. E.: 1998a, 'Homogeneity adjustments of *in situ* atmospheric climate data: A review', *Internat. J. Climatol.* **18**, 1493–1517.

Peterson, T. C., Karl, T. R., Jamason, P. F., Knight, R. and Easterling, D. R.: 1998b, 'First difference method: maximizing station density for the calculation of long-term global temperature change', *J. Geophys. Res.* **103**, 25967–25974.

Peterson, T. C., Gallo, K. P., Lawrimore, J., Owen, T. W., Huang, A., and McKittrick, D. A..: 1999, 'Global rural temperature trends', *Geophys. Res. Lett.* **26**, 329–332.

Quayle, R. G., Easterling, D. R., Karl, T. R., and Hughes, P. Y.: 1991, 'Effects of recent thermometer changes in the Cooperative Station Network', *Bull. Amer. Meteorol. Soc.* **72**, 1718–1723.

Rayner, N. A., Horton, E. B., Parker, D. E., Folland, C. K., and Hackett, R. B.: 1996, 'Version 2.2 of the Global sea-Ice and Sea Surface Temperature data set, 1903-1994', Hadley Centre Climate Research Technical Note CRTN 74. Available from the Hadley Centre, Met. Office, Bracknell, UK.

Reynolds, R. W.,: 1993, 'Impact of Mount Pinatubo aerosols on satellite-derived sea surface temperatures', *J. Climate* **6**, 768–774.

Reynolds, R. W. and Smith, T. M.: 1994, 'Improved global sea surface temperature analyses using optimum interpolation', *J. Climate* **7**, 929–948.

Reynolds, R. W., Folland, C. K. and Parker, D. E.: 1989, 'Biases in satellite-derived sea-surface-temperature data', *Nature* **341**, 728–731.

Santer, B. D., Hnilo, J. J., Wigley, T. M. L., Boyle, J. S., Doutriaux, C., Fiorino, M., Parker, D. E., and Taylor, K. E.: 1999, 'Uncertainties in observationally based estimates of temperature change in the free atmosphere', *J. Geoph. Res.* **104**, 6305–6333.

Spencer, R. W., and Christy, J. R.: 1992, 'Precision and radiosonde validation of satellite gridpoint temperature anomalies. Part II: A tropospheric retrieval and trends during 1979-90', *J. Climate* **5**, 858–866.

Spencer, R. W., and Christy, J. R.: 1993, 'Precision lower stratospheric temperature monitoring with the MSU: technique, validation and results 1979-1991', *J. Climate* **6**, 1194–1204.

Wallace, J. M., Zhang, Y., and Bajuk, L.: 1996, 'Interpretation of interdecadal trends in Northern Hemisphere surface air temperature', *J. Climate* **9**, 249–259.

Wentz, F. J., and Schabel, M.: 1998, 'Effects of orbital decay on satellite-derived lower-tropospheric temperature trends', *Nature* **394**, 661–664.

LONG TERM CLIMATE RECORDS FROM POLAR ICE

BERHARD STAUFFER
University of Bern, Switzerland

Received: 20 October 1999; Accepted in final form: 29 December 1999

Abstract. One of the great challenges in climate research is to investigate the principal mechanisms that control global climatic changes and an effective way to learn more about it, is the reconstruction of past climate changes. The most important sources of information about such changes and the associated composition of the atmosphere are the two large ice caps of Greenland and Antarctica. Analysis of ice cores is the most powerful means we have to determine how climate has changed over the last few climatic cycles, and to relate this to changes in atmospheric composition, in particular to concentrations of the principal greenhouse gases – CO_2, CH_4 and N_2O (carbon dioxide, methane, and nitrous oxide).

Transitions from cold ice age climates to warmer interstadials have always been accompanied by an increase of the atmospheric concentration of the three principal greenhouse gases. This increase has been, at least for CO_2, vital for the ending of glacial epochs. A highly simplified course of events for the past four transitions would then be as follows: first, changing orbital parameters initiated the end of the glacial epoch; second, an increase in greenhouse gases then amplified the weak orbital signal; third, in the second half of the transition, warming was further amplified by decreasing albedo, caused by melting of the large ice sheets in the Northern Hemisphere going parallel with a change of the ocean circulation.

The isotopic records of Greenland ice cores show evidence for fast and drastic climatic changes during the last glacial epoch. Possible causes and mechanisms of such changes and their significance as global climatic events are discussed here. Ice core results also enable the reaction of the environment to past global changes to be investigated.

It will also be discussed how reliable stable isotope records are as a local temperature proxy and how representative paleoclimatic results from Greenland and Antarctica are in relation to global climate.

1. Introduction

The snow deposited in the central parts of polar ice sheets is buried by subsequent snow falls and compacted under the pressure of the overlaying snow. It is densified to ice by a sintering process and penetrates with time deeper and deeper into the ice sheet as it also moves toward the coast. Figure 1 shows schematically a cross section through the Greenland ice sheet. It is in equilibrium concerning volume and shape if the annual loss of ice, melted at the coast or breaking off as ice bergs, is compensated by the annual snow deposition in the interior of the ice sheet. It is obvious from Figure 1 that the age of the ice is increasing with depth. Therefore, core drilling into an ice sheet allows the collection of a well stratified and complete set of precipitation samples from the past few tens to several hundred thousand

Space Science Reviews **94:** 321–336, 2000.

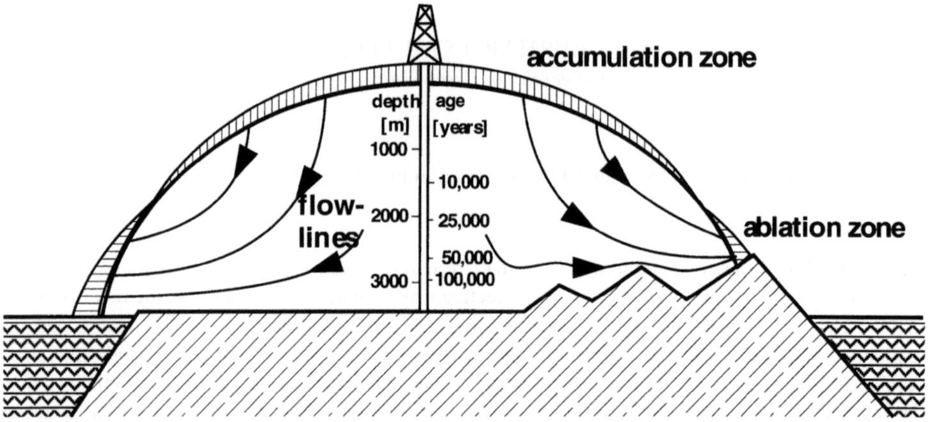

Figure 1. Schematic vertical cross section of an ice sheet resting on bedrock. Snow deposited at the surface of the accumulation zone will follow the flow lines and melt or break off as ice bergs in the ablation zone. In the central part of the ice sheet the age of the ice is getting continuously older with increasing depth. The annual layer thickness is getting smaller due to the plastic flow of ice. The given depth and ages are valid for a cross section through Summit in Greenland.

years. The analyses of these samples provide a variety of important proxies for climate and environmental conditions at the time of snow deposition or ice formation respectively.

– The stable isotope ratio in sea water is about 0.002 for $^{18}O/^{16}O$ and about 0.00015 for $D/^{1}H$. Deviations from these mean values given as δ-values (relative deviation in per mille) are a proxy for the local mean annual surface temperature. From annual snow samples collected at locations with different annual surface temperature in Greenland the following empirical relation was derived (Johnsen *et al.*, 1992):

$$\delta^{18}O = 0.67 \times \bar{T}_s[^\circ C] - 13.7‰ \quad (1)$$

However, this relation is not necessarily the same for past temperature changes. Indeed, borehole temperature data show that it was far from this relation in the past as will be discussed in section 4. For a temperature record it is not relevant whether it is based on $\delta^{18}O$ or δD values.

– Volcanic eruptions emit large amounts of dust, vapour and gases into the atmosphere. Large explosive eruptions around the world are characterised in ice cores mainly by an elevated sulphate concentration (Hammer, 1980). It is possible to look in such selected layers for volcanic fragments which allow an identification of the volcano. Until now no significant global cooling lasting longer than a few years could be observed after volcanic eruptions (Hammer *et al.*, 1980).

– The concentration of various chemical compounds depends on the activity and distribution of corresponding sources, the transport mechanisms to the polar ice sheets and the transfer from the atmosphere into the snow cover. Measurements allow the determination of pre industrial atmospheric concentrations.

Longer records show that several compounds showed systematic variations before human impacts. Ice from the last glacial epoch is enriched in most aerosols (Petit *et al.*, 1981). It is also possible to get information about the oxidation capacity of the atmosphere in the past by measuring the concentration of reactive species like hydrogen peroxide and formaldehyde (Staffelbach *et al.*, 1991). In Greenland, changes of ammonium concentration and of the concentrations of carboxylic acids give information about changes of the bioactivity of the North American continent (Fuhrer and Legrand, 1997).

- In the central parts of large polar ice sheets the surface temperature does not reach the melting point, not even during summer days. Ice is formed at such places by a dry sintering process. Atmospheric air, filling the pore space between firn grains, is more and more driven out with the densification of the firn. A final amount, about 10% by volume, is finally enclosed in bubbles. This air has the composition of the atmosphere at the time of enclosure. It has to be kept in mind that the enclosed air has a younger age than the surrounding ice (Schwander *et al.*, 1993). By the analysis of air extracted from well dated ice samples the atmospheric composition in the past can be determined (Raynaud *et al.*, 1993).
- Radioisotopes like ^{10}Be and ^{36}Cl are produced by cosmic radiation in the atmosphere and deposited to the earth and, therefore, also to the snow surface. The concentration of such isotopes in ice can be used to reconstruct a record of past cosmic radiation as discussed by J. Beer in this volume.

Information from all these proxies is only valuable if the age of the ice (time since snow was deposited) and the age of the enclosed air can be determined. An age scale can be obtained by:

- Counting annual layers: isotopic ratios and the concentration of several chemical compounds show seasonal variations, which can be identified by measurements and counted if the annual accumulation is sufficiently high. For the ice cores from central Greenland it was possible to count annual layers down to about 1750 m, corresponding to 14500 years BP (Johnsen *et al.*, 1992).
- Comparison with reference horizons like volcanic eruptions, variations in gas and dust concentrations (Hammer, 1989; Blunier *et al.*, 1998).
- Calculations based on ice flow properties and based on estimated accumulation rates. The accumulation rate is expected to depend on mean annual air temperature (Johnsen and Dansgaard, 1992).

2. The Last Four Climatic Cycles

Late Quaternary climate (about the last 1 Million years) is characterised by changes between glacial epochs and warm interstadials. Glacial epochs are characterised in Antarctica by much colder temperatures, reduced precipitation rates and enhanced storminess. Interstadials are considered to be similar to our present climate. The

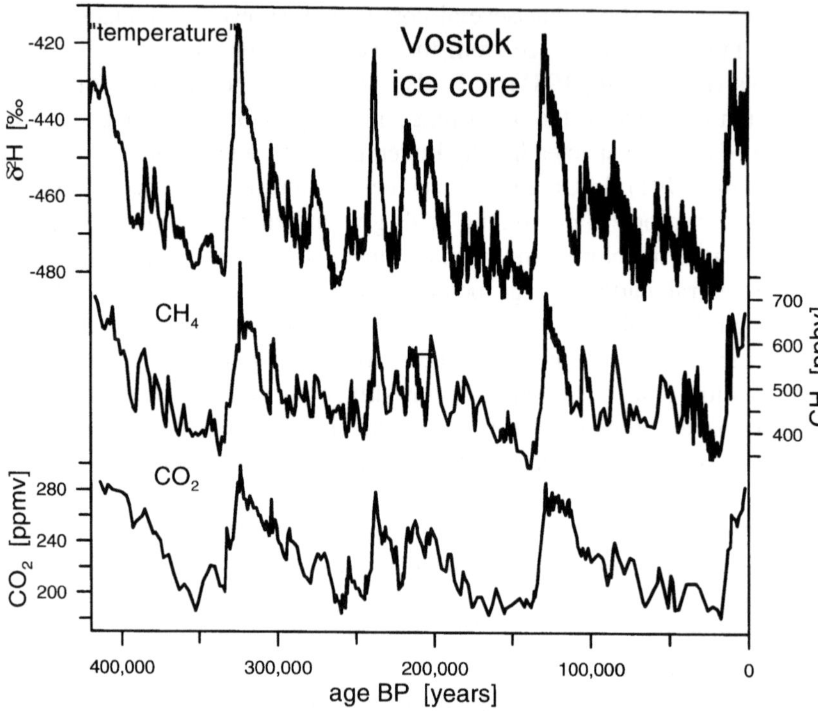

Figure 2. Deuterium, carbon dioxide and methane record of the Vostok (East Antarctica) ice core on a linear age scale according to Petit *et al.* (1999). The age resolution for the CO_2 measurements are between 1,500 and 6,000 years, for CH_4 measurements between about 100 and 4,500 years. The accuracy is better than 1‰ for Deuterium, about 3 ppmv for CO_2 and 20 ppbv for CH_4.

δD record in Figure 2 representing a temperature proxy for Vostok (Antarctica) (Petit *et al.*, 1999) shows that the main transition between glacial to warm epochs started at about 335,000 years, 245,000 years, 135,000 years and 18,000 years BP. From this one would infer a roughly 100,000 year periodicity. Time-series analysis confirm this periodicity and there is general agreement that it is caused by changing orbital parameters (in this case especially eccentricity) of the Earth. However, the nature of the interaction between the orbital parameters, which cause only a minor variation of the total solar irradiance and its latitudinal distribution, are still unclear.

Figure 2 shows as well the records of the two most important greenhouse gases CO_2 and CH_4 (Petit *et al.*, 1999). The records show that the atmospheric concentration of both gases was never as high as today during the past 420,000 years (present values: 370 ppmv for CO_2 and 1,700 ppbv for CH_4) and that there is a strong correlation between the temperature represented by the δD record and the two greenhouse gas records. The interplay of temperature and the atmospheric greenhouse gas concentration is of course of special interest.

The atmospheric methane concentration was varying between 350 and 550 ppbv during glacial epochs and increased up to 700 ppbv during interstadials. Methane

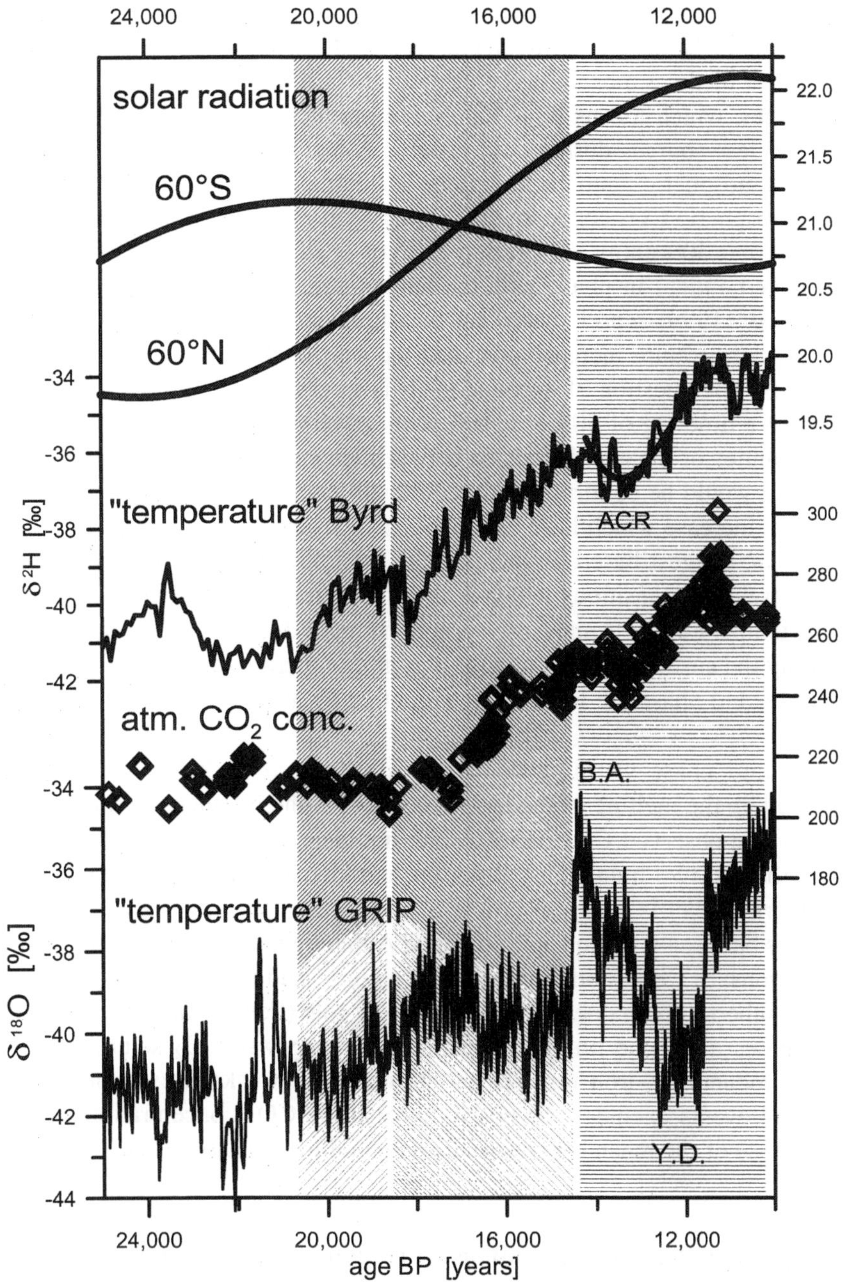

Figure 3. Solar radiation for the summer month (JJA for Northern Hemisphere, DJF for Southern Hemisphere), Antarctic temperature, atmospheric CO_2 concentration measured on Byrd ice core and Greenland temperature (Johnsen *et al.*, 1992) on a common time scale to illustrate a possible mechanism for the transition from the last glacial epoch to the Holocene (Petit *et al.*, 1999). The time intervals with different hatchings are explained in the text.(ACR: Antarctic Cold Reversal; B.A.: Bølling/Allerød; Y.D.: Younger Dryas)

was produced in pre industrial times mainly by wetlands (Chappellaz *et al.*, 1993). The main sink is the reaction with OH radicals in the atmosphere. It is assumed that the variations of the atmospheric concentrations are mainly caused by variations of the sources, which means larger and more productive wetlands during warmer climate and less sources during cold epochs. A feedback on climate by radiative forcing of atmospheric methane variations is small.

The atmospheric CO_2 concentration increases at each transition from about 180 ppmv to 280 – 300 ppmv. According to Petit *et al.* (1999) the CO_2 increase is (within uncertainties) in phase with the temperature increase. On the other hand Fischer *et al.* (1999) report, based on measurements on the same core that the start of the temperature increase preceded the start of the CO_2 increase by 500 – 1, 000 years. The question of lags and leads is a very important one, but the identification of a time lag of 500 – 1, 000 years is very difficult taking into account the uncertainties of the age difference between ice and enclosed air bubbles. Even if a time lag would be confirmed CO_2 could still be an important amplifier for the temperature increase which lasts itself several thousand years.

Petit *et al.* suggest based on their results the following simplified scenario for glacial-interglacial transitions which is shown schematically in Figure 3:

- The temperature in Antarctica starts to increase due to a high solar irradiance during the austral summer months at high Southern latitudes.
- In a second phase the warming is amplified by an increasing CO_2 concentration.
- In a third phase warming due to an albedo feed back becomes dominant caused by the melting of the large ice sheets in the Northern Hemisphere.

Causes for the observed CO_2 increases are not clear yet. From the Vostok results covering the last two transitions, Broecker and Henderson (1998) concluded that the Southern Ocean is likely to be the main player regulating atmospheric CO_2. Similarities between CO_2 concentration and Antarctic temperature for the previous two transitions, as well as other parts of the records, add further support to the idea that the Southern Ocean does indeed have a key role (Petit *et al.*, 1999). However, all hypotheses about the linking mechanisms (changes in CO_2 solubility, phytoplankton productivity, iron fertilization etc.) all have weaknesses at present. The sequence of events excludes the possibility that the CO_2 concentration increased just as a cause of the mean ocean surface temperature or a rise in sea level.

3. Fast Climatic Variations During the Past Glacial Epoch

Figure 4 shows part of the $\delta^{18}O$ record from the GRIP ice core drilled at Summit in the centre of the Greenland ice sheet. Only the last 50,000 years are presented to allow more details to be shown (Dansgaard *et al.*, 1993).

Clearly visible is the transition from the last glaciation to the Holocene. The difference of the $\delta^{18}O$ value of 8‰ would correspond to a local temperature in-

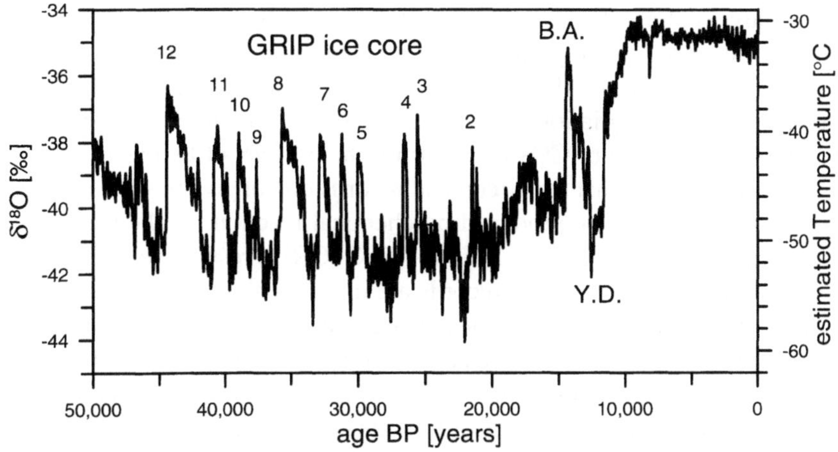

Figure 4. Part of the continuous $\delta^{18}O$ record from the GRIP ice core along a linear time scale, covering the last 50,000 years. Glacial interstadials, so called Dansgaard/Oeschger events are numbered 12 – 2. B.A. is event 1, known in Europe as the Bølling-Allerød interstadial. The cold period Younger Dryas is marked by Y.D.. (Dansgaard *et al.*, 1993)

crease of 12°C according to the relation found for the spatial dependence between $\delta^{18}O$ and mean annual surface temperature given in the introduction. However, a different relation has to be applied for the transition as will be discussed in the next chapter. So the local temperature increase is about 24°C according to our present knowledge (Johnsen *et al.*, 1995; Cuffey *et al.*, 1995). While temperature increases slowly and continuously in Antarctica at the end of the last glaciation, according to the Vostok record, the increase in Greenland shows large fluctuations. Temperature also started to increase slowly around 20,000 years BP, but a first clear, abrupt and fast temperature increase is observed at $14,450 \pm 200$ years BP. It is especially astonishing that the local temperature increased in two decades by more than 10°C. The following warm period, known in Europe as the Bølling/Allerød, gave way after less than two millennia to a cold period, known as Younger Dryas in Europe. The Younger Dryas period was terminated again by an abrupt temperature increase at 11,550±70 years BP the final transition from the last glacial epoch to the Holocene in Greenland (Johnsen *et al.*, 1992; Severinghaus *et al.*, 1998). The two fast temperature increases and the Younger Dryas period as a whole are an important key for understanding the mechanisms responsible for large global climatic changes.

Fast and drastic temperature variations are also seen in the GRIP record during the last glacial epoch. This fast variations, called Dansgaard/Oeschger events, have been a confirmation of the same variations observed already in the Dye-3 (South Greenland) ice core (Dansgaard *et al.*, 1984). All Dansgaard/Oeschger events start with an abrupt temperature increase of several degrees in a few decades, followed by a slow temperature decrease to low glacial temperatures again.

CO_2 analyses along the Dye-3 and the GRIP ice core show large concentration variations of the order of 50 ppmv parallel to Dansgaard/Oeschger events (Stauffer *et al.*, 1984; Anklin *et al.*, 1997). However, measurements on Antarctic ice cores did not confirm these variations (Neftel *et al.*, 1988), which have to be global due to the short mixing time of CO_2 in the atmosphere of about two years. New measurements show indeed that the Greenland records do not give the true atmospheric record because CO_2 can be produced by chemical reactions in the ice despite the low temperatures of $-20°C$ (Dye-3) and $-32°C$ (GRIP) respectively (Tschumi and Stauffer, in press). Antarctic CO_2 records are more reliable due to the lower impurity concentrations in the ice. Recently it became possible to synchronise the age scale of Greenland and Antarctic ice core records to about 200 years uncertainty as will be discussed in section 6. Therefore, it is now possible to compare Antarctic CO_2 records with Greenland δ^{18} records. This comparison shows definitively that the large concentration variations with an amplitude of 50 ppmv are an artefact. On the other hand they suggest that there have been variations going parallel with the Dansgaard/Oeschger events which have a long duration but with an amplitude of only $10 - 20$ ppmv (Stauffer *et al.*, 1998).

Most of the Dansgaard/Oeschger events recorded in Greenland ice cores have parallels in the record of ice-rated debris found in deep sea sediments from the North Atlantic (Bond *et al.*, 1993; Bond and Lotti, 1995). This correlation provides some clues concerning possible mechanisms which cause the fast variations observed in Greenland but not in Antarctica (or at least only strongly attenuated there). Based on observations and model calculations there is general agreement today that the Dansgaard/Oeschger events are correlated with variations of the deep water formation in the North/Atlantic which controls to a great deal the deep ocean circulation, the so called conveyor belt (Broecker, 1994). The deep water formation is caused at present by very cold and saline and, therefore, dense surface water. An input of fresh water caused, for example, by large ice masses calving into the North Atlantic from the large glacial ice sheets covering North America and North Europe, could decrease or stop the deep water formation. This would have lead to a cooling in the North Atlantic region. Resumption of deep water formation after a certain time would cause the abrupt temperature increases observed in Greenland ice cores (Paillard, 1994). If this mechanism can be confirmed, there still remains the question about the cause of large ice berg discharges at irregular intervals. The break off could be caused by pure ice dynamic properties (a growing ice sheet gets unstable when its margin reaches the continental borders), or there could also be an influence of extraterrestrial quantities like Earth orbital parameters or variations of the solar luminosity. It is obvious that these fast variations are also an important key for the understanding of mechanisms governing global climate.

4. How Reliable are Stable Isotope Records as Proxy for the Local Temperature?

The $\delta^{18}O$ and δD values of precipitation, based on simple physical laws, depend on the condensation temperature (Dansgaard, 1964). However, the good correlation between the mean $\delta^{18}O$ value of an annual precipitation sample and the mean annual temperature is an empirical finding based on spatial temperature distributions and corresponding variations of the stable isotope ratios (Dansgaard *et al.*, 1973). The question remains, as in the case of other proxies like tree ring width, pollen and foraminifera distributions, whether this spatial dependence ("modern analogue method") is the same for the climatic variations observed in the past. For ice cores it was recently possible to test this hypothesis by two new tools, inverse modelling based on bore hole temperature measurements and based on the thermal diffusion of gases in the permeable firn layer.

The firn temperature at 10 m depth corresponds, in regions where surface melting can be neglected, to the mean annual surface temperature (Paterson, 1994). Changes in the mean annual air temperature propagate into the ice by diffusive heat flow and by ice flow. The temperature profile through an ice sheet thus provides a record of past surface temperature, modified by heat diffusion, ice flow and a small amount of heat generated by ice deformation, and in deeper strata by the geothermal heat flux (Dahl-Jensen *et al.*, 1998). The reconstruction of the surface temperature based on the temperature profile is difficult. The record of high frequency events is lost entirely by heat diffusion especially in deeper layers. The reconstruction is not unambiguous, several temperature records can lead to the same temperature profile. However, temperature records e.g. based on stable isotope profiles can be tested by comparing with the temperature profile. Such tests have been made independently for the GISP-2 and the GRIP bore hole in Central Greenland. Both tests showed that the temperature shift between the glacial epoch and the Holocene had to be larger as predicted by the $\delta^{18}O$ record assuming a relationship of 0.67‰/°C (Johnsen *et al.*, 1995; Cuffey *et al.*, 1995). The best fit is obtained if for this temperature variation a relation of 0.33‰/°C is applied. This leads to a temperature increase of 24°C from the last glacial maximum to the Holocene as shown on the right scale in Figure 4. However, this relation is not necessarily valid for other temperature variations like the Dansgaard/Oeschger events.

For fast temperature variations there is another tool to test temperature differences, the thermal diffusion. In a large part of the permeable firn layer, air is only transported by molecular diffusion. This causes a slight enrichment of heavier molecules at the bottom of the firn layer due to gravitation (Schwander *et al.*, 1988). If there is a temperature difference between the bottom and the top of the firn layer with diffusive mixing, there is in addition a separation of components with different masses by thermal diffusion (Severinghaus *et al.*, 1998). This process slightly alters the isotopic as well as the elemental composition of the air, which gets enclosed in bubbles. The measurement of records of the $^{15}N/^{14}N$ ratio in N_2 along an ice core section representing a fast temperature increase, allows the temperature difference to be determined. Severinghaus *et al.* determined the temperature increase at the end of the Younger Dryas period to have been 15°C. Lang *et al.* (1999) determined the temperature difference at the onset of Dansgaard/Oeschger event nr. 19 (about 70,000 years BP) to be 16°C. The comparison of these temperature increases with the corresponding increases of stable isotope δ-values allows the relations between temperature changes and e.g. $\delta^{18}O$ values to be re-assessed. For the two measured events 0.3‰/°C have been obtained for the Younger Dryas - Holocene transition and 0.42‰/°C for the onset of Dansgaard/Oeschger event 19. It is not evident that all Dansgaard/Oeschger events show the same relation. The reason for the differences in these relations seems to be changes in the seasonal precipitation distribution. In the cold parts of the last glacial epoch less winter snow was deposited.

5. How Representative are Polar Ice Core Results?

In the previous chapters some interesting and partly spectacular results from polar ice cores have been discussed. However, how representative are these results? - The Dansgaard/Oeschger events are only observed in Greenland but not in Antarctic ice cores. Therefore, they cannot be global events in the sense of synchronous temperature changes in all parts of the earth. On the other hand they would not be very interesting if they would just characterise a local phenomenon. For several events like the Younger Dryas period similar paleo records have been found in Europe and other parts of the world. However, similar records can be a pitfall as shall be shown later.

A very convincing argument that Dansgaard/Oeschger events are of global significance comes from methane records measured on ice cores. The GRIP methane record shows variations during the last glacial epoch which go parallel with Dansgaard/Oeschger events (Chappellaz *et al.*, 1993). Methane variations are caused, as mentioned already, mainly by variations of the methane sources which have been wetlands in pre industrial times. Wetlands have not existed in the neighbourhood of Greenland during the last glaciation. If the extension and activity of methane sources far away from Greenland changed synchronous with Dansgaard/Oeschger events, it is evident that it is a phenomenon which is not restricted to Greenland and not to the North Atlantic region either. The same methane variations are also observed in Antarctic ice core records despite the fact, that Dansgaard/Oeschger events are not (or only faintly) observed in the stable isotope records. Almost synchronous variations are expected due to the short mixing time of methane in the atmosphere. However, small differences of the atmospheric methane concentration between high Northern and high Southern latitudes are caused by an asymmetric distribution of the methane sources. The sources are larger in the Northern

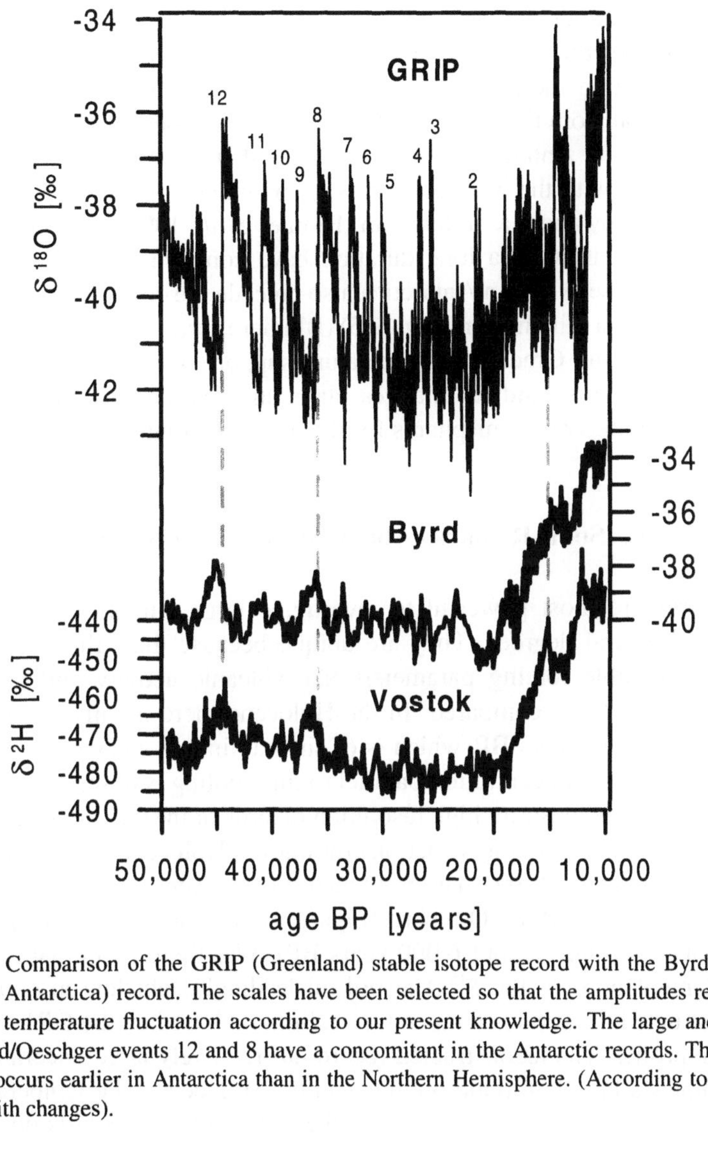

Figure 5. Comparison of the GRIP (Greenland) stable isotope record with the Byrd and the Vostok (both Antarctica) record. The scales have been selected so that the amplitudes represent about the same temperature fluctuation according to our present knowledge. The large and long lasting Dansgaard/Oeschger events 12 and 8 have a concomitant in the Antarctic records. The temperature increase occurs earlier in Antarctica than in the Northern Hemisphere. (According to Blunier *et al.* (1998) with changes).

hemisphere causing at present a concentration difference of about 140 ppbv. The concentration differences can be reconstructed for past climatic epochs. Ice core records from Greenland and Antarctica show concentration differences between North and South of −3 ppbv to +50 ppbv for the last 50,000 years (Chappellaz *et al.*, 1997; Dällenbach *et al.*, submitted). This record allows a rough estimate about the latitudinal distribution of methane sources in the past. This is a first step to reconstruct not only climate changes based on ice core records but also reaction of the biosphere to such changes. Absolute time scales of polar ice cores are not accurate enough to investigate time lags between the two hemispheres by a millennium or less. The parallel methane variations allow the age scales of Greenland and Antarctic ice core records to be synchronised. Best suited for this purpose are fast methane concentration increases at the beginning of Dansgaard/Oeschger events and at the transition from the last glacial epoch to the Holocene. The synchronisation of Greenland and Antarctic ice cores showed that the Antarctic cold reversal is not synchronous with the Younger Dryas, as previously assumed, but preceded the Younger Dryas by about a millennium (Blunier *et al.*, 1997). The largest Dansgaard/Oeschger events with a long duration have concomitants in Antarctic ice cores. However, these events in Antarctica also precede the corresponding events in Greenland by about a millennium (Figure 5) (Blunier *et al.*, 1998). This asynchrony between Antarctic and Greenland temperature supports the idea that the thermohaline ocean circulation and its changes plays an important role in the climate coupling between the two hemispheres for this kind of variations.

6. Some Remarks About Shorter Time Scales

Polar ice cores are most powerful to investigate large climatic variations with amplitudes of several degrees. They are unique because they allow temperature records with possible forcing parameters like volcanic activity and greenhouse gas concentrations to be compared. In the Holocene there is only one event, the cooling around 8,250 years BP, which is of such a kind. A clear minimum of the atmospheric methane concentration parallel to this cooling is observed. Recently it became possible to reconstruct high resolution records of the CH_4 and CO_2 concentrations which show surprising variations during the Holocene (Chappellaz *et al.*, 1993; Indermühle *et al.*, 1999). The $\delta^{18}O$ record is less spectacular. It is especially astonishing that the GRIP $\delta^{18}O$ record (Figure 4) does not show any clear signal of the climatic optimum (about 6,000 years BP). Was there no climatic optimum in central Greenland or is the $\delta^{18}O$ record misleading? Dahl-Jensen *et al.* (1998) have reconstructed the surface temperature at Summit based on the GRIP bore hole temperature profile. The reconstructed record has relatively large uncertainties but it clearly shows a broad climatic optimum in the Holocene. The temperature was 2.5°C warmer than today. The reason that the stable isotope records do not show this large temperature deviation is most probably again a change in the seasonal distribution of precipitation. It is assumed that there was a higher contribution of winter snow, and that the low δ-values of the winter snow compensated the increased mean annual surface temperature effect.

There is ample evidence that the stable isotope records are reliable proxies for the temperature variations caused by the changes between glacial epochs and interstadials and by variations like the Dansgaard/Oeschger events or the 8,200 year BP cooling (Leuenberger *et al.*, 1999). Such evidence is missing at present for more detailed variations. This does not imply that more detailed fluctuations do not provide important climatic information but there is no guarantee for it. If remarkable fluctuations are found in two ice cores or in an ice core and another paleoclimate record, this can still be accidentally. A third and fourth record will have to show if the signal is consistent. Even if a signal in stable isotope records turns out to be consistent, there still remains the question whether it is a temperature signal or something else.

Fischer *et al.* (1996) investigated various ice core records from Greenland and Canada covering the last 3,000 years. They used normalised records and applied empirical orthogonal functions to separate local noise from a common signal. For the last few centuries they compared the signal also with measurements from surrounding meteorological stations. About 50% of the variance of single records is local noise, but about 50% seems to represent a "temperature" signal for the last two centuries. Going further back in time Fisher *et al.* observe that the records from Canada and North Greenland show a "little ice age" but in records from Central and South Greenland a clear signal is missing. Fischer *et al.* (1998) stacked three ice cores from a traverse in North Greenland and obtained with the stacked record, covering the last 500 years, a clear signal of a "little ice age" with temperature minima in the late 17th century and a stronger one in the early 19th century. However, if this cooling is a large scale phenomenon, there is no reason that Central and South Greenland should be excluded. Johnsen *et al.* (1997) investigated a stacked record from Summit in Central Greenland. There is a small minimum at around 1700 AD and 1900 AD but a much stronger one in the late 14th century. Dahl-Jensen *et al.* (1998), as mentioned already, have reconstructed a temperature record based on the GRIP bore hole temperature profile at Summit. It is for methodical reasons a very smoothed record. It shows a broad significant minimum at 1600 AD and one at about 1850 AD. Fischer *et al.* have an explanation for the different behaviour of δ-records in Central and North Greenland. They suspect that the δd-records in South and Central Greenland are masked because relatively warm cyclones bring most of the winter precipitation to the Southern half of the Greenland ice sheet. In this case we would expect that the stacked record from North Greenland correlates well with the temperature record by Dahl-Jensen *et al.*. The general trend is indeed similar, however, the first minimum is about one century later in the North Greenland record and the second minimum about 25 years earlier.

The goal of this discussion was not to give a final answer about the temperature history during the past thousand years in Greenland but to demonstrate that there is certainly information about short time climatic variations in polar ice cores. However, an extraction of such information is laborious and difficult. It should be a warning about overinterpretation of records from one single ice core.

7. Conclusions

Polar ice cores are unique as a source of information about large scale climatic variations and about variations of the atmospheric concentration of greenhouse gases.

This allows the influence of greenhouse gas concentrations on past global climatic changes to be investigated. In addition polar ice cores provide information about past volcanism and about variations of cosmic radiation. There is hope that the measurement of additional chemical parameters will also allow information about changes of the global environment in relation to climatic changes to be obtained. The most powerful proxy for short term temperature changes are the stable isotope records. However, the quantitative relation between δ-values and mean annual surface temperature has to be used with caution. The stable isotope records show a relatively large local noise which makes it difficult to obtain information about temperature variations in the order of 1°C or less.

References

Anklin, M., Schwander, J., Stauffer, B., Tschumi, J., Fuchs, A., Barnola, J. M., and Raynaud, D.: 1997, 'CO$_2$ record between 40 and 8 kyr B.P. from the Greenland Ice Core Project ice core', *Journal of Geophysical Research* **102** (C12), 26539–26546.

Blunier, T., Schwander, J., Stauffer, B., Stocker, T., Dällenbach, A., Indermühle, J., Tschumi, J., Chappellaz, D., Raynaud, D., and Barnola, J.-M.: 1997, 'Timing of the Antarctic Cold Reversal and the atmospheric CO$_2$ increase with respect to the Younger Dryas event', *Geophysical Research Letters* **24** (21), 2683–2686.

Blunier, T., Chappellaz, J., Schwander, J., Dällenbach, A., Stauffer, B., Stocker, T., Raynaud, D., Jouzel, J., Clausen, H. B., Hammer, C. U., and Johnsen, S. J.: 1998, 'Asynchrony of Antarctic and Greenland climate change during the last glacial period', *Nature* **394**, 739–743.

Bond, G. C., and Lotti, R.: 1995, 'Iceberg discharges into the North Atlantic on millennial time scales during the last deglaciation', *Science* **267**, 1005–1010.

Bond, G., Broecker, W., Johnsen, S., McManus, J., Labeyrie, L., Jouzel, J., and Bonani, G.: 1993, 'Correlations between climate records from North Atlantic sediments and Greenland ice', *Nature* **365**, 143–147.

Broecker, W. S.: 1994, 'An unstable superconveyor', *Nature* **367**, 414–415.

Broecker, W. S. and Henderson, G. M.: 1998, 'The sequence of events surrounding Termination II and their implications for the cause of glacial-interglacial CO$_2$ changes', *Paleoceanography* **13**, 35–364.

Chappellaz, J., Blunier, T., Raynaud, D., Barnola, J. M., Schwander, J., and Stauffer, B.: 1993, 'Synchronous changes in atmospheric CH$_4$ and Greenland climate between 40 and 8 kyr BP', *Nature* **366**, 443–445.

Chappellaz, J., Blunier, T., Kints, S., Dällenbach, A., Barnola, J.-M., Schwander, J., Raynaud, D., and Stauffer, B.: 1997, 'Changes in the atmospheric CH$_4$ gradient between Greenland and Antarctica during the Holocene', *Journal of Geophysical Research* **102** (D13), 15987–15999.

Cuffey, K. M., Clow, G. D., Alley, R. B., Stuiver, M., Waddington, E. D., and Saltus, R. W.: 1995, 'Large arctic temperature change at the Wisconsin-Holocene glacial transition', *Science* **270**, 455–458.

Dahl-Jensen, D., Mosegaard, K., Gundestrup, N., Clow, G. D., Johnsen, S. J., Hansen, A. W., and Balling, N.: 1998, 'Past Temperatures Directly from the Greenland Ice Sheet', *Science* **282**, 268–271.

Dällenbach, A., Blunier, T., Flückiger, J., Stauffer, B., Chappellaz, J., and Raynaud, D.: 2000, 'Changes in the atmospheric CH$_4$ gradient between Greenland and Antarctica during the last glacial and the transition to the Holocene', *Geophysical Research Letters*, submitted.

Dansgaard, W.: 1964, 'Stable isotopes in precipitation', *Tellus* **16**, 436–468.

Dansgaard, W., Johnsen, S. J., Clausen, H. B., and Gundestrup, N.: 1973, 'Stable isotope glaciology', *Meddelelser om Grønland* **197** (2), 1–53.

Dansgaard, W., Johnsen, S. J., Clausen, H. B., Dahl-Jensen, D., Gundestrup, N., Hammer, C. U., and Oeschger, H.: 1984, 'North Atlantic Climatic oscillations revealed by deep Greenland Ice Cores', In *Climatic Processes and Climate Sensitivity, Geophysical monograph 29, Maurice Ewing.*, Vol. **5** (eds. J. E. Hansen and T. Takahashi), pp. 288–298.

Dansgaard, W., Johnsen, S. J., Clausen, H. B., Dahl-Jensen, D., Gundestrup, N. S., Hammer, C. U., Hvidberg, C. S., Steffensen, J. P., Sveinbjörnsdottir, A. E., Jouzel, J., and Bond, G.: 1993, 'Evidence for general instability of past climate from a 250-kyr ice-core record', *Nature* 364, 218–220.

Fischer, H., Werner, M., Wagenbach, D., Schwager, M., Thorsteinnson, T., Wilhelms, F., Kipfstuhl, J., and Sommer, S.: 1998, 'Little ice age clearly recorded in northern Greenland ice cores', *Geophysical Research Letters* **25** (10), 1749–1752.

Fischer, H., Wahlen, M., Smith, J., Mastroianni, D., and Deck, B.: 1999, 'Ice core records of atmospheric CO_2 around the last three glacial terminations', *Science* **283**, 1712–1714.

Fisher, D. A., Koerner, R. M., Kuivinen, K., Clausen, H. B., Johnsen, S. J., Steffensen, J.-P., Gundestrup, N., and Hammer, C. U.: 1996, 'Inter-Comparison of Ice Core δ^{18}O and Precipitation Records from Sites in Canada and Greenland over the Last 3500 Years and over the Last Few Centuries in Detail Using EOF Techniques', In *Climatic Variations and Forcing Mechanisms of the Last 2,000 Years. NATO ASI Series I 41 Global Environmental Change*, Vol. **41** (eds. P. D. Jones, R. S. Bradley, and J. Jouzel), Springer-Verlag, pp. 297–328.

Fuhrer, K. and Legrand, M.: 1997, 'Continental biogenic species in the Greenland Ice Core Project ice core: Tracing back the biomass history of the North American continent', *Journal of Geophysical Research* **102** (C12), 26735–26746.

Hammer, C. U.: 1980, 'Acidity of polar ice cores in relation to absolute dating, past volcanism, and radio-echoes', *Journal of Glaciology* **25**, 359–372.

Hammer, C. U.: 1989, 'Dating by physical and chemical seasonal variations and reference horizons', In *The environmental record in glaciers and ice sheets*, Vol. Physical, Chemical, and Earth Sciences Research Report **8** (ed. H. Oeschger and C. C. Langway, Jr.), John Wiley & Sons, pp. 99–121.

Hammer, C. U., Clausen, H. B., and Dansgaard, W.: 1980, 'Greenland ice sheet evidence of post-glacial volcanism and its climatic impact', *Nature* **288**, 230–235.

Indermühle, A., Stocker, T. F., Joos, F., Fischer, H., Smith, H. J., Wahlen, M., Deck, B., Mastroianni, D., Tschumi, J., Blunier, T., Meyer, R., and Stauffer, B.: 1999, 'Holocene carbon-cycle dynamics based on CO_2 trapped in ice at Taylor Dome', Antarctica. *Nature* **398**, 121–126.

Johnsen, S. J., and Dansgaard, W.: 1992, 'On flow model dating of stable isotope records from Greenland ice cores', In *The Last Deglaciation: Absolute and Radiocarbon Chronologies. NATO ASI series*, Vol. **I 2** (eds. E. Bard and W. S. Broecker), Springer Verlag, pp. 13–24.

Johnsen, S. J., Clausen, H. B., Dansgaard, W., Fuhrer, K., Gundestrup, N., Hammer, C. U., Iversen, P., Jouzel, J., Stauffer, B., and Steffensen, J. P.: 1992, 'Irregular glacial interstadials recorded in a new Greenland ice core', *Nature* **359**, 311–313.

Johnsen, S., Dahl-Jensen, D., Dansgaard, W., and Gundestrup, N.: 1995, 'Greenland palaeotemperatures derived from GRIP bore hole temperature and ice core isotope profiles', *Tellus* **47B**, 624–629.

Johnsen, S. J., Clausen, H. B., Dansgaard, W., Gundestrup, N. S., Hammer, C. U., Andersen, U., Andersen, K. K., Hvidberg, C. S., Dahl-Jensen D., Steffensen, J. P., Shoji, H., Sveinbjörnsdóttir Á., White, J., Jouzel, J., and Fisher, D.: 1997, 'The δ^{18}O record along the Greenland Ice Core Project deep ice core and the problem of possible Eemian climatic instability', *Journal of Geophysical Research* **102** (C12), 26397–26410.

Lang, C., Leuenberger, M., Schwander, J. and Johnsen, S.: 1999, '16°C Rapid Temperature Variation in Central Greenland 70,000 Years Ago', *Science* **286**, 934–937.

Leuenberger, M., Lang, C., and Schwander, J.: 1999, 'Delta ^{15}N measurements as a calibration tool for the paleothermometer and gas-ice age differences: A case study for the 8200 B.P. event on GRIP ice', *Journal of Geophysical Research* **104**, 22,163–22,170.

Neftel, A., Oeschger, H., Staffelbach, T., and Stauffer, B.: 1988, 'CO_2 record in the Byrd ice core 50, 000 – 5, 000 years BP', *Nature* **331**, 609–611.

Paillard, D.: 1994, 'Role of the thermohaline circulation in the abrupt warming after Heinrich events', *Nature* **372**, 162–164.

Paterson, W. B. S.: 1994, *The physics of glaciers*, Pergamon press.

Petit, J.-R., Briat, M., and Royer, A.: 1981, 'Ice age aerosol content from East Antarctic ice core samples and past wind strength', *Nature* **293**, 391–394.

Petit, J. R., Jouzel, J., Raynaud, D., Barkov, N. I., Barnola, J.-M., Basile, I., Bender, M., Chappellaz, J., Davis, M., Delaygue, M., Delmotte, M., Kotlyakov, V. M., Legrand, M., Lipenkov, V. Y., Lorius, C., Pépin, L., Ritz, C., Saltzman, E., and Stievenard, M.: 1999, 'Climate and atmospheric history of the past 420,000 years from the Vostok ice core', Antarctica. *Nature* **399**, 429–436.

Raynaud, D., Jouzel, J., Barnola, J. M., Chappellaz, J., Delmas, R. J., and Lorius, C.: 1993, 'The ice record of greenhouse gases', *Science* **259**, 926–933.

Schwander, J., Stauffer, B., and Sigg, A.: 1988, 'Air mixing in firn and the age of the air at pore close-off', *Annals of Glaciology* **10**, 141–145.

Schwander, J., Barnola, J.-M., AndriĆ C., Leuenberger, M., Ludin, A., Raynaud, D., and Stauffer, B.: 1993, 'The age of the air in the firn and the ice at Summit Greenland', *J. Geophys. Res.* **98**, 2831–2838.

Severinghaus, J. P., Sowers, T., Brook, E. J., Alley, R. B., and Bender, M. L.: 1998, 'Timing of abrupt climate change at the end of the Younger Dryas interval from thermally fractionated gases in polar ice', *Nature* **391**, 141–146.

Staffelbach, T., Neftel, A., Stauffer, B., and Jacob, D.: 1991, 'A record of the atmospheric methane sink from formaldehyde in polar ice cores', *Nature* **349**, 603–605.

Stauffer, B., Hofer, H., Oeschger, H., Schwander, J., and Siegenthaler, U.: 1984, 'Atmospheric CO_2 concentration during the last glaciation', *Annals of Glaciology* **5**, 160–164.

Stauffer, B., Blunier, T., Dällenbach, A., Indermühle, A., Schwander, J., Stocker, T., Tschumi, J., Chappellaz, J., Raynaud, D., Hammer, C. U., and Clausenm H. B.: 1998, 'Atmospheric CO_2 concentration and millennial-scale climate change during the last glacial period', *Nature* **392**, 59–62.

Tschumi, J., and Stauffer, B.: 2000, 'Reconstructing the past atmospheric CO_2 concentration based on ice core analyses: open questions due to the in situ production of CO_2 in the ice', *Journal of Glaciology*, in press.

ANTHROPOGENIC AND NATURAL CAUSES OF TWENTIETH CENTURY TEMPERATURE CHANGE

P. A. STOTT[†] (pastott@meto.gov.uk), S. F. B. TETT[†], G. S. JONES[†], M. R. ALLEN[‡], W. J. INGRAM[†], J. F. B. MITCHELL[†]
[†] *Hadley Centre for Climate Prediction and Research, Met Office, UK Met Office, Bracknell, Berks, UK RG12 2SY*
[‡] *Rutherford Appleton Laboratory*

Received: 18 October 1999; Accepted in final form: 6 December 1999

Abstract. We analyse spatio-temporal patterns of near-surface temperature change to provide an attribution of twentieth century climate change. We apply an "optimal detection" methodology to seasonal and annual data averaged over a range of spatial and temporal scales. We find that solar effects may have contributed significantly to the warming in the first half of the century although this result is dependent on the reconstruction of total solar irradiance that is used. In the latter half of the century, we find that anthropogenic increases in greenhouses gases are largely responsible for the observed warming, balanced by some cooling due to anthropogenic sulphate aerosols, with no evidence for significant solar effects.

1. Introduction

Twentieth century temperature change has been far from uniform. In the global mean, near-surface temperatures rose sharply between 1910 and 1940, and then flattened off up until the late 1960s before rising rapidly again, such that 1998 was the warmest year in the instrumental record, almost 0.6K warmer than the 1961-1990 average. In addition the warming has not been spatially uniform, with land areas warming faster than ocean areas, and greatest recent warming in high latitudes. In this study we aim to understand what caused these changes by means of a sophisticated coupled general circulation model of the atmosphere and ocean that includes forcings due to both anthropogenic and natural causes. The model we use, HadCM2 (Johns et al., 1997), has a horizontal resolution of 2.5° in latitude by 3.75° in longitude with 19 atmospheric and 20 oceanic levels and uses prescribed adjustments to surface heat and water fluxes to the ocean. We account for the effects of natural internal variability by a 1700 year control run of HadCM2 in which external climate forcings, except for the diurnal and seasonal cycle, are kept constant.

We have considered four possible external forcings; changes in well-mixed greenhouse gases (**G**), changes in greenhouse gases plus changes in surface albedo due to anthropogenic sulphate aerosols (**GS**), changes in stratospheric volcanic aerosols (**Vol**) and changes in total solar irradiance. For solar changes, we used

Space Science Reviews **94:** 337–344, 2000.

two reconstructions of total solar irradiance, one based on proxy data until 1991 (Hoyt and Schatten, 1993) and extended to 1996 with satellite data (Willson, 1997) (**HS**), and an alternative reconstruction whose long term component is based on the amplitude rather than the length of the solar cycle (Lean et al., 1995), extended to 1997 (Lean, pers. comm.) (**LBB**). We do not include changes in spectral composition nor resultant changes in stratospheric ozone. Each ensemble contains four runs, where for the **G**, **GS** and **Vol** ensembles, each ensemble member was started from different initial conditions taken from the long control simulation. For the two solar ensembles, one member of each ensemble was started with initial conditions taken from the control and run with changing forcing from 1700 for **HS** and 1750 for **LBB**. The remaining three ensemble members were started by adding random perturbations to the atmospheric state of the first ensemble member in 1890.

We compare observed and model near-surface temperature data by means of standard "optimal detection" methodology (Hasselmann, 1993). "Optimal detection" is essentially generalised linear regression (Allen and Tett, 1999). We estimate the amplitudes, β_i, of the model-simulated signals, $\mathbf{x}_i$, in the observations, $\mathbf{y}$:

$$\mathbf{y} = \sum_i \beta_i \mathbf{x}_i + \mathbf{u}, \tag{1}$$

where $\mathbf{u}$ is the internal climatic variability. A signal is formally detected in our analysis if the 5-95% uncertainty range for the signal amplitude, β_i, is entirely positive.

In our case, rather than analyse trends (Hegerl et al., 1997), we investigate space-time patterns spanning the 50 year periods, 1906–56, 1916–66, 1926–76, 1936–86 and 1946–96. In the annual analysis, each 50-year space-time pattern consists of 5 decadal mean fields, where the modelled data are masked by the equivalent observational data mask and converted to anomalies against the 50-year period. Observed decadal means are calculated from a $5° \times 5°$ dataset of near-surface temperature anomalies (Parker et al., 1994) by averaging annual means where there were at least 8 out of 12 months in at least half the years. In the seasonal analysis, each 50-year space-time pattern consists of the 5 decadal mean fields for each of the four seasons, September, October, November; December, January, February; March April, May; and June, July, August.

All data are projected onto spherical harmonics retaining only those wave numbers representing spatial scales larger than 5000km, as a perfect model study (Stott and Tett, 1998) had shown that the model has no detectable skill at spatial scales smaller than this. In optimal detection, spatio-temporal patterns of low natural variability are weighted up. It is therefore important not to include spatio-temporal patterns that are poorly represented by the model, especially since the model underestimates climate variability at small spatial scales (less than 2000km) (Stott and Tett, 1998). We retain only the first ten patterns in both the annual and seasonal analyses. This truncation has been chosen by applying a consistency test (Allen and Tett, 1999) to the residuals of regression using all four signals, and taking the minimum truncation found in all five 50-year periods. The methodology is explained in more detail elsewhere (Stott et al., 1999), (Allen and Tett, 1999).

If different patterns resemble each other, or resemble linear combinations of each other, they are degenerate. In this case, regression results may become unreliable and very sensitive to noise in the data, even when all relevant patterns are included on an equal footing. We only consider decadal means and apart from forcing due to sulphate aerosols, all the forcings we consider have a similar annual mean geographical distribution. As the climate system's response will tend to smooth these patterns and add noise, and our analysis only considers the response's projection onto the climate system's favoured patterns of variability, it is not surprising that we find there are no more than two independent pieces of information available in each half-century, so in the annual analysis we consider combinations of up to two signals. With decadal seasonal means the degeneracy is lost and we may legitimately consider results from regressing up to all four signals at once.

2. Annual Analysis

For a signal combination to explain temperature change over the century, the residuals of regression should be consistent with internal climate variability as estimated from the control run (Allen and Tett, 1999). We find that no combination of natural signals or internal variability can explain the observed warming since 1946, even though the consistency test we use is a relatively weak test.

Greenhouse gases, sulphate aerosols and solar changes (both **HS** and **LBB**) are detected in at least one 50-year period and including the influence of sulphate aerosols in 1946-96 provides a better fit to observations than other explanations. Volcanoes are not detected in any of the five 50-year periods considered here, although volcanic effects are detected on annual scales (Stott et al., 1999) and are sometimes detected using decadal averages for 50 year periods with different starting years than used here (Stott et al., 1999). However, we aim to provide a necessary but sufficient explanation of recent climate change (Tett et al., 1999) and include only signals detected in at least one of the five 50-year periods considered and no others providing that this combination is consistent with the observations. Due to degeneracy, we can consider only two signals simultaneously. The best explanations of twentieth century temperature change, based on annual data, are combinations involving **GS** and solar changes, and changes in greenhouse gases and sulphate aerosols alone (denoted **GHG** and **SUL**). **GHG** is given by the **G** simulation and **SUL** is derived from the difference of the **GS** and **G** simulations.

The best-estimate pattern amplitudes, β, and their uncertainties for these combinations are shown in Fig 1. The **GHG&SUL** ellipse for 1946-96 is above both the $\beta_{\mathrm{GHG}} = 0$ and the $\beta_{\mathrm{SUL}} = 0$ axes, indicating that both greenhouse gases and sulphate aerosols are detected in this period. The **GHG&SUL** ellipse for 1906-56 lies below the $\beta_{\mathrm{GHG}} = \beta_{\mathrm{SUL}}$ line, indicating that we need to reduce the ratio of

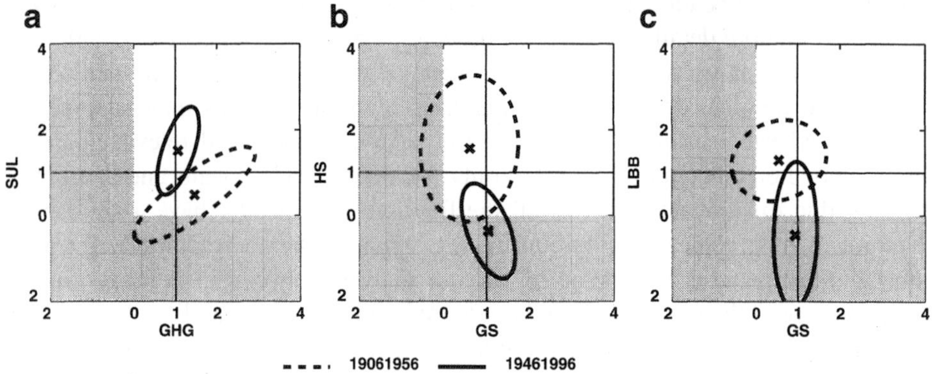

Figure 1. Ellipses containing 90% of the estimated joint distribution of signal amplitudes for a) **GHG&SUL**, b) **GS&HS** and c) **GS&LBB** signals.

the sulphate to greenhouse gas signal amplitudes below that specified in the **GS** experiment to remain consistent with the observations. Including solar changes in addition to the **GS** signal leads to model consistency with the observations whichever solar reconstruction is used (the point (1,1) now lies within the **GS&HS** and **GS&LBB** ellipses for 1906-56). Both the **GS&HS** and **GS&LBB** ellipses are above the x axis in 1906-56 indicating that solar changes are detected in combination with **GS** early in the century regardless of which reconstruction of total solar irradiance is used. However detection of solar changes is dependent on the relative amplitude of the anthropogenic signals. For HS, this relative amplitude must be approximately the same as in the GS experiment. If we reduce the amplitude of sulphates by 33% or more, we no longer detect the HS solar signal in 1906-56 whereas we would need to reduce the amplitude of sulphates by more than half to no longer detect the LBB solar signal in 1906-56.

Although solar changes contribute detectably to the early century warming, they make a relatively small contribution to the overall warming during the twentieth century (Fig 2). However, for consistency with previous work, we have used standard estimates of pattern amplitudes which are biased towards zero when there is noise in the estimated signals. This will affect weaker signals more than stronger signals and therefore might be expected to affect the solar signal more since it is weaker than the anthropogenic ones.

3. Seasonal Analysis

By including seasonal information, the signals become less degenerate and up to four signal amplitudes can be estimated simultaneously. Greenhouse gases, sulphate aerosols and solar changes are all detected during at least one of the 50-year periods. Using the HS reconstruction (Hoyt and Schatten, 1993), solar changes are

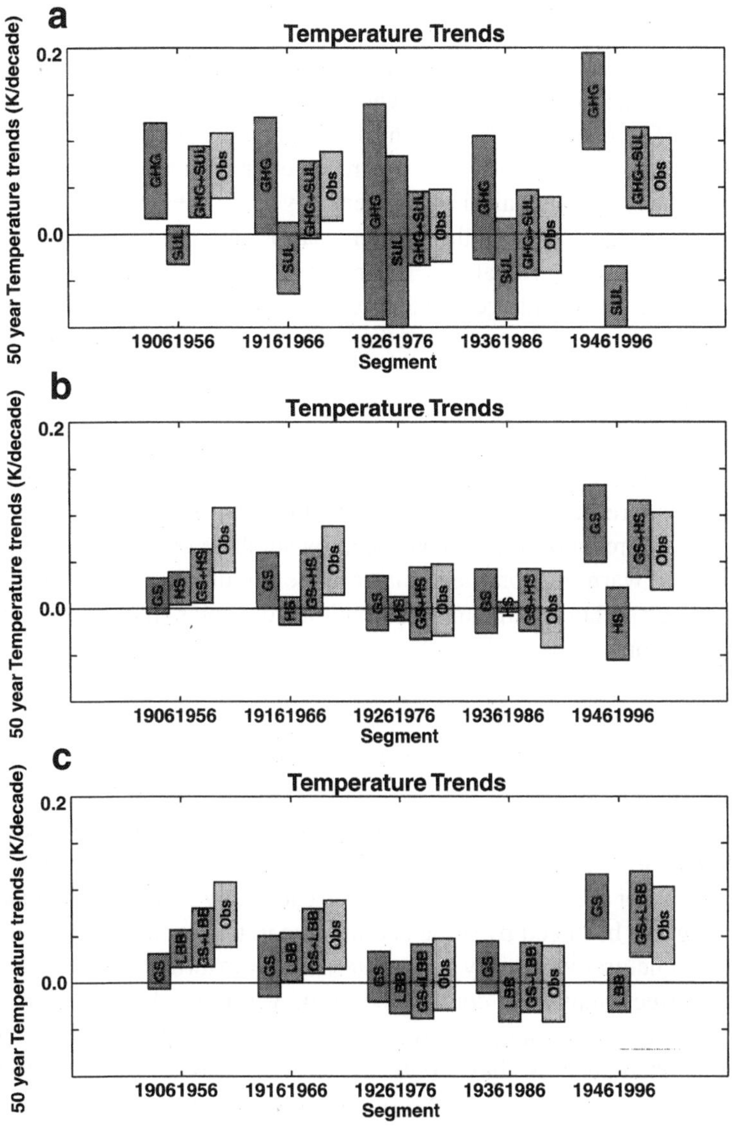

Figure 2. 50-year temperature trends and their uncertainties due to a) **GHG&SUL**, b) **GS&HS** and c) **GS&LBB** from annual analysis.

detected in the three-way regression with greenhouses and sulphate aerosols; in this case their amplitudes are not constrained to be the same as in the **GS** simulation as in the annual case. However, using the LBB reconstruction (Lean et al., 1995), solar changes are not detected in combination with **GHG** and **SUL**, and when combined with **GS**, the residual of the best fit of **GS** and **LBB** to the observations fails the consistency test (Allen and Tett, 1999). Although both solar reconstructions show a long-term increase in the early part of the twentieth century, then a dip of approximately two solar cycles followed by another increase, the LBB reconstruction shows the dip in the 1960s whereas the HS reconstruction shows the dip in the 1950s. The observed decadal mean temperatures show a peak in the 1940s followed by a dip in the 1950s which agrees better with the HS reconstruction in this period. The seasonal analysis has potentially four times the temporal resolution as the annual analysis but only the first ten spatio-temporal patterns are retained in both annual and seasonal analyses. Therefore the seasonal analysis concentrates largely on global mean changes which explains why the discrepancy between the observations and the LBB reconstruction shows up more clearly using seasonal data.

The contributions of greenhouse gases, sulphate aerosols and changes in total solar irradiance to the trends in global-mean temperatures for each of the 50-year periods considered in the seasonal analysis are shown in Fig 3. The overall pattern is of steadily increasing contributions from the anthropogenic components with warming due to greenhouse gases balanced by sulphate cooling in the mid century and greenhouse warming dominating later in the century. Early in the century, there is a relatively small but significant contribution from solar changes according to the HS reconstruction.

4. Conclusions

We have analysed space-time patterns of near-surface temperature patterns using an "optimal detection" methodology. The results indicate that the warming observed in the latter half of the twentieth century was caused by anthropogenic increases in greenhouse gases balanced by some cooling due to tropospheric sulphate aerosols, rather than by natural variability either internally or externally forced. We also find that natural effects may have contributed to the warming observed in the first half of the century. By optimising over seasons and using one particular reconstruction of total solar irradiance, we have found evidence that solar effects contributed significantly to this warming, regardless of any error in the relative amplitudes of the forcing of sulphates and greenhouse gases prescribed in the model. However, this was not the case when an alternative reconstruction of total solar irradiance was used, based more on the amplitude than the length of the solar cycle.

There are a number of caveats with this work. The modelling simulations we used have a number of limitations; in particular we did not consider changes in the spectral distribution of solar irradiance or possible associated changes in ozone (Haigh, 1999). We do not include the effects of observational sampling error, although this effect is believed to be small (Hegerl et al., 1999). Future work addresses these limitations in a model that unlike HadCM2 does not need flux adjustments to maintain a stable climate. A limitation of our analysis procedure is

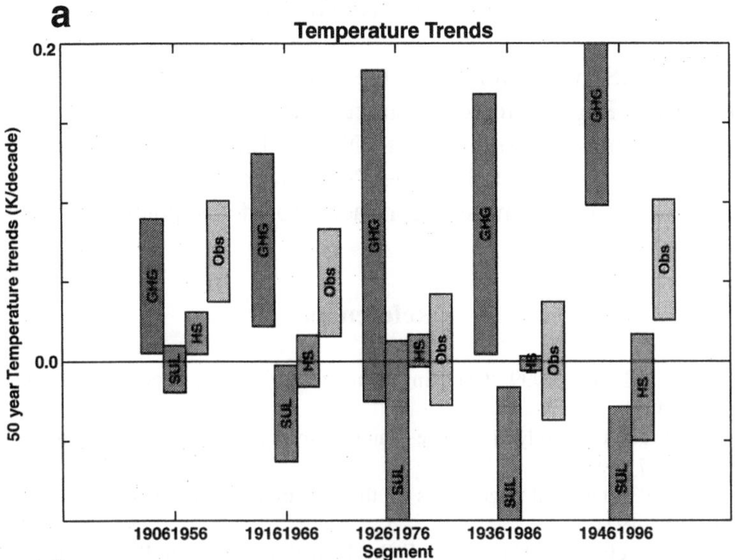

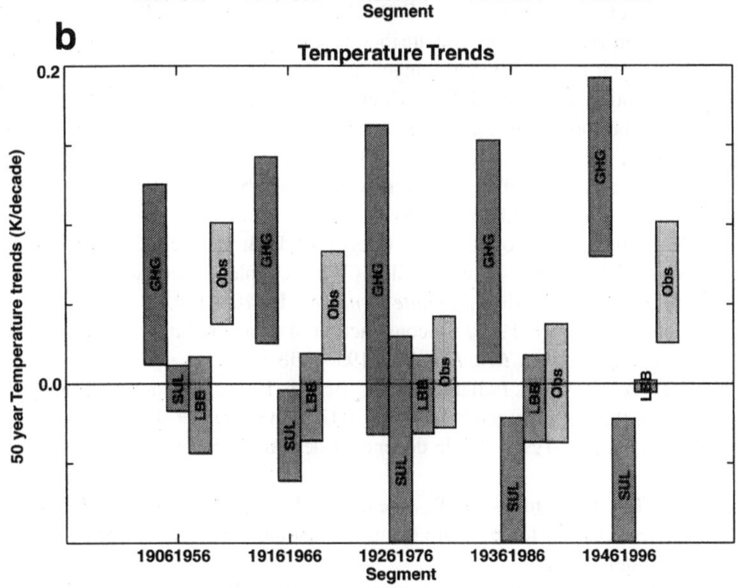

Figure 3. 50-year temperature trends and their uncertainties due to a) **GHG&SUL&HS** and b) **GHG&SUL&LBB** from seasonal analysis.

that we use standard regression which may lead to a low-bias in pattern amplitudes that is likely to be more serious for weak signals. This is currently under investigation and future work will report on the implications of taking account of noise in model-predicted signals for the results reported here.

Acknowledgements

PAS, SFBT and GSJ were funded by the U.K. Department of the Environment, Transport and the Regions under contract PECD 7/12/37. MRA was supported by a Research Fellowship from the U.K. Natural Environment Research Council. WJI and JFBM were supported by the Public Meteorological Service Research Programme. European Commission contracts provided supplementary support.

References

Allen, M. R. and S. F. B. Tett: 1999, 'Checking for model consistency in optimal fingerprinting', *Cli. Dyn.* **15**, 419–434.

Haigh, J. D.: 1999, 'A GCM study of climate change in response to the 11-year solar cycle', *Q. J. R. Met. Soc.* **125**, 871–892.

Hasselmann, K.: 1993, 'Optimal fingerprints for the detection of Time-dependent Climate Change', *J. Clim.* **6**, 1957–1971.

Hegerl, G. C., K. Hasselman, U. Cubasch, J. F. B. Mitchell, E. Roeckner, R. Voss, and J. Waskewitz: 1997, 'Multi-fingerprint detection and attribution analysis of greenhouse gas, greenhouse gas-plus-aerosol and solar forced climate change', *Cli. Dyn.* **13**, 613–634.

Hegerl, G. C., P. A. Stott, S. F. B. Tett, J. F. B. Mitchell, M. R. Allen, and U. Cubasch: 1999, 'Optimal detection and attribution of climate change : sensitivity of results to climate model differences', *Climate Dynamics*, in press.

Hoyt, D. V. and K. H. Schatten: 1993, 'A discussion of plausible solar irradiance variations, 1700–1992', *J. Geophys. Res.* **98**, 18,895–18,906.

Johns, T. C., R. E. Carnell, J. F. Crossley, J. M. Gregory, J. F. B. Mitchell, C. A. Senior, S. F. B. Tett, and R. A. Wood: 1997, 'The second Hadley Centre coupled ocean-atmosphere GCM: model description, spinup and validation', *Cimate Dynamics* **13**, 103–134.

Lean, J., J. Beer, and R. Bradley: 1995, 'Reconstruction of solar irradiance since 1610: implications for climate change', *Geophys. Res. Lett.* **22**, 3195–3198.

Parker, D. E., P. D. Jones, C. K. Folland, and A. Bevan: 1994, 'Interdecadal changes of surface temperature since the late nineteenth century', *J. Geophys. Res.* **99**, 14,373–14,399.

Stott, P. A. and S. F. B. Tett: 1998, 'Scale-dependent detection of climate change', *J. Climate* **11**, 3282–3294.

Stott, P. A., S. F. B. Tett, G. S. Jones, M. R. Allen, W. J. Ingram, and J. F. B. Mitchell: 1999, 'Attribution of twentieth century temperature change to natural and an thropogenic causes', *Climate Dynamics*, submitted.

Tett, S. F. B., P. A. Stott, M. R. Allen, W. J. Ingram, and J. F. B. Mitchell: 1999, 'Causes of twentieth century temperature change near the Earth's surface', *Nature* **399**, 569–572.

Willson, R. C.: 1997, 'Total solar irradiance trend during solar cycles 21 and 22', *Science* **277**, 1963–1965.

Address for Offprints: P. A. Stott, Hadley Centre for Climate Prediction and Research, Met Office, UK Met Office, Bracknell, Berks, UK RG12 2SY pastott@meto.gov.uk

OBSERVATIONS OF CLIMATE VARIABILITY

Discussion Session 3a

PETER A. STOTT
Hadley Centre, The Met. Office, Bracknell, Berkshire RG12 2SY, UK

KEITH P. SHINE
Dept. of Meteorology, University of Reading, Earley Gate, Reading RG6 6BB, UK

Received: 28 January 2000; Accepted in final form: 30 May 2000

1. Introduction

Before any attempt is made to attribute any observed climate change to a particular mechanism, be it solar variability or changes in greenhouse gas concentration, it is necessary to be sure that the observations used to deduce the climate change are reliable. Most of the data that is used to construct our picture of climate change are either from instrument networks that were not specifically designed for the detection of climate change or, prior to the instrumental period, from proxies that provide some kind of indirect information on atmospheric conditions.

The instrumental record from land and sea-surface observations is only available on anything approaching a global-scale for 150 years (although in specific locations, such as central England, records extend back for about 350 years). The distribution of observations is anything but uniform over this period. The type of instrumentation used, or the way it is used, has also changed with time, and from country to country. To build up a picture of real climate variability, it is important to appreciate when such changes occurred, and their impact. Otherwise it is difficult to have confidence that any observed climate change is indeed real, rather than an artifact of the observation system. For observations away from the surface, the record is even shorter (radiosonde observations only became sufficiently widespread to build up a large-scale picture around 40 years ago) and the instrumentation has been prone to even more changes than the surface network. Satellite data, for the detection of trends, has only been available since around the late-1970s. Although individual instruments can give a near-global view of the climate system, often trend analyses are built up from a succession of similar instruments on different satellites. Careful corrections have to be made to intercalibrate different instruments and account for drift (either due to deterioration of the instrument or changes in the satellite's orbital characteristics).

Not only is the quality of the record of importance, but so is its length, particularly if long-period cycles are sought from the data. A particular example might

Space Science Reviews **94:** 345–348, 2000.

be attempts to detect 11-year solar periodicity from datasets which are of the same length, or sometimes shorter, than the 11-year cycle itself.

Even if climate change signals are "real" (i.e. not artefacts of the observing system) this is but a first step. To know why the climate has changed we need information on the size of the unforced "chaotic" variability of the climate system, the natural climate change mechanisms (such as variability in solar output, or the impact of volcanic eruptions) and the anthropogenic climate change mechanisms. Longer period data sets can generally give more information on the size of the "chaotic" variations, but it is difficult to be sure that these changes are truly "chaotic" rather than having an identifiable external cause.

The situation for climate series developed from proxies, such as tree rings and ice cores, has its own dangers. Normally the data come from a few specific locations and so there is a danger of interpreting local signals as truly global signals. There are also a host of other factors, not directly related to temperature, which may lead to a signal in the proxy.

This discussion session aimed at exploring some of these issues.

2. Instrumental Record

For all data sources, the quality of data affects the robustness of conclusions that can be drawn from the data. Instrumental records are available on a global scale for the last 150 years. Two specific concerns were raised concerning the quality of instrumental data that relate to topics of considerable current interest.

The first concerns the early century warming that is seen in near-surface temperatures. In this volume, Stott et al. (2000) show, using a coupled ocean atmosphere general circulation model, that the large warming seen in the first half of this century cannot be explained by internal variability alone nor can it be explained entirely by anthropogenic forcings. This indicates a role for natural forcings, especially changes in solar output, as contributing to this warming. However, how confident are we that this early century warming is not an artefact of instrumentation changes such as the move over from measuring sea surface temperatures (SST) from buckets to engine intakes? In this volume, Parker et al. (2000), use SSTs adjusted using a physical model to compensate for this problem and argue that support for both marine and land surface air temperature records comes from the agreement between their global trends. Air temperatures from islands and coasts have also been compared with collocated SSTs and this work lends confidence to the SST records.

The second question of considerable interest and debate concerns differences between temperature trends at the near-surface and in the free troposphere. The warming seen at the surface has not been seen in satellite observations made by the Microwave Sounding Unit instrument since 1979. There has been cooling in the upper troposphere since 1979 according to radiosonde records, in contradiction with increasing-greenhouse gas simulations using climate models that instead typically show greater warming in the tropical upper troposphere than at the surface. Parker et al. show that there has been an apparent increase in global and tropical lower tropospheric lapse rate during the last 20 years. This may be due to spurious changes due to biases in any of the surface, radiosonde or satellite datasets or may arise from real changes in tropospheric atmospheric circulation, thermal forcing from stratospheric ozone loss or explosive volcanic eruptions. A US National Research Council Panel has considered this subject in depth and in a report published in January 2000 ascribes much of the lapse rate changes to real physical effects, though uncertain contributions from observational deficiencies remain. In addition, the panel recommended that extensive simulations be made with improved models with more complete and realistic forcings. However, longer term surface and tropospheric temperature trends from radiosondes going back to the mid 1960s are in closer agreement and indicate a general warming of the lower atmosphere that is inconsistent with model estimates of natural internal variability of the climate system.

3. Data on Millennium Timescales

To build up a picture of large scale climate variation prior to the instrument period, it is necessary to start using proxies for climate change. Perhaps the most noteable attempt to develop such a picture is the recent work of Mann et al. (1998, 1999). The proxies used include data from ice core, ice melt, tree ring and coral records. Mann et al. (1998) develop a series back to 1600 and, using far less data, Mann et al. (1999) extend the northern hemispheric record back to the year 1000. They discuss in detail the difficulties in merging such records and the associated uncertainties, which amplify the further back in time one goes. However, their best-estimate indicates that recent climate is unusually warm. With what they describe as "moderately high levels of confidence", the 20th century was the warmest century of the millenium, the 1990s were the warmest decade and 1998 was the warmest individual year. An interesting aspect of this data set is that prior to 1900 it appears that the climate was cooling at a rate of 0.02°C/century; this makes the recent warming even more marked. It is data sets such as these that can, if used with care, help put some kind of limits on the size of both unforced and natural variability, and increase our confidence in our attributions of the causes of the recent warming.

4. Ice Core Data

Ice cores can be used to provide climate information going back over 400,000 years. Clearly however, ice cores can only be made where ice is present, principally in Antarctica and Greenland. How representative are these locations of global and large scale patterns of climate change? In his contribution to this volume, Stauffer (2000) discusses the representativeness of polar ice core results. He notes that methane records indicate that methane varies in parallel with Dansgaard-Oeschger events seen in the GRIP Greenland ice core. Wetlands, the main source of methane, have not existed in the vicinity of Greenland during the last glaciation and the same methane variations are seen in Antarctic ice cores, indicating that Dansgaard-Oeschger events are of global significance.

The mid-holocene optimum is not seen in the GRIP $\delta^{18}O$ record. However, there is a warming seen in the GRIP bore-hole temperature record and Stauffer proposes that this discrepancy could be a result of an increase in winter snowfall in Greenland compensating for the effect of increased annual surface temperatures.

More recently, $\delta^{18}O$ records from ice cores in Canada and North Greenland show a little ice age with temperature minima in the late 17th century and the early 19th century which are much weaker in Central and South Greenland $\delta^{18}O$ records. Stauffer discusses evidence that relatively warm cyclones bring most of the winter precipitation to the southern half of the Greenland ice sheet and mask the Central and South Greenland $\delta^{18}O$ records.

These results indicate that local meteorological effects can influence ice core records and thus mask large scale hemispheric climate changes. Nevertheless there is also evidence that ice core records do provide data representative of large scale patterns of climate change.

5. Concluding Remarks

The characterisation of climate variability remains a high priority topic in climate research. Not only is there the continued challenge of continuing the analysis of past data and understanding its strengths and weaknesses, but also there is a need to look to the future. Decisions to change instrumentation or, as is happening in some countries, reduce the number of observations being made, will have an adverse impact on our ability to monitor future changes. It is important that the consequences of any changes in the current network, of both ground-based and satellite-based instruments, are well understood, ahead of the time that the changes actually are made.

References

Parker, D. E., *et al.*: 2000, *Space Sci. Rev.*, this volume.

Mann, M. E., Bradley, R. S., and, Hughes, M. K.: 1998, 'Globale-scale temperature patterns and climate forcing over the past six centuries', *Nature* **393**, 779–787.

Mann, M. E., Bradley, R. S., and Hughes, M. K.: 1999, 'Northern hemisphere temperatures during the past millenium: inferences, uncertainties and limitations', *Geophys. Res. Lett.* **26**, 759–762.

Stauffer, B.: 2000, *Space Sci. Rev.*, this volume.

Stott, P. A., *et al.*: 2000, *Space Sci. Rev.*, this volume.

THE SUN'S ROLE IN LONG-TERM CLIMATE CHANGE

JAMES E. HANSEN
NASA Goddard Institute for Space Studies

Received: 23 February 2000; Accepted in final form: 9 June 2000

Abstract. Evidence suggests that changes of solar irradiance in recent centuries have provided a significant climate forcing and that the sun has been one of the principal causes of long-term climate change. During the past two decades the solar forcing has been much smaller than the climate forcing caused by increasing greenhouse gases. But it is incorrect to assume that the sun necessarily will be an insignificant player in climate change of the 21st century. Indeed, I argue that moderate success in curtailing the growth of anthropogenic climate forcings could leave the sun playing a pivotal role in future climate change.

1. Introduction

The sun has long been suspected of being an important cause of decade-to-century climate changes. The classical paper of Eddy (1976) presented a strong case that the luminosity of the sun was lower during the Maunder minimum of solar activity in the seventeenth and early eighteenth centuries, when much of the Earth experienced notably low temperatures of the "little ice age". There are a large number of empirical relations between solar activity and climate variations that offer support for the contention that the sun has been a significant factor in climate change, even though such correlations are unsatisfactory for cause and effect analysis. The principal difficulty in quantifying the role of the sun in climate change has been the absence of precise measurements of solar irradiance that avoid the interfering effects of the Earth's atmosphere. Although rather precise data has been obtained from satellites in the past two decades (Willson and Hudson, 1991), there is disagreement as to whether these data reveal an underlying trend over the last two solar cycles (Willson, 1997; Fröhlich and Lean, 1998). For earlier times it has been necessary to estimate long-term change from assumed relationships between solar irradiance and measured solar properties, for example, the extent of convective activity visible on the Sun's surface. Hoyt and Schatten (1993) and Lean et al. (1995) have constructed such solar irradiance histories for the past two centuries using several proxy measures of solar activity. Their estimates of solar irradiance changes have significant differences, but both estimates have an increase of solar irradiance by a few tenths of a percent over the past two centuries.

An earlier, similar, version of Hoyt's solar irradiance time series was used by Hansen et al. (1981) in climate model simulations of global mean temperature for

Space Science Reviews **94:** 349–356, 2000.

the period since 1880. This assumed solar variability was found to contribute to the overall global warming and to shorter period fluctuations of global temperature, but the predominant cause of long-term warming in the model was increasing atmospheric carbon dioxide. Similar conclusions have been reached in more recent studies, such as those of Rind et al. (1999) that employed the irradiance time series of Lean et al. (1995) for simulations from 1600 to the present. White et al. (1997) find evidence in global ocean temperature change of a response to changing solar irradiance.

Several studies with three-dimensional global climate models (Haigh, 1996; Balachandran et al., 1999; Shindell et al., 1999) have simulated effects of changing solar irradiance on tropospheric climate patterns that resemble observed changes. These studies present further evidence of the Sun's role in climate variability and change. But evaluation of the contribution of solar variability to past and future global temperature change can be made best on the basis of quantitative comparisons of solar and competing climate forcings. This evaluation must include both the direct and indirect climate forcings.

2. Climate Forcings, 1850–2000

Climate fluctuates even in the absence of any forcings, because of chaotic dynamical fluctuations of the atmosphere and ocean. But long-term climate changes primarily in response to climate forcings, which are imposed perturbations of the Earth's energy balance with space (Hansen et al., 1997). An example of a direct climate forcing is a change of the solar irradiance incident on the Earth. An example of an indirect climate forcing is the change of atmospheric ozone induced by solar ultraviolet changes, which thus further alters the heating of the troposphere. Other indirect solar forcings, such as cloud changes induced by solar modulation of cosmic ray fluxes, have been suggested, but in the absence of convincing observational data these remain hypotheses.

Climate forcings in the industrial era, i.e., in the past two centuries, have been estimated and discussed by IPCC (1996) and Hansen et al. (1998). The presentation here (Figure 1) is a slight modification of that of Hansen et al. (1998). Here we break down the well-mixed greenhouse gases into different categories, and order these according to their contribution to the global climate forcing. Also we associate with CH_4 its indirect effects on tropospheric ozone and stratospheric water vapor, and we associate with the chlorofluorocarbons their indirect effect on stratospheric ozone. The indirect effect of CH_4 on tropospheric ozone is assumed to be one-quarter of the total tropospheric ozone change in the industrial era, based on simulations of Lelieveld et al. (1998). The forcings due to stratospheric water vapor and stratospheric ozone changes are based on simulations with the GISS model (Hansen et al., 2000).

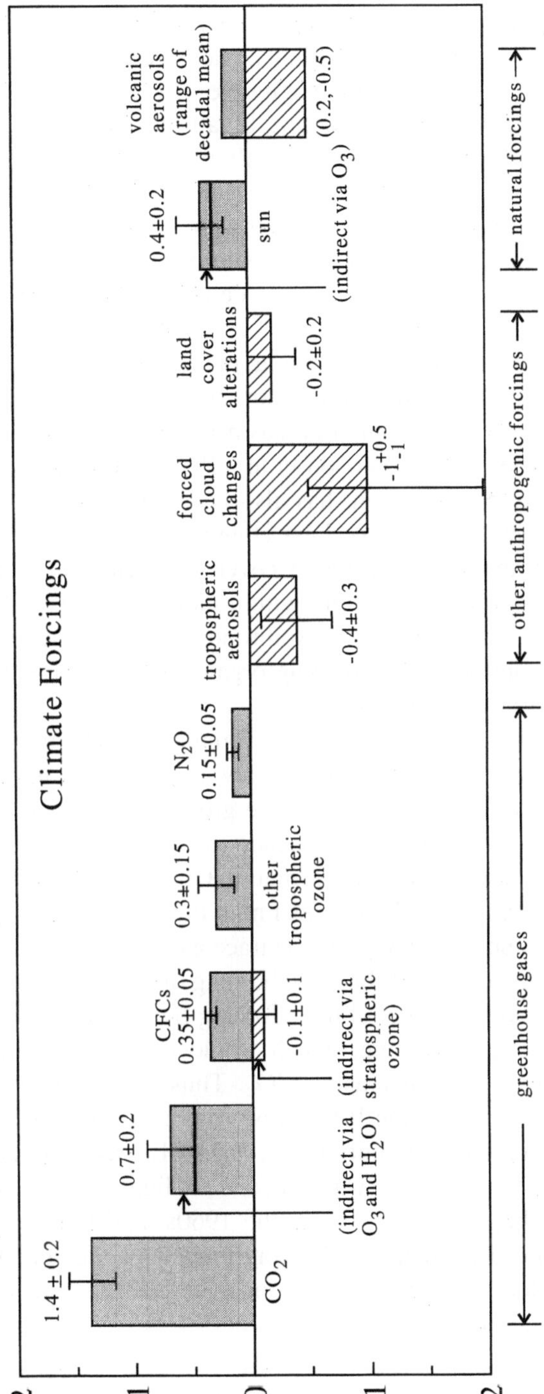

Figure 1. Estimated global climate forcings between 1850 and 2000.

A common misconception that is perhaps promoted by comparisons such as in Figure 1, especially when the greenhouse gas forcings are stacked one on top of another, is that the solar forcing is negligible. But a more relevant comparison of the solar forcing is with the net forcing by all mechanisms. Current estimates of the net forcing for the past century are only of the order of 1 to 1.5 W/m^2. In that event a contribution of 0.4 W/m^2 from the sun is far from negligible. Indeed, if the solar forcing had instead been –0.4 W/m^2, the global temperature change in the past century might have been so small as to be insignificant. A similarly pivotal role for the sun is possible next century, depending upon the balance among the competing forcings.

Note that assessment of the net forcing, in the past and future, depends critically on better understanding of the contribution of aerosols, including their indirect effect on clouds. Considerable attention has been paid to sulfate aerosols, especially subsequent to the paper of Charlson et al. (1992). The direct effect of absorbing aerosols, primarily black carbon, has also been addressed (Haywood and Shine, 1995, 1997; Penner et al., 1998). Less attention has been paid to the semi-direct effect of aerosol absorption on cloud cover (Hansen et al., 1997) and the indirect effects of absorbing inclusions within cloud particles (Chylek et al., 1995; Heintzenberg and Wendisch, 1996). Anthropogenic carbonaceous aerosols, including black carbon and weakly absorbing organic aerosols, are produced mainly by the burning of fossil fuels and biomass.

One reason to break down the climate forcings as in Figure 1 is that the growth rates of the different greenhouse gases vary, as illustrated in Figure 5 of Hansen et al. (1998). Between 1960 and 1990 the growth rate of climate forcing by CO_2 increased from about 0.1 W/m^2 per decade to just over 0.2 W/m^2 per decade. But the additional forcing by trace gases, mainly CFCs and CH_4, increased even more rapidly, to almost 0.2 W/m^2, before declining in the 1990s.

The distinction between CO_2 and the trace gases is important, because the same activities that produce most of the CO_2, burning of fossil fuels and land conversion, also produce aerosols. The net climate forcing by aerosols, direct plus indirect, is almost certainly one of cooling, which would tend to at least partially obscure globally warming due to increasing CO_2. Thus I suggest that the sharp global warming trend that began in the 1960s was primarily a consequence of the activities producing the trace gases, mainly CFCs and CH_4 , as these gases produce only warming. Prior to the 1960s the warming and cooling forcings were closer to being in balance than they have been since the 1960s, and thus the net forcing prior to 1960 was more dependent on the contribution of the sun. We must consider whether such a closer balance (of anthropogenic positive and negative forcings) is also possible in the future.

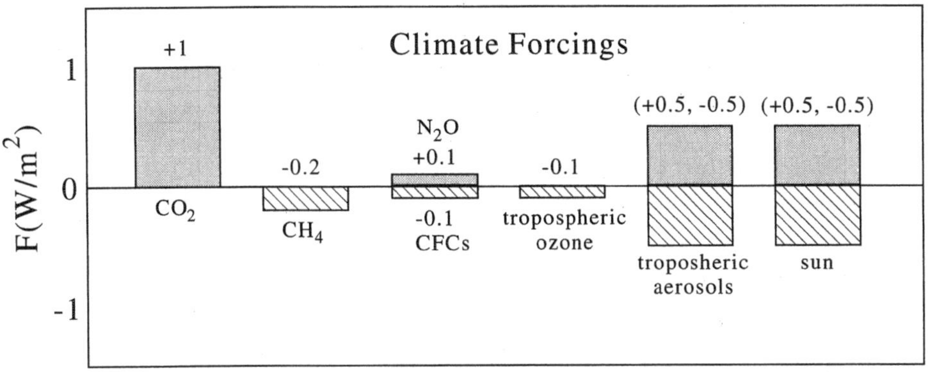

Figure 2. A scenario for additional climate forcings between 2000 and 2050. The sense and magnitudes of the uncertain aerosol and solar forcings strongly influence the net forcing.

3. Future Climate Forcings

Climate change scenarios for the future commonly have rapidly increasing greenhouse gases. The IPCC (1996) scenario that is used most frequently in climate simulations, IS92a, is sometimes called "business-as-usual" and has a rapid growth of greenhouse gases equivalent to almost 1% CO_2 increase per year or about 0.5 W/m^2 per decade. In that case most other climate forcings, including plausible variations of solar irradiance, are unimportant.

But such a trend is only one possible scenario, and it is not necessarily representative of actual trends. Indeed, as shown in Figure 5 of Hansen et al. (1998), the growth rate of the greenhouse gas climate forcing has declined by about 30% since the late 1970s and now falls below the lowest IS92 scenario. Future growth rates will depend not only on CO_2, but also on the continuing changes of CFCs and CH_4.

Let us assume, for example, that attempts to constrain CO_2 growth rates are partially successful, such that increased emissions in some countries are approximately balanced by decreases in other countries. In that case the CO_2 climate forcing increase is about 0.2 W/m^2 per decade or 1 W/m^2 in 50 years (Figure 2). CFC-11 and CFC-12, because of regulations already enacted, are expected to decline in abundance over the next 50 years. However, other CFCs are expected to increase over that period. Thus the net climate forcing by halocarbons over the next 50 years is likely to be near zero, but it could be slightly negative if regulations were extended to include additional gases, such as HFC-134a, and destruction of the accessible bank of CFC-12.

CH_4 has the potential to either increase or decrease. Its growth rate has declined by more than half in the past decade (Dlugokencky et al., 1998; Etheridge et al., 1998). A fraction of the decrease in its growth rate may be associated with stratospheric ozone depletion (Lelieveld et al., 1998), in which case there may be a CH_4 increase associated with future stratospheric ozone recovery. However, the principal reason for the slowdown is probably a reduced growth rate of CH_4 sources (Etheridge et al., 1998). Because the lifetime of CH_4 is short, about 8–10 years, the CH_4 abundance rapidly reflects changes in sources and sinks.

There are many opportunities to reduce the sources of atmospheric CH_4 (Hogan et al., 1991), including the tapping of landfills for CH_4 fuel, capture of CH_4 in mining operations, reducing pipeline losses, and more efficient agricultural practices. The indirect effects of CH_4 change will have the same sign as the direct effect. Thus a 30% decrease in the methane source would cause a negative forcing of about $-0.2\ W/m^2$. Reduction of CH_4 would have the added benefit of increasing atmospheric OH and reducing tropospheric O_3, a pollutant that is harmful to human health and agriculture (McKee, 1994).

A global strategy that links the air pollution and global warming issues should be attractive to both the developed and developing worlds. Technology for reducing air pollution, including black carbon, already exists. We suggest that a climate forcing scenario at least as likely as "business as usual" is one with concerted introduction of modern efficient energy systems and with special attention paid to reducing trace gas amounts and black carbon emissions. In this scenario the net change of anthropogenic climate forcing over the next 50 years could plausibly be less than $1\ W/m^2$.

Assuming that anthropogenic climate forcings follow a path similar to this alternative scenario, then the contributions of competing climate forcings, particularly solar irradiance, become important. If the change of solar forcing in the next 50 years, direct and indirect, is say $-0.5\ W/m^2$, as opposed to say $+0.5\ W/m^2$, that would have a large effect on the net climate forcing. The Maunder Minimum of solar activity has been estimated to represent a reduction of solar radiation of at least 0.2%, equivalent to a direct solar forcing of about $-0.5\ W/m^2$. Given our present understanding of the sun, we can not rule out a solar climate forcing of either $+0.5\ W/m^2$ or $-0.5\ W/m^2$ in the next 50 years. This suggests the need for research to understand the nature of solar variability and perhaps develop predictive capabilities for future solar change. Finally, I note that I am not disputing the merits of "business as usual" scenarios. It is appropriate to use these as a warning of potential climate effects should worst case practices be followed. At the same time, for the sake of comparison, we should consider less extreme alternative scenarios that may be achievable. Better understanding of all climate forcings is needed for reliable prediction of where climate is headed and for determination of the most effective ways to influence the direction of future climate change.

References

Balachandran, N.K., Rind, D., Lonergan, P., and Shindell, D.T.: 1999, 'Effects of solar cycle variability on the lower stratosphere and the troposphere', *J. Geophys. Res.* **104**, 27,321–27,339.

Charlson, R.J., Schwartz, S.E., Hales, J.M., Cess, R.D., Coakley, J.A., Hansen, J.E., and Hofmann, D.J.: 1992, 'Climate forcing by anthropogenic aerosols', *Science* **255**, 423–430.

Chylek, P., Videen, G., Ngo, D., Pinnick R.G., and Klett, J.D.: 1995, 'Effect of black carbon on the optical properties and climate forcing of sulfate aerosols', *J. Geophys. Res.* **100**, 16,325–16,332.

Dlugokencky, E.J., Masarie, K.A., Lang, P.M., and Tans, P.P.: 1976, 'Continuing decline in the growth rate of the atmospheric methane burden', *Nature* **393**, 447–450.

Eddy, J.: 1976, 'The Maunder Minimum', *Science* **192**, 1189–1202.

Etheridge, D.M., Steele, L.P., Francey, R.J., and Langenfelds, R.L.: 1998, 'Atmospheric methane between 1000 AD and present: evidence of anthropogenic emissions and climatic variability', *J. Geophys. Res.* **103**, 15,979–15,993.

Fröhlich, C., and Lean, J.: 1998, 'The sun's total irradiance: cycles, trends and related climate change uncertainties since 1976', *Geophys. Res. Lett.* **25**, 4377–4380.

Haigh, J.D.: 1996, 'The impact of solar variability on climate', *Science* **272**, 981–984.

Hansen, J., Johnson, D., Lacis, A., Lebedeff, S., Lee, P., Rind, D., and Russell, G.: 1981, 'Climate impact of increasing atmospheric carbon dioxide', *Science* **213**, 957–966.

Hansen, J., Sato, M., Ruedy, R., Lacis, A., and 39 co-authors: 1997, 'Forcings and chaos in interannual to decadal climate change', *J. Geophys. Res.* **102**, 25,679–25,720.

Hansen, J.E., Sato, M., Lacis, A., Ruedy, R., Tegen, I., and Matthews, E.: 1998, 'Climate forcings in the industrial era', *Proc. Natl. Acad. Sci. USA* **95**, 12,753–12,758.

Hansen, J., et al.: 2000, 'Climate forcings in GISS SI2000 simulations', *J. Geophys. Res.*, to be submitted.

Haywood, J.M., and Shine, K.P.: 1995, 'The effect of anthropogenic sulfate and soot aerosol on the clear sky planetary radiation budget', *Geophys. Res. Lett.* **22**, 603–606.

Haywood, J.M., and Shine, K.P.: 1997, 'Multi-spectral calculations of the direct radiataive forcing of tropospheric sulphate and soot aerosols using a column model', *Q. J. R.. Meteorol. Soc.* **123**, 1907–1930.

Heintzenberg, J., and Wendisch, M.: 1996, 'On the sensitivity of cloud albedo to the partitioning of particulate absorbers in cloudy air', *Beitr. Phys. Atmos.* **69**, 491–499.

Hogan, K.B., Hoffman, J.S., and Thompson, A.M.: 1991, 'Methane on the greenhouse agenda', *Nature* **354**, 181–182.

Hoyt, D.V., and Schatten, K.H.: 1993, 'A discussion of plausible solar irradiance variations, 1700–1992', *J. Geophys. Res.* **98**, 18,895–18,906.

Intergovernmental Panel on Climate Change (IPCC): 1996, *Climate Change 1995*, J.T. Houghton, L.G. Meira Filho, B.A. Callander, N. Harris, A. Kattenberg and K. Maskell (eds.), Cambridge University Press, Cambridge, England, 572 pp.

Lean, J., Beer, J., and Bradley, R.: 1995, 'Reconstruction of solar irradiance since 1610: implications for climate change', *Geophys. Res. Lett.* **22**, 3195–3198.

Lelieveld, J., Crutzen, P.J., and Dentener, F.J.: 1998, 'Changing concentration, lifetime and climate forcing of atmospheric methane', *Tellus* **50B**, 128–150.

Logan, J.A., Megretskaia, I.A., Miller, A.J., Tiao, G.C., Choi, D., Zhang, L., Stolarski, R.S., Labow, G.J., Hollandsworth, S.M., Bodeker, G.E., Claude, H., De Muer, D., Kerr, J.B., Tarasick, D.W., Oltmans, S.J., Johnson, B., Schmidlin, F., Staehelin, J., Viatte, P., and Uchino, O.: 1999, 'Trends in the vertical distribution of ozone: a comparison of two analyses of ozonesonde data', *J. Geophys. Res.* **104**, 26,373–26,399.

McKee, D.J.: 1994, *Tropospheric Ozone*, Lewis, Boca Raton, Florida, 333 pp.

Penner, J.E., Chuang, C.C., and Grant, K.: 1998, 'Climate forcing by carbonaceous and sulfate aerosols', *Clim. Dyn.* **14**, 839–851.

Rind, D., Lean, J., and Healy, R.: 1999, 'Simulated time-dependent climate response to solar radiative forcing since 1600', *J. Geophys. Res.* **104**, 1973–1990.

Shindell, D., Rind, D., Balachandran, N., Lean, J., and Lonergan, P.: 1999, 'Solar cycle variability, ozone, and climate', *Science* **284**, 305–308.

White, W.B., Lean, J., Cayan, D.R., and Dettinger, M.D.: 1997, 'Response of global upper ocean temperature to changing solar irradiance', *J. Geophys. Res.* **102**, 3255–3266.

Willson, R.C., and Hudson, H.S.: 1991, 'The sun's luminosity over a complete solar cycle', *Nature* **351**, 42–44.

Willson, R.C.: 1997, 'Total solar irradiance trend during solar cycles 21 and 22', *Science* **277**, 1963–1965.

Address for Offprints: James Hansen, NASA Goddard Institute for Space Studies, 2880 Broadway, New York, NY 10025; jhansen@giss.nasa.gov

CLIMATE MODELLING

Discussion Session 3b

R.J. HAARSMA (haarsma@knmi.nl)
Royal Netherlands Meteorological Institute, P.O. Box 201, 3730 AE De Bilt, The Netherlands

Received: 31 March 2000; Accepted in final form: 18 April 2000

1. Introduction

The central theme of this discussion was the role of numerical models in understanding the effect of variations in solar irradiance on the climate. Observational data are, of course, the basis of any knowledge of the relation between solar irradiance and climate. However, there are many problems concerning the availability and quality of data. Accurate data of the variability of solar irradiance only exists for about two decades since the introduction of satellite measurements (Fröhlich, 2000). It is believed that the largest changes in solar irradiance occur on longer time scales, varying from decades to centuries (Lean, 2000). There exist estimates of solar irradiance variability based on sunspot characteristics which have been measured over a much longer period. The quality of these estimates is however hard to judge. The situation for climate data is slightly better. A global surface temperature dataset spanning the past century is available. The quality of this dataset is, however, questionable, especially if we go further backwards in time. It is not sure that a small climate response to variations in solar irradiance can be detected with confidence from the available surface temperature data. Paleo reconstructions give valuable information about the climate, but dating, spatial structure and quantitative information on climate parameters are becoming increasingly difficult for data from earlier periods. These data problems would be not so severe if there was a clear signal to noise ratio. That is, if the climate response to changes in solar irradiance would be much larger than the internal variability of the climate system (noise) and other external factors affecting the climate system like for instance changing greenhouse gases. In that case the main question would be not the detection but the understanding of the response. However, present evidence suggests that variations in solar irradiance only have a small impact, comparable or smaller than other factors affecting the variability of the climate system.

The two theses posed during the discussion were first that the poor quality of the data together with the low signal to noise ratio implies that the question about the impact of variations of solar irradiance on the climate system cannot be answered by solely analysing the data and second that the use of climate models is inevitable

Space Science Reviews **94:** 357–361, 2000.

and beneficial for the detection and understanding of the effect of variations in solar irradiation on climate.

2. The Use of Climate Models

2.1. Methodology

These two theses were posed because the presentations during this workshop clearly revealed the deficiencies of current climate models. In simulating the present climate these models show, in certain variables, deviations from the observations which are of the order of the signal to be detected – such as the response to variations in solar irradiance. Sometimes these deviations are corrected by introducing artificially correction terms, e. g. radiant flux correction at the atmosphere-ocean boundary. These problems are well known within the climate modelling community and attempts have been made to design a coherent approach to improve the climate models. An example of such a attempt is the comparison of the simulation of clouds in different GCMs (Cess et al., 1996). It was intended that this comparison would then reveal, within the context of a well defined experiment, the different deficiencies of the models and would indicate how to improve the parameterization of clouds in climate models. Evaluation of this comparison revealed a wide variety in results. Deducing from these results which cloud scheme was the best and how to improve future cloud schemes, however, proved to be difficult. This was because, in general, if a cloud scheme was successful in certain aspects of the effect of clouds on climate its quality in other aspects was less. In addition a number of models was tuned to give the best results, which clearly diminished the relevance of this experiment. The conclusion is therefore that climate models are rather imperfect and that at present there are no clear short cuts to improve them. It should be stressed, however, that the ability of climate models to simulate the climate has been substantially improved over the last decades. This is partly due to the increase in spatial resolution and partly due to a better understanding and simulation of the physical processes involved, especially subgrid scale processes. It is therefore not unreasonable to assume that this will continue into the future due to increasing computer power and understanding of the climate system.

Realising the imperfections of climate models it is an important question how climate models can be used for increased understanding of the climate and its variability. The great advantage of climate models is that they can be used in a controlled environment. In these "laboratory" experiments they can be used for the investigation of specific questions. Usually more than one physical mechanism is involved in a specific aspect of climate variability. By careful designed experiments climate models can provide information of the relative importance of these mechanisms. Similarly the different steps between cause and effect can be analyzed with the help of climate models. Also in cases of low signal to noise ratio this can be increased by artificially increasing the source of the signal, such as the amplitude of the variations in solar irradiance. In addition extending the time integration facilitates the detection of the signal against the noise. Because of the complexity of the physical processes involved a hierarchy of models can be used (Arnold and Robinson, 2000). With simple models laboratory experiments can be designed, performed and analyzed with more ease and a larger part of the parameter space can be explored. By comparing the results of simpler models with more advanced models hypotheses emerging from the former can be tested.

In this workshop several examples of the use of climate models for the understanding of the impact of solar irradiance variability on climate have been presented. Some of these will be discussed below.

2.2. Examples

The climate is a chaotic system and its dynamics can be described by a strange attractor. One of the consequences of this is that the time evolution is sensitively dependent on the initial state. This implies that the actual evolution of the climate system as observed is only one of the possible realizations. A slightly different initial condition would have resulted in another trajectory in phase space. It is therefore important to know the possible range of different time evoulutions of the climate system under the same boundary conditions. In other words what is the probability density distribution of the time evolution? A way to obtain an estimate of this distribution is to perform ensemble integrations. Ensemble integrations performed by Cubasch (2000) have shown that the 11-year cycle in the solar irradiance with an amplitude of about 0.5 Wm^{-2} does not give rise to a signal which surpasses the internal noise level of the climate system, but that longer fluctuations of about 80 years in the solar irradiance with larger amplitudes of about 2 Wm^{-2} do give a clear significant signal. From these experiments it is however not completely clear whether it is the longer time scale or the larger amplitude which is responsible for the clear signal. Due to the inertia of the climate system, mainly caused by the oceans, lower frequencies in the solar forcing will result in a stronger signal. To investigate this effect Drijfhout et al. (1999) have performed an number of idealized experiments in which they modulated solar irradiance with a fixed frequency and amplitude. From this they concluded that for variations in solar irradiance with a period of 80 years already amplitudes of 0.5 Wm^{-2} would give a significant signal above the internal noise of the climate system.

Due to nonlinearities in the climate system variations in solar irradiance could give rise to time scale interactions and resonance. For a clear understanding of these processes the use of climate models is crucial. For certain regions in parameter space the dynamical behaviour of a nonlinear system is sensitive dependent on the parameter values. Climate models can be used to explore these so-called bifurcation points. Tobias and Weiss (2000) and Haarsma et al. (2000) have used climate models in this way to explore the sensitivity of the climate system to variations in solar irradiance.

Another question is the relative importance of solar forcing compared to other external forcings such as anthropogenic changes in greenhouse gases and aerosols concentrations or volcanic eruptions. Within a climate model this can be investigated by isolating each of these forcings. On the basis of these experiments Hansen (2000) and Stott et al. (2000) concluded that variations in solar forcing have significantly contributed to the observed variations in climate.

Apart from the hypothesis that changes in solar irradiance affects the earth climate by direct heating of the earth surface and the troposphere there exist other hypotheses which point to a more indirect way of influencing the climate. For testing these hypotheses climate models are a very valuable tool, especially for unravelling the various steps in the cause and effect. Larkin et al. (2000) showed within the context of a general circulation model that changes in stratospheric ozone caused by changes in solar irradiance have a significant impact on the earth's climate.

Sensitivity studies with climate models indicate that the suggested strong link between the length between of the solar cycle and the Northern Hemisphere temperature (Friis-Christensen and Lassen, 1991) only can be explained if an amplifying mechanism is operating in the climate system. Solar induced changes in cloud cover are proposed as a possible mechanism (Marsh and Svensmark, 2000). In addition to observations climate models could used for testing this hypothesis.

Another application of climate models is the assimilation of data. Data coverage is usually sparse and the quality of the data is often hampered by imperfections. These problems become more serious if we go further backwards in time in particular for paleo data. By the assimilation of data climate models can contribute to a better quality of the data. In this way numerical weather forecast centres like ECMWF and NCEP have recently constructed atmospheric datasets over the past 15 years. Assimilated data sets covering almost half a century will be available in the near future. The assimilation of paleo data is a promising but yet hardly explored possibility.

3. Conclusions

These examples show how climate models are and can be used in testing hypotheses about the impact of variations in solar irradiance on climate. An important issue during the discussion was, however, whether the quality of climate models is good enough to perform these experiments as outlined above. Stated bluntly are they not just (expensive) noise generators and as such of no real value? It was clearly stated in the discussion by the modellers and accepted by most participants of the workshop that climate models in general are based on sound physical principles. The main problems arise because of the low resolution due to the lack of computer power and the problems in correctly simulating the subgrid scale processes, partly due to a lack of understanding of these processes. Comparison of the output of numerical simulations with observations reveals that climate models are able to simulate qualitatively correct the climate and can therefore be used as a tool in the investigation of the impact of variations in solar irradiance on climate. A notion of concern is the use of tunable parameters outside the physically accepted range of values and artificial correction terms in order to remain sufficiently close to the observed climate. However, recent progress in climate modelling shows that the latest generation climate models is less hampered by these undesirable deficiencies. For instance an increasing number of climate models now run without flux correction. The main conclusion is therefore that if used in a sensible way climate models are a very valuable tool in understanding climate variability and that, based on present evidence, one might expect that the quality of the climate models, and therefore their contribution to the understanding of the climate system, will increase in the future.

Acknowledgements

R.J. Haarsma was supported by the Scale INTeracion EXperiments (SINTEX) project of the European Commission, contract number ENV4-CT98-0714.

References

Arnold, N. F., and Robinson, T. R.: 2000, *Space Sci. Rev.,* this volume.

Cess, R. D., et al.: 1996, 'Cloud feedback in atmospheric general circulation models: an update', *J. Geophys. Res.* **101**, 12,791–12,794.

Cubasch, U.: 2000, *Space Sci. Rev.,* this volume.

Drijfhout, S. S., Haarsma, R. J., Opsteegh, J. D., and Selten, F. M.: 1999, 'Solar-induced versus internal variability in a coupled climate model', *Geophys. Res. Lett.* **26**, 205–208.

Friis-Christensen, E., and Lassen, K.: 1991, 'Length of the solar cycle: an indicator of solar activity closely associated with climate', *Science* **254**, 698–700.

Fröhlich, C.: 2000, *Space Sci. Rev.,* this volume.

Haarsma, R. J., Drijfhout, S. S., Opsteegh, J. D., and Selten, F. M.: 2000, *Space Sci. Rev.,* this volume.

Hansen, J. E.: 2000, *Space Sci. Rev.,* this volume.

Larkin, A., Haigh J. D., and Djavidnia, S.: 2000, *Space Sci. Rev.,* this volume.

Lean, J.: 2000, *Space Sci. Rev.,* this volume.

Marsh, N., and Svensmark, H.: 2000, *Space Sci. Rev.,* this volume.

Stott, P. A., Tett, S. F. B., Jones, G. S., Allen, M. R., Ingram, W. J., and Mitchell, J. F. B.: 2000, *Space Sci. Rev.,* this volume.

Tobias, S. M., and Weiss, N. O.: 2000, *Space Sci. Rev.,* this volume.

Address for Offprints: R.J. Haarsma, Royal Netherlands Meteorological Institute,P.O. Box 201, 3730 AE De Bilt, The Netherlands, haarsma@knmi.nl

RADIATIVE FORCING OF CLIMATE CHANGE

KEITH P. SHINE (k.p.shine@reading.ac.uk)
Department of Meteorology, University of Reading, Earley Gate, Reading RG6 6BB, UK.

Received: 18 October 1999; Accepted in final form: 17 January 2000

Abstract. Our current understanding of mechanisms that are, or may be, acting to cause climate change over the past century is briefly reviewed, with an emphasis on those due to human activity. The paper discusses the general level of confidence in these estimates and areas of remaining uncertainty. The effects of increases in the so-called well-mixed greenhouse gases, and in particular carbon dioxide, appear to be the dominant mechanism. However, there are considerable uncertainties in our estimates of many other forcing mechanisms; those associated with the so-called indirect aerosol forcing (whereby changes in aerosols can impact on cloud properties) may be the most serious, as its climatic effect may be of a similar size as, but opposite sign to, that due to carbon dioxide. The possible role of volcanic eruptions as a natural climate change mechanism is also highlighted.

1. Introduction

Variations in the output of the sun, which are considered in detail elsewhere in this volume, are but one of many potential mechanisms that can cause climate change. The purpose of this paper is to provide a perspective by briefly reviewing our current knowledge of mechanisms other than solar variability, concentrating particularly on those that are due to human activity. This is an area which has been regularly reviewed over the past decade or so, often as part of international assessments (e.g. Ramanathan *et al.*, 1987; IPCC, 1990, 1992, 1994, 1995). This paper is largely based around the author's own recent review (Shine and Forster, 1999) in which more detail and a much more extensive list of references can be found. The Intergovernmental Panel on Climate Change (IPCC) is also extensively reviewing this subject for its Third Assessment Report, due for completion in 2001.

Our understanding of the range of potential mechanisms has evolved rapidly over the past decade. As an example, IPCC (1990) mentions only briefly the role of aerosols and ozone change, but later reports consider these at great length. The rate of the development in this area warns against overconfidence that we have yet identified all the major climate change mechanisms.

The conceptual framework used in this paper is so-called "radiative forcing". Radiative forcing is a measure in Wm^{-2} of the change in the Earth-atmosphere system's radiation budget when some parameter is changed such as, for example, the total solar irradiance or the atmospheric concentration of carbon dioxide.

For small perturbations, climate model experiments (e.g. IPCC, 1994; Hansen *et al.*, 1997) indicate that the equilibrium global-mean surface temperature change

Space Science Reviews **94:** 363–373, 2000.

(ΔT_s) is related to the radiative forcing (ΔF) (if imposed for a time period long compared to the multi-decadal response time of the climate system) so that

$$\Delta T_s \approx \lambda \Delta F \tag{1}$$

where λ is the climate sensitivity (in $K(Wm^{-2})^{-1}$). The value of λ is determined by the direct response of the surface temperature to the change in energy balance and by a number of feedbacks. The feedbacks, which arise as part of the response to climate change, include changes in water vapour, sea-ice and snow extent, and cloud properties.

Early climate model experiments indicated that the climate sensitivity, λ, depends on the size, but not the nature, of the global-mean radiative forcing. Thus, a change in total solar irradiance would have approximately the same climate impact as a change in carbon dioxide concentration if they caused the same radiative forcing. This is despite the fact that they impact on the climate system in quite different ways. For example, the effect of a change in total solar irradiance is initially felt primarily through a change in the surface shortwave ($0.2 - 4\,\mu$m) radiation balance. The change in carbon dioxide is initially felt via changes in the mid-infrared ($4 - 200\,\mu$m) radiation budget in the upper troposphere.

An important reason for using radiative forcing rather than the resulting change in climate is that there is currently a large uncertainty in the value of λ (e.g. IPCC, 1995). Different models give values in the range from about 0.3 to 1.1 $K(Wm^{-2})^{-1}$; most of this uncertainty is due to model differences in cloud feedbacks (IPCC, 1995). Hence for a particular climate change mechanism, it is more straightforward to compare different estimates of ΔF than ΔT_s. There are other reasons for calculating radiative forcing. It is computationally much less demanding than using a sophisticated climate model; the sensitivity to poorly-known parameters can be more easily explored, and more sophisticated radiative transfer models can be used.

The applicability of the concept of radiative forcing is now under intense scrutiny (see particularly Hansen *et al.*, 1997). Part of the reason is that its initial development occurred when relatively "simple" climate change mechanisms were being considered (such as changes in carbon dioxide and total solar irradiance). Now the changes being considered (such as complex height dependent changes in ozone, or spatially inhomogeneous changes in aerosol parameters) are much less simple, and the conclusions from earlier studies may not carry over to these new mechanisms. An additional important reason for the scrutiny is that the concept of radiative forcing is now implicitly enshrined in the United Nations Framework Convention on Climate Change (the Kyoto Protocol), as part of the means for comparing the climatic effect of emissions of different gases.

The current situation is that it appears that radiative forcing remains a useful "first-order" method of comparing climate change mechanisms. At the moment, experiments with sophisticated climate models (the General Circulation Models) do not present a consistent picture as to how the climate sensitivity might vary amongst the different climate change mechanisms. Until such a time that a con-

TABLE I

Comparison of the ratio of the climate sensitivity for ozone changes, λ_{O_3}, to the climate sensitivity for carbon dioxide changes, λ_{CO_2}, from recent general circulation model experiments. Also shown is the absolute climate sensitivity λ_{CO_2} (in $K(Wm^{-2})^{-1}$) for the experiments.

Study	Brief Description of Experiment	λ_{CO_2} $[K(Wm^{-2})^{-1}]$	$\lambda_{O_3}/\lambda_{CO_2}$
Hansen *et al.* (1997)	Ozone change based on observations, fixed clouds	0.69	1.25
Hansen *et al.* (1997)	As above, but with varying clouds	0.92	0.7
Christiansen (1999)	Perpetual January calculations of idealised lower stratospheric ozone change, varying clouds	0.4	1.0
Christiansen (1999)	Perpetual January calculations of idealised upper stratospheric ozone change, varying clouds	0.4	1.4
Forster and Shine (1999)	Stratospheric ozone changes based on observations (and mostly in the lower stratosphere), varying clouds	0.5	1.4

sistent picture does develop, it would be unwise to conclude that the concept of radiative forcing is flawed. Indeed, if a consistent picture does develop, it may be possible to enhance the concept of radiative forcing by explicitly taking account of the fact that a given Wm^{-2} from certain mechanisms is more or less effective than the same Wm^{-2} from other forcings.

As an illustration of the current lack of consensus in general circulation model experiments, Table 1 shows a comparison of recent calculations of the climate sensitivity, λ , due to changes in ozone. (Ozone is chosen as several results are now available to compare its effect with that of carbon dioxide increases.) Since there is a wide variation in model calculations of the absolute value of λ , Table 1 shows the sensitivity for ozone changes, λ_{O_3}, relative to that for carbon dioxide, λ_{CO_2}, from the same model. Recall that, ideally, λ would be independent of radiative forcing mechanism, so that $\lambda_{O_3}/\lambda_{CO_2}$ would be unity. Table 1 shows that the ratio *does* depart from unity in several models, but there is no consensus as to how it does so; in some models, ozone is more effective at changing climate than is indicated by its radiative forcing, whilst in others it is less sensitive. The reason for these differences presumably lies in the physical parameterisations used in the models, for example surface energy transfer, or, perhaps, the methodology adopted to calculate the radiative forcing. These need to be disentangled for further progress to be made in understanding the relationship between radiative forcing and climate change. (The table also shows the variation in the value of λ_{CO_2}, to emphasize the point that there is significant disagreement in the absolute value of this parameter.)

In the following sections, the background behind our knowledge of different mechanisms is discussed. In the final section, the radiative forcing from each mechanism is compared and the possible role of volcanic eruptions on climate is briefly discussed. Note that only climate change mechanisms that act via a change in radiative budget are discussed. It is possible, as an example, that changes in land surface properties (such as surface roughness or evapotranspiration) may also act to alter climate (see e.g. Henderson-Sellers, 1995; Chase *et al.*, 1996).

2. Well-mixed Greenhouse Gases

"Well-mixed greenhouse gases" refers to gases that are relatively long-lived (lifetimes of more than a few years) so that increases are on a global scale and can be characterised by just a few measurements across the globe. The increases are well documented and much of the increases can be attributed to the effects of human activity. The most notable of these is carbon dioxide, but the cumulative contributions of others, such as methane, nitrous oxide and the chlorofluorocarbons (CFCs) is also significant. Of the mechanisms considered here, the effect of these gases is believed to be best understood. This is for two main reasons. First is the quality of observations indicating the change in concentration over the past century or so. Second, it is believed that the radiative properties of these gases in the mid-infrared are well-characterised so that we can calculate the impact of changes in concentration on the Earth's energy budget within about 10%. Nevertheless, as discussed in recent assessments (e.g. IPCC, 1994, 1995; WMO, 1999) our understanding of the reasons for increases for some of these gases is incomplete, and radiative transfer calculations will continue to be revised.

In addition, the mix of gases that are being emitted into the atmosphere is changing. Most particularly, emissions of CFCs are being reduced to zero in response to the Montreal Protocol process, aimed at protecting the stratospheric ozone layer. Concentrations of these gases has begun to, or soon will, fall. Other "new" gases, some used as replacements to the CFCs, are now being detected in the atmosphere (e.g. IPCC, 1995; WMO, 1999). Although the current concentrations of these new gases remain individually small, emissions of many are increasing. This necessitates continued monitoring of concentrations and assessments of their potential climate impact via laboratory studies of their infrared spectroscopy and their reactivity in the atmosphere, in conjunction with numerical modelling studies. (WMO, 1999) lists more than 80 species in their assessment.

3. Ozone

The calculation of the radiative forcing due to changes in ozone is much more difficult than that of the well-mixed greenhouse gases for a variety of reasons (see e.g. IPCC, 1992; WMO, 1999; Shine and Forster, 1999). First, human activity is changing ozone in different ways in different regions of the atmosphere. In the troposphere (at heights below around 12 km), emissions of carbon monoxide, methane and oxides of nitrogen from (amongst other factors) burning of fossil fuels and biomass, lead to an increase in ozone. In the stratosphere, the halocarbons, such as the CFCs, are leading to a depletion of ozone. Even the sign of the climatic effect of changes in ozone depends on the altitude of ozone change. A decrease in ozone in the upper stratosphere (i.e. above about 30 km) leads to a surface *warming* of climate (at least in models!), as more solar ultraviolet and visible radiation reaches the troposphere. A decrease in the lower stratosphere is believed to lead to a surface *cooling*. Whilst more solar radiation still reaches the troposphere, causing a warming effect, this is more than balanced by the decreased emission of infrared radiation from stratosphere to troposphere. This is because: (i) ozone is an effective absorber and emitter of infrared radiation so that less ozone means less emission and (ii) the loss of ozone leads to a cooler lower stratosphere and hence less emission by the other gases, such as carbon dioxide and water vapour. The balances between these processes are subtle and there is considerable spread in the values of radiative forcing reported in the literature.

A second reason for the uncertainty in the ozone effect on climate is due to our limited knowledge of the nature of the ozone change (e.g. WMO, 1999). The increase in tropospheric ozone has gone on over the past hundred years and yet the even today tropospheric ozone is not well observed. Hence there is a heavy reliance on numerical models to simulate the change in ozone due to the changes in emissions, and there are inadequate data to fully evaluate these models. In the stratosphere, the most reliable measurements are of the change in the so-called "total ozone column", i.e. the height-integrated change. Measurements of the trends as a function of height are much less reliable, particularly in the lower stratosphere, although much progress has been made recently (e.g. WMO, 1999). As mentioned above, even the sign of the climatic effect of ozone change depends on the height of the change. The consensus is that the observed loss of stratospheric ozone has led to a cooling of the climate system, while the tropospheric increases have caused a warming.

A third reason relates not to the direct effect of the ozone change on climate but due to consequential impacts via photochemistry (e.g. WMO, 1999). A decrease in stratospheric ozone allows more ultraviolet radiation into the troposphere. This directly destroys ozone in the troposphere. It also leads to an increase in concentrations of the hydroxyl radical (OH) which is an important "cleanser" of the troposphere. Increased OH leads to a more rapid destruction of gases such as methane and the hydrofluorocarbons. It has been further suggested that the conversion of sulphur gases to aerosol particles results in an impact on cloud properties. The sign of all these additional chemical impacts appears to accentuate the "direct" cooling due to the stratospheric ozone loss, although the size remains uncertain.

Uncertainties in the effect of ozone loss are at least 50%. The net impact of ozone change over a 100 year period is believed to be a warming, as the tropospheric increases have had a larger impact than the stratospheric losses. Over the past twenty years, though, when the stratospheric depletion has been more marked, it is likely that the cooling effect has dominated.

Since one important mechanism by which the climatic effect of solar variations might be amplified is via changes in ozone (see Larkin *et al.*, this volume), many of the uncertainties referred to above are as relevant to solar effects as they are to human effects on climate.

4. Aerosols

Since the late 1980s, there has been a massive increase in interest in the effects of increased concentrations of aerosols (i.e. small airborne particles generally in the 0.1 to 1.0 μm radius range) on climate. There are two basic mechanisms by which they influence climate. The so-called "direct effect" results from the increased scatter and absorption of radiation in the atmosphere. The "indirect effect" results from the crucial role that aerosols play in cloud formation. Cloud particles form around aerosol particles. An increase in aerosol concentrations has been observed to lead to more, but smaller, cloud droplets which make the clouds more effective at scattering solar radiation back to space — this has a cooling effect on climate. It has also been proposed that the presence of smaller droplets leads to an inhibition of the production of rain, leading to a thickening of the clouds (see IPCC, 1994, 1995).

Evaluations of the direct effect of the aerosols suffer from some of the same problems discussed for tropospheric ozone above. Even today, the distributions and properties of aerosols are poorly observed, particularly away from the surface, and there is a heavy reliance on numerical models, which are difficult to evaluate. The problem is made worse because there are many different types of aerosols. Some, such as sulphate and organic carbon aerosols are so-called secondary aerosols which result from gaseous emissions, often associated with fossil fuel or biomass burning. Following chemical processing in the atmosphere, aerosols are formed. Others, such as soot and mineral dust (due to changes in land use), are so-called primary aerosols, emitted into the atmosphere as particles. To complicate matters, soot, because it is highly absorbing, leads to a warming of the climate system, sulphate and organics cool, as their scattering effect dominates; the sign of the climatic impact of mineral dust is uncertain. A further serious uncertainty is that these aerosol species do not necessarily exist as separate entities, but individual particles can be a mixture of types. This has a significant impact on the climatic effect, as, for example, mixing a small amount of soot into a sulphate aerosol can flip the sign of its climate impact. The current consensus is that, overall, increased amounts of aerosols have led to a cooling effect, but confidence in these assessments is low.

Confidence in the assessments of the climate impact of the indirect effect of aerosols is much lower. Although individual field observations confirm the basic mechanism that more aerosols lead to smaller droplets, scaling these up to the global scale has formidable problems. The majority of studies of the indirect effect have been based on the same type of numerical model referred to above. In addition to the difficulties in evaluating these models, there are further difficulties. To date, models have predicted the *mass* of the aerosol. For the indirect forcing, this mass has to be related in some manner to the number of cloud droplets. The nature of this relationship is likely to be very dependent on the pre-existing aerosol amount and type in the atmosphere and hence geographically highly variable.

5. Other Mechanisms

A large range of other mechanisms has been proposed which may play a role in climate change. These include changes in land use, which cause a change in surface albedo, and aircraft-induced contrails.

Recently, our own work has highlighted the possible role of stratospheric water vapour changes (Forster and Shine, 1999) in climate change. There is patchy evidence (partly because either it is geographically limited, or it is from short-duration observations) that there has been an increase in stratospheric water vapour over the past couple of decades. Some increase is anticipated from the increase in emissions of methane; oxidisation of methane in the stratosphere leads to water vapour production. However, the observed increase apparently far exceeds that expected from methane. If changes in stratospheric water vapour are indeed global and sustained, they would add to the greenhouse effect and cause a surface warming; they also lead to a cooling of the stratosphere. Previous assessments (e.g. WMO, 1999) had concluded that the observed long-term cooling of the lower stratosphere could be attributed to ozone depletion; Forster and Shine (1999) show that the water vapour increase could be responsible for a cooling similar to that due to ozone. There are many uncertainties in this area, not least a conceptual one. The increase may be due to, for example, a change in the transport of water vapour across the tropopause as a result of climate change from some other source. If this is the case then the effect of the stratospheric water vapour should be viewed as a feedback (and hence contributing to λ in Equation 1), rather than a forcing.

6. Comparison of the Size and Sign of Different Mechanisms

Figure 1 shows a comparison of the global-mean radiative forcings since the 19th century due to many of the mechanisms mentioned above, taken from Shine and Forster (1999). It is their attempt to update a figure presented in recent IPCC reports (1994, 1995). There are several important aspects of this figure. First, the

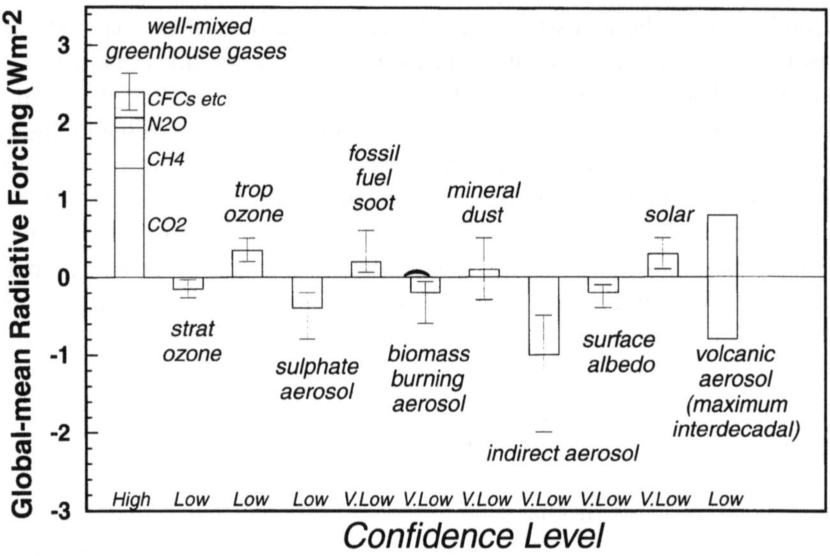

Figure 1. The global-mean radiative forcing due to individual mechanisms for the period from pre-industrial times to the present taken from Shine and Forster (1999). The "error bars" indicate the range of forcings inferred, where possible, from the recent literature. The confidence level indicates a subjective confidence level that the forcings actually lie within these "error bars". Any cancellation between mechanisms in terms of global-mean forcing must not be taken to indicate a reduction in climate impact, as regional scale changes can still be significant. (Reproduced from Shine and Forster, 1999, with permission from Elsevier Science.)

"error bars" are meant solely as a qualitative indication of the spread of values in the literature, although some studies will have reported values outside this range. The "confidence level" is again a subjective assessment of the likelihood that the true value lies within the "error bars" based on the difficulties in deriving those estimates. In addition to the caveats on radiative forcing discussed in Section 1, it must be remembered that global-mean values are a crude tool. Compensation between radiative forcing mechanisms at the global level must not be taken to imply an absence of climate change, as the regional forcings and regional responses can still be substantial. Shine and Forster (1999) present plots of the geographical distribution of many of these mechanisms.

Since the purpose of this paper is to place the radiative forcing due to solar variability into a wider perspective, we briefly discuss the solar term here. The solar radiative forcing of 0.3 ± 0.2 Wm^{-2} in Figure 1 is taken from IPCC (1995) and represents an estimate of the effect of the change in total solar irradiance since around 1850. The many difficulties in deriving such a value are discussed by Solanki and Fligge (this volume) and Lean (this volume) who also show time series of the inferred variation in solar variation in recent centuries. The change since the Maunder Minimum may have been double the value given by IPCC (1995) (see Lean, this volume). (It must be recalled that the change in solar radiative forcing is

0.175 times the change in total solar irradiance; this takes into account the fact that the incident global-mean solar radiation is one-quarter of the total solar irradiance (the ratio of the cross-sectional area of the Earth to its actual surface area) and the fact that, because the planetary albedo is 0.3, only 70% of the incident radiation is absorbed.) The solar value in Figure 1 does not include the effects of ozone changes, nor does it include the much more speculative cosmic ray impacts on clouds.

Figure 1 shows that the contribution from the well-mixed greenhouse gases causes a forcing of about 2.4 Wm^{-2}, 60% of which is due to carbon dioxide. The stratospheric ozone forcing of -0.15 Wm^{-2} looks small in comparison, but this has all occurred over the past two decades; over this time period it may have offset the well-mixed greenhouse forcing over the same period by around 30%. This estimate neglects the indirect photochemical impact of the ozone loss, which was discussed in Section 3. The tropospheric ozone forcing of 0.35 Wm^{-2} could be one of the larger forcings. It is for historical reasons that stratospheric and tropospheric ozone have been treated largely separately, In the future it is hoped that studies will start to consider them together.

Of the aerosol forcings, the direct impact of sulphate aerosols is by far the most studied with an estimated effect of -0.4 Wm^{-2}, which is about the same size as, but the opposite sign to, the tropospheric ozone forcing. The other forcings are much more uncertain; fossil fuel soot, by lowering the planetary albedo, has been estimated to cause a small positive forcing. Even the sign of the mineral dust forcing is not known with confidence. As discussed above, the interrelationships between the different aerosol types may mean that it is not appropriate to show them disaggregated into separate entities, as has been done in Figure 1. Future studies should present a more "integrated" view of the impact of aerosols.

The size of the indirect aerosol forcing is a matter of much debate. IPCC (1994, 1995) refrained from even giving a central estimate, giving the range of forcing as being (albeit with very low confidence) between 0 and -1.5 Wm^{-2}. Shine and Forster (1999) considered that the range of studies that had been brought to bear on this subject were such that a central estimate of -1 Wm^{-2} was justified, although they retained the "very low confidence". It will be interesting to see how IPCC's Third Assessment Report deals with this issue. Whatever, the indirect aerosol forcing has the potential to offset, at least in the global mean, much of the forcing due to the well-mixed greenhouse gases and clarification of its size remains a priority.

The stratospheric water vapour forcing discussed in Section 5 (if it indeed should be viewed as a forcing) was estimated by Forster and Shine (1999) to be 0.2 Wm^{-2} over the period 1979-1997. This would make it as significant as (although of the opposite sign to) the stratospheric ozone loss.

Previous IPCC reports have excluded the effect of volcanic eruptions on climate in such a figure but it is worth dwelling briefly on this here, as it is not covered in detail elsewhere in this volume. Explosive volcanic eruptions can inject huge quantities of sulphur gases into the stratosphere which condense to sulphate aerosols which act to increase the planetary albedo and cool the climate. The aerosols from each individual eruption (at least of the size that has occurred in recent centuries) generally only last a few years, before they are removed from the stratosphere.

The eruption of Mount Pinatubo in 1991 presented a real test of climate models as it was, for the first time, possible to both predict a climate change and clearly see it occurring in the years following the eruption. Indeed, measurements from satellite were able to detect the large negative radiative forcing (around $-4\ \mathrm{Wm}^{-2}$) at the peak of the aerosol loading. Comparison of the modelling studies made soon after the eruption (Hansen *et al.*, 1992) with the subsequently observed climate change (see e.g. IPCC, 1994, 1995) affirm that our climate models do have predictive capability and that the concept of radiative forcing is, at least for this forcing, useful. The role of volcanoes in longer term climate change is less clear, again due to lack of good observations. Figure 1 shows the change in volcanic radiative forcing between a decade with high loading and a decade with a low loading. The volcanic effect is shown as having either a positive or a negative sign. This is because moving from a "volcanic" decade to a non-volcanic decade would constitute a relative warming effect, whilst moving from a non-volcanic to a volcanic decade would constitute a relative cooling effect.

The role of volcanoes in century-scale climate change is a subject that has been revisited many times; one recent example, using a simple energy balance model, is Rowntree (1998). Although individual eruptions only have an impact for a few years, the large eruptions have not been evenly distributed and the impacts of one eruption can merge into the impacts of another. There was a period of relatively frequent (climatically significant) eruptions in the latter half of the 19th century, followed by a quiescent period between around 1920 and 1960, after which there have been several large eruptions. Rowntree (1998) shows that this distribution of eruptions may have contributed significantly to the observed surface warming between the 1920s and 1940s. And the recent eruptions may have been depressing the global mean surface temperatures over the past decades, despite the fact that they are the warmest in the 150 year record. Certainly volcanic eruptions deserve at least as much attention as solar variability in considerations of the natural mechanisms of climate change, and in attempts to understand variations in climate over the past few centuries.

References

Chase, T. N., Piekle, R. A., Kittel, T. G. F., Nemani, R., Running, S. W.: 1996, 'The sensitivity of a general circulation model to changes in leaf area index', *J. Geophys. Res.* **101**, 7393–7408.

Christiansen, B.: 1999, 'Radiative forcing and climate sensitivity: the ozone experience', *Quart. J. Roy. Meteorol. Soc.* **125**, 3011–3035.

Forster, P. M. de F., and Shine, K. P.: 1999, 'Stratospheric water vapour changes as a possible contributor to observed stratospheric cooling', *Geophys. Res. Lett.* **26**, 3309–3312.

Hansen, J., Lacis, A., Ruedy, R. and Sato, M.: 1992, 'Potential climate impact of Mount Pinatubo eruption' *Geophys. Res. Lett.* **19**, 215–218.

Hansen, J., Sato, M. and Ruedy, R.: 1997, 'Radiative forcing and climate response', *J. Geophys. Res.* **102**, 6831–6864.

Henderson-Sellers, A.: 1995, 'Human effects on climate through the large-scale impacts of land use change', In: A. Henderson-Sellers (Editor), *Future Climates of the World: A Modelling Perspective*. World Survey of Climatology Volume **16**, Elsevier, Amsterdam, pp. 433–475.

IPCC: 1990, Climate Change, Intergovernmental Panel on Climate Change, Cambridge University Press.

IPCC: 1992, Climate Change 1992, Intergovernmental Panel on Climate Change, Cambridge University Press.

IPCC: 1994, Climate Change 1994, Intergovernmental Panel on Climate Change, Cambridge University Press.

IPCC: 1995, Climate Change 1995, Intergovernmental Panel on Climate Change, Cambridge University Press.

Larkin, A., Haigh, J. D., and Djavidnia, S.: 2000, *Space Sci. Rev.*, this volume.

Lean, J. L.: 2000, *Space Sci. Rev.*, this volume.

Ramanathan, V., Callis, L., Cess, R., Hansen, J., Isaksen, I., Kuhn, W., Lacis, A., Luther, F., Mahlman, J., Reck, R., Schlesinger, M.: 1987, 'Climate-chemical interactions and effects of changing atmospheric trace gases', *Rev. Geophys.* **25**, 1441–1482.

Rowntree, P. R.: 1998, 'Global average climate forcing and temperature response since 1750', *Int. J. Climatol.* **18**, 355–377.

Shine, K. P., and Forster, P. M. de F.: 1999, 'The effect of human activity on radiative forcing of climate change: a review of recent developments', *Global and Planetary Change* **20**, 205–225.

Solanki, S. K., Fligge, M.: 2000, *Space Sci. Rev.*, this volume.

WMO: 1999, Scientific Assessment of Stratospheric Ozone Depletion: 1998: World Meteorological Organisation Global Ozone Research and Monitoring Project — Report No 44, Geneva Switzerland.

Address for Offprints: Department of Meteorology, University of Reading, Earley Gate, Reading RG6 6BB, UK., k.p.shine@reading.ac.uk

WHAT CAN CLOUD OBSERVATIONS TELL US ABOUT CLIMATE VARIABILITY?

JOEL R. NORRIS
National Center for Atmospheric Research, Boulder, Colorado * †

Received: 21 October 1999; Accepted in final form: 1 December 1999

Abstract. Clouds have a large impact on the Earth's radiation budget and hence have the potential to exert strong feedbacks on climate variability and climate change. These feedbacks are not well-understood, so it is essential to investigate observed relationships between cloud properties and other parameters of the climate system. Suitable cloud datasets based on surface observations and satellite observations are described and various advantages and disadvantages of each are discussed. In particular it is noted that significant inhomogeneities likely exist in the datasets which have important implications for studies of climate variability. Recommendations are made for the use of cloud data in future investigations.

1. Introduction

Cloudiness is one of the most important variables influencing the Earth's radiation budget, but our present understanding of its role in the climate system is limited. Current general circulation models do not consistently and correctly simulate cloud feedbacks internal to the climate system (Cess et al., 1996). Other large uncertainties are possible cloud responses to external forcings such as anthropogenic aerosol (Twomey et al., 1984; Albrecht 1989) or cosmic ray flux (Marsh and Svensmark, 2000). For these reasons it is important to examine past cloud variability and potential connections to other internal and external climate parameters. The two primary sources of cloud data are surface synoptic observations and satellite radiance measurements, each with its own advantages and disadvantages. Surface observations extend over a much longer period of record, but lack the quantitative radiative information and global coverage provided by satellites. Unfortunately, artificial inhomogeneities appear to exist in all cloud datasets, thus making interpretation and attribution of observed long-term changes in global cloud cover problematic.

* The National Center for Atmospheric Research is sponsored by the National Science Foundation.

† Current affiliation: Program in Atmospheric and Oceanic Sciences, Princeton University, Princeton, New Jersey

TABLE I
Cloud Datasets

	EECRA	ISCCP
source	synoptic weather observations	weather satellites
identification	visual, by synoptic code	radiance threshold
time period	1971–96 (land), 1952–95 (ocean)	Jul 1983 to Dec 1993
coverage	irregular	global
resolution	user-defined averaging	3-hourly, 250km x 250 km
parameters	total cloud amount low cloud amount low, mid, high cloud type	cloud amount cloud optical thickness cloud top pressure
advantages	identification of cloud type multidecadal time record unobstructed view of low clouds	quantitative radiative information global regular sampling unobstructed view of high clouds
disadvantages	no quantitative radiative information irregular and incomplete sampling obstructed view of high clouds cloud measured in sky-cover inhomogeneities	uncertain identification of cloud type short time record obstructed view of low clouds uncertain threshold for cloud inhomogeneities

2. Cloud Datasets

2.1. Surface-Based

Surface-based cloud datasets suitable for studying climate variability are constructed from synoptic weather observations. Human observers on ships or at land stations visually measure the amount of total cloudiness and low cloudiness by eighths or tenths of sky-cover and identify cloud type and low, mid, and high levels following the synoptic code (WMO 1974). Note that the fraction of cloud covering the sky dome may not be the same as the fraction of cloud covering the earth. Warren et al. (1986; 1988) have averaged these observations to seasonal means at $5° \times 5°$ (land) and $10° \times 20°$ (ocean) resolution for the years 1971–1981 (land) and 1952–1981 (ocean). Much of the Southern Ocean and some land regions are poorly sampled. No global collection of land observations is available prior to 1971, and obvious inhomogeneities are present in the ocean observations prior to 1952. These data have been updated by Hahn and Warren (1999) in the Extended Edited Cloud Report Archive (EECRA) to 1971–1996 (land) and 1952–1995 (ocean). The EECRA is a quality-controlled collection of the original individual cloud observations and thus allows the user to define the averaging regions and time periods. Because the observers view from the surface, the EECRA is especially useful for examining changes in low cloud cover. Further information is summarized in Table I.

2.2. Satellite-based

Satellite-based cloud datasets are typically constructed from visible and infrared radiance measurements. The International Satellite Cloud Climatology Project (ISCCP) (Rossow and Schiffer, 1991) has the longest period of record available (1983–1993). It is derived from a succession of geostationary and polar-orbiter weather satellites and provides global coverage at 3-hourly and 250 km $\times$ 250 km resolution. The threshold method is used to identify cloudiness. Since clouds are almost always brighter and colder than the surface, the darkest and warmest pixels are identified as clear sky and pixels brighter and colder beyond a certain threshold are identified as cloudy. This has some uncertainty, and a decrease in the threshold from the C-series to the D-series of ISCCP caused observed global cloud cover to increase by 5% (Doutriaux-Boucher and Sèze, 1998). Cloud optical thickness and cloud top pressure can be calculated using a radiative transfer model. Due to their top-down view, satellite data are the best resource for examining high cloudiness. Since the threshold method used by ISCCP has difficulty identifying thin cirrus, the High Resolution Infrared Radiation Sounder (HIRS) dataset (Wylie and Menzel, 1999) may be a better resource for examining this cloud type. Because different datasets use different thresholds and other retrieval methods, information from two datasets cannot be combined into a homogeneous time series. Further information is summarized in Table 1.

3. Possible Inhomogeneities

In studies of cloud variability it is important to be aware of possible inhomogeneities in the data. These can arise from systematic biases that become more prominent as natural variability of opposite signs is averaged out. Thus, inhomogeneities are worst for the largest averaging intervals in space and time (i.e. multi-year changes in global cloud cover). Norris (1999) documented that changes in EECRA zonal mean total cloud cover and low cloud cover between 1952 and 1995 were positive at every latitude with sufficient sampling, but expressed uncertainty regarding the reliability of the trends. On the one hand no likely cause for an artifact was found, but on the other hand it was not clear what underlying physical mechanism would cause substantially different processes in the Tropics, subtropics and midlatitudes to all produce increasing cloudiness. For this reason it is important to corroborate observed changes in cloud cover with physically consistent changes in related climate parameters (e.g., Norris et al., 1998; Norris, 1999). EECRA variations can also be compared to ISCCP variations for the time period the two datasets overlap. Figure 1 shows that they do not agree, suggesting that the apparent increase in surface-observed global ocean cloud cover since the early 1950's may be largely spurious.

In order to gain insight into possible causes of the ISCCP time series presented in Figure 1, correlations between the Figure 1 time series and the time series of

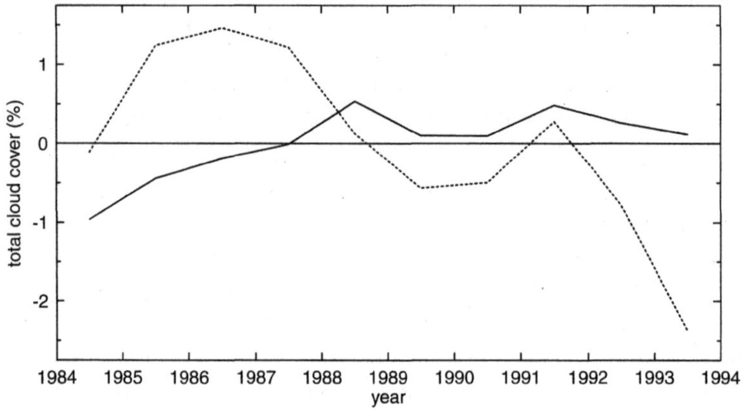

Figure 1. Annual mean ocean cloud cover averaged over 40°S - 60°N (where the EECRA has sufficient sampling). The solid line is EECRA and the dotted line is ISCCP D-series.

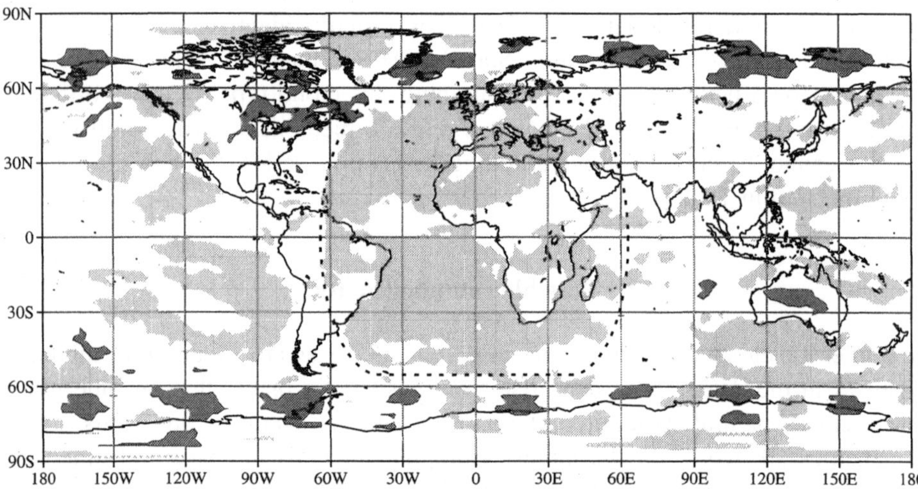

Figure 2. Correlation of ISCCP D-series gridbox time series with ISCCP time series in Fig. 1. Lighter gray indicates correlations above 0.4; darker gray indicates correlations below -0.4. The dotted line indicates the approximate footprint of Meteosat.

annual means at each grid box were calculated in hope that the spatial pattern of correlation coefficients (shown in Fig. 2) would suggest meteorological processes responsible for the variation in near-global ocean cloud cover. The most striking feature of Figure 2 is the circular patch of positive correlation centered on 0°N, 0°E. It seems unlikely that a natural process would produce such a regular pattern, but it is interesting to note that the pattern almost exactly resembles the footprint of Meteosat data incorporated into ISCCP (Meteosat is the European geostationary satellite). In particular, note the discontinuities in the western North Atlantic and South Atlantic. Some of the correlation patterns elsewhere around the globe are suggestive of other geostationary satellite footprints. The semi-regular patterns of alternating correlation sign seen between 60° and 70° latitude are also suspicious and may be related to polar-orbiter swaths. Similar features are observed for correlations with the time series of ISCCP monthly mean global cloud cover (not shown).

It has been known that instrument degradation and miscalibration between successive satellites created inhomogeneities in the previous version (C-series) of ISCCP data (Klein and Hartmann, 1993). The D-series ISCCP data used in the present study was generated using new methods to avoid the problems of the C-series, but it now appears that the attempt was not completely successful. The correlation patterns evident in Figure 2 suggest that the time series of ISCCP global mean cloudiness does not result from any natural process but is instead somehow related to systematic biases in the retrieval and/or processing of satellite radiances, though further investigation will be required to identify the exact problems. Accordingly, the ISCCP time series presented in Figure 1 is probably spurious, and any resemblance to time series of other parameters (e.g., Marsh and Svensmark, 2000) is merely coincidental.

4. Recommendations

The above results do not invalidate all uses of surface-observed or satellite cloud data, but rather enjoin caution. Natural variability is much greater on small space and time scales; hence, investigations of synoptic and regional variability in cloudiness should not greatly suffer from inhomogeneities. It is only when natural variability is averaged out in large space and time domains that the systematic biases become relatively strong. Some recommendations for studies of cloud-climate variability follow.

- Use extreme caution when interpreting cloud variability averaged over large space and time domains.
- Always check for the presence of unphysical variability in cloud data.
- Corroborate observed cloud variability with variability in other physically-related climate parameters.
- Look at geographical patterns of cloud variability since they provide more insight than large-scale means.
- Examine cloud type as well as total cloud cover since cloud type is a better measure of processes and radiative impacts.

Acknowledgements

This work was supported by the National Science Foundation while the author was an Advanced Study Program postdoctoral fellow at the National Center for Atmospheric Research.

References

Albrecht, B. A., 1989: Aerosols, cloud microphysics, and fractional cloudiness. Science, 245, 1227-1230.

Cess R. D., et al., 1996: Cloud feedback in atmospheric general circulation models: an update. J. Geophys. Res., 101, 12791-12794.

Doutriaux-Boucher, M., and G. Sèze, 1998: Significant changes between the ISCCP C and D cloud climatologies. Geophys. Res. Lett., 25, 4193-4196.

Hahn, C. J., and S. G. Warren, 1999: Extended edited synoptic cloud reports from ships and land stations over the globe, 1952-1996. NDP026C, Carbon Dioxide Information Analysis Center, Oak Ridge National Laboratory, Oak Ridge, TN.

Klein, S. A., and D. L. Hartmann, 1993: Spurious changes in the ISCCP dataset. Geophys. Res. Lett., 20, 455-458.

Marsh, N., and Svensmark, H.: 2000, 'Cosmic Rays, Clouds, and Climate', *Space Sci. Rev.*, this volume.

Norris, J. R., 1999: On trends and possible artifacts in global ocean cloud cover between 1952 and 1995. J. Climate, 12, 1864-1870.

Norris, J. R., 1999: Interannual and interdecadal variability in the storm track, cloudiness, and sea surface temperature over the summertime North Pacific. J. Climate, in press.

Norris, J. R., Y. Zhang, and J. M. Wallace, 1998: Role of low clouds in summertime atmosphere-ocean interactions over the North pacific. J. Climate, 11, 2482-2490.

Rossow, W. B., and R. A. Schiffer, 1991: ISCCP cloud data products. Bull. Amer. Meteor. Soc., 72, 2-20.

Twomey, S. A., M. Piepgrass, and T. L. Wolfe, 1984: An assessment of the impact of pollution on global cloud albedo. Tellus, 36B, 356-366.

Warren, S. G., C. J. Hahn, J. London, R. M. Chervin, and R. L. Jenne, 1986: Global distribution of total cloud cover and cloud type amounts over land. NCAR/TN-273+STR, NCAR Boulder, CO, 29 pp. + 200 maps.

Warren, S. G., C. J. Hahn, J. London, R. M. Chervin, and R. L. Jenne, 1988: Global distribution of total cloud cover and cloud type amounts over the ocean. NCAR/TN-317+STR, NCAR Boulder, CO, 42 pp. + 170 maps.

WMO, 1974: Manual on Codes. WMO Publ. 306, Vol. 1 [Available from World Meteorological Organization, Avenue Giuseppe-Motta 41, CP2300, 1211 Geneva 2, Switzerland.]

Wylie, D. P., and W. P. Menzel, 1999: Eight years of high cloud statistics using HIRS. J. Climate, 12, 170-184.

CLOUD FORMATION AND THE POSSIBLE SIGNIFICANCE OF CHARGE FOR ATMOSPHERIC CONDENSATION AND ICE NUCLEI

R. GILES HARRISON
Department of Meteorology, The University of Reading P.O. Box 243, Reading RG6 6BB, UK

Received: 11 November 1999; Accepted in final form: 7 March 2000

Abstract. Cloud formation in the atmosphere is related to the presence of water vapour, cloud condensation nuclei (CCN) and ice nuclei (IN). Ionisation in the atmosphere is caused by a variety of sources, but the contribution from cosmic rays is always present and is modulated by the solar cycle. Methods of investigating the variability in ionisation are described. The mechanisms proposed by which (1) ionisation could influence cloud formation, and (2) by which changes to the CCN and IN could occur are discussed. Direct formation of sulphate CN is conceivable in atmospheric air by radioactivity, and charging of molecular clusters leads to greater collisions rates than for neutral clusters. Modification of the ice nucleation efficiency of aerosol could also have atmospheric effects through latent heat release. However in both cases definitive atmospheric experimental work is lacking and therefore any link between solar variability and clouds remains unproven.

1. Introduction

Clouds are principally composed of ice and water, although high and low altitude clouds of limited vertical extent may contain principally just one phase of liquid water or ice. Clouds may be stratiform or convective, and in some cases contain electrified particles. The bulk properties of clouds have a marked influence on climate, although the sign of the effect depends on cloud properties. It is conceivable that the microphysical processes leading to droplet and ice formation could also be influenced by electrical effects, and it is this question which is discussed further here.

Water vapour is effectively a gaseous constituent of atmospheric air, and its concentration can consequently be determined by its partial pressure. At any given temperature there is an associated maximum value of partial pressure due to water vapour, the *saturation vapour pressure*. Air containing sufficient water vapour to generate the saturation vapour pressure is said to be *saturated*, and has a relative humidity of 100%. Occasionally it is possible for the saturation vapour pressure to be exceeded in localised regions, known as *supersaturation*. In the atmosphere this is never greater than an excess relative humidity of few percent, due to the abundance of small particles (or *aerosol*) on which condensation can occur. Many different kinds of aerosol are capable of acting as *condensation nuclei* CN but it is the subset able to permit condensation at atmospheric supersaturation (*cloud condensation nuclei*, CCN) which are of greatest interest.

Space Science Reviews **94:** 381–396, 2000.

If the temperature is below 0°C, liquid water droplets may persist without actually freezing *i.e.* 0°C is the temperature at which ice melts, but not necessarily that temperature at which a water droplet will freeze. Any water droplet with a temperature below 0°C is *supercooled*, and in a thermodynamically unstable state in which freezing can be initiated. Freezing occurs by *heterogeneous* or *homogeneous nucleation*. In heterogeneous nucleation, the supercooled water freezes as a result of the presence of a suitable *ice nucleus*. Homogeneous nucleation becomes more probable at low temperatures, and all supercooled water in atmospheric clouds turns to ice at about −40°C by this process. Suitable Ice Nuclei (IN) are very rare in the atmosphere, and only a very small (but variable) fraction of the atmospheric aerosol is able to initiate ice. Consequently a significant fraction of atmospheric cloud contains supercooled water. At temperatures between about −5°C and −10°C, *ice multiplication* processes can occur. When a particle freezes at these temperatures, mechanical stresses during freezing lead to the ejection of small ice fragments, which are highly efficient ice nuclei. These processes are addressed in the cloud physics text books, *e.g.* Rogers and Yau (1989), Mason (1971) and Pruppacher and Klett (1997).

Although a possible relation between cosmic rays and climate has attracted interest due to the correlation published in Svensmark and Friis-Christensen (1997), their work is certainly not the first to suggest that cosmic rays could influence cloud processes, and indeed there is other work in which more detailed mechanisms are proposed. Svensmark and Friis-Christensen (1997) suggested that electrical charging of atmospheric aerosol influences the particles' efficacy as cloud nuclei. Pudovkin and Veretenenko (1995) found correlations between cloud coverage and Forbush decreases in a narrow latitude band, which they suggested occurs via ice cloud. A more detailed mechanism involving ice clouds has been proposed in Tinsley and Dean (1991). Their work develops an atmospheric electrical mechanism to explain observed correlations between sunspot number and various indices of mid-latitude cyclone intensity. It depends on the electrification of atmospheric aerosol enhancing the aerosol particles' efficacy as ice nuclei.

1.1. Atmospheric Particles

Atmospheric air is rarely regarded as an electrical conductor, but it is able to pass a small electric current because of the presence of singly-charged particles known as molecular *small ions*. These have a high electrical mobility, and are consequently easily accelerated by the electric fields which exist in the atmosphere. Such a high mobility conventionally distinguishes them from *Aitken Nuclei* and the larger *aerosol particles* (AP), although they strictly are all part of the atmospheric aerosol (see Table 1).

The interaction between ions and aerosol in low electric fields is essentially due to diffusive interactions, and charge exchange is invariably associated.

TABLE I
Typical Properties of Atmospheric Particles and Droplets

Particle	Radius/μm	Mobility/10^{-6} $m^2V^{-1}s^{-1}$
Small ions	0.005 to 0.008	0.5–100
Aitken Nuclei	0.008 to 0.03	0.05–0.5
Aerosol	0.01 to 10	
Cloud droplets	2 to 30	
Rain drops	30 to 3000	

1.2. The Atmospheric Electrical System

Electric fields present in the atmosphere vary between fair weather values of typically 10^2 Vm^{-1} at the surface to about 10^5 Vm^{-1} in thunderstorms before a lightning discharge (McGorman and Rust, 1998). The atmospheric electric circuit is thought to be sustained by thunderstorms continually active around the tropics, in which a current of 2,000 A typically flows. This total current flows throughout the atmosphere, in regions of disturbed and undisturbed weather, and the conduction current is carried by vertical electrical migration of small ions.

2. Atmospheric Ionisation

2.1. Ionisation and Ion Formation

Ion-pairs are continually produced in the atmosphere by radiolysis of air molecules. The ions produced are rarely single species but clusters of water molecules around a central ion. Typical atmospheric ion concentrations in unpolluted air and fine weather are about 500 ions cm^{-3}, and vertical charge fluxes caused by conduction in fair weather conditions are typically 2 pAm^{-2} ($\sim$12 ions cm^{-2} s^{-1}).

Atmospheric radiolysis occurs from three principal sources: radon, cosmic rays and terrestrial gamma radiation (Chalmers, 1967), and the partitioning between the sources varies vertically. Near the surface, ionisation from turbulent transport of radon and other radioactive isotopes is important, together with gamma radiation from isotopes below the surface. Ionisation from cosmic rays is always present, comprising about 20% of the ionisation at the surface. The cosmic fraction increases with increasing height in the atmosphere and dominates above the planetary boundary layer. Ions are removed by their own self-recombination, attachment to existing aerosol and, possibly as discussed in Section 3.1, ion-induced aerosol formation.

Atmospheric small ions of both signs with number concentrations n_+ and n_- are governed by

$$\frac{dn_\pm}{dt} = q - \alpha n_\pm n_\mp - n_\pm \int_{a=0}^{\infty} \sum_{j=-\infty}^{\infty} \beta_{\pm 1,j}(a) N_j(a) da \tag{1}$$

where the ions are produced at a rate q per unit volume. Ions (which are assumed to carry unit charges) are removed by ion- ion recombination (with recombination coefficient α), and by attachment to aerosol particles, which causes charge transfer to the aerosol. The aerosol attachment rate $\beta_{\pm 1,j}(a)$ depends on aerosol particle radius a and the number of elementary charges j present on the aerosol particle (Gunn, 1954). In Equation 1, the size and charge distributions of atmospheric aerosol particles are accounted for by the integral of number concentration $N(a)$ over all particle radii, and by a sum across all possible particle charges at each radius.

Small ions consist of clusters of water molecules collected around a singly charged ion. They have a lifetime of the order of a hundred seconds. Clusters such as $H_3O^+(H_2O)_n$, $H^+(H_2O)_n$, $NO^+(H_2O)_n$ and $NO_2^+(H_2O)_n$ are common for the positive ions and $O_2^-(H_2O)_n$, $CO_4^-(H_2O)_n$, $NO^-(H_2O)_n$ or $NO_2^-(H_2O)_n$ for the negative ions (Volland, 1984). The chemical difference between the species in the positive and negative ions leads to some physical asymmetries in the ion properties, and the negative ions are more mobile. The ratio of mobilities $\mu_-/\mu_+ \sim 1.2$. Collisions between the ions and atmospheric aerosol lead to charge-exchange and electrification of the aerosol, and the ion asymmetry ensures that the collisions do not lead to an average charge of zero. Local electric fields can cause further asymmetries, by depletion of one sign of ion concentration, and consequently substantial aerosol electrification in that region.

In considering how natural atmospheric ionisation might influence cloud formation, there seem two more possible routes to discuss:

1. *direct* processes, such as production of new aerosol (*e.g.* sulphate) by gas-to-particle conversion (GPC) or homogeneous nucleation
2. *indirect* processes, such as the modification of existing heterogeneous nucleation processes by affecting the condensation nuclei (CN) or ice nuclei (IN).

2.2. Temporal Variation in Atmospheric Ionisation

The ion-pair production rate per unit volume q, shows temporal and spatial variability. The increasing fractional contribution of cosmic ray ionisation with height has already been mentioned, but the temporal variations in q at the surface are also substantial. Figure 1 shows the count rates of two Geiger counters operated at Reading over a ten hour period. One was operated above the grassy surface of the Department of Meteorology's Field Site, and the other within the Department building itself, 500 m distance, and sampled at 1 Hz using a ratemeter system. At all

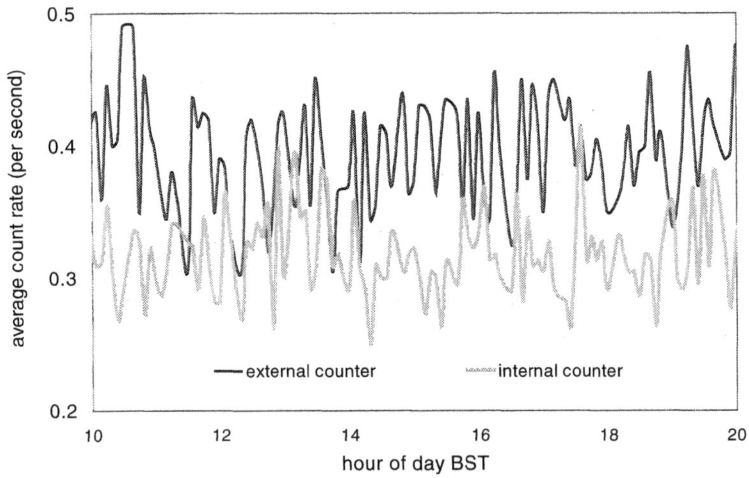

Figure 1. Five-minute averaged count rates from two Geiger counters sampled using ratemeters, one placed within the basement of the Meteorology Department building at Reading, and the other exposed on the Department's field site 500 m distant, on 16th October 1999.

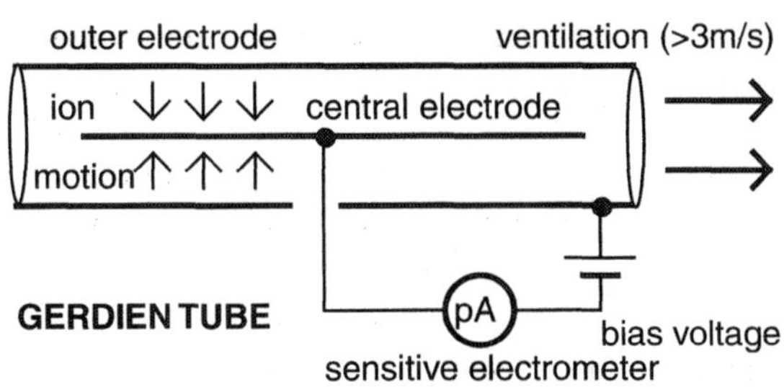

Figure 2. Essential features of a Gerdien condenser for air ion measurements, operating in the current mode. A bias potential on the central electrode permits unipolar collection of ions up to a critical value of mobility, and the total current determined from the ions captured is measured using a sensitive ammeter, from which the ion number concentration can be determined.

time scales there are occasional peaks which coincide, perhaps related to energetic air showers. Whatever the precise origin, it seems appropriate to regard q, which has a frequently quoted surface value of 10 ionpairs cm^{-3} s^{-1} as a long-term mean. Micrometeorological processes, such as the turbulent transport of Radon isotopes, also contribute to the variability.

2.3. Measurement of Atmospheric Ionisation

Atmospheric ions can be studied by measuring the conductivity, if a fixed mobility fraction is selected. Gerdien developed a method of measuring air conductivity using a cylindrical condenser (Gerdien, 1905). A voltage is applied between two cylindrical electrodes and the inner one is earthed via a sensitive ammeter (*e.g.* Harrison, 1997), Figure 2. In the tube, ions of the same sign as the polarising voltage are repelled by the outer electrode, and move in the electric field to meet the inner electrode, where they cause a small current. In the miniature Gerdien used by Aplin and Harrison (1999) the current was of order 10^{-15} A. The conductivity σ is given by

$$\sigma = \frac{i\varepsilon_0}{CV} \tag{2}$$

where C is the capacitance of the Gerdien and V the polarising voltage across the electrodes, as long as the output current is proportional to the polarising voltage, indicating that a fixed fraction of air ions are collected by the central electrode.

2.4. Aerosol Electrification

The net charge present in the air is partitioned between atmospheric small ions and aerosol particles, between which there are interactions. Particle charges can be calculated by solving Equation 1 and the associated aerosol charging equations. It is not possible to find analytical time- dependent solutions, but it is possible in the steady-state. The number concentration N_j of monodisperse aerosol particles carrying j elementary charges is given by the Modified Boltzmann Distribution (Clement and Harrison, 1992) as

$$\frac{N_j}{N_0} = \left[\frac{n_+\mu_+}{n_-\mu_-}\right]^j \frac{8\pi\varepsilon_0 akT}{je^2} \sinh\left[\frac{je^2}{8\pi\varepsilon_0 akT}\right] \exp\left[\frac{-j^2e^2}{8\pi\varepsilon_0 akT}\right] \tag{3}$$

with a mean charge (Gunn, 1954)

$$\bar{j} = \frac{4\pi\varepsilon_0 akT}{e^2} \ln\left[\frac{n_+\mu_+}{n_-\mu_-}\right] \tag{4}$$

where $\mu_\pm$ are the positive and negative small ion mobilities, $n_\pm$ their number concentrations, T the temperature, e the modulus of the electronic charge, k Boltzmann's constant and ε_0 the permittivity of free space.

The ion asymmetry factor $x(= \mu_+n_+/\mu_-n_-)$ determines the mean of the charge distribution. x is not unity for two reasons. Firstly, chemical difference between positive and negative ions cause $\mu_+ \neq \mu_-$. Secondly, $n_+ \neq n_-$ because the ion concentrations n_+ and n_- vary considerably in the atmosphere, especially on the boundaries of aerosol and water droplet regions (Gringel, 1978). Paltridge (1965) simultaneously measured the small-ion density and electrical conductivity and found, for the altitude range 4 to 24 km, n_+/n_- was 1.43 ± 0.14, and $\langle x \rangle = 1.36$. Measurements of polar conductivities in clouds were also reported by Scott and Evans (1969), who found an increase in the relative value of the positive polar conductivity in a region of high relative humidity, giving $x > 3$. They observed that the ratios of the polar conductivities inside the cloud were approximately equal to those outside and noted that in fair weather the negative conductivity exceeded the positive conductivity. Continuous mountain recordings of n_+ and n_- have been carried out by Reiter (1974) and Dolezalek *et al.* (1985), giving variations of n_+/n_- from 1.75 to 1.9 in various weather conditions.

3. Processes Directly Related to Atmospheric Ionisation

3.1. Aerosol Nucleation

Horrak *et al.* (1998) reported the spontaneous formation of intermediate size ions in atmospheric air, by observing changes in the ion mobility spectrum of urban air. Aerosol can be formed spontaneously in unpolluted marine air (O'Dowd *et al.*, 1996) and theoretical work on clustering reactions of ions (Castleman, 1982; Raes *et al.*, 1986) suggests ions are sometimes critical in gas-to-particle conversion. The chemical composition of carrier gases is important in ion-induced nucleation, and Emi *et al.* (1998) report laboratory experiments in which alpha-particles produce nanometre particles in clean air containing 5 ppm SO_2 and 3500 ppm water vapour. Adachi *et al.* (1998) report that the presence of NH_3 enhances the effect, and find particle concentrations of $\sim 10^4\,cm^{-3}$ with $\sim$6 MBq α-activity. Yu *et al.* (1998) find that there is significant ion- induced sulphate CN formation in aircraft plumes and hypothesise its occurrence in the atmosphere (Turco *et al.*, 1998).

In clean air, the principal removal mechanism for ions is self- recombination. As a typical positive or negative ion will, in general, be hydrated by a few water molecules (because of its electric charge), the recombination process will lead to a neutral cluster consisting of a number of water molecules. Neutral water molecules combine and dissociate to form clusters in a dynamic equilibrium continuously: the significance of electrification might be that collisions between the relatively rare large clusters are made much more likely if they carry opposite signs.

Simplifying Equation 1 by ignoring the effect of polarity ($n_+ \approx n_- = n$), then the variation of ion concentration with time is

$$\frac{dn}{dt} = q - \alpha n^2 - \beta n Z - \gamma n \tag{5}$$

where Z is the equivalent monodisperse aerosol number concentration and γ an ion-induced nucleation coefficient. It has been usual to regard the aerosol and ions as distinct systems of particles, which interact purely by collisions, however the mobility bursts observed by Horrak *et al.* (1998) now suggest that the γ-term could sometimes be significant, and would couple the time dependence in the ion concentration equations to time dependence in the aerosol concentration. Ion-induced formation of water droplets is not possible at the low levels of supersaturation found in the atmosphere (Mason, 1971).

If, however, the small ion concentration was limited only by recombination, and *all* the ions were converted to CCN by some hitherto unknown atmospheric mechanism, typical *surface* values of q and α would suggest a *maximum* concentration of 2500 cm^{-3}. This would cause a measurable change in continental CCN concentrations, and a significant CCN change in oceanic air.

3.2. Formation of Molecular Clusters by Recombination

The recombination term in Equation 5 leads to the formation of neutral clusters. Positive and negative ions are typically clusters of a few water molecules around a central ion, thus recombination could typically yield neutral clusters of up to ten water molecules. Large neutral molecular clusters can be formed by a variety of possible collisions between smaller charged and uncharged clusters. Rate equations can be written to investigate this, as the collisions which can occur are:

1. single-charged clusters with neutral clusters (to give single-charged larger clusters)
2. single-charged clusters with oppositely-charged single charged clusters (to give larger neutral clusters)
3. neutral-neutral collisions (to give larger neutral clusters)

A simplified analysis of this problem is illuminating. To do this, molecular clusters are assumed to contain k identical molecules, and the charged clusters are assumed to carry single charges only. P_k is the number concentration of positive clusters containing k molecules, M_k the number concentration of negative clusters containing k molecules and N_k the number concentration of neutral clusters containing k molecules. Neglecting the aerosol terms, Equation 5 can be extended to consider larger clusters formed by recombination. For single-molecule clusters (produced at a rate q per unit volume), the concentration variation is

$$\frac{dP_1}{dt} = q - \sum_{j=1}^{\infty} \alpha_R(1, j)P_1 M_j - \sum_{j=1}^{\infty} \alpha_C(1, j)P_1 N_j \tag{6}$$

and for k-molecular positive clusters

$$\frac{dP_k}{dt} = \sum_{j=1}^{k-1} \alpha_C(k-j, j)P_{k-j}N_j - \sum_{j=1}^{\infty} \alpha_C(k, j)P_k N_j - \sum_{j=1}^{\infty} \alpha_R(k, j)P_k M_j \tag{7}$$

where $\alpha_R(k, m)$ is the *recombination coefficient* between positive clusters containing k molecules and negative clusters containing m molecules and $\alpha_C(k, m)$ the *combination coefficient* between positive k-clusters and neutral m-clusters. Similar equations can be written for negative clusters M_k.

For the neutral clusters, the equation for their number concentration is

$$\begin{aligned} \frac{dN_k}{dt} &= \sum_{j=1}^{k-1} \alpha_R(j, k-j)P_j M_{k-j} + \frac{1}{2}\sum_{j=1}^{k-1} \alpha_C(k-j, k)N_{k-j}N_k \\ &\quad - \sum_{j=1}^{\infty} \alpha_C(k, j)N_k P_j - \sum_{j=1}^{\infty} \alpha_C(k, j)N_k M_j - \sum_{j=1}^{\infty} \alpha_C(k, j)N_k N_j \end{aligned} \tag{8}$$

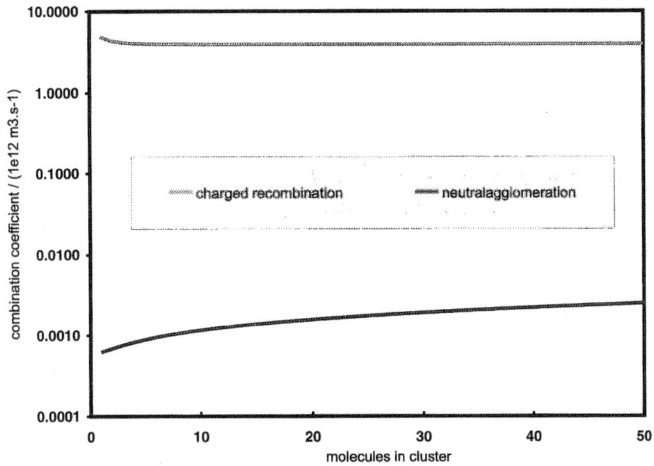

Figure 3. The combination coefficients α_R and α_C between charged clusters (recombination) and neutral clusters (kinetic collisions). Combination rates between charged molecular ions are typically 10^4 greater than those between neutral molecular clusters of comparable size.

The collision coefficients above differ depending on whether there is charged or uncharged interaction. In the case of *uncharged interactions* the coefficient is that for collisional agglomeration (Landau and Lifshitz, 1980)

$$\alpha_C(m,n) = 2\left(\frac{3}{4\pi}V\right)^{\frac{2}{3}}\left(\frac{2\pi k_B T}{m_L}\right)^{\frac{1}{2}}\left(m^{\frac{1}{3}}+n^{\frac{1}{3}}\right)^2\left(\frac{m+n}{mn}\right)^{\frac{1}{2}} \tag{9}$$

where T is temperature, k_B Boltzmann's constant, m_L the molecular mass of a unit cluster, V the molecular volume (m_L/ρ), ρ the density of liquid in cluster and there are considered to be m and n molecules in the two clusters combining. For *charged interactions* (essentially ion-ion recombination) Thomson's theory (Thomson, 1924) gives

$$\alpha_R(m,n) = \pi s^2\sqrt{c_m^2+c_n^2}\left[2-\frac{\lambda_m}{2s}\left(1-e^{-2s/\lambda_m}\right)-\frac{\lambda_n}{2s}\left(1-e^{-2s/\lambda_n}\right)\right] \tag{10}$$

where the m and n clusters have opposite signs and c is their root mean-square speed, λ their mean free path and s a critical radius depending on their electrical energy.

Figure 3 compares the charged and uncharged coefficients: charged interactions are able to occur much more rapidly but, as a number fraction of the water molecules present, are expected to be exceedingly rare.

Using these combination coefficient expressions, the cluster equations have been solved as a function of time to give their variation in concentration, Figure 4. Up to ten molecules per cluster were permitted, although in practice including additional loss terms (*e.g.* through aerosol attachment or direct evaporation) would also

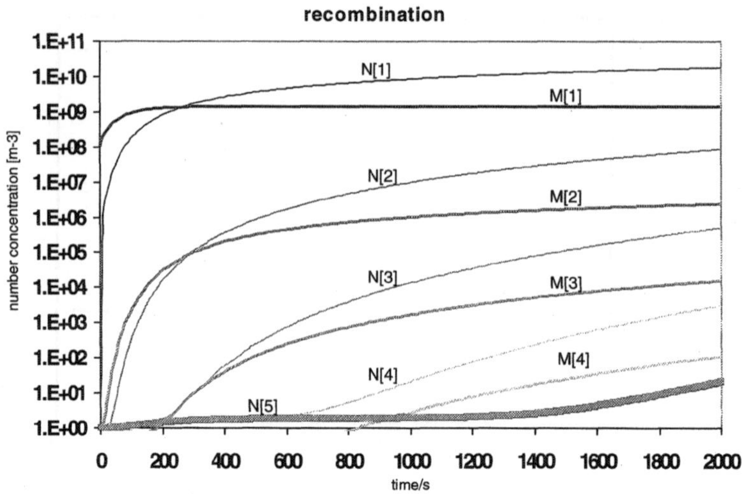

Figure 4. Number concentration of negative and neutral clusters containing k molecules $M[k]$ and $N[k]$ as a function of time.

be important. From these calculations it appears possible that collisions between large oppositely charged clusters could enhance the production rate of large neutral clusters. This charge-assisted route might offer a more rapid source of a few large neutral clusters than neutral-neutral interactions.

4. Processes Indirectly Related to Atmospheric Ionisation

4.1. Ice Nucleation

The study of ice nuclei (IN) is only partially conclusive (Mason, 1971, Pruppacher and Klett, 1997). It is clear that only a small fraction of aerosol particles (AP) can act as IN although this fraction increases with decreasing temperature (see also Figure 7b). This is usually explained in terms of a chemical or shape effect, depending on the precise nucleation mechanism considered. Atmospheric concentrations of IN show diurnal variations (*e.g.* Rosinski *et al.*, 1995), some of which have their origin as CN (Rosinski, 1995).

Much of the early work on ice nucleation (Fletcher, 1966) suggests that typical IN number concentrations are between 10^{-6} and 10^{-3} cm^{-3} at warmer than −20°C, rising by several orders of magnitude at the colder temperatures from −20 to −35°C. Although there appear to be grossly inadequate IN concentrations to account for the amount of ice observed, *ice multiplication* mechanisms (where splintering occurs producing ice particles which themselves act as nuclei) will certainly contribute to ice formation at the warmer temperatures (−3 to −8°C) (Mossop, 1985). The work of Pruppacher (1973), and to a lesser extent that of

Gabarashvili and Gliki (1967) suggested that aerosol electrification could affect ice nucleation, however the work of Katz (1968) emphasises that trace impurities or physical differences are an important confounding factor. Enhancing heterogeneous ice nucleation by electrification has been proposed by Tinsley and Dean (1991) but it is currently supported by very little definitive experimental work (Pruppacher, 1973). Large particle charges are considered necessary to cause the enhancement of aerosol as IN, and it is therefore important to establish if such highly-charged aerosol particles actually exist in the atmosphere.

Charge distributions on atmospheric particles can be found using the modified Boltzmann distribution, Equation (3). The mean charge varies with x and for values of x typical of cloudless atmospheric air ($x = 1.2$), calculations show that in a monodisperse number concentration of 1000 particles cm^{-3} there will only be a few particles carrying large ($j > 10$) charges. This is possible even though the *mean* charge for the particle *population* is close to zero, but it is unlikely that any highly charged particle will maintain its charge state for anything more than a short transient, as ions will tend to neutralise such particles.

4.2. Freezing of Supercooled Water

Although some qualitative experiments suggest that the electrical interactions between a supercooled droplet and a suitable ice nucleus might be implicated in freezing, quantitative laboratory or atmospheric experimental support is lacking. In studying heterogeneous nucleation, Harrison and Lodge (1998) described apparatus to supercool a single droplet, and detect unambiguously (by temperature measurement) when the droplet freezes. Their apparatus suspended a 1microlitre water droplet on an ultra-miniature thermistor, using a Peltier cooling device to supercool the droplet. Electrical connections to the microthermistor allowed the temperature of the droplet to be constantly monitored, and they also permitted direct electrical heating of the droplet after freezing occurred. Deionised water droplets could be readily supercooled to −20°C before freezing, and repeated determinations of the freezing temperature for the same droplet were possible. The thermal characteristics of the phase change are very well-defined, Figure 5, with rapid latent heat release on freezing, allow a distribution of freezing temperatures to be found (Figure 6). Tinsley and Dean (1991) argue that if charged aerosol are more effective IN than neutral aerosol, small electrical changes in the aerosol could modulate significant amounts of latent heat within clouds. There would then be an amplifying effect, possibly allowing small electrical energies to have greater significance for cloud development than would otherwise be expected.

4.3. Electrofreezing

Electrofreezing is a term used inconsistently by different authors and here it will be used to describe any electrical-enhancement of the phase transition between

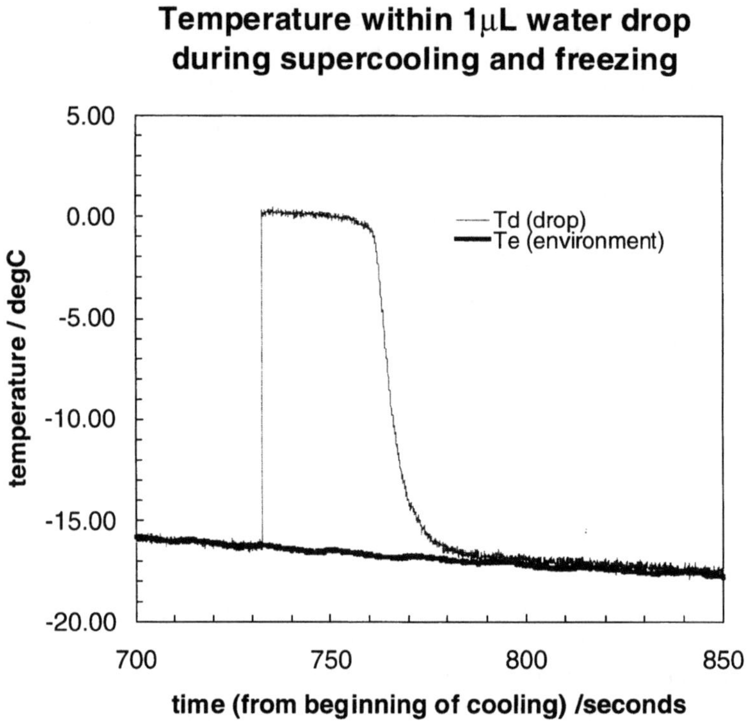

Figure 5. Time series (obtained at 12 Hz) of droplet temperature T_d in an environment at T_e during supercooling and freezing process. The initial supercooling phase continues until the droplet begins to freeze, when its temperature instantaneously rises to about 0°C as latent heat is released. After it has completely frozen it cools, initially rapidly, to the temperature of its surroundings (from Harrison and Lodge, 1998).

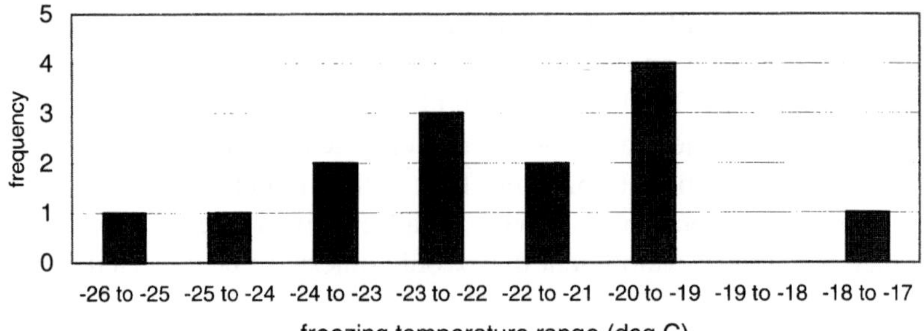

Figure 6. Distribution of freezing probabilities of deionised 1 μlitre water droplets, found using the supercooling apparatus of Harrison and Lodge (1998).

supercooled water and ice, (Dawson and Cardell, 1973). The different mechanisms suggested are now briefly discussed.

In *Field-Induced Electrofreezing*, a (usually large) electric field causes the freezing of a supercooled water drop. A significant series of experiments was described by Smith *et al.* (1971), in which it was suggested that freezing was due to mechanical disruption of the drop by the large fields, and that this disruption was the direct cause of the freezing rather than a direct electrical interaction with the water molecules comprising the drop. *Charge-enhanced contact nucleation* has also been described as electrofreezing. Pruppacher (1973) described experiments in which micron- diameter sulphur particles (which usually acted as poor ice nuclei, hardly inducing freezing at all above −20°C) were able to cause ice formation at −8°C. Pruppacher suggested this was due to negative electrification of the sulphur particles. *Electroscavenging* is also a possible electrofreezing mechanism. It is known that electrical charging of water drops and aerosol particles increases their collision efficiency (Pruppacher and Klett, 1997), and therefore the rate at which interactions between supercooled drops and aerosol occur. Only a small fraction of atmospheric aerosol particles can act as ice nuclei, but the Coulomb and image-force enhanced capture probabilities would also increase the capture rate of the rare aerosol particles suitable as ice nuclei.

4.4. Electroscavenging

To investigate the sensitivity of electrical enhancement of freezing by contact nucleation alone, Tripathi (2000) calculated the aerosol particle (AP) collection efficiency of a charged cloud droplet, taking full account of inertial and electrical (coulomb and image) forces. Previous particle collection calculations using the Coulomb force (Wang *et al.*, 1978) and those including the electrical image (Tinsley *et al.*, 2000) were extended to include gravitational and drag forces, for particle charges typical of the atmosphere in non- thunderstorm conditions. If some of the aerosols collected are active as ice nuclei, and the collecting droplet is supercooled, electrically-induced changes in collection efficiency could conceivably change freezing rates by a change in the contact nucleation rate.

Tripathi (2000) defined the electrical enhancement of aerosol collection as

$$P = E(a, j, A, J)/E_0 \tag{11}$$

where E is the droplet's efficiency in collecting aerosol particles, which varies with the particle radius a and charge j, and the droplet's radius and charge A and J. E_0 is the droplet's efficiency in collecting aerosol particles when both aerosol and drop are neutral. In Figure 7(a) the computed electrical enhancement P is shown for three cloud droplets of radii 40, 32 and 26 μm for charged collisions, for three aerosols of radii 0.4, 0.5 and 0.6μm respectively. These sizes optimise the electrical effect over inertial effects, but are nevertheless fairly typical of atmospheric ice nuclei. There is a three-fold increase in P for an aerosol carrying ten charges. In

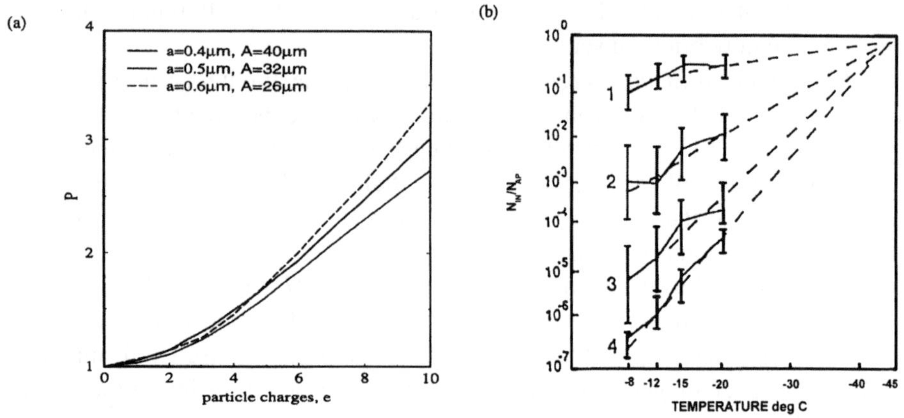

Figure 7. (a) Electrical enhancement of aerosol collection rate P, calculated as a function of number of aerosol particle charges j, for supercooled droplets of radii 40, 32 and 26 μm carrying 300 charges collecting aerosol particles of radii 0.4, 0.5 and 0.6 μm respectively (from Tripathi, 2000). (b) Temperature dependence of active fraction of natural atmospheric aerosol particles active as ice nuclei for aerosol particles sizes (1) 100 μm (2) 10 μm (3) 1 μm (4) 0.1 μm. (from Pruppacher and Klett, 1997, with changes)

Figure 7(b) (from Pruppacher and Klett, 1997), the fraction of natural atmospheric aerosols particles active as ice nuclei (N_{IN}/N_{AP}) is plotted as a function of temperature. Depending on which experimental curve is chosen, it is clearly possible to compare the change in contact nucleation rates due to charging with the change in ice-nuclei active aerosol particles with temperature. This requires the assumption that all the ice production occurs by contact nucleation.

5. Conclusion

From the viewpoint of the solar physicist, the modulation of cosmic rays by solar activity is an established fact which clearly leads to variations in ionisation in the terrestrial atmosphere. Given the relative scales of star and planet, it might seem likely that there has to be a response in the climate to these changes. From the viewpoint of the atmospheric physicist however, it is unclear if the changes in ionisation can actually lead to changes in cloud, although electrical mechanisms certainly seem possible. The discussions here have highlighted the fact that local variability in condensation and ice nuclei concentrations are important factors in cloud formation and distribution, and that electrification is one of the properties of most atmospheric particles. It is conceivable that ionisation changes could lead to changes in the formation of molecular clusters, or influence aerosol acting as CCN or IN. It does also seem possible that there could be direct ion-assisted formation of CCN, although this would account for only some of the atmospheric variability in CCN observed. These possibilities are all very intriguing, however there does *not* yet appear to be a physical mechanism, theoretically and experimentally established to actually occur in the terrestrial atmosphere which can explain a correlation between solar-induced cosmic ray changes and variations in clouds.

References

Adachi, M., Ishida, T., Kim, T. O., and Okuyama, K.: 1998, 'Effects of NH_3 on particle formation from NO_2/H_2O/air and SO_2/H_2O/air mixtures by α-ray radiolysis', *J. Aerosol. Sci.* **29**, S1, 58,339–58,340.

Aplin, K. L., and Harrison, R. G.: 1999, 'The interaction between air ions and aerosol particles in the atmosphere', *in* D. M. Taylor (ed.), *Proceedings of* 10th *International Electrostatics Conference*, Cambridge, April 1999, Institute of Physics Conference series **163**, 411–414.

Castleman, A. W.: 1982, *in* D. R. Schryer (ed.), *Heterogeneous Atmospheric Chemistry*, AGU, Washington.

Chalmers, J. A.: 1967, *Atmospheric Electricity*, 2nd edition, Pergamon, Oxford.

Clement, C. F. and Harrison, R. G.: 1992, 'The charging of radioactive aerosols', *J. Aerosol. Sci.* **23**, 5, 481–504.

Dawson, G. A., and Cardell, G. R.: 1973, Electrofreezing of supercooled waterdrops', *J. Geophys. Res.* **78**, 36, 8864–8866.

Dolezalek, H., Reiter, R. and Kroling: 1985, 'Basic comments on the Physics, occurrence in the atmosphere, and possible biological effects of air ions', *Int J. Biometeorology* **29**, 207–242.

Emi, H., Shintani, E., Namiki, N., and Otani, Y.: 1998, 'Measurement of the ion mobility distribution at a new mobility analyser with separation in axial direction to the flow', *J. Aerosol. Sci.* **29**, S1, S1247–S1248.

Fletcher N. H.: 1966, 'The physics of rainclouds', CUP.

Gabarashvili, T. G., and Gliki N. V.: 1967, 'Origination of the ice phase in supercooled water under the influence of electrically charged crystals of cholesterol and naphthalene', *Izv. Atmos. and Oceanic Physics* **3**, 5, 570–574.

Gerdien, H.: 1905, 'Die absolute Messung der Spezifischen Leitfähigkeit und der Dichte des vertikalen Leitungstroms in der Atmosphäre', *J. Terr. Mag.* **65**, 10.

Gringel, W.: 1978, 'Untersuchungen zur elecktrischen Luftleitfähigkeit unter Berücksichtigung der Sonnenaktivität und der Aerosolteilchenkonzentration bis 35 km Höhe', PhD Dissertation, Eberhard-Karls-Universität zu Tübingen.

Gunn, R.: 1954, 'Diffusion charging of atmospheric droplets by ions and the resulting combination coefficients', *J. Meteor* **11**, 339–347.

Harrison, R. G.: 1997, 'Climate change and the global atmospheric electrical system', *Atmospheric Environment* **31**, 20, 3483–3484.

Harrison, R. G., and Lodge, B. N.: 1998, 'A calorimeter to detect freezing in supercooled water droplets, *Rev. Sci. Inst.* **69**, 11, 4004–4005.

Horrak, U., Salm, J., and Tammet, H.: 1998, 'Bursts of intermediate ions in atmospheric air' *J. Geophys. Res.* **103** D12, 13909–13915.

Katz, U.: 1968, 'The ice nucleation activity of electrically charged and uncharged CuS particles', *in Proc. Int. Conf. on Cloud Physics*, August, 26 – 30, 1968, Toronto, pp. 183–187.

Landau, L. D., and Lifshitz, E. M.: 1980, *Statistical Physics*, Part 1, 3rd edition, Pergamon.

Mason, B. J: 1971, *The Physics of Clouds*, OUP.

McGorman, D. R., and Rust, W. D.: 1998, *The electrical nature of storms*, OUP.

Mossop, S. C.: 1985, 'The origin and concentration of ice crystals in clouds', *Bull. Am. Meteor. Soc.* **66**, 264–273.

O'Dowd, C. D., Smith, M. H., Lowe, J. A., Harrison, R. M., Davison, B., and Hewitt, C. N.: 1996, 'New particle formation in the marine environment', *in* M. Kulmala and P. E. Wagner (eds.), *Proc. 14th Int. Conf. on Nucleation and Atmospheric Aerosols*, pp. 925–928.

Paltridge, G. W.: 1965, 'Experimental measurements of the small-ion density and electrical conductivity of the stratosphere', *J. Geophys. Res.* **70**, 2751–2761.

Pruppacher, H. R.: 1973, 'Electrofreezing of supercooled water', *Pure Appl. Geophys.* **104**, 623–634.

Pruppacher, H. R., and Klett, J. D.: 1997, *Microphysics of clouds and precipitation*, 2nd edition, Kluwer.

Pudovkin, M. I., and Veretenenko, S. V.: 1995, 'Cloudiness decreases associated with Forbush–decreases of galactic cosmic rays', *J. Atmos. Terr. Phys.* **75**, 1349–1355.

Raes, F., Janssens, A., and Van Dingenen, R.: 1986, 'The role of ion-induced aerosol formation in the lower atmosphere', *J. Aerosol. Sci.* **17**, 3, 466–470.

Reiter, R.: 1974, *in* H. Dolezalek and R. Reiter (eds.), *Electric Processes in Atmospheres*, Steinkopff, Darmstadt, FRG, 1977.

Rogers, R. R., and Yau, M. K.: 1989, *A short course in cloud physics*, Pergamon, 3rd edition.

Rosinski, J., Nagamoto, C. T., and Zhou, M. Y.: 1995, 'Ice forming nuclei over the East China Sea', *Atmospheric Research* **36**, 95–105.

Rosinski, J.: 1995, 'Cloud condensation nuclei as a real source of ice forming nuclei in continental and marine air masses', *Atmospheric Research* **38**, 351–359.

Smith, M.H., Griffiths, R.F., and Latham, J.: 'The freezing of raindrops falling through strong electric fields', *QJRMS* **97**, 495–505.

Scott, J. P., and Evans, W. H.: 1969, 'The electrical conductivity of clouds', *Pure and Appl. Geophysic.* **75**, 219–232.

Svensmark, H., and Friis-Christensen, E.: 1997, 'Variations of cosmic ray flux and global cloud coverage — a missing link in solar-climate relationships', *J. Atmos. Solar-Terrestrial Phys.* **59**, 1225–1232.

Thomson, J.J.: 1924, 'Recombination of gaseous ions, the chemical combination of gases and monomolecular reactions', *Phil. Mag* **47**, 337–378.

Tinsley, B. A., and Dean, G. W.: 1991, 'Apparent tropospheric response to MeV-GeV particle flux variations: a connection via electrofreeezing of supercooled water in high-level clouds?' *J. Geophys. Res.* **96**, 22,283–22,296.

Tinsley, B. A., Rohrbaugh, R. P., Hei, M., and Beard K. V.: 2000, 'Effects of image charges on the scavenging of aerosol particles by cloud droplets, and on droplet charging and possible ice nucleation processes', *J. Atmos. Sci.*, in press.

Tripathi, S. N.: 2000, 'Removal of Charged Aerosols', PhD Thesis, University of Reading, U.K., submitted.

Turco, R. P., Zhao, Jing-Xia, and Yu, F.: 1998, ' A new source of tropospheric aerosols: Ion-ion recombination' *Geophys. Res. Lett.* **25**, 5, 635–638.

Volland, H.: 1984, *Atmospheric Electrodynamics*, Springer-Verlag, Berlin.

Wang, P. K., Grover, S. N., and Pruppacher, H. R.: 1978, 'On the effect of electric charges on the scavenging of aerosol particles by clouds and small raindrops', *J. Atmos. Sci.* **34**, 1735–1743.

Yu, F., Turco, R. P., Kärcher, B., and Schroder, F. P.: 1998, 'On the mechanisms controlling the formation and properties of volatile particles in aircraft wakes', *Geophys. Res. Lett.* **25**, 20, 3839–3842.

SOLAR VARIABILITY AND CLOUDS

Discussion Session 3c

JASPER KIRKBY
CERN, Geneva, Switzerland

ARI LAAKSONEN
University of Kuopio, Kuopio, Finland

Received: 14 April 2000; Accepted in final form: 24 July 2000

Abstract. Satellite observations have recently revealed a surprising imprint of the 11-year solar cycle on global low cloud cover. The cloud data suggest a correlation with the intensity of galactic cosmic rays. If this apparent connection between cosmic rays and clouds is real, variations of the cosmic ray flux caused by long-term changes in the solar wind could have a significant influence on the global energy radiation budget and the climate. However a direct link between cosmic rays and clouds has not been unambiguously established and, moreover, the microphysical mechanism is poorly understood. New experiments are being planned to find out whether cosmic rays can affect cloud formation, and if so how.

1. Introduction

There seems little doubt that a link exists between solar activity and climate change (Reid, 2000). The best known example is the Maunder Minimum, the Little Ice Age between 1645 and 1715, when there was almost a complete absence of sunspots. This and other climate changes during the Holocene often seem to have coincided with anomalies in the production of light radio-isotopes such as ^{14}C and ^{10}Be. Since these radio-isotopes are produced by the interactions of galactic cosmic rays (GCRs) with nuclei in the atmosphere, their palaeo-reconstructed abundances directly measure the prevailing GCR intensity. A high GCR intensity seems to be associated with cool temperatures, and a low GCR intensity with warm temperatures. Since the GCR flux reaching Earth is modulated by variations in the strength of the solar wind (as well as variations of the geomagnetic field), this provides clear evidence for a connection between solar activity and climate variability.

Whereas there is little disagreement that such a link exists, the *size* of the effect on the Earth's temperature is not well established. For example, the Mann *et al.* (1999) temperature reconstruction for the past millennium shows relatively small temperature variations, except during the last century. However, these and other data taken together are not inconsistent with systematic temperature variations during the Holocene of at least $\pm 0.5°C$ about the mean value.

Another uncertain aspect of climate variability is whether the effects have been *global* or only *local*. The latter could, for example, be explained by internal climate

Space Science Reviews **94:** 397–409, 2000.

re-distributions. Although the reconstructions of past climate changes have relied strongly on ice cores, which are polar-centric, there exist complementary proxies such as tree rings and lake sediments that are experimentally and geographically independent. These proxies frequently confirm the qualitative patterns seen in the ice cores. A recent example was presented by Verschuren *et al.* (2000) who show that the "European" climate pattern of the last 1100 years extended into the equatorial regions. Their study revealed a close correlation of the ^{14}C record with the depth and salinity of a lake in equatorial east Africa. It therefore appears likely that these climate anomalies were indeed largely global phenomena.

If the Sun is inducing climate variability then what mechanism or mechanisms are responsible? The obvious first thing to suspect is a variation of the solar irradiance. High-precision satellite measurements over the last 2 decades (Fröhlich, 2000) have revealed that the solar irradiance is indeed varying, but only by about 0.1% from sunspot maximum to minimum. The solar irradiance is slightly higher at sunspot maximum; although sunspots are cooler and have reduced emission, this is more than compensated by an associated increase in the nearby bright *plages* and *faculae*. This is a small, but non-negligible, variation: 0.3 Wm^{-2}, globally-averaged.

An important, but as yet unanswered, question is whether this represents simply an 11-year periodic variation of the Sun's output or whether there are longer-term systematic changes. In the absence of any direct measurements, the sunspot record is generally used as a proxy for estimating the changes in solar irradiance over the past 300–400 years (Lean, 2000; Solanki, 2000), in order to allow comparisons with the temperature records. The input parameters for these models generally include the sunspot number, the demodulated amplitude and the cycle length. As an example, Lean (2000) estimates an irradiance increase since 1700 of about 3.3 Wm^{-2} (0.8 Wm^{-2}, globally averaged), i.e. 0.24% of the present irradiance.

Clear correlations with climate changes are observed in these exercises. However it is important to keep in mind that there is at present no direct experimental evidence of long-term variations of the irradiance of the Sun. What is actually observed is a correlation between the temperature record and *sunspot activity*. The latter is known to be associated with the GCR intensity, for which there *is* unambiguous evidence of long-term variability—in the light radio-isotope record. Is it possible that the GCRs themselves could be influencing the climate and that the solar irradiance itself is not varying beyond the established 0.1% periodic effect due to sunspot darkening and facular brightening?

But how could cosmic rays affect the Earth's climate? The energy deposited by cosmic rays is only a few parts per billion compared with the incident solar energy, so a strong amplifying mechanism would be necessary. The answer may have been found by Svensmark and Friis-Christensen (1997) who, in 1997, reported the observation of an unexpected correlation between global cloud cover and the incident cosmic ray intensity. In the remainder of this paper we will follow this important new clue for a mechanism that may underlie solar-climate variability.

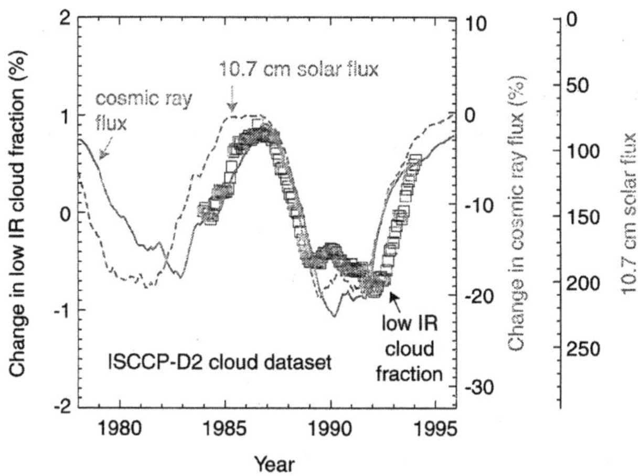

Figure 1. Variation over the period 1983-1994 of the IR (λ = 10–12 μm) low cloud fraction (below about 3.2 km) in the ISCCP-D2 dataset, in comparison with the changes of GCR intensity (solid line) and 10.7 cm solar irradiance (dashed line), (Marsh and Svensmark, 2000a, 2000b). The cloud data have complete global coverage, day and night. The data are smoothed using a 12-month running mean.

2. Cloud Variation Over the Solar Cycle

The satellite data that revealed the first evidence for a possible GCR-cloud link had several limitations. In particular the data comprised a composite of four individual datasets from different satellites, each covering different periods of time and different regions of the globe. This introduced uncertainties in the inter-calibration between the datasets. Furthermore the data analysed were restricted to oceans, to latitudes between 60N and 60S, and mostly to the daytime.

Kernthaler *et al.* (1999) and others have pointed out these and other uncertainties in the Svensmark and Friis-Christensen observation. A recurrent criticism has also been the absence of any microphysical mechanism that could connect cosmic rays to cloud formation. This criticism is curious for two reasons. Firstly, a puzzling observation should be reason for optimism; if the observation is correct and yet there is no known physical mechanism, then this is surely a sign that an important advance is at hand. Secondly, it would in fact be more precise to say that a possible microphysical mechanism is not widely known, rather than to say it does not exist (see §4).

The limitations of the original observation have largely been removed in a new analysis reported at this meeting by Marsh and Svensmark (2000a), based on the recently-released International Satellite Cloud Climate Project (ISCCP) D2 dataset, which covers the period from July 1983 to September 1994. The new data indicate the presence of a significant correlation between GCR intensity and the frequency of *low* clouds, below about 3.2 km (Fig. 1), but none with clouds at higher altitudes.

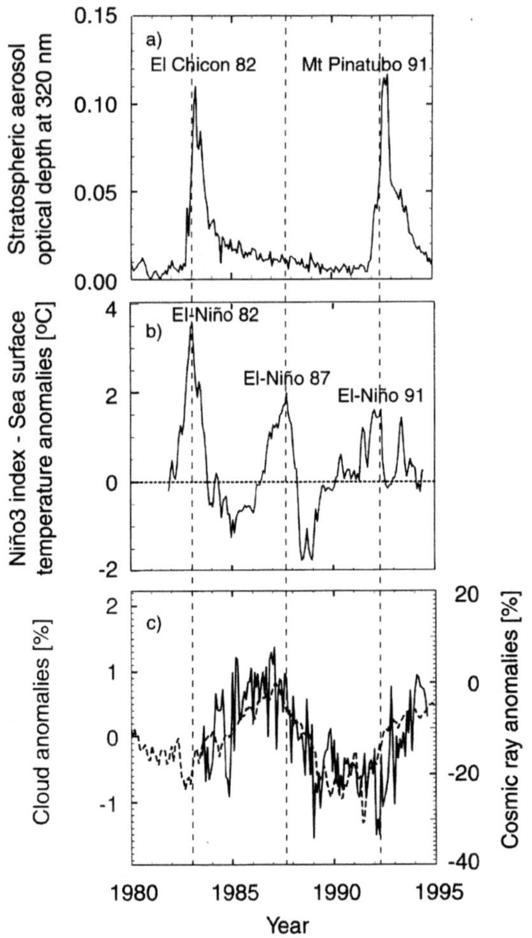

Figure 2. Comparison of the occurrences of volcanic and El-Niño events with the cloud anomalies: a) Stratospheric aerosol optical depth. The two peaks are caused by injections of aerosols into the stratosphere by the two large volcanic eruptions (El Chichon, Mexico, 1982, and Pinatubo, Philippines, 1991). b) Niño3 index which indicates 3 El-Niño events (1982, 1987 and 1991) during the indicated period. c) Low cloud cover anomaly (solid curve) and GCR flux from the Climax neutron monitor (dashed curve) (Marsh and Svensmark, 2000a).

Following the original observation, questions were raised concerning the influence of volcanoes and ENSO events on cloud cover during the measurement period. However these show a poor correlation with the observed cloud modulation in the new data (Fig. 2) and it seems unlikely that they could explain the major features of the cloud variability.

Marsh and Svensmark (2000b) find the GCR intensity is also strongly correlated with the cloud-top temperatures of low IR clouds. The signs of the two correlations are that more cosmic rays give more clouds and a higher cloud-top temperature.

However the spatial distributions of the correlations across the globe are different in the two cases; the regions of high correlation of cloud frequency are widely scattered whereas the cloud-top temperature correlation is high and essentially continuous over the entire tropics. This is a surprising result since the solar modulation of the cosmic ray intensity is a minimum near the geomagnetic equator, with a peak variation of only about 5%. The reason for these different spatial distributions is not understood, although it should be noted that they are two separate cloud properties. Cloud frequency measures the presence or absence of a cloud, reflecting changes in the cloud lifetimes, cloud sizes, or cloud number. On the other hand, the inferred cloud-top temperature depends on its altitude and on the microphysical properties of existing clouds.

The net radiative properties of a cloud are mainly dependent on its altitude and optical thickness. Optically-thin clouds at high and middle altitudes cause a net warming due to their relative transparency at short wavelengths but opacity in the IR region, whereas thick clouds produce a net cooling due to the dominance of the increased albedo of shortwave solar radiation. Since the new data indicate that the solar modulation appears in the thick low clouds, the *sign* of the GCR-climate effect is now known: increased cosmic rays are associated with increased low-clouds and therefore with a cooler climate. This is consistent with the sign inferred from the light radio-isotope record.

Estimates from the Earth Radiation Budget Experiment (ERBE) indicate, overall, that clouds reflect more energy than they trap, leading to a net cooling of about 28 Wm^{-2} from the mean global cloud cover of 63% (Hartmann, 1993). Figure 1 indicates an absolute variation in low cloud cover of about 2% over a solar cycle, or about 7% relative variation since low clouds constitute a mean global cloud fraction of 28%. Since low clouds have a net radiative forcing of about -17 Wm^{-2}, this would imply a solar maximum-to-minimum change in the Earth's radiation budget of about 1.2 Wm^{-2} (0.3% of the total irradiance).

This is a significant effect. During the course of the last century the coronal source flux carried out into the heliosphere by the solar wind increased by 140% (Lockwood *et al.*, 1999), causing a reduction of the GCR intensity on Earth by about 15%, globally averaged. (A variation of 15% is approximately equal to a solar maximum-to-minimum change of the globally-averaged present GCR intensity.) If the GCR-cloud link is real then this suggests a positive radiative forcing from reduced cloud cover during the last century that could be comparable to the estimated radiative forcing of 1.5 Wm^{-2} from the increase in CO_2 concentration.

3. The Causes of Solar-cloud Variability

Other solar activity also follows the sunspot cycle, so in principle the cloud variations might be attributed to (a) a direct effect of GCRs on clouds, (b) a direct effect of other solar activity on clouds, or (c) an effect of solar activity on global weather that indirectly results in a change in cloud cover. Although the other explanations mentioned here cannot be ruled out, Fig. 1 contains evidence favouring (a), the direct cosmic ray effect.

The solar 10.7 cm microwave flux is a good proxy both for variations in the sunspot count and for variations in the solar output of other electromagnetic radiation, namely visible light, soft X-rays and ultraviolet rays. In Fig. 1 the change in cloud fraction followed the cosmic rays fairly closely but sometimes lagged as much as two years behind the change in the 10.7 cm flux. Delays in cosmic ray variations occur because disturbances in the solar wind, which scatter the cosmic rays, involve events such as coronal mass ejections (CMEs) that are only loosely related to the sunspot count. Moreover the heliosphere is so large that disturbances can take a year or more to reach its boundary, the heliopause. The delays show that the cloud variations correlate more closely with the cosmic rays than with the solar electromagnetic flux.

Could the solar wind have a direct effect on clouds, while coincidentally modulating the cosmic rays? This is unlikely because solar wind protons have very little energy (a few keV) and if they penetrate the Earth's magnetosphere in the auroral zone, the outer atmosphere stops them at altitudes above 100 km. Shock waves associated with occasional CME disturbances in the solar wind can accelerate protons to 100 MeV or more, creating "solar cosmic rays", but only rarely do they have sufficient energy to reach the tropopause (8–18 km).

The sign of the effect is a consideration in assessing possible solar mechanisms influencing the cloud fraction. While the cloud fraction correlates positively with the cosmic ray intensity, it correlates negatively with solar electromagnetic radiation and with the strength of the solar wind. If enhancements in electromagnetic radiation or the solar wind were responsible for a direct effect on clouds, they would be in the counter-intuitive sense of reducing the cloud fraction.

Perhaps changes in cloud fraction might be an indirect result of changes in global weather due to solar effects that coincide with the cosmic ray modulation? Other mechanisms have been proposed for a solar role in climate change, although none suggests an obvious effect on cloud fraction. One example is the warming of the upper stratosphere during the solar cycle due to increases in ultraviolet (UV) radiation, which is absorbed by stratospheric ozone. This has the potential to influence large-scale dynamics of the troposphere (Larkin *et al.*, 2000), although vertical mixing between the stratosphere and troposphere is weak.

Solar activity has many manifestations, and all of the means by which it may influence the Earth's climate deserve further investigation. It is no part of our discussion to suggest that the link between cosmic rays and clouds is the only important solar-climatic mechanism. For the reasons mentioned here we nevertheless consider that the preferred interpretation of Figure 1 is the simplest, namely that cloud behaviour is directly influenced by cosmic rays. As to the lack of a clear microphysical explanation, this is discussed in the next section.

4. The Microphysical Influence of Cosmic Rays on Clouds

If cosmic rays can influence clouds, it is likely to be through their effect on aerosols. Aerosols are found throughout the atmosphere and constitute efficient cloud condensation nuclei (CCN) for activation of cloud droplets. The presence of a largely abundant supply of CCN ensures that the maximum water vapour supersaturations in the atmosphere rarely exceed values of about 1% since higher values are arrested by the removal of water vapour during droplet growth. These values are far below the supersaturations (~500%) required to directly activate small ions into droplets.

Despite the abundance of CCN, however, extra aerosols can influence cloud radiative properties, lifetimes and precipitation. For example a recent study (Rosenfeld, 2000) with NOAA satellite data looked at clouds that formed downwind of industrial sites located in pristine areas. The otherwise uniform cloud data from these regions was streaked with bright (highly reflective) plumes from the industrial sites. The droplets in these plumes were found to be more numerous than the nearby regions and of a smaller diameter—typically less than 10 μm and therefore below the threshold size for them to coalesce efficiently and precipitate. In contrast the droplets outside the plumes measured more than 25 μm in diameter. The high reflectivity of the plumes resulted from a higher droplet number density at fixed liquid water content. Analysis of independent satellite data over the same regions from the Tropical Rainfall Measuring Mission confirmed that these plumes did indeed produce less rain and therefore had a longer lifetime than clouds in the nearby regions.

Cosmic rays may affect atmospheric aerosols either through the *ions* or the *radicals* they create. The latter primarily involve N, O, and OH radicals created by the dissociation of the main active constituents of air: N_2, O_2 and H_2O. There are three distinct processes to consider: 1) the creation of new condensation nuclei (CN) from condensable vapours, 2) the growth of CN into CCN, and 3) the creation of ice nuclei. These are illustrated in Fig. 3 and described in more detail below.

4.1. Creation of New CN

The ions or condensable vapours created by cosmic rays may increase the nucleation rate of new aerosols *in situ* (gas-to-particle conversion). It has long been speculated that atmospheric sulphate particles are formed by ion-induced nucleation (see, for example, refs. Raes and Janssens (1985), Raes *et al.* (1986) and references quoted therein). Also processes such as ion-ion recombination have been proposed to lead to the production of stable clusters in the lower stratosphere (Arnold, 1980) and in the troposphere (Turco *et al.*, 1998; Turco, 2000). Since the nucleation rate of new aerosols from these processes is proportional to ion concentration, variations of the cosmic ray flux may translate into variations of CN and CCN concentrations.

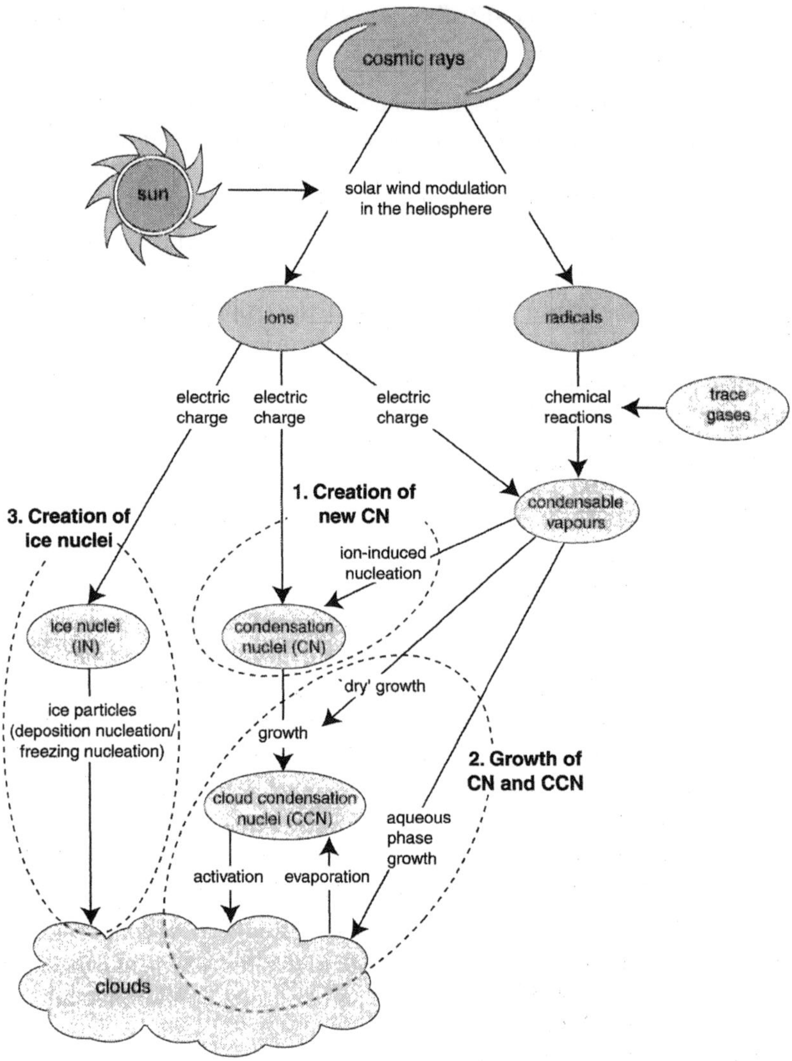

Figure 3. Paths that may connect galactic cosmic rays to clouds and hence link solar and climate variability through the solar wind modulation.

There is very little experimental data on the effect of ions in the atmosphere on new particle formation (Harrison, 2000). Observations have been made of nucleation bursts of CN in the atmosphere that cannot be explained by classical theories. For example, Hõrrak *et al.* (1998) reported the spontaneous formation of bursts of intermediate size ions in urban air, which they suggest may be due to ion-induced nucleation. Also, Clarke *et al.* (1998) observed new particle formation in the tropical marine boundary layer that could not be explained by classical binary nucleation theory at the low measured concentrations of sulphuric acid (2–

5×10^7 molecules cm^{-3}). However an ion-ion recombination model study reported recently by Yu and Turco (2000) *is* able to successfully reproduce the observations of Clarke *et al.* This model indicates that the nucleation rate of fresh CN in the marine boundary layer is generally limited by the available ion production rate from GCRs, in contrast with the upper atmosphere, where the nucleation rate is limited by the trace vapour concentration. This provides a possible explanation why the solar modulation is only seen in low clouds (Marsh and Svensmark, 2000a).

Direct experimental evidence that ions are involved in the nucleation of new particles under atmospheric conditions is lacking. However two experiments provide indications of positive effects. Bricard *et al.* (1968) observed new particle production in filtered (aerosol-free) Paris air exposed to very high radiation doses (3×10^8 Bq $m^{-3} \times 300$ s). On the other hand, Vohra *et al.* (1984) carried out experiments at naturally-occurring ionisation levels of 1–10 Bq m^{-3} and observed new particle production proportional to ionisation rate, but in artificial air containing high concentrations of trace gases (300 ppb SO_2, 100 ppb C_2H_4 and 160 ppb O_3).

4.2. Growth of CN Into CCN

Ions may also play a role in the growth of CN into CCN by enhancing the growth rates and reducing the evaporation rates. The coagulation and condensation processes during growth may be enhanced by the electrostatic interactions of charged clusters, giving them a competitive advantage over neutral clusters. Recently a search for massive ions in the upper troposphere and lower stratosphere was started by MPIK Heidelberg (Eichkorn *et al.*, 1999) using aircraft-based ion mass spectrometers. Preliminary results indeed indicate the presence of massive positive and negative ions.

Growth of aerosols can lead to a higher CCN number concentration, and to CCN that activate into droplets at lower supersaturations. Clouds that form in air containing high CCN concentrations tend to have high droplet concentrations, which enhances the shortwave (solar) albedo. Increase in the CCN concentration also inhibits rainfall and therefore increases cloud lifetimes (cloud coverage). These effects—which are due to more, smaller droplets at a fixed liquid water content—are particularly significant in marine air, where the CCN concentrations are generally quite low.

The possible effect of charges on aerosol *activation* into cloud droplets has usually been ignored since the conventional electrostatic interaction term for small charges is expected to be appreciable only for very small aerosols of ~1 nm diameter. However, a recent study (Cheng *et al.*, 2000) of heterogeneous nucleation of n-butanol vapour on charged and uncharged insoluble particles shows a surprisingly large charge effect. The charges have a clear effect on the nucleation process even when the seed particles are as large as 90 nm in diameter and the supersaturation is only 0.5%. For these experiments the classical theory predicts that the difference in the nucleating ability of charged and uncharged seed particles should vanish for particle diameters above 20 nm, corresponding to supersaturations of about 100%. If this result is correct, it indicates that charges may play a role in atmospheric cloud drop activation processes, at least when the CCN are partially insoluble (e.g. carbon) and carry multiple charges.

Galactic cosmic rays also create free radicals in the atmosphere and these may be able to influence the atmospheric chemistry of condensable vapours under certain conditions, especially if catalytic reactions are involved. The most important are OH and NO; they have production rates of about 1–2 OH molecule per ion-pair and 1.5 NO molecule per ion-pair. These rates imply mixing ratios of about 0.7 pptv OH or NO are generated per day by GCRs directly in the upper troposphere.

4.3. Creation of Ice Nuclei

The presence of ice in clouds has an important influence on their radiative properties and it can also lead to precipitation. However the apparent lack of ice nuclei (IN) in the atmosphere is at present a mystery: measurements indicate the presence of about 1 IN per litre at $-20°C$ (compared with about 10^6 aerosol particles per litre) whereas the observed concentrations of ice particles in clouds at these temperatures vary between 10 and 300 per litre. Two nucleation processes can lead to ice formation: a) direct sublimation of vapour to the solid phase (deposition nucleation) and b) freezing of a supercooled liquid droplet (freezing nucleation). It has been suggested that the ionisation produced by cosmic rays may be able to affect either of these processes (Tinsley, 2000).

The possible effect of charges on ice nucleation has been studied to some extent (see Pruppacher and Klett (1997) for a review), and effects have been seen in a number of experiments. In particular, the ice nucleating ability of various types of seed particles seems to be influenced by ionisation. For example, cloud chamber experiments at the University of Missouri-Rolla in 1980 (Anderson *et al.*, 1980) found an enhancement of frozen droplets in the regions where cosmic rays traversed the cloud chamber.

Enhancing heterogeneous ice nucleation by electrification (Tinsley, 2000) is supported by very little experimental work so far. If the electrofreezing effect is real, possible influences of cosmic rays on cirrus formation and on the properties of middle clouds are implied. Interestingly, cirrus cloudiness at high latitudes has been linked to the GCR intensity (Veretenenko and Pudovkin, 1999). Electrofreezing could also affect glaciation (which triggers rain) in convective clouds. Possibly related to this effect, Stozhkov *et al.* (1995) have found a correlation between short-term, large (Forbush) decreases of cosmic rays and reduced precipitation in the Amazonian region.

Finally we mention that it has been proposed that solid polar stratospheric clouds form by deposition nucleation of nitric acid and water directly onto cosmic ray-generated ions or ion clusters. These persistent and optically very thin clouds are often seen over several thousand kilometer regions and cannot be explained by any recognised mechanism. However, to date, there have been no experiments that can confirm or dispute the possibility of formation by GCRs. An understanding of such particles appears to be critical to a complete understanding of denitrification and subsequent ozone loss in the polar regions (Waibel *et al.*, 1999).

5. Summary and Outlook

Satellite observations suggest that solar variability may be linked to climate variability by a chain that involves the solar wind, cosmic rays and clouds. The weak link is the connection between cosmic rays and clouds. This has not been unambiguously established and, moreover, the microphysical mechanism is poorly understood. Cosmic rays are the dominant source of ions in the free troposphere and stratosphere and they also create free radicals. It has been proposed that ions may be important both in the nucleation of new aerosols and in their growth via clustering to ultimately become cloud condensation nuclei and thereby seed clouds. In addition to their effects on aerosol formation and growth, cosmic rays may also influence the formation of ice particles in clouds.

New experimental data and better models are critically needed to investigate the apparent link between solar and cloud variability—and in particular to test the GCR hypothesis. In addition to on-going atmospheric studies, a new experiment has recently been proposed (Cloud collaboration, 2000) to test the GCR-cloud link under controlled laboratory conditions in a beam at the CERN Proton Synchrotron (PS), which provides an adjustable source of "cosmic rays". The experiment, which is named CLOUD (Cosmics Leaving OUtdoor Droplets), is based on a cloud chamber that is designed to duplicate the conditions prevailing in the atmosphere. This requires the capability of producing the very low water vapour supersaturations—typically a few tenths of a percent—found in clouds, so that the precise cloud-forming properties of the air parcel under study can be measured.

The advantage of this approach is that all the experimental conditions can be precisely controlled and measured. In particular, by comparing measurements at several beam intensities near atmospheric levels, the existence of a causal relationship between ionising radiation and cloud formation can be explored. Such measurements are difficult to perform with cosmic rays in the atmosphere since the natural intensity variations are modest and, moreover, they follow the slow 11-year solar cycle. Furthermore, the CLOUD experiment provides full control of the trace gases and initial aerosol content and, in addition, complete physical and chemical analysis of the final products after beam exposure. This will greatly facilitate an understanding of the microphysics of any observed effects.

Where positive effects are found, detailed simulations will be made to assess their importance for cosmic rays and clouds in the atmosphere. This will not be easy. Natural clouds, and the processes that affect them, are highly variable in both space and time. Furthermore only a limited understanding exists of the processes that affect cloud development. Attribution of effects due to variations of the GCR flux will need to demonstrate that they are significant within the natural variations of other parameters.

If the link between cosmic rays and clouds is real, it may provide a significant new mechanism for climate change. During the 20th Century the cosmic rays reaching the Earth diminished by about 15% as a result of increasing strength of the solar wind. The inferred reduction in cloud cover could have warmed the Earth by a large fraction of the amount currently estimated to be due to man-made carbon dioxide. In that case, the effect of carbon dioxide may have been overestimated. If, on the other hand, the link to cosmic rays proves to be illusory, present diplomatic efforts to curb emissions of carbon dioxide will be more strongly supported scientifically. Settling the issue, one way or the other, is therefore an urgent task.

References

Anderson, R. J., Miller, R. C., Kassner, J. L. Jr., and Hagen, D. E.: 1980, 'A study of homogeneous condensation-freezing nucleation of small water droplets in an expansion cloud chamber', *J. Atmospheric Sciences* **37**, No. 11, 2508–2520.

Arnold, F.: 1980, 'Multi-ion complexes in the stratosphere: implications for trace gases and aerosols', *Nature* **284**, 610.

Bricard, J. *et al.*: 1968, 'Formation and evolution of nuclei of condensation that appear in air initially free of aerosols', *J. Geophys. Res.* **73**, 4487.

Cheng, C.-C., Tao, C.-J., and Shu, H.-J.: 2000, 'Heterogeneous nucleation of n-butanol vapour on sub-micrometer charged and neutral particles of lactose and monosodium glutamate', *J. Colloid Interface Sci.* **224**, 11.

Clarke, A. D. *et al.*: 1998, 'Particle nucleation in the tropical boundary layerand its coupling to marine sulfur sources', *Science* **282**, 89.

CLOUD collaboration (B. Fastrup *et al.*): 2000, CERN proposal SPSC/P317, SPSC 2000-021.

Eichkorn, S., Wohlfrom, K. H., and Arnold, F.: 2000, 'Massive ion detection in the upper troposphere: implications for aerosol particle formation via ion-induced nucleation', poster presented at the *EGS General Assembly*, The Hague.

Fröhlich, C.: 2000, 'Observations of Irradiance Variability', *Space Sci. Rev.*, this volume.

Harrison, R. G.: 2000, 'Cloud formation and the possible significance of charge for atmospheric condensation and ice nuclei', *Space Sci. Rev.*, this volume.

Hartmann, D. L.: 1993, 'Radiative effects of clouds on Earth's climate, *in* P. V. Hobbs (ed.), *Aerosol-Cloud-Climate Interactions*, International Geophysics Series, Vol. 54, Academic Press Inc., San Diego, 151.

Hõrrak, U., Salm, J., and Tammet, H.: 1998, Bursts of intermediate ions in atmospheric air, *J. Geophys. Res.* **103** D12, 13909.

Kernthaler, S. C., Toumi, R. and Haigh, J.D.: 1999, 'Some doubts concerning a link between cosmic ray fluxes and global cloudiness', *Geophys. Res. Lett.* **26**, 863.

Larkin, A., Haigh, J. D., and Djavidnia, S.: 2000, 'The effect of UV Irradiance Variations on the Earth's Atmosphere', *Space Sci. Rev.*, this volume.

Lean, J. L.: 2000, 'Short term, indirect indices of solar variability', *Space Sci. Rev.*, this volume.

Lockwood, M., Stamper, R., and Wild, M. N.: 1999, 'A doubling of the Sun's coronal magnetic field during the past 100 years', *Nature* **399**, 437.

Mann, M. E., Bradley, R. S., and Hughes, M. K.: 1999, 'Northern hemisphere temperatures during the past millennium: inferences, uncertainties and limitations', *Geophys. Res. Lett.* **26**, 759.

Marsh, N., and Svensmark, H.: 2000a, 'Cosmic Rays, Clouds and Climate', *Space Sci. Rev.*, this volume.

Marsh, N., and Svensmark, H.: 2000b, 'Low cloud properties influenced by solar activity', *Geophys. Res. Lett.*, submitted.

Pruppacher, H. R., and Klett, J. D.: 1997, *Microphysics of clouds and precipitation*, 2nd ed., Kluwer Academic Publishers, The Netherlands.

Raes, F., and Janssens, A.: 1985, 'Ion-induced aerosol formation in a H_2-H_2SO_4 system - I. Extension of the classical theory and search for experimental evidence', *J. Aerosol Sci.* **16**, 217.

Raes, F., Janssens, A., and Dingenen, R. V.: 1986, 'The role of ion-induced aerosol formation in the lower atmosphere', *J. Aerosol Sci.* **17**, 466.

Reid, G. C.: 2000, ' Solar Varibility and the Earth's Climate: Introduction and Overview', *Space Sci. Rev.*, this volume.

Rosenfeld, D.: 2000, 'Suppression of rain and snow by urban and industrial air pollution', *Science* **287**, 1793.

Solanki, S. K.: 2000, 'Reconstruction of past solar irradiance', *Space Sci. Rev.*, this volume.

Stozhkov, Yu. I., *et al.*: 1985, 'Rainfalls during great Forbush decreases', *Il Nuovo Cimento* **18C**, 335.

Svensmark, H.: 1998, 'Influence of cosmic rays on the Earth's climate', *Phys. Rev. Lett.* **81**, 5027.

Svensmark, H., and Friis-Christensen, E.: 1997, 'Variation in cosmic ray flux and global cloud coverage—a missing link in solar-climate relationships', *Journal of Atmospheric and Solar-Terrestrial Physics* **59**, 1225.

Tinsley, B. A.: 2000, 'Influence of solar wind on the global electric circuit, and inferred effects on cloud microphysics, temperature and dynamics in the troposphere', *Space Sci. Rev.*, this volume.

Turco, R. P., Zhao, J.-K., and Yu, F.: 1998, 'A new source of tropospheric aerosols: ion-ion recombination', *Geophys. Res. Lett.* **25**, 635.

Turco, R. P., Zhao, J.-K., and Yu, F.: 2000, 'Ultrafine aerosol formation via ion-mediated nucleation', *Geophys. Res. Lett.* **27**, 883.

Verschuren, D., Laird, K., and Cumming, B.: 2000, 'Rainfall and drought in equatorial East Africa during the past 1100 years', *Nature* **403**, 410.

Veretenenko, S. V., and Pudovkin, M. I.: 1999, 'Variations of solar radiation input to the lower atmosphere associated with different helio/geophysical factors', *J. Atmos. Solar-Terrestrial Phys.* **61**, 521.

Vohra, K. G., Subba Ramu M. C., and Muraleedharan, T. S.: 1984, 'An experimental study of the role of radon and its daughter products in the conversion of sulphur dioxide into aerosol particles in the atmosphere', *Atmospheric Environment* **18**, 1653.

Waibel, A. E., Peter, T., Carslaw, K. S., Oelhaf, H., Wetzel, G., Crutzen, P. J., Pöschl, U., Tsias, A., Reimer, E., and Fischer, H.: 1999, 'Arctic ozone loss due to denitrification', *Science* **283**, 2064-2069.

Yu, F., and Turco, R. P.: 2000, 'From molecular clusters to nanoparticles: the role of ambient ionization in tropospheric aerosol formation', *J. Geophys. Res.*, submitted.

SOLAR VARIABILITY AND CLIMATE

A Summary

EIGIL FRIIS-CHRISTENSEN
Danish Space Research Institute, Juliane Maries Vej 30, DK-2100 Copenhagen, Denmark

Received: 19 June 2000; Accepted in final form: 16 August 2000

1. Introduction

During the last decades our understanding of the space environment of the Earth has increased dramatically. This is primarily due to access to observations made outside our limiting atmosphere. The research that has been stimulated by these completely new data has had a great impact in many scientific fields, including astrophysics, planetary physics and geophysics. In addition to the extension of our knowledge in the basic sciences we have also learned how to utilise our space environment for a wide range of applications including navigation, communication, resource management and monitoring as well as day-to-day weather prediction. All these applications have changed the operational mode of our global society in a very profound way during the last couple of decades and it is now very difficult to imagine our society without them.

In this process of relying more and more on space technology we have also realised that the Earth and its environment is subject to varying influences from outside, mainly from the ultimate source of energy on Earth, namely the Sun. It has been realised that our Sun is not just a constantly radiating source of energy but a variable star, displaying different levels of variability in different bands of the electromagnetic spectrum. These changes are related to solar activity with the irradiance being higher during high activity. From time to time the Sun also emits huge masses of plasma as a coronal mass ejection. All these processes affect in some way or another the Earth's environment and its atmosphere down to the surface of the Earth.

Also during the last couple of decades we have become conscious about the effect of mankind itself on the environment. The growth of the World's population is associated with distinct and observable changes in living conditions. One of the primary challenges of our time is to assess the risk for humanity that is associated with such changes and take adequate measures to mitigate the effects or to adapt to the changes in appropriate ways for instance by developing new technology. In order to do so, we need to detect the changes and to identify and understand their causes. Observations from space are one of the most promising tools in this respect.

Space Science Reviews **94:** 411–421, 2000.

From space we can monitor the Earth and its environment and from space we can also observe the varying conditions in the outer environment, to which the Earth is subjected.

One of the issues of global change that has attracted most attention during the last decades is the increasing amount of greenhouse gases, in particular carbon dioxide, which is one of the results of human activity. The associated change in the composition of our atmosphere may change the radiative properties so that the climate may be significantly affected. Much effort has therefore been devoted to assess the possible effects on climate and its impact on society. This task is extremely difficult because the Earth's climate is the result of a number of complex processes that are only rudimentarily understood. The issue is even more complicated because climate changes occur due to a number of natural processes as historic recordings clearly demonstrate. During the last couple of centuries many reports have suggested that some of the major climate changes are associated with changes in solar activity although without being able to point at any distinct physical mechanism that could account for such variations.

Policy measures to mitigate the effects of climate changes associated with human activity will have a tremendous impact on society. Therefore it is important that such measures are based on a good scientific understanding of the problem and the best possible estimate of the extent of its influence. In order to achieve that, we need to understand also the range of natural variations in climate and their causes. Solar activity variations comprise one of the proposed causes that can not easily be dismissed. On the other hand a thorough analysis of the complex problem and the fact that no mechanism has yet been accepted calls for a truly interdisciplinary approach involving the climate research community as well as solar physicists.

Realising its responsibility to explore the huge implications of the knowledge of our solar affected environment that we have acquired during the last decades, the International Space Science Institute (ISSI, Bern) decided to host a workshop devoted to the issue of "how does the solar variability affect climate change". The format of the workshop was based on invitations to a number of leading scientists in various scientific disciplines spanning from atmospheric physics to solar physics. The aim was to have an open and unbiased discussion about the issues that hopefully could result in a common understanding of the facts and of the remaining questions that could be of inspiration to the scientific community in its strive to understand the Earth's climate and its changes.

It is evident that the present foundation of the whole issue is based on observational data that indicate a good correlation between solar activity variations and climate changes. It is also evident that the significance of the found correlation can not be assessed until sufficiently detailed physical mechanisms can be suggested for further investigations. For these reasons the workshop was organised in a set of sessions that critically assessed observations of solar and climate variability and in particular focussed on those aspects of solar variability that could have a direct or indirect influence on climate. In order to contribute to the understanding of the physical mechanisms and to assess the significance of the empirical relationships some time was devoted to the physical causes of solar and climate variations. Due to the complexity of the issue of climate change most work is based on numerical climate models. Therefore some talks were devoted to the specific question of the capability of present climate models to predict future climate.

2. Solar Variability

It has been known for a long time that the Sun varies with solar activity, the most obvious manifestation of which is the varying number of dark spots on the solar surface. With the invention of the telescope around 1600 regular, although not quite systematic observations of the solar surface were carried out at various observatories in Europe from which Wolf introduced around 1850 the sunspot number. There is also a relationship between the solar surface activity and the occurrence of aurora, which could also be seen in the short-term fluctuations in the Earth's magnetic field, called geomagnetic activity, which was systematically observed at magnetic observatories from around 1870. Radio astronomers found that the intensity of the 10.7 cm radio signal from the Sun was well correlated with the sunspot number and in the 1950ies Forbush discovered that the intensity of the cosmic ray flux was noticeably reduced during high solar activity. This latter finding pointed at a completely new aspect of solar activity, namely the solar wind consisting of plasma blowing out from the solar surface and carrying with it a magnetic field from the solar surface. This varying modulation of the cosmic ray flux by the solar wind is also the reason for the important discovery of the modulation of ^{14}C production due to the changing geomagnetic field and the varying solar activity. The ^{14}C content in organic material and the radioactive isotope ^{10}Be in precipitated snow comprise the major source of knowledge regarding past variations of solar activity.

The most important solar parameter is the total power as irradiance received from the Sun at the Earth. Until 20 years ago this was believed to be constant, as the term "solar constant" indicates. For the last two decades it has been possible to make high-precision measurements of the total solar irradiance outside the Earth's atmosphere. This has confirmed that there are indeed small but systematic variations in phase with the solar activity with an increase in irradiance from sunspot minimum to sunspot maximum of about 0.1% corresponding to a global average of $0.3\ W\,m^{-2}$. To derive this number is not an easy task as explained by Fröhlich (2000) because it has to rely on the intercalibration of a set of radiometers on different space platforms during different times. The next question is how the spectral distribution of the solar output varies with changes in total irradiance. Lean (2000) explained how a model can be build from proxies which is able to explain both the observed spectral and total irradiance during the last 20 years reasonably well.

But can we expect larger changes of the solar irradiance over longer time periods? This question was addressed by the presentation of reconstructions of past irradiance variations based on models using various proxies of irradiance like sunspot number, solar cycle length, geomagnetic activity, and the cosmic ray flux (Solanki and Fligge, 2000). From these reconstructions which are 'calibrated' with the observations of the last 20 years one may deduce a total solar irradiance during the Maunder minimum around 1700 about 0.25% lower than today (Lean et al., 1998). It was also emphasised, however, that such estimates have limited confidence since they are not based on a detailed physical understanding.

Bazilevskaya (2000) discussed another aspect of solar variability, namely the modulation of the cosmic ray flux. Although this modulation closely mirrors the sunspot record and therefore might be used as a proxy for total solar irradiance there are also important differences, in particular the long-term behaviour. For example the cosmic ray flux was considerably more depressed in solar cycle 22 than in cycle 21 although the sunspot maxima for these two cycles were similar in amplitude. This difference in the long-term trend may be related to yet another index of solar activity, the geomagnetic activity index, aa, which shows a nearly monotonous upward long-term trend during this century. Based on a comparison with recent satellite data Lockwood et al. (1999) estimated that the open magnetic flux of the Sun has increased by 130% since the beginning of the century.

An estimate of the long-term activity variations of the Sun may most directly be derived from the radio-isotopes ^{14}C and ^{10}Be that are produced by the interaction of cosmic rays with atmospheric gases. An interesting aspect is that during the Maunder minimum where nearly no sunspots were observed there still is a distinct 11-year cycle in the ^{10}Be concentration in the Greenland ice cores (Beer, 2000). Again, this demonstrates the problems associated with an appropriate proxy for solar activity to be used in studies of Sun-climate relationships. It also demonstrates the need of improved physical models of solar activity and its different manifestations. The Ulysses mission and in particular the SOHO mission has provided some of the most promising data that will eventually improve our understanding of solar variability. Solar dynamo theory and the results from helioseismology have also provided a wealth of information about the Sun and possible sources for the variability (Weiss and Tobias, 2000; Kuhn and Schüssler, 2000; Spruit, 2000).

3. Climate Variability

A number of climate parameters have been used in the published reports on Sun-climate relationship. Some reports are based on local data only. This is in particular the case when dealing with old data sets, and those reports may be subject to the criticism that the results may just correspond to a climate redistribution caused by internal oscillations without any need for any external forcing. Although this may be true in many cases this does not necessarily mean that the correlation is accidental. If the correlation is significant, even on a local or regional scale, it may indicate some kind of triggering mechanism that may have a large regional effect without necessarily relying on a major change in power input.

Examples of that kind, although on a much greater scale, are the repeated occurrence of glaciation with a period of approximately 100,000 years. On top of this dominant quasi-cycle, shorter quasi-cycles of roughly 41,000 and 21,000 years are superimposed. The apparent regularity of this phenomenon and the calculated periods led Milankovitch to propose that the climatic cycles are related to the variations in the received solar energy and in particular its distribution over the Earth's surface. Model calculations show that the response to the orbital forcing is non-linear and that internal processes and feedback mechanisms play a crucial role. Whether the external orbital forcing parameters drive the internal processes or whether they just act as a pacemaker for the free oscillations in the internally driven system is still an open question. Stauffer (2000) discussed the importance of the feedback mechanisms in the long-term changes in climate. The ice core data from Antarctica and Greenland show a significant increase of the amount of the three principal greenhouse gases, CO_2, CH_4, and N_2O at the end of each of the glacial periods.

The modern instrumental record, although short in climatic sense, provides the most comprehensive set of global data that can be used for climate studies. Even though measurements of for example the temperature have been performed at a sufficient number of sites to provide a Northern Hemisphere Land average temperature since 1860, a large effort had to be made in order to calibrate and correct the individual time series of temperature. Parker (2000) described the various sources of errors that may affect these data, including urbanisation, change in land use, change in instrumentation and measurement procedures. He also described the various tests that had been performed, including the derivation of vertical profiles for various sites based on radiosondes (during 40 years) and Microwave Sounding Unit (MSU) measurements from satellites (during the last 20 years).

The largest disagreement concerns the reconstructed temperature series for the last millennium, until the beginning of the instrumental temperature series. Reconstruction of the global or Northern Hemisphere average temperature for this interval is based on different proxies like tree rings, measurements of the $\delta^{18}O$ ratio in ice cores (Stauffer, 2000), and advances and retreats of glaciers. Several different reconstructed temperature series have been proposed, all based on different data sets, different calibrations and different assumptions. Although similarities exist between the various proposed series there are also large differences. Two of the most recent reconstructed temperature series were discussed in particular, namely the temperature series by Jones et al. (1998) and the one by Mann et al. (1998). The comparison between these curves is particularly interesting because they are derived by two different methods. Jones et al. (1998) used a classical calibration with modern temperature measurements for each site and calculated a usual average temperature curve based on these data. Mann et al. (1998), however, based their global average on data that were attributed with weighting factors corresponding to how well a given temperature time series followed the expected trends based on spatial correlation maps. The latter methodology may have the advantage of putting less weight on data that may be erroneous. On the other hand this method assumes that the basic climate patterns have remained unchanged during the whole period. This assumption may be challenged and perhaps it accounts for the fact that the amplitude of the temperature oscillations are much less, nearly a factor of two compared to the classical temperature reconstruction (Jones et al., 1998). The assumption of an unchanged circulation pattern during the whole period including the modern instrumental time is also inconsistent with the claim that a global warming caused by an enhanced greenhouse effect has a specific "fingerprint". If that were the case then the climate pattern during the present era would indeed be different than during previous times, and temperature series that were not consistent with present day patterns would automatically receive less weight in the averaging.

But good observations are not sufficient. Numerical climate models provide the ultimate test of any theory concerning causes of climate change. But how good are present day climate models? And how can we test the models? The obvious way to test a Sun-climate relationship is to use an ocean-atmosphere general circulation model to simulate the effect on climate of a change in the total solar irradiance. Cubasch and Voss (2000) used as input two different reconstructed time series of total solar irradiance and calculated the corresponding time series of the global temperature. A major result of the modelling is that the atmospheric response to the 11-year variation in reconstructed solar irradiance data is heavily dampened. On the other hand, the long-term variations in solar activity are simulated quite well in the temperature response. The dampening of the 11-year signal might be expected because of the large thermal inertia of the ocean and it is also confirmed by the fact that very few surface temperature records show a clear 11-year variation in phase with solar activity variations. White et al. (1997) however, do show decadal modulation of sea surface temperatures with amplitude about 0.1 K in phase with solar activity.

In the lower stratosphere, however, van Loon and Labitzke (2000) have found a very clear variation of the height of the 30 hPa pressure level in phase with the solar activity variation. This correlation has been consistent during the last 4 solar cycles. Recently their results, which were related to the Northern Hemisphere data, have been shown to be valid also in the Southern Hemisphere. Interesting enough, this clearly observed signal is not found in the model experiments by Cubasch and Voss (2000). A possible explanation for this fact is that the stratospheric signal is not primarily due to changes in total solar radiation but rather to changes in the ultraviolet part of the spectrum.

A major uncertainty in all climate models is the treatment of clouds. This uncertainty is related both to the cloud formation processes and to the impact on the Earth's radiation budget. Cess (1996) made a comparison between a number of different general circulation models and showed that they did not in a consistent way simulate the cloud radiative feedback. Furthermore, even when two models did show the same net result they might differ significantly when calculating the result for short- and long-wave radiation separately. This indicates that the physics in the models may be different but also that the output of the climate models is tuned to the data sets. The overall impression of his presentation was that today's models are quite useful as educational tools but they are presently not sufficiently credible to be used as prediction tools for future climate.

The ultimate test of climate models is the comparison with observations. In this respect cloud data are necessary but also extremely difficult. Ground data lack the quantitative radiative information and global coverage while the satellite data rely on thresholds and retrieval algorithms that are different for different data sets, which means that they cannot easily be combined. Inhomogeneities in the data therefore represents a problem when such data are used in comparison with other data series (Norris, 2000). But spatial inhomogeneities related to the fact that different satellites have different instrumentation does not necessarily imply that the time series of global average cloudiness cannot be trusted and used in climate studies relating to trends in solar activity or El Niño. When dealing with cloud data it is recommended, however, also to examine the individual cloud types because cloud type is a better measure of processes and radiative impacts (Norris, 2000).

4. The Physical Mechanisms

The claim of a direct cause and effect relationship needs a plausible physical mechanism. The most obvious candidate is the direct effect of the varying total solar irradiance (the solar "constant") with solar activity. But even if changes in total irradiance may amount to ~0.25% over longer periods as has been suggested to be the case for the Maunder minimum, they may only be part of the explanation of the observed climate change. Whereas most of the global warming during this century may be due to the enhanced greenhouse effect, this cannot be the case for the significant past climate variations. Since many of the past climate changes have been demonstrated to be correlated with changes in solar activity parameters, one obviously has searched for indirect mechanisms by which solar activity variations could modulate climate.

One possibility concerns the ultraviolet part of the solar irradiance. Solar cycle variations in this spectral band are much larger than for the visible light. Haigh showed the results of Larkin et al. (2000) indicating that the effect in the stratosphere may be simulated by taking into account also the changes in the UV part of the spectrum. The UV-radiation is absorbed by the stratospheric ozone and this gives rise to a stratospheric warming during high solar activity. The model simulations further show that the whole vertical structure of the atmosphere is affected and this can even influence the tropospheric circulation. Although such model simulations can explain many of the observed features in the stratosphere (van Loon and Labitzke, 2000) the amplitude of the simulated changes are still too small compared to the observations.

Even more pronounced manifestations of solar cycle effects in the solar energy output are associated with the solar wind, the emission of particles and fields from the solar surface. Although the energy in the solar wind is negligible compared to the energy in the ultraviolet and visible spectral bands, the relative variations are much larger. And we know from the record of the radio-isotopes ^{14}C and ^{10}Be that there is a clear signal in the troposphere. But how could these small energy fluctuations possibly affect climate? The cosmic ray flux interacts in a complicated way with the whole atmosphere (see Bazilevskaya, 2000). The cosmic ray flux is in fact the main cause of the ionisation of the atmosphere below 55–60 km, except near the ground level below 3–4 km over land. And since we know that the ionisation rate changes significantly with solar activity, we could imagine that physical processes in the atmosphere that are directly affected by ionisation could contribute to climate change.

Condensation of vapour in terms of cloud formation is a complex process that was discusses by Tinsley (2000). He in particular focussed on the higher troposphere where the cosmic ray flux is largest and where clouds consist of ice particles. In particular he emphasised the vertical electric current that is dependent on the atmospheric conductivity, which in turn is affected by the ionisation. The electric current is, however, also modulated by the precipitation of energetic electrons from the magnetosphere. Large gradients in the conductivity near the interface between the cloud top and the clear air are established and lead to large space charges that may influence the production of ice forming nuclei and thereby the formation of clouds.

Another process related to cloud formation is discussed in Marsh and Svensmark (2000). They build upon the work by Svensmark and Friis-Christensen (1997) and Svensmark (1998) who discovered that the global cloud cover measured by satellites during a considerable part of a solar cycle was highly correlated with the cosmic ray flux. Using an improved and recently released data set from the International Satellite Cloud Climate Project (ISCCP) covering the period from 1983 to 1994 they were able to demonstrate that only the low clouds are responsible for the found correlation. This is a very important information because it gives a much better hint regarding the microphysical processes that may be involved.

A change in cloud cover would indeed be a very effective amplifying mechanism for climate forcing. The result that mainly low clouds are affected by the modulation of the cosmic rays is also in line with the long term oscillations in temperature. An increase in the amount of low clouds will mean a cooling whereas an increase in high clouds would result in a warming according to present understanding of the radiative properties of clouds.

An effect of cosmic rays on low clouds is most likely to be through their effect on aerosols. Marsh and Svensmark (2000) point at the importance of aerosols in cloud formation by referring to observations of the effect on low clouds of ship exhaust which show that small perturbations in the aerosol production may have a large effect on clouds. They further referred to recent computer simulations (Turco et al., 1998) of aerosol formation by ion-ion recombination. These results indicate that indeed some unexplored mechanisms may contribute to the understanding of how ionisation by cosmic rays may enhance cloud condensation, in particular at low altitudes. More recent computer simulations indicate that even quantitatively the results correspond to the observations, although there still is much work to do in this area before a complete physical understanding is available.

5. Conclusion

The issue of a Sun-climate relationship is gaining more and more respect these days. Our understanding of the Sun and solar processes has increased dramatically during recent years. Moreover, it is understood that the Sun affects the Earth's environment in a much more complicated way than we have imagined. And we realise that it is impossible to describe the effect of the Sun on Earth by just a few parameters.

The Earth's climate is a very complex system with a great number of interacting elements and with many feedback mechanisms. A number of different processes including solar variability may cause climate changes. Since many data indicate the possibility of a significant impact on terrestrial climate of solar variability it seems a most relevant and urgent task to quantify the effect. This is a task that cannot be accomplished by a single workshop. But a workshop like this may serve as a forum where representatives from different scientific disciplines can meet and express the status of knowledge, both those aspects that are relatively well understood and in particular those problems that are still unresolved.

The workshop was centred on three types of possible mechanisms of a solar-climate relationship. A direct effect through changes in the total irradiance, an indirect effect through changes in the solar ultraviolet irradiance, and finally an indirect effect caused by changes in the fluxes of energetic charged particles impinging on the Earth's atmosphere. Due to the complexity and our limited understanding of some of the basic atmospheric processes and the fact that the effects of the different mechanisms do overlap, it is not an easy task to separate the role of each potential mechanism.

In the early days of climate modelling it was implicitly assumed that the response of climate to a certain amount of global-mean radiative forcing (the climate sensitivity) was independent on the exact nature of the forcing. Shine (2000) demonstrated that this concept of a simple climate sensitivity parameter would need to be modified in order to take into account the various feedback mechanisms and their different spatial and vertical distribution. The task of assessing the relative importance of the different mechanisms is therefore even more difficult.

The fundamental question we want to answer is of course: Is there a relationship between solar activity variation and climate or is there none? The associated scientific question is a little more multifaceted. In order to quantify the effect we must understand the physical mechanism or mechanisms. Perhaps one of the most conspicuous results of the workshop was the realisation of the fact that we may not only be faced with one single dominating mechanism. Solar activity effect on climate may work simultaneously through all the different possibilities, total irradiance variations together with UV variations and variations in the flux of energetic particles. There seemed to be sufficient evidence presented at the workshop indicating that this might indeed be the case. And since these mechanisms are different, in their temporal as well as in their spatial response, this may also explain the reason for some of the discrepancies that are often present in a single study.

The scientific process based on a variety of concerted approaches in addition to a strict peer review process has demonstrated its efficiency in reaching robust conclusions, at least in the long run. New ideas that tend to change the "current wisdom" are more difficult to introduce. Ideas that are not sufficiently well described to be included in physical or numerical models have a hard time being accepted. This is not surprising because a very crucial part of the scientific process is to create sufficient documentation wherein new ideas are put on solid ground. The result of this workshop represents just one of the endeavours in this direction. But the results of the workshop also suggest that this topic is getting more and more mature and that a break-through in the understanding of the solar-climate relationship may not be too far away.

References

Bazilevskaya, G.A.: 2000, *Space Sci. Rev.*, this volume.

Beer, J.: 2000, *Space Sci. Rev.*, this volume.

Cess, R., et al.: 1996, 'Cloud feedback in atmospheric general circulation models: An update', *J. Geophys. Res.* **101**, 12,791–12,794.

Cubash, U., and Voss, R.: 2000, *Space Sci. Rev.*, this volume.

Fröhlich, C.: 2000, *Space Sci. Rev.*, this volume.

Jones, P. D., Briffa, K. R., Barnett, T. P., and Tett, S. F. B.: 1998, 'High-resolution palaeoclimatic records for the last Millennium: Interpretation, integration and comparison with general circulation model control-run temperatures', *The Holocene* **8**, 455–471.

Kuhn, J.R., and Schüssler, M.: 2000, *Space Sci. Rev.*, this volume.

Larkin, A., Haigh, J. D., and Djavidnia, S.: 2000, *Space Sci. Rev.*, this volume.

Lean, J., Beer, J., and Bradley, R. S.: 1998, 'Reconstruction of solar irradiance since 1610: Implications for climate change', *Geophys. Res. Lett.* **22**, 3195–3198.

Lean, J.: 2000, *Space Sci. Rev.*, this volume.

Lockwood, M., Stamper, R., and Wild, M. N.: 1999, 'A doubling of the Sun's coronal magnetic field during the past 100 years', *Nature* **399**, 437.

Mann, M. E., Bradley, R. S., and Hughes, M. K.: 1998, 'Global-scale temperature patterns and climate forcing over the past six centuries', *Nature* **392**, 770–787.

Marsh, N., and Svensmark, H.: 2000, *Space Sci. Rev.*, this volume.

Norris, J. R.: 2000, *Space Sci. Rev.*, this volume.

Parker, D. E., Basnett, T., Brown, S., Gordon, M., Horton, B., and Rayner, N.: 2000, *Space Sci. Rev.*, this volume.

Shine, K: 2000, *Space Sci. Rev.*, this volume.

Solanki, S. K., and Fligge, M.: 2000, *Space Sci. Rev.*, this volume.

Spruit, H.: 2000, *Space Sci. Rev.*, this volume.

Stauffer, B.: 2000, *Space Sci. Rev.*, this volume.

Svensmark, H.: 1998, 'Influence of Cosmic rays on Climate', *Phys. Rev. Lett.* **81**, 5027.

Svensmark, H., and Friis-Christensen, E.: 1997, 'Variation in Cosmic Ray Flux and Global Cloud Coverage – a Missing Link in Solar-Climate relationships', *J. Atmos. Solar-Terr. Phys.* **59**, 1225–1232.

Tinsley, B.: 2000, *Space Sci. Rev.*, this volume.

Turco, R. P., Zhao, J.-K., and Yo, F.: 1998, 'A New Source Of Tropospheric Aerosols: Ion-Ion Recombination', *Geophys. Res. Lett.* **25**, 635.

van Loon, H., and Labitzke, K.: 2000, *Space Sci. Rev.*, this volume.

Weiss, N. O., and Tobias, S. M.: 2000, *Space Sci. Rev.*, this volume.

White, W. B., Lean, J., Cayan, D. R., and Dettinger, M. D.: 1997, 'Response of global upper ocean temperature to changing solar irradiance', *J. Geophys. Res.* **102**, 3255–3266.

PRESS RELEASE

THE SUN'S INFLUENCE IS AN IMPORTANT FACTOR IN GLOBAL CLIMATE CHANGE

However, the rapid warming during the last 20 years is dominated by other factors

2 July 1999

The variability of the solar radiation is an important factor affecting the climate on Earth. Even though its magnitude merely varies by about a permille over the solar activity cycle, it causes direct variations of the global temperature by about half a degree. But the solar influence is not the only important factor affecting the climate. The rapid temperature increase during the last 20 years is probably not due to a changing Sun, but is most likely caused by man-made greenhouse gases (such as carbon dioxide), the concentration of which was never as high during the last 15'000 years as it is today in the atmosphere. These conclusions were drawn by scientists from around the world this week in Bern.

Discussions about the influence of the solar radiation on global warming are still ongoing. At this week's international workshop of the International Space Science Institute (ISSI) in Bern leading scientists evaluated the corresponding state of knowledge. It is clear that the solar influence is an important factor affecting the Earth's surface temperature, but it is also clear that it is not the only one. The variability of solar irradiance of about 0.1% during a cycle is too small to directly cause the observed changes on the Earth's surface temperature. However, solar variability also influences the climate indirectly: Variability in the ultraviolet radiation affects the ozone in the upper atmosphere and thus may lead to a temperature change. Solar variability also modulates the cosmic rays, which in turn may affect the cloud cover and thus also leads to a temperature change. These indirect effects are highly nonlinear and therefore may well be larger than the direct effect, but the corresponding mechanisms as well as the range of their influence are unclear.

Changes of solar irradiance have been directly observed since 1978 and are mainly linked to the variability of magnetic activity which is e. g. manifested by the number of sunspots. The amplitude of the irradiance variations is approximately 0.1% over the 11-year cycle. There is also evidence for longer-term variability. Observations of sunspots since the invention of the telescope in the 16th century show variations of the strength of the 11-year cycle. The so-called "Maunder Minimum" in the 17th century is an example when very few sunspots were seen over 30 years and the climate was called the "little ice age".

Space Science Reviews **94:** 423–424, 2000.

The mechanisms which could cause such a relation are difficult to investigate because they cannot be studied separately but have to be distilled from large data sets of weather observations (which are less and less reliable the further back one looks) or even from indirect data.

Several proposed mechanisms have been discussed extensively:

- Total irradiance: Larger irradiance changes are expected on time scales larger than 11 years.
- Solar magnetic field: During periods of high solar activity, the increased solar magnetic field enhances the shielding of cosmic rays. It is postulated, based on empirical correlations, that less cosmic rays reduce the global cloud cover and therefore increase the temperature.
- UV flux: The changes in UV flux are much higher (more than 10 times) than the changes of the total irradiance in the upper atmosphere. Both absorption and the enhanced production of ozone lead to a temperature change of the stratosphere and therefore also of its dynamics. The coupling of stratosphere and troposphere dynamics could lead to changes in tropospheric wind regimes and thus to temperature changes at the Earth's surface.

The influence of solar variability has to be compared to other forcing factors. The main forcing factors in the decadal to centennial timescale besides solar variability are volcanic eruptions, the concentration of aerosols as well as of greenhouse gases (CO_2, methane, etc.). In the past 300'000 years there were several rapid climate changes, as we know from ice cores, that are larger than what could be explained by solar variability.

Existing and reconstructed data sets allow statistical analysis of the influence of these different factors on climatic changes during the last centuries. These analysis using different data sets all show similar results: Changes of global temperatures in the 17th and 18th centuries can be explained quite well by changes of solar activity, i. e. there is a good statistical correlation. With the onset of fossil fuel combustion in the last century, the contribution of the greenhouse effect increases and becomes dominant during the last two decades. During this period, the best agreement between statistical as well as physical dynamical models and actually observed global temperature changes is obtained if all factors, i. e. changes in solar radiation, volcanic activity, aerosols, and greenhouse gas concentrations are taken into account.

Author Index

List of Participants

Neil Arnold, *Univ. of Leicester, UK,* nfa1@ion.le.ac.uk
Galina Bazilevskaya, *Lebedev Physical Inst., Moscow, Russia,* bazilevs@fiand.msk.su
Jürg Beer, *Eawag, Dübendorf, Switzerland,* beer@eawag.ch
Raymond Bradley, *Univ. of Massachusetts, Amherst, USA,* rbradley@geo.umass.edu
Alfred Bürgi, *Arias, Bern, Switzerland,* arias-bern@bluewin.ch
Robert Cess, *SUNY, Stony Brook, USA,* cess@atmsci.msrc.sunysb.edu
Ulrich Cubasch, *Deutsches Klimarechenzentrum, Hamburg, Germany,* cubasch@dkrz.de
Paul Evenson, *National Science Foundation, Washington DC, USA,* pevenso@nsf.gov
Len Fisk, *Univ. of Michigan, Ann Arbor, USA,* lafisk@umich.edu
Marcel Fligge, *ETH-Zentrum, Zürich, Switzerland,* fligge@astro.phys.ethz.ch
Linton Floyd, *Naval Research Lab., Washington DC, USA,* floyd@susim.nrl.navy.mil
Eigil Friis-Christensen, *Danish Space Res. Inst., Copenhagen, Denmark,* efc@dsri.dk
Claus Fröhlich, *PMOD/WRC, Davos, Switzerland,* c.frohlich@pmodwrc.ch
Johannes Geiss, *ISSI, Bern, Switzerland,* geiss@issi.unibe.ch
Reindert Haarsma, *Royal Netherlands Met. Inst., The Netherlands,* haarsma@knmi.nl
Joanna Haigh, *Imperial College, London, UK,* j.haigh@ic.ac.uk
James Hansen,*Goddard Inst. for Space Studies, New York, USA,* jhansen@giss.nasa.gov
Giles Harrison, *Univ. of Reading, UK,* r.g.harrison@reading.ac.uk
Jasper Kirkby, *CERN, Geneva, Switzerland,* jasper.kirkby@cern.ch
Jeffrey Kuhn, *National Solar Observatory, Sunspot, USA,* jkuhn@sunspot.noao.edu
Erkki Kyrölä, *Geofysiikan Tutkimus, Helsinki, Finland,* erkki.kyrola@fmi.fi
Ari Laaksonen, *Univ. of Kuopio, Finland,* alaakson@messi.uku.fi
Karin Labitzke, *Freie Univ. Berlin, Germany,* labitzke@strat01.met.fu-berlin.de
Judith Lean, *Naval Research Lab., Washington DC, USA,* lean@demeter.nrl.navy.mil
Werner Mende, *Berlin-Brandenburgische Akad. der Wiss., Germany,* mende@bbaw.de
Joel Norris, *NCAR, Boulder, USA,* jnorris@ucar.edu
Judit Pap, *UCLA, Los Angeles, USA,* pap@pulsator.astro.ucla.edu
David Parker, *Hadley Centre, Bracknell, UK,* deparker@meto.gov.uk
George Reid, *NOAA, Boulder, USA,* reid@ucar.edu
Gary Rottman, *Univ. of Colorado, Boulder, USA,* gary.rottman@lasp.colorado.edu
Werner Schmutz, *PMOD/WRC, Davos, Switzerland,* w.schmutz@pmodwrc.ch
Manfred Schüssler, *MPAe, Katlenburg-Lindau, Germany,* schuessler@linmpi.mpg.de
Keith Shine, *Univ. of Reading, UK,* k.p.shine@reading.ac.uk
Sami Solanki, *MPAe, Katlenburg-Lindau, Germany,* solanki@linmpi.mpg.de
Henk Spruit, *MPI für Astrophysik, Garching, Germany,* hspruit@mpa-garching.mpg.de
Bernhard Stauffer, *Univ. of Bern, Switzerland,* stauffer@climate.unibe.ch
Rudolf von Steiger, *ISSI, Bern, Switzerland,* vsteiger@issi.unibe.ch
Oskar Steiner, *KIS, Freiburg, Germany,* steiner@kis.uni-freiburg.de
Thomas Stocker, *Univ. of Bern, Switzerland,* stocker@climate.unibe.ch
Peter Stott, *Hadley Centre, Bracknell, UK,* pastott@meto.gov.uk
Henrik Svensmark, *Danish Space Res. Inst., Copenhagen, Denmark,* hsv@dsri.dk

Brian Tinsley, *Univ. of Texas, Dallas, USA,* tinsley@utdallas.edu
Steve Tobias, *Univ. of Cambridge, UK,* smt11@damtp.cam.ac.uk
Yvonne Unruh, *Univ. Wien, Austria,* ycu@venus.astro.univie.ac.at
Heinz Wanner, *Univ. of Bern, Switzerland,* wanner@giub.unibe.ch
Nigel Weiss, *Univ. of Cambridge, UK,* now@damtp.cam.ac.uk
Oran White, *High Altitude Observatory, Boulder CO, USA,* orw@hao.ucar.edu